International Shipping

The sudden collapse of the world's complete cruise industry in March 2020, caused by the spread of the COVID-19 pandemic, became obvious by the fate of the so far largest cruise liner from German production, built under a contract of UK's P&O Cruises. The builders, Meyer Werft in Papenburg, decided to bring forward the transfer of the uncompleted 184,700 gt IONA from the shipyard site to the North Sea. But its final fitting out and testing had to be delayed temporarily. In a desperate move, the largest cruise operator and P&O parent company, US-headquartered Carnival Group, offered their ships to be used as hospital ships for COVID-19 patients (Ralf Witthohn)

Ralf Witthohn

International Shipping

The Role of Sea Transport in the Global Economy

Ralf Witthohn
Schiffdorf-Spaden, Germany

ISBN 978-3-658-34275-3 ISBN 978-3-658-34273-9 (eBook)
https://doi.org/10.1007/978-3-658-34273-9

This Springer imprint is published by the registered company Springer Fachmedien Wiesbaden GmbH, part of Springer Nature.
The registered company address is: Abraham-Lincoln-Str. 46, 65189 Wiesbaden, Germany

Contents

Abbreviations

AHC	Active heave compensation, fitted to balance ship movements during crane operation
AIS	Automatic identification system, unencrypted and therefore publicly accessible VHF-based system for the recognition of ships in the close range, prescribed for merchant ships
BSU (*Bundesamt für Seeunfalluntersuchung*)	Federal Bureau for Marine Accident Investigation, located at Hamburg, Germany
ceu	car equivalent unit, describes the cargo capacity of vehicle carriers
ECA	Emission Control Area, special zone for the restriction of harmful exhaust gases from marine fuels such as sulphur compounds (Sulphur Emission Control Area, SECA)
feu	40 ft equivalent unit, standard container of 40 ft length (12,192 mm), 8 ft width, and 8 ft height (2435 mm)
FPSO (Floating production storage and offloading unit)	Ship anchored at an oil well to store and process crude oil and transfer it to a tanker for transport
gt	gross tons (tonnage), indicates the total volume of a ship with all superstructures exceeding 1 m^3 of volume, calculated by using a formula on the basis of drawing and survey and recorded in the ship's tonnage certificate

IACS	International Association of Classification Societies
IMDG Code	International Maritime Dangerous Goods Code international regulations for the carriage of dangerous goods on seagoing vessels
IMO	International Maritime Organization, the United Nations suborganisation responsible for shipping, issuing regulations from its headquarters in London
LARS	Launch and Recovery System, launching and recovery device for underwater work equipment
LNG	Liquefied natural gas, natural gas (methane) cooled at -163 °C for transport and storage purposes
LR1, LR2	Long Range 1 and Long Range 2 designations for the two largest tanker classes regularly transporting oil products of up to 75,000 dwt (LR1) and 115,000 dwt (LR2), resp.
LRIT	Long-Range Identification and Tracking, encrypted system for the recognition of ships over long distances by satellite
MCR	Maximum continuous rating, maximum continuous power of an engine
NCR	Normal continuous rating, continuous power of an engine, 85% of MCR
nm	Nautical mile = 1.852 km
nt	Net tons (tonnage), indicates the commercially usable tonnage of a ship determined on the basis of drawings, survey of the ship, and a formula
PTI	Power take-in mechanical auxiliary drive on the gearbox, e.g. for emergency operation of the propulsion system if the main engine fails
PTO	Power take-off mechanical auxiliary drive on the gearbox for operation of a shaft generator
rfeu	Refrigerated forty-foot equivalent unit, standard container used for the carriage of reefer cargoes
ROV	Remotely operated vehicle, cable-controlled underwater vehicle for construction or research tasks

SOLAS	Safety of Life at Sea, UN Convention on Safety Regulations, especially on passenger ships
SWATH	Small Waterplane Area Twin Hull, double hull ship with high stability and low water resistance
SWL	Safe working load, maximum permissible load capacity of a hoist
teu	20 ft equivalent unit, standard container of 20 feet length (6055 mm), 8 feet width, and 8 feet height (2435 mm)
VLCC (very large crude carrier)	Large tankers from 200,000 dwt to 400,000 dwt for the exclusive transport of crude oil
VLOC (very large ore carrier)	Large dry cargo ships of between 200,000 dwt and 400,000 dwt for the exclusive transport of iron ore

List of Figures

Introduction

Up to 90% of the world's cargo transport volume is carried across the sea, which extends over 71% of the earth' surface, making most regions accessible by ships and many more by navigating canals and rivers. At the same time, only the application of the physical properties of water according to the Archimedes' principle allows a transport performance that exceeded 10 billion tonnes in 2015 for the first time and reached 11 billion tonnes in 2018, when growth predictions of annually 3.4% until 2024 were made. Such forecasts became obsolete in the wake of the COVID-19 outbreak, which disrupted maritime transport and caused a decrease by 3.8% in 2020 to a total volume of 10.7 million tonnes. Shortly before this book went to press, international sea transports were severely affected as a result of the Russian war of aggression against Ukraine. Major shipping companies such as Denmark's Maersk Line terminated container shipments to Russian ports in the Baltic Sea and the Far East, with the exception of food and humanitarian goods. By a total volume of 24.1 million tonnes, 45% of which were fossil fuels, the Russian Federation had been, for example, the largest trading partner for goods imported or exported via German seaports in the first nine months of 2021. In course of the war, the blockade of the Ukrainian ports at the Black Sea and the Azov Sea caused an interruption of grain exports with the potential of creating a world-wide food crisis.

The world's annual GDP, which is regularly superseding a value of USD 80 trillion, depends fundamentally on the performance of sea transport. Oil, gas, and coal are shipped to produce energy and ores to be refined into steel, copper, or aluminium, which in turn are refined to industrial products. Their distribution is again carried out by transport on specially developed ship types including car, trailer, and plutonium transporters as well as in chemical, vegetable oil, and orange juice tankers. Only the standardised container made it possible to economically ship all types of conventional general cargo by sea, even before globalisation generated a volume growth that the old conventional cargo carriers could not have dealt with.

Shipping was the main actor in the globalisation process, and still keeps it going, despite increasing criticism from a number of institutions and even governments. In the light of the climate discussion, it has become even clearer that shipping is responsible for a substantial share of environmental destruction. The mass overseas transport of raw materials has led to industrialisation and all its consequences of air, water, and soil pollution. At the same time, ships are intensive sources of harmful emissions themselves. Toxic gases are emitted from the burning of fuel oil, of which ships consume several hundred million tonnes every year. Toxic paints are applied on ship hulls to reduce fouling, and harmful coolants are used in reefer containers. Waste and oil residues pollute the oceans, noise from the propulsion systems disturbs marine life, and ballast water imports invasive species.

Climate change and its resulting environmental consequences have created the urgent need to create a less polluting economy. Above all, the restructuring will entail the renunciation of fossil fuels, the growing use of renewable energies, and the conservation of resources. Particularly the transport of raw materials used as energy source will be the shipping segment most affected by the decarbonisation process. Consequently, bulk and tanker shipping performing about one-third of the total global sea transport volume has to expect a considerable downturn. Already hit in recent years was the offshore sector, in which less auxiliary vessels for the exploitation of oil and gas were employed.

How vulnerable the global transport chains are became evident when the COVID-19 pandemic increasingly paralysed societies and industries, not least the shipping sector, from the beginning of 2020 on. Particularly cruise shipping, which had shown strong growth rates in the years before, came to a sudden stop, in a way that had never occurred in a shipping branch before and that puts up the question whether such maritime leisure activities will ever return to the old state. Hard hit were the crews who suffer from social isolation under the conditions of modern shipping anyhow. They were not anymore allowed to leave their ships in port, while the mandatory crew changes often had to be postponed, thus prolonging the times the seamen had to spend on board. Another event that revealed the sensitivity of shipping was the accident of the ultra-large container ship EVER GIVEN in the Suez Canal in March 2021, resulting in the temporary interruption of the exchange of goods on the Far East–Europe route as well as of oil shipments from the Middle East.

The spectrum of industrial goods shipped is wide. Besides vehicles of the most varied types, aircraft or even ships and floating docks can be carried. At the same time, transport services include bizarre offers such as the transport of North Sea crabs to Morocco for peeling and return transport across the Mediterranean Sea. Although supply purposes predominate, shipping is still assigned problematic disposal tasks, like the removal of chemical warfare agents from Syria.

In addition to being used as a transport lane, the sea is as well a workplace, first for fishermen, but also for the extraction of mineral resources such as oil, gas, sand, and precious stones or for the production of renewable energies. The range of special offshore vessel types including those for seismographic research, drilling and production, pipe and cable laying, anchor handling, or pipe transports is immense. Crane ships, wind farm installation ships, salvage cranes, or diamond prospecting ships are further highly sophisticated vessels, whose design and construction require great knowledge of naval architecture and whose use under often adverse weather conditions requires a high degree of proven seamanship and technical experience.

Despite the dominant share of freight transport in shipping and the vast diversity of all kinds of work done on the oceans, the safe carriage of people remains one of the most important tasks—for the driver who accompanies his truck from Turkey to the Ukraine, for the voyager from Berlin to Copenhagen who takes the ferry across the Fehmarn Belt, or for the cruise passenger who is offered to relax and to experience nearly any foreign country. In addition to the well-known cruise liners and ferries, there are special passenger ships that perform humanitarian tasks as hospital or rescue vessel.

As advocates of free trade, shipowners and operators often see themselves as service providers striving for economic success and rarely question the background of the transports they undertake. In order to distract from the high environmental impact, they hide behind alibi measures, when for example vehicle carriers are fitted with solar collectors that generate only a fraction of the energy required to transport vehicles that are equipped with harmful combustion engines. The same applies to the cruise industry, which has caused high environmental pollution in port cities and natural areas. After the British classification society Lloyd's Register had already determined the pollutant emissions on ships in practice during an initial research programme in the 1990s, it took more than two decades, until January 2020, to suspend the long-time use of heavy oil as the dominant fuel without adequate exhaust gas cleaning.

In shipping, the oil burning diesel engine still remains almost the only form of propulsion chosen, and thus is an essential contributor of greenhouse gases, figured by the International Maritime Organization (IMO) at around 940 million tonnes of CO_2 annually, equivalent to 2.5% of global emissions. The IMO has therefore envisaged their reduction by at least 40% until 2030 and by 50% until 2050 first of all by technical innovations and the introduction of alternative fuels or energy sources. In July 2017, the German Ministry of Transport announced the promotion of fuel cell technology for merchant ships. A short time later, it promoted the use of LNG as marine fuel.

Much of information on maritime logistics and the work at sea collected in this book could only be obtained by extensive on-spot research and by interviews with crews and dock workers. Shipping companies, agents, port operators, and processing companies tend to withhold data on their activities for competitive or security reasons. The latter are particularly cited since the terrorist events of September 2001. As well, the companies do not want to be a target for criticism of questionable transports. Important sources of information are the AIS system (Automated Identification System) on worldwide ship movements, the Equasis electronic information system developed by the European Commission and the French Ministry of Transport, registers of classification societies, and the private New Zealand-based online register Miramar for historical data of ships, for container shipping the France-based industry service Alphaliner, and for transport statistics UNCTAD and FAO data, as well as information from shipping companies and shipyards.

Understanding maritime logistics as the base of global economy is the primary prerequisite for mastering the upcoming challenges, particularly the inevitable need for decarbonisation. This book is the first which, by illustrating actual on-scene reports and detailed descriptions of ships, ports, and facilities, paints a comprehensive picture of the facets connected to overseas goods flows and the special industry that is carrying it out. Only a few months after the publication of the work *Transport, Arbeit und Erholung auf dem Meer, Die Rolle der Schifffahrt in der globalen Wirtschaft* by Springer Fachmedien Wiesbaden in early 2019, the publisher declared its friendly willingness to issue this substantially extended English edition.

In July 2019, MSC GÜLSÜN was the first unit in a series of 16 container vessels of 23,756 teu capacity, then the world's largest of their type, from South Korean shipyards Samsung and Daewoo for operation by Switzerland/Italy-based Mediterranean Shipping Co. The picture reveals the breadth of the funnel casing, designed to incorporate the scrubber systems prescribed to reduce the harmful effects of the exhaust gases. At the same time, the ship was the first container carrier fitted with monitors able to fight burning deck containers, after explosions in containers had severely hit a number of container ships (Ralf Witthohn)

Part I

Transport by Sea

Developing countries are the main contributors to the international exchange of goods by sea. In 2019 they accounted for 59% of exports and 66% of imports, and in 2020 their shares increased to 60% and 70%, resp. While those countries' portion in exports has remained rather similar over the past decades, their quota of import volumes has increased many times over. Developing countries are no longer merely suppliers of raw materials, but also participants in global production processes and importers of raw materials such as oil.

Among the types of cargo, dry and liquid bulk goods have the largest shares in terms of weight. In 2018, 2019 and 2020 major dry bulk cargoes (iron ore, coal, grain) accounted for around 3.2 billion tonnes. Oil and gas transports yet shrinked by 7.7% from 3.2 in 2019 to 2.9 billion tonnes in 2020. Steel products remained among the declining types of cargo by a 5% decrease to 0.35 billion tonnes, while forest products lost 3% to a similar weight of nearly 0.37 billion tonnes in 2020. In 2020, container ships moved 149.2 million 20-foot equivalent units (teu), down from 150.9 million in the year before.

The total number of merchant vessels was figured at 99,800 units in early 2021, meaning a 3% annual plus, though the ship deliveries had declined by 12% during 2020. Calculated in dwt, bulk carriers had at this time a share of 42.8% (913 million dwt), oil tankers of 29% (619 million dwt), container ships of 13.2% (282 million dwt) and general cargo ships of 3.6% (77 million dwt). The total deadweight tonnes stood at 2.13 billion. Bulk carriers had, by 3.8%, the second largest growth rate next to gas tankers, which achieved a 5.1% plus. By number the largest share was contributed by 16,500 tankers with a total tonnage of 411 million gt, followed by 16,200 general cargo ships

(58.4 million gt), 11,700 bulk carriers (448 million gt) and 6,700 container and ro-ro vessels (283 million gt).

Based on deadweight capacities, the major shipowning countries as of 2021 were Greece (4,700 ships under national and foreign flags), China (7,300 ships), Japan (4,000), Singapore (2,800), Hong Kong (1,800), Germany (2,400), Republic of Korea (1,600) and Norway (2,000), accounting for more than 50 per cent of the world fleet's total deadweight. More than 70 per cent of the fleet is registered under a foreign flag. Based on gross tons, only three shipbuilding countries—Korea, China and Japan—accounted for 93% of the newbuilding activities in 2020.

1

Agricultural and Forestry Products, Animals, Feedingstuffs

The major share of agricultural products such as wheat, coarse grains (barley, maize, oats, millet), rice, oilseeds or sugar, is transported in bulk. The size range of bulk carriers used extends from coastal freighters and minibulkers with carrying capacities of up to 10,000 dwt to capesize carriers of around 180,000 dwt.

1.1 Cereals

Next to iron ore and coal, grain holds the third rank in the weight balance of dry bulk cargoes. Its transport volume reached a total of 512 million tonnes in 2020, substantially up by 7.1% from 478 million tonnes in 2019. It thus accounted for about 10% of all dry bulk goods. According to UNCTAD figures, the most important grain exporter in 2020 was the United States by a share of 26%, before Brazil (23%), Argentina (11%), Ukraine (10%), European Union (9%), Russia (7%) and Canada (6%). Countries in East and South Asia were the importers of 49% of the grain volume, Africa of 14%, followed by South and Central America (10%) and the European Union (9%). As, like Russia, an important player in agrifood markets, Ukraine in 2020 accounted for a 13% share in the global corn and of 36% in the sunflower oil and seeds trade. The export of grain through Ukrainian ports, which had reached 5 million tonnes monthly, was severely disrupted after Russia's invasion in February 2022. Individual shipments resumed only in August 2022.

About 70% of grain transport refers to wheat and coarse grain, predominantly winter wheat in the countries of cultivation in the northern hemisphere. Canada, Kazakhstan, Russia and the United States are also involved in

R. Witthohn, *International Shipping*, https://doi.org/10.1007/978-3-658-34273-9_1

Fig. 1.1 As a major shipment place for grain, the port of Rouen/France has storage capacities of 1.2 million tonnes and achieves yearly bulk export figures of around 10 million tonnes. In January 2021, the port, which is marketed together with Le Havre and Paris under the name HAROPA, was called at by the bulk carrier ANDROS to take a record load of 57,000 tonnes of barley (Ralf Witthohn)

spring production, which starts later. Subsequently, in countries of the southern hemisphere like Australia and Argentina wheat is sown, so that nearly permanent harvest times require corresponding transport efforts.

The most important grain export ports in Europe are Rouen (Fig. 1.1), Rostock and Hamburg, in the United States New Orleans, Houston, Baltimore and Norfolk, Vancouver in Canada and the ports on the St. Lawrence River, Novorossiysk in Russia, Yuzhniy and Odessa in Ukraine, Rosario and Buenos Aires in Argentina, Santos and Rio Grande do Sul in Brazil, Fremantle, Bunbury and Adelaide in Australia. The largest exporter of wheat, based on its value, in 2018 was Russia by 20.5%, followed by Canada (13.8%), United States (13.2%) and France (10%).

1.1.1 Canadian Wheat from the Thunder Bay

After having passed the city of Sault Ste. Marie at the western entrance of the St. Lawrence Seaway and having crossed Lake Superior, the bulk carrier BAIE ST. PAUL is berthed late in the evening at the pier in Thunder Bay. There, the

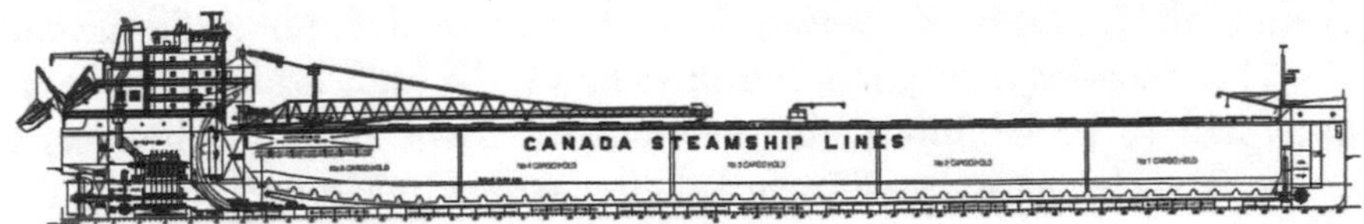

Fig. 1.2 While the cargo is loaded by port facilities onto the Laker BAIE ST. PAUL, it is discharged using the ship's own system (Chengxi)

grain trading company Viterra operates a 362,000 tonnes storing silo, from which the so-called laker is loaded within only 27 h. The port of Thunder Bay has the largest grain storage capacity in North America. Eight terminals achieve handling capacities of 1000 t/h to 3400 t/h and can accommodate a total of 1.2 million tonnes. They are operated by the companies Superior Elevator of the Cargill group, Parrish & Heimbecker, Canada Malting, Richardson International, G3, Western Grain By-Products Storage and Viterra (Fig. 1.2).

By handling performances of up to 6000 t/h, self-discharging cargo vessels offer the opportunity to increase unloading rates significantly. At the same time, the number of persons involved in the handling operations is reduced. The technically sophisticated automated discharge systems are not only used for grain transport on the Great Lakes of North America. They have also become familiar in the transport of ore and building materials in Northern Europe, as well as in international cement trades. Medium-sized bulk carrier types of conventional design are instead regularly equipped with revolving cranes, which afford considerably longer stays of several days in port.

Worked out by Canada's engineering firm Cooke Naval Architect Consultants the design of BAIE ST. PAUL has been adapted to the maximum dimensions of the St. Lawrence Seaway and is thus meeting the criteria of a typical laker. At 225.5 m length, 23.8 m breadth and 9 m draught the ship's parameters result in a deadweight capacity of 37,690 tonnes, while the volume of its five cargo holds is figured at 41,700 m^3. They are emptied with the help of a self-unloading system supplied by the company EMS-Tech from Belleville/Ontario. It achieves a capacity of 5450 t/h. According to the gravity principle, the cargo slides through the W-shaped hold bottom onto two belts travelling underneath in two tunnels. It is then transported upwards in buckets via an elevator and further ashore with the aid of a slewable boom.

BAIE ST. PAUL, delivered by Chengxi Shipyard in Jiangyin/China to Montreal-based CSL Group (Canada Steamship Lines) in 2012, is accelerated to a speed of 12.5 knots by an MAN B&W two-stroke engine putting out 8750 kW. The ship was built in accordance with the regulations of the British classification society Lloyd's Register of Shipping for navigation on the Great

Lakes and the St. Lawrence Seaway, its hull structure does not fulfill the conditions for deep-sea navigation. For this reason, the hull, loaded with ballast stones, had to be reinforced for the transfer voyage from the builders to Canada. The elements were dismantled after arrival in Montreal. In addition to this temporary constructional measure, the 8 weeks lasting sailing across the Pacific, through the Panama canal to Canada's Atlantic coast was planned with the support of weather consultants, in order to reduce stresses on the hull eventually caused by heavy seas. Exceptional for a cargo ship is the equipment with a dynamic positioning system in combination with transversal thrusters forward and aft. They allow safe positioning during waiting times off and in the ports and locks of the seaway. The lead vessel BAIE ST. PAUL was succeeded by sister vessels WHITEFISH BAY, THUNDER BAY and BAIE COMEAU. The owner's newbuilding programme benefitted from Canada's abolition of a 25% import tax on ships. Interestingly, the supplying country of the ships, China, had imported two similar-sized bulk carriers in 1994, when Bremer Vulkan in Germany delivered the 37,532 dwt self-dischargers HAI WANG XING and TIAN LONG XING for China's domestic bulk trade.

Canada Steamship Lines, claiming to be the world's largest operator of self-discharging bulk carriers, extended their activities and set up offices in America, Europe, Asia and Australia, so that the self-discharging principle is meanwhile used worldwide. In 2019, the entire CSL fleet comprised 58 vessels that were self-discharging or transshipment units, with the exception of six handysize gearless bulk carriers operated on the Great Lakes. The smaller self-dischargers CSL ELBE (Fig. 1.3), CSL RHINE, CSL TRIMNES and

Fig. 1.3 In contrast to the crane-equipped, 2019-built conventional 37,340 dwt bulk carrier AQUAMARINE STAR, the self-unloader CSL ELBE of Canada Steamship Lines is fitted with an automated discharge system (Ralf Witthohn)

TERTNES, mainly used for granite transport, are operated from the European headquarters in UK-based Richmond.

While BAIE ST. PAUL is being loaded in Thunder Bay, the previously handled bulk carrier ALGOMA DISCOVERY has completed its voyage on the St. Lawrence Seaway and reached Port Cartier at the mouth of the waterway. More than 160 million tonnes are moved annually on the Seaway, which has been used commercially since 1679. The main grain ports through which US and Canadian farmers load their crops are Hamilton, Windsor, Sarnia, Goderich, Owen Sound, Johnstown, Port Colborne and Oshawa. These ports are also used for the supply with fertilizer. The cargoes from Ontario are destined for Toledo, Buffalo, Montreal, Sorel, Trois Rivières, Quebec, Baie Comeau and Port Cartier, from where the grain is exported overseas. Thunder Bay is mainly used to load the wheat variety Marquis, specially bred for the short Canadian harvest season and cultivated in the prairies.

The Canadian Wheat Board, which has been responsible for marketing and sales since 1935, was partially privatised in 2015 through the sale of a 50.1% stake to a joint venture of the US agricultural group Bunge Limited and the Saudi Agricultural and Livestock Investment Company (SALIC) from Riyadh and renamed G3 Canada Limited. The company markets cereals and oilseeds in more than 70 countries. It has 18 locations close to the St. Lawrence seaway at Bloom, Glenlea, Colonsay, Pasqua, Alexander, Plenty, Prairie West, Kindersley, Luseland, Dodsland, Leader, St-Hyacinthe, Ste-Madeleine, St-Lambert, Ste-Élisabeth and the ports of Thunder Bay, Hamilton, Trois-Rivières and Québec. An export terminal with a storage capacity of 180,000 tonnes was opened in Vancouver in 2020.

ALGOMA DISCOVERY belongs to the fleet of the Algoma Central Corporation in St. Catharines/Canada, which in 2019 operated 11 self-dischargers, eight gearless bulk carriers and eight product tankers (Fig. 1.4). Seven more self-dischargers are ocean-going, including WESER STAHL, which was specially designed for ore transport to the steelworks of ArcelorMittal at Bremen/Germany. The gearless bulk carriers ALGOMA DISCOVERY, ALGOMA EQUINOX, ALGOMA GUARDIAN, ALGOMA HARVESTER, ALGOMA SPIRIT, ALGOMA STRONGFIELD, G3 MARQUIS and TIM S. DOOL are primarily used for eastbound grain transport on the St. Lawrence, while iron ore is transported westward. But they are also employed on overseas routes, including ALGOMA DISCOVERY, which was in 2009 engaged in the transport of steel from Bremen. The eight newbuildings of the ALGOMA EQUINOX class that came into operation from 2013 to 2019 have been equipped with scrubber systems from Wärtsilä to reduce sulphur in the exhaust gases. They wash the emissions of the main engine, the auxiliary diesel engines and the oil-fired boiler in order to comply

Fig. 1.4 Usually operated on the Great Lakes the 34,750 dwt Laker ALGOMA DISCOVERY ran aground in 2009 on the Lower Weser near Brake with a cargo of steel products from the Arcelor-Mittal works at Bremen. The ship was only refloated after lightering of the cargo (Ralf Witthohn)

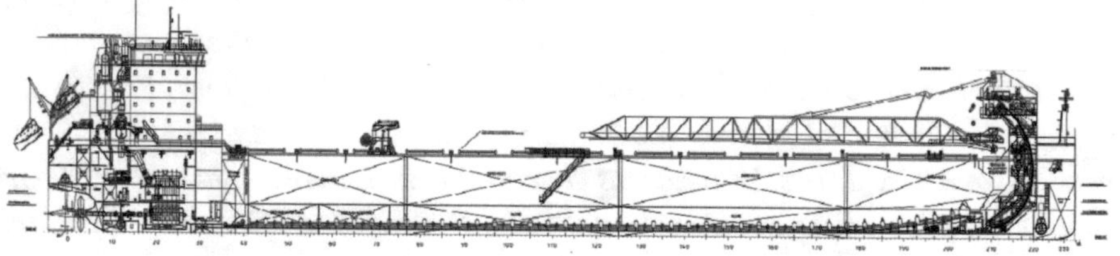

Fig. 1.5 To achieve the same speed, Algomas smaller, but equally wide 25,000 dwt newbuilding from Rijeka was fitted with the identical main engine type chosen for the 35,900 dwt version (Uljanik)

with the regulations established by the International Maritime Organization (IMO) for the Emission Control Areas (ECAs), including that on the Great Lakes in Canada and the United States (Fig. 1.5).

1.1.2 Croatian-Built Self-dischargers for Canada

When the builder of the 37,367 tdw self-dischargers ALGOMA NIAGARA, ALGOMA SAULT and ALGOMA CONVEYOR, China's Yangzijiang Shipyard, went bankupt during the construction period, Algoma subsequently contracted a new order with Nantong Mingde Heavy Industries in China, which was yet cancelled. Algoma then ordered five newbuildings of two different size classes from Croatia's Uljanik Group, headquartered in Pula, for delivery in 2017/18. In December 2017, the 198.4 m length, 23.8 m breadth ALGOMA INNOVATOR was delivered as the first of two ships of the smaller four-hatch type from shipyard 3 Maj in Rijeka. Due to adverse weather conditions on the North Atlantic, the transfer voyage was postponed until February the next year. The vessel achieves a deadweight of 25,054 tonnes. The larger five-hatch type of 225.6 m length and 23.8 m breadth was designed to carry 35,900 tonnes. Although both types were to be equipped with an identical Wärtsilä RT-flex 50-D main engine of 8725 kW, the speed forecast of 14 knots applied to both. At the same breadth thus, no higher propulsion power was expected to be necessary for the longer ship. Different from the newbuildings from China, the discharge tower and the slewing boom of the Croatian buildings were to be mounted forward. Due to financing problems at Uljanik, the remaining three Algoma newbuildings were yet cancelled. In September 2019 Algoma agreed with the Croatian builder to continue the construction of a sister ship to ALGOMA INNOVATOR, delivered as ALGOMA INTREPID in September 2020, with financing provided by the Croatian Bank for Reconstruction and Development (HBOR) (Fig. 1.6).

1.1.3 Grain Port Churchill

Grain handling activities in Thunder Bay, the northwesternmost of the Great Lakes ports, were strengthened from August 2016 on by the temporary closure of Churchill, a port 1200 km north of Thunder Bay on the Hudson Bay. After controversial discussions with local residents about the planned handling of crude oil in the sensitive region, the transport company Omnitrax Canada abandoned its plans to expand the port and in 2018 sold it to the Arctic Gateway Group representing a consortium of First Nations, local governments and investors. In addition to exporting wheat, which volumes had already fallen from around 500,000 tonnes a year to less than 200,000 tonnes in 2015, Churchill served as a loading site for wood, vehicles, general cargo, breakbulk and fuel, as well as import place for fertilisers from Russia. Shipments of grain from the initially state-owned Arctic port had taken place

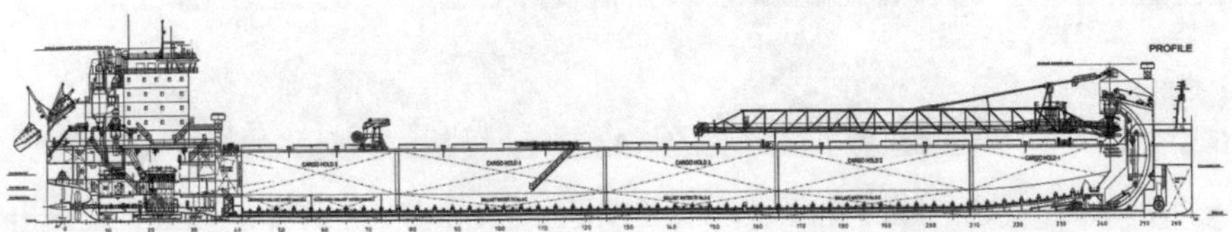

Fig. 1.6 Rijeka-based shipard 3rd. Maj stopped the realisation of this 35,900-dwt-self-discharger for Algoma due to financing problems (Uljanik)

since 1929. The navigation period is limited to the period from 15 July to 31 October, the length of the ships to 225 m.

One of the bulk carriers loaded at Churchill with 27,500 tonnes of the wheat variety No. 2 Canadian Western Red Spring was the German bulk carrier PUFFIN. The ship transported the cargo to Colombia via the mouth of the Churchill River on the western bank of the Hudson Bay over 500 nm to the Hudson Strait and another 400 nm to the Atlantic Ocean. In 2019 the port was again used by ocean-going ships, such as FEDERAL ASAHI of Canada's shipping companyFednav. The largest permissible draught at the deepest of the three grain quays in Churchill is 11.5 m at low tide, the 166 m long tanker pier can accept 9.9 m deep-going ships with the assistance of tugs. The advantage of the Canadian freshwater port compared to Thunder Bay at the Great Lakes is the almost three-day time advantage on the route to northern Europe and the avoidance of the high fees on the St. Lawrence Seaway. In November 2021, the closure of the port for 2 years until 2023 was announced due to the replacement of rail tracks leading to the site (Fig. 1.7).

Handysize bulk carriers such as PUFFIN are predestined for wheat shipments from ports with limited fairway conditions. Delivered by Shanghai

Fig. 1.7 Among the ships that exported wheat through the port of Churchill at the Hudson Bay was the German-owned handysize bulk carrer PUFFIN, which was in 2017 converted into the transshipment vessel GREENDALE for operation in the Black Sea (Ralf Witthohn)

Shipyard to Bremen-based Harren & Partner in 2003 it was time-chartered by Canada's Canfornav shipping company. At 199.9 m length, 23.7 m breadth and 11.3 m draught the ship has a deadweight capacity of 37,300 tonnes. The volume of its five holds totals to 42,779 m^3. Two months after delivery in May 2003, Harren & Partner took over the sister ship POCHARD. It was renamed POCHARD S. in 2014 after being sold to Istanbul shipping company Armador. In June 2004, another unit of the type, to be named PADANG, was delivered as BLUEBILL to the Greek shipping company Navarone.

In 2017, PUFFIN was taken over by Transship Co. in Odessa. The owner is engaged in transshipment activities and employed the ship for this task in the Black Sea. Renamed GREENDALE it was equipped with two special Liebherr cranes. Its classification switched from the Norwegian-German company DNV GL to the Russian Register of Shipping, the flag from Antigua & Barbuda to Bélize. The GREENDALE took a position in the Kerch Strait, and was in February 2022 reported to be still anchoring there. The blockade of the Ukrainian ports as consequence of Russia's military attack in February 2022 yet halted all transshipment activities.

1.1.4 Standard Designs from Shanghai

The design of the PUFFIN as described is named Odely 35,200, according to an engineering office based in Shanghai. The naval architects developed a series of bulk carrier classes along with other ship types. Four of them, of 20,000 dwt, 25,500 dwt, 30,000 dwt and 35,200 dwt, were laid out for possible operation on the Great Lakes, four others have deadweight capacities of 35,500, 56,600, 67,000 and 80,000 tonnes. Among the Odely designs is one of the first ultramax bulk carrier types, at 36 m breadth and 199.9 m length clearly exceeding the former panamax breadth of 32.3 m. The ship's midship section shows a conventional bulk carrier with hopper and wing tanks. Its deadweight capacity amounts to 67,000 tonnes, five holds have a volume of 83,000 m^3 (grain). The hatches are uniformly 20.6 m wide and 23 m long, with the exception of the 21.3 m measuring hatch 1. The holds are closed by folding covers and served by four 36 tonnes lifting revolving cranes of 28 m outreach. The ballast water capacity, including hold 3, is figured at 34,500 m^3. As main engine chosen was a two-stroke MAN B&W motor of type 6S60ME-BC8.2 putting out 9660 kW for a speed of 14 knots (Fig. 1.8).

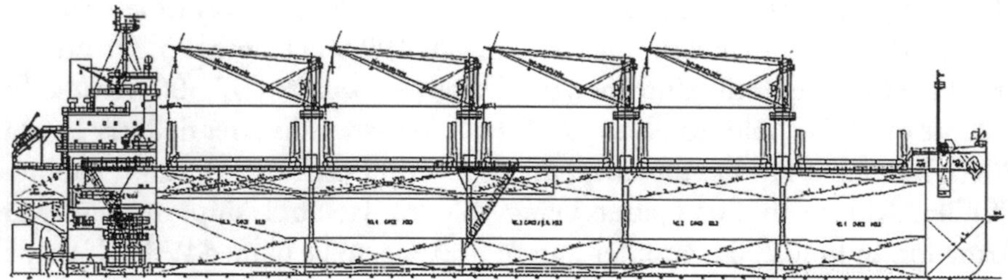

Fig. 1.8 A Shanghai design bureau offers this postpanamax wide 67,000 dwt supramax carrier designated Odely 67 (Odely)

1.1.5 Grain Transport in Handysize, Supramax and Panamax Carriers

The majority of bulk carriers engaged in grain transport, but also in other types of bulk cargo, vary in size from handysize vessels of 28,000 dwt to panamax/kamsarmax ships of 80,000 dwt. Postpanamax and capesize units with carrying capacities up to 180,000 tonnes are also employed. In 2021, the world's active bulk carrier fleet reached a number of about 11,300 units, comprising 2300 units of the capesize (180,000 dwt), 2300 of the panamax (80,000 dwt), 3900 ships of the supramax class (up to 67,000 dwt) and 2500 vessels of the handysize class (up to 40,000 dwt). The longer the voyage distance, the more economical large ships become, for example on sailings from the Americas to Asia. The share of shipping costs of cargoes in relatively large, simply constructed bulk carriers sailing at moderate speeds, accounts for about 10%.

Due to their often standardized designs, bulk carrier construction has become one of the favourite activities of Asian shipyards, especially in Japan, China and South Korea. In smaller numbers, builders in Indonesia, Vietnam and Singapore participate in this market as well. The shipyards developed different designs, though the basic concept is very much uniform across all countries featuring an aft positioned main engine, hatch arrangements with generally five holds and four cranes for the handysize to the supra/ultramax class, seven holds for panamax carriers—most of which are built without cranes—and nine hatches for the capesize class.

Thanks to the equipment with revolving cranes usually mounted in way of the transverse bulkheads, bulk carriers of the smaller size classes are independent of handling facilities, which are often lacking at smaller ports. Almost without exception, the 14 knots to 16 knots achieving ships are powered by slow-running crosshead engines of the MAN B&W or Wärtsilä/WinGD types, which have outputs of between 6000 and 8000 kW. At 100 to 130 rpm, they act directly—without reduction gear—on the propulsion shaft turning a non-variable pitch propeller. Such relatively modest vessel speeds are considered sufficient in bulk transport and are rarely superseded. The main engines are regularly manufactured under licence of European manufacturers at Asian shipyards or in mechanical engineering companies that are often part of large industrial conglomerates such as Mitsubishi, Mitsui, Kawasaki, Hyundai or Daewoo (Fig. 1.9). Originally laid out to burn heavy fuel oil they had, due to new IMO regulations, to switch to low sulfur fuel oil (LSFO) with a maximum 0.5% sulphur content from January 2020 on. As consequence, a large number of bulk carriers have since then been equipped with scrubber systems, which allow the continued burning of heavy fuel oil (Fig. 1.10).

Fig. 1.9 The handysize standard type Seahorse 35 is represented by the 35,000 dwt bulk carrier LONE STAR, in 2012 delivered by Nantong Jinghua shipyard in China to Nordic Bulk Carriers in Hellerup/Denmark (Ralf Witthohn)

Fig. 1.10 The 58,000 dwt supramax carrier GRANDE ISLAND of Hong Kong-based Pacific Basin Shipping, in 2009 delivered by Japan's Tsuneishi shipyard, is one of numerous bulkers retroactively fitted with a scrubber system in the therefore extended funnel casing (Ralf Witthohn)

1.1.6 Wheat from the Parana

In January 2017, the German bulk carrier RENATE is sailing for 16 h on the meandering Parana before reaching the La Plata north of Buenos Aires. In Puerto General San Martin, the river's last deep-water harbour, the bulk carrier has been loaded with wheat up to the maximum fresh water draught of 10.2 m. This corresponds to the maximum fairway depth to which the Parana has been dredged up to the city of 11,000 inhabitants.

The 34,850 dwt handysize bulk carrier is a product of Zchi Shipbuilding in Zhoushan, which from 2010 to 2014 laid the keels of a small series of eight newbuildings of this type designated Seahorse 35. In 2012/2013, the shipyard delivered the 180 m length, 30 m breadth, 14.7 m depth and 10.1 m draught NINA-MARIE and RENATE to Germany's MST Mineralien Schiffahrt (Minship) in Schnaittenbach, Bavaria. Lübeck shipping company F. H. Bertling received four newbuildings named AQUITANIA, PATAGONIA, ALENTEJO and DALARNA. The German owner had already taken over the 50,000 dwt ships NAVARRA and DALMATIA from the same shipyard in 2010/2011 (Fig. 1.11).

Fig. 1.11 For two established German shipping companies the keel of six newbuildings of the RENATE type was laid in Zhoushan (Ralf Witthohn)

Another Seahorse 35 type unit came into German hands, when Hamburg-based shipping company Vogemann purchased HANZE GRONINGEN from Dutch owners Hanzvast and renamed it VOGE JULIE in 2018. HANZE GRONINGEN was in 2010 built at Qidong Daoda shipyard, which afterwards delivered the sister vessels HANZE GOSLAR, HANZE GENDT, HANZE GDANSK, HANZE GOTEBORG and HANZE GENUA. VOGE JULIE is managed by company Ahrenkiel Vogemann Bolten, majority-owned by the MPC Capital Group. In its fleet of 14 handy-size carriers in 2022, there were three more Seahorse 35 units, CARLOTA BOLTEN, CAROLINA BOLTEN and, since 2019, AMORGOS BOLTEN. ANDALUCIA of type Seahorse 375, in 2013 built for Hamburg owners at Yangzhou Guoyu shipyard in Yizheng/China, switched to Hamburg-based Nordic Hamburg Shipmanagement.

1.1.7 Open Hatch Bulk Carriers

The midship section of handysize and supramax bulk carriers usually features a double bottom and triangular hopper and in the lower and wing tanks in the upper corners of the cargo hold. This kind of tank configuration is to avoid the shifting of bulk cargoes and to ease their discharge. At the same time, the tanks are, together with the double bottom tanks, serving as ballast tanks. Only rarely is new cargo available at the port of discharge, so that ballast voyages to the next port of loading are to be undertaken. The arrangement of the top tanks is yet restricting the maximum possible opening of the hatches. Because this is particularly required on breakbulk carrying ships, so-called open hatch bulk carriers (OHB) are being constructed at increasing numbers. In contrast to conventionally designed bulk carriers, the open hatch ships have a double hull formed of the shell plating and an inner longitudinal bulkhead below the hatchway coaming. The side compartments created serve as ballast water tanks.

A number of shipyards have switched from the building of classic single-hull bulk carriers to such open hatch carriers, which midship section resembles that of container ship. Such bulk carriers are suitable for breakbulk are often employed within scheduled liner services. One of the more modern bulk carrier types with a deadweight capacity of 39,000 tonnes was designed by the Shanghai-based design office Bestway under the designation Emerald 39. The standard type is laid out for the transport of cargoes like grain, coal, sulphur, cement, ammonium, fertiliser and steel rollers in the holds as well as for logs on the hatches. At a ship's breadth of 30 m and hatch openings of 25.3 m, the side tanks' breadth results in 2.3 m each. At 179 m length, 14.8 m depth, a design draught of 9.5 m and a maximum draught of 10.6 m draught, the ships achieve a deadweight of about 38,900 tonnes on the latter. The capacity of the five holds served by four revolving cranes amounts to 48,150 m^3. Main drive is a 4810 kW developing MAN B&W diesel engine of type 5S50ME-B9.3 Tier II. At a speed of 14 knots it consumes 18.8 tonnes daily securing a range of 18,000 nm. Almost all handysize and supramax types are offered by the shipyards in a variation as log carriers. These are equipped with supports at the bulwark, which are used to secure logs or sawn timber with the aid of ropes and chains (Fig. 1.12).

Among other shipyards, 2005-founded Jiangmen Nanyang Ship Engineering in Jiangmen/China realized as its first Emerald 39 unit JIANGMEN TRADER in 2012. Sister vessels are JAMES BAY, JAMAICA BAY, JERICHO BEACH, JERVIS BAY, JULES POINT, CASCADE, JURA,

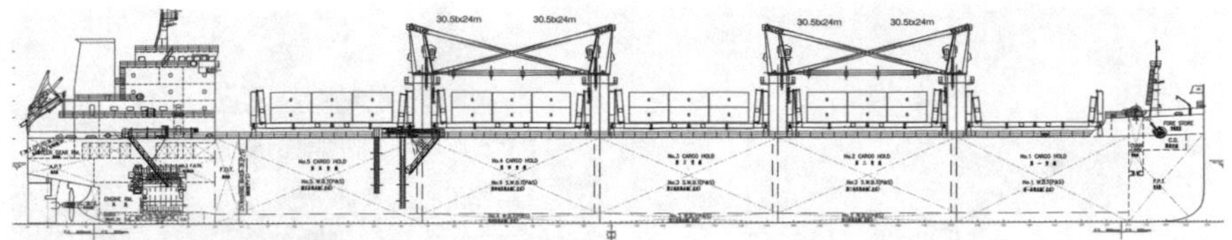

Fig. 1.12 Bestway's Emerald 39 design is an open hatch bulk carrier (Zhejiang Zengzhou Shipbuilding)

WESTERN MIAMI, WESTERN LONDON, WESTERN PANAMA, WESTERN LIMA, WESTERN DURBAN, XING ZUN HAI, XING RONG HAI, HAMBURG WAY, XING RU HAI, XING YI HAI, BERGE BANDAI, HAMBURG PEARL, CIELO DE MIZUSHIMA, BRIGHTEN TRADER, ACTION TRADER, ASTON TRADER, LA FRESNAIS, LA LOIRAIS, NORD SANTIAGO, NORD ANNAPOLIS, WESTERN DONCASTER, CANADIAN BULKER, CHILEAN BULKER, HAMBURG TEAM, TAIKOO TRADER, PAULA TRADER, HAMBURG CITY, ULTRA FOREST, ULTRA BOSQUE, ULTRA SILVA, TAC ODESSA, TAC SUZUKA and TAC IMOLA, the latter commissioned in November 2021. In the course of the series construction, the deadweight capacity was increased from 35,000 dwt to 40,200 dwt without a change of the ship's main dimensions.

38,000 dwt is the typical key figure for numerous handysize carrier designs. In Japan, Naikai Zosen took part in the supply of a handymax design by delivering MILAU BULKER, CRIMSON PRINCESS, NICOLINE BULKER, ANNE METTE BULKER, EVA BULKER, JIN YU, POS OVELIA, ECO SPLENDOR, AFRICAN JAY, AFRICAN GROUSE, BERGE HALLASAN and GLENPARK until 2017. The deadweight of the 183 m length, 30.6 m breadth and 14.5 m depth ships is figured at around 38,000 tonnes related to a draught of 10 m.

China's Huatai Heavy Industry Nantong at Rugao participated since its founding in 2010 in the realization of such a handysize type. The builder supplied Hamilton/Bermudas based Interlink Maritime with the 190 m length, 28.5 m breadth 37,070 dwt ships INTERLINK ACUITY, INTERLINK PARITY, INTERLINK VERITY, INTERLINK EQUITY and INTERLINK LEVITY. The basic design was named Dolphin 37 (Fig. 1.13) and developed by the Shanghai Merchant Ship Design & Research Institute (SDARI). Of this standard open hatch bulk carrier type more than 100 units have been built by over a dozen Chinese shipyards. The further developed INTERLINK ABILITY, INTERLINK DIGNITY, INTERLINK PROBITY and

Fig. 1.13 The Dolphin 37 class unit INTERLINK EQUITY achieves a deadweight of 37,300 tonnes (Ralf Witthohn)

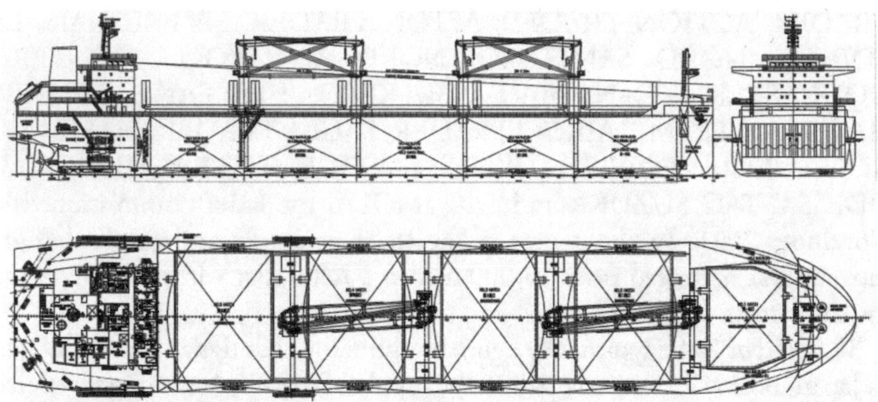

Fig. 1.14 SDARI's Dolphin 38 class shows an open hatch configuration (SDARI)

INTERLINK QUALITY as well as the INTERLINK FIDELITY, INTERLINK MOBILITY, INTERLINK SAGACITY, INTERLINK ACTIVITY, INTERLINK PRIORITY, INTERLINK TENACITY, INTERLINK VERACITY, INTERLINK CAPACITY and INTERLINK EQUALITY of type Dolphin 38 achieve 38,800 dwt at 180 m length and 32 m breadth (Fig. 1.14).

1.1.8 d'Amico's Open Hatch Ships

One of the Jiangmen Nanjang newbuildings of the Emerald 39 type was CIELO DI MIZUSHIMA delivered to Italy's shipowner d'Amico in 2016. In the following year, the Rome-headquartered company employed the 39,388 dwt bulk carrier for some time in a liner service connecting the Brazilian ports of Munguba and Outeiro with Singapore, Taichung, Changshu and Qingdao. The service was later joined by the CIELO DI TOCOPILLA, which features nearly identical dimensions. Its design, yet, is based on the Deltamarin 37 type of the Finnish engineering firm Deltamarin. D'Amico placed a total of ten orders of this design with China's Yangfan shipyard and took delivery between 2014 and 2016 of GIULIA I, CIELO DI MONACO, CIELO DI TOCOPILLA, LENTIKIA, CIELO DI VIRGIN GORDA, CIELO DI VALPARAISO, CIELO DI CARTAGENA, CIELO DI ANGRA, CIELO DI TAMPA and CIELO DI JARI.

Some of the ships were temporarily used, among other vessels, in a second d'Amico service from the Mediterranean to the Pacific Northwest and US West Coast. CIELO DI TAMPA, CIELO DI TOKYO and CIELO DI MONACO sailed from Tarragona, Savona, Livorno, Salerno and Manfredonia via Panama Canal to Lynnterm and Fibreco. The 39,200 dwt ships have a cargo hold volume of 50,800 m^3. The 2650 m^3 larger volume compared to the Emerald 39 type was achieved by extending hold 5 further aft. In 2019, d'Amico operated a total of 21 open hatch bulk carriers of the 37/39,000 dwt class, seven buildings from the Japanese shipyards Saiki, Kanda and Minami Nippon as well as four from Korea's Hyundai Mipo shipyard, in addition to the Yangfan-built ships.

1.1.9 Finnish Design for Chinese Builders

Of the 37,000 dwt open hatch Deltamarin design twelve units were ordered at the Yangfan group by Szczecin/Poland-based Polska Zegluga Morska (PZM, Polsteam). Yangfan Shipyard, together with Yangzijiang Shipbuilding, delivered ARMIA KRAJOWA, LEGIONY POLSKIE, SZARE SZEREGI, TCZEW, GARDNO, DRAWNO, SOLIDARNOSC, JAMNO, NARIE and KARLINO from 2016 to 2019. 13 units were contracted by Oldendorff Carriers from Lübeck/Germany for construction at Jinling shipyard in China. Of the conventional bulk carrier version, Deltamarin signed a contract with China Shipbuilding & Offshore International to build four units, plus four ice-strengthened ships, at Tianjin Xingang Shipbuilding Heavy Industry for China's COSCO shipping company. Two Yangfan units were built for Transbulk shipping company in Tunis/Tunesia. China Navigation Co. in Singapore ordered 24 box-shape versions of the 37,000 dwt type, of which 16 were delivered by Chengxi and eight by Zhejiang Ouhua shipyards from 2013 on, starting with WUCHANG (Figs. 1.15 and 1.16).

Fig. 1.15 The open hatch carrier CIELO DI JARI was in 2016 the last of ten Deltamarin 37 type units delivered by China's Yangfan shipyard to Italian owner d'Amico (Ralf Witthohn)

Fig. 1.16 Polsteam's 39,072 dwt open hatch bulk carrier SZARE SZEREGI was in 2017 built by the Zhoushan shipyard site of the Yangfan group according the Deltamarin 37 design (Ralf Witthohn)

1.1.10 Ukrainian Grain for the World

In addition to the transshipment vessel GREENDALE, converted from a bulk carrier, and floating cranes, the Ukrainian shipping company Transship employs the crane ship ATLAS DOUBLE, specially designed for the handling of grain from the Ukrainian ports of the Azov Sea. In the Kerch Strait between the Azov Sea and the Black Sea, four transshipment zones have been established to transfer, among other bulk cargoes, grain originating from the ports of Berdjansk, Mariupol and from the Don estuary in barges and smaller ships into larger overseas vessels up to the capesize class. The crane ship is operating about 30 nm south of Kerch in the Black Sea. During the transshipment process, it is moored with the aid of eleven 20 tonnes pulling automatic

Rolls-Royce winches between the lighter and the cargo ship to be loaded. The grabs of two 30/35 tonnes lifting articulated Liebherr cranes with an outreach of 46 m can achieve up to 90 hoists per hour. Thus, the ATLAS DOUBLE achieves a maximum handling capacity of 48,000 tonnes per day. The grabs can work up to 29.5 m above and 20.6 m below the waterline. The 78.4 m length, 2 m breadth and 5.1 m depth pontoon construction together with a heeling compensation system allows the crane ship to carry out its handling tasks at wind speeds of up to 20 m/s (7 knots) without the help of a tug. To increase stability during the handling work, the draft is increased from 2.45 to 3.5 m by a maximum of 2495 m^3 ballast. As a result, the displacement of the ship, which amounts to 3379 tonnes during transit, is increased to 5160 tonnes (Fig. 1.17).

The diesel-electric propulsion system fed by three Caterpillar C32 diesel engines of 940 kW each supplies the energy for the two cranes and the power for two Rolls-Royce azimuth propellers of 450 kW output each. They are mounted under the fore and the aft ship enabling a speed of 6.5 knots. There is accommodation for 25 people occupied with the operation of the ship and its cranes. The ATLAS DOUBLE is classified by the Russian Maritime Register of Shipping and flys the Bélize flag. The crane ship was completed by Craneship Shipyard in Kerch in 2014. In 2019, the Transship fleet comprised the bulk transshipment complexes AFINA, ALINA, BK-1 and GREENDALE, which were all converted from bulk carriers, the crane ship ATLAS DOUBLE, seven floating cranes equipped with a revolving crane, six feeder vessels and barges and 19 tugs. In addition to the units for the transshipment activities, Transship Bulk employed in worldwide service the 33,800 tdw freighters

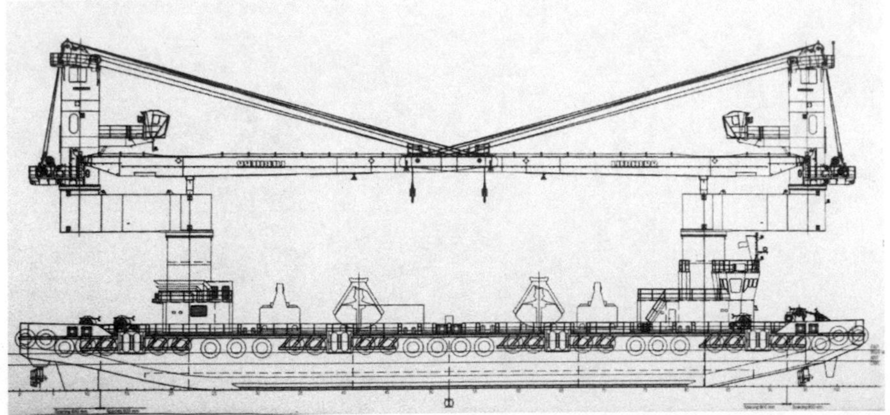

Fig. 1.17 The craneship ATLAS DOUBLE transfers grain cargoes into large bulk carriers off the port of Kerch (Craneship)

ZINA and FLORIANA, built in 2012 by the twenty-first Century shipyard in Tongyeong/Korea, and the 35,000 dwt bulk carrier TIZIANA, delivered in 2016 by Tsuneishi shipyard in Zhoushan/China (Fig. 1.18).

The fairways of the Azov Sea offer sufficient water depth only to smaller ships, the port of Berdyansk for ships with a maximum of 7.9 m. The port on the northern shore of the sea invested in the construction of a new grain terminal with a capacity of 18,000 tonnes and an annual handling volume of 1.5 million tonnes in the first phase. The terminal is equipped with a wagon discharging facility with a capacity of 350 tonnes per hour. The investments reflect the strong role of the agricultural sector, which grain production was increased to 60 million tonnes in 2017, with a two third export share. In order to cope with the increase, the expansion of ports such as Yushny is being pushed forward and water depths are being increased, for example in the Dnjepr estuary. The loss of terminals on the Crimea peninsular annexed by Russia reduced Ukraine's handling capacities. Since the construction of a bridge across the Kerch Strait to Crimea, even the sea route to the Azov Sea is restricted. In August 2017, the signal mast of the Bélize-flagged 18,000 dwt bulk carrier COPAN destined for Mariupol had to be shortened to enable its passage. Mariupol can be approached by ships at a maximum draught of 12 m, though all handling activities at Ukrainian ports were stopped after Russia's invasion and the blockage of ports in February 2022.

Fig. 1.18 Ukraine's shipping company Transship, which is primarily active in the transshipment business, operates three bulk carriers in worldwide trade, including FLORIANA (Ralf Witthohn)

1.1.11 German Wheat for South Africa

At P. Kruse companie's huge silo plant in the Hamburg district of Wilhelmsburg, the Panama-flagged bulk carrier APEX BULKER loads 10,000 tonnes of German-harvested wheat for South Africa. Afterwards, the 177 m length and 28 m breadth ship of 33,000 dwt moves through the Reethe bridge into the Reiherstieg canal to the quay of the feedingstuff company Habema to complete its cargo there.

Ships with a maximum draught of 13 m can berth at the silo facility once built by the company in 1940. Two pneumatic systems and the grab bridge achieve an hourly output of 550 tonnes. This means that 18,000 tonnes of grain or 12,000 tonnes of feed can be discharged daily, while the loading capacity amounts to 20,000 tonnes per day. The companie's conveyor system is connected with Kali-Transport on the western part of the quay, so that the maximum daily loading capacity reaches 100,000 tonnes. When over 70,000 tonnes carrying bulk carriers leave the Reethe quay, they occasionally supplement their export cargo in Rotterdam. The Hamburg bulk terminal can load and unload grain, oilseeds and animal feed at the same time. 200 cells of its silo plant take up 100 tonnes to 7000 tonnes amounting to a total of 80,000 tonnes. 1500 m trough chain conveyors, 250 m conveyor belts, 25 elevators, forklifts, wheel loaders and trim tracks are used for transport. If the moisture content of the cargo is too high, a drying system is used which can remove 4% of the water from 25 tonnes every hour. An aspiration plant cleans 120 tonnes of grain per hour (Fig. 1.19).

Fig. 1.19 18,000 tonnes of grain can daily be discharged into P. Kruse's silo in Hamburg (Ralf Witthohn)

1.1.12 Wheat from Vancouver for Japan in Clean Holds

Only weeks after delivery by Japan's Iwagi Zosen to Taiwan's Kuang Ming Shipping in November 2017, the Liberian-flagged bulk carrier KM LONDON proceeds to Vancouver to load a cargo of wheat for Japan. The holds of the newbuilding are not yet contaminated by other cargoes, so that they are best suited to be filled with sensitive agricultural goods. Iwagi has built the ultramax carrier as the third unit within a quartet of newbuildings started by KM VANCOUVER and KM WEIPA and succeeded by KM JAKARTA. That brandnew ships do not only offer exclusive cargo opportunities but may bear operational risks, becomes evident on the day the ship leaves Vancouver on 14 December 2017. Due to a technical failure in the steering system, the bulk carrier grounds in the Columbia River off Longview/Washington. The grounding causes leaks in the forepeak and in a ballast tank. The next day, KM LONDON is refloated by tugs and able to be berthed in the port of Longview for inspection and repair.

Iwagi shipyard started to deliver supramax carriers by the QUEEN SAPPHIRE in 2011 and afterwards built nearly hundred units of this type designated I-Star 61 and I-Star 63, including the KM LONDON (Fig. 1.20). The type was as well built by Imabari shipyard. The 199.9 m length, 32.2 m breadth, 19.2 m depth and 13.4 m draught KM LONDON is strengthened for the carriage of heavy cargoes, such as ore and steel coils, and allowed to sail with the cargo holds 2 and 4 empty. All five holds, which are laid out for grab operation, have a capacity of 80,500 m^3. In January 2020, the 14.5 knots

Fig. 1.20 High newbuilding numbers were achieved by the 63,386 dwt ultramax carrier type KM LONDON from the Japanese builders Iwagi and Imabari (Ralf Witthohn)

achieving bulk carrier, delivered a cargo of Swedish iron ore from Narvik at Bremen. After founding in 1990 as subsidiary of Taiwan's shipping company Yang Ming, Kuang Ming Shipping entered bulk carrier operation in 2008 and built up a fleet of own and time-chartered ultramax, panamax, postpanamax and capesize carriers numbering 18 units in 2022.

1.2 Fruit, Coffee and Cocoa

The first shipment of 50 bunches of bananas from Bélize to the United States on a sailing ship took place in 1869, the first shipment of citrus fruit from Sicily and Spain to New Orleans a year later. It was organized by immigrants of Sicilian descent and members of the mafia, who first imported bananas, pineapples, coconuts and limes from Honduras on a steamboat in 1878. Chiquita Brands, one of the largest shipping companies in banana transport today under the name Great White Fleet, attributes its beginnings to the delivery of 160 tufts of bananas on the TELEGRAPH fishing schooner from Jamaica to Jersey City in 1870 and the founding of the Boston Fruit Company in 1885, later to become United Fruit. From 1900 onwards, the introduction of refrigerated ships considerably reduced transport losses caused by early ripening of the bananas (Fig. 1.21).

In 2019, Antwerp-based Great White Fleet, a subsidiary of Chiquita, shipped bananas on three main routes, from Central America to the east and west coast of the US and to Europe. For the latter, newly established in January

Fig. 1.21 In 2014, Chiquita returned four CHIQUITA SCHWEIZ ships to the lessor, 2 years later the pictured Bremerhaven fruit terminal was cancelled from the companie's sailing schedule in favour of Vlissingen (Ralf Witthohn)

2019, were container vessels chartered, named HANNAH SCHULTE ex MAERSK NEEDHAM, COSMOS, EM CORFU and employed them together with the owned CHIQUITA EXPRESS and CHIQUITA TRADER (Fig. 1.22). Their port rotation included calls at Vlissingen, Moin, Almirante and Manzanillo with transshipment opportunities at Manzanillo to and from Puerto Cortes, Puerto Barrios and Santo Tomas de Castilla. The chartered vessels were replaced by the Japanese-owned COOL EXPRESS, COOL EAGLE and COOL EXPLORER built from 2018 on. Before, Chiquita had chartered the conventional refrigerated ships CHIQUITA BELGIE, STAR FIRST STAR SERVICE 1, STAR QUALITY and STAR TRUST for the service connectiong Santa Marta, Puerto Limon and Almirante with Sheerness and Vlissingen (Fig. 1.23). In January 2016, Bremerhaven, which had been served by reefer vessels since the 1920s, was removed from the Central America-Europe service timetable.

The German-built container vessels CHIQUITA VENTURE (Fig. 1.24) and CHIQUITA PROGRESS operating together with the former German-owned GARWOOD from Port Hueneme on the US West Coast to Manzanillo, Puerto Quetzal and Puerto Madera are equipped with 400 reefer connections. This Great White Fleet service was set up in 2017 in replacement of a slot arrangement on Hamburg Süd container ships.

1.2.1 Bananas for St. Petersburg

On a Sunday noon the German-owned reefer vessel HANSA BREMEN operated by the Dutch shipping company Seatrade leaves the fruit terminal of Guayaquil

Fig. 1.22 From Januar 2019 until 2021 Chiquita chartered five container vessels including HANNAH SCHULTE ex MAERSK NEEDHAM for its newly launched North Europe-Central America service (Ralf Witthohn)

Fig. 1.23 Only in 2014 Chiquita had replaced four vessels of the CHIQUITA SCHWEIZ reefer class by ships of the pictured conventional STAR TRUST type, which were, substituted by container vessels and, in their turn, partly by Japanese-owned reefers (Ralf Witthohn)

Fig. 1.24 The German container vessel CONTI SALOMÉ transports bananas from Guatemala to California under the name CHIQUITA VENTURE (Ralf Witthohn)

and runs seawards at speeds of only a few knots on a tributary of the Guayas river in front of the Isla Trinitaria. During 3 days, the four holds of the ship have been filled with grass-green bananas, until the draught marks indicated an immersion of 8.5 m, one meter below the maximum draught. Now the ship has still to sail 3 h, before it passes at 13 km through the narrowness between the mainland tip and the Isla Puna. The humming vessel's ventilators cool the sensitive fruit load at 12 °C, the refrigerating process is automatically monitored. A total of 6450 pallets of fruit packed in cartons are stored on four cargo decks. On the hatch covers is space for 75 40-ft-long reefer containers to be handled by the ship's four revolving cranes.

Two days after departure from Guayaquil, HANSA BREMEN anchors off Balboa and, after 2 days of waiting, is granted permission to enter the Panama Canal. Under the Bridge of the Americas, the ship sails to the western one of the old Miraflores locks, in the east of which the container ship MAERSK BUTON is heading for the Atlantic. 12 h later, the reefer arrives at Concepcion at the eastern exit of the Panama canal and, at a voyage speed of over 17 knots, sets course for the European ports of St. Petersburg, where the ship calls 17 days later, and Helsingborg, reached 3 days on. The bananas are still green and brought into a saleable condition under temperature control in a ripening facility.

The building of HANSA BREMEN and its sister vessels was sponsored by private investors' capital available in abundance on the ship investment market. It helped the Hamburg-based finance company Hansa Treuhand to secure the financing of the reefer vessels built by Bremer Vulkan between 1989 and 1991 and managed by Leonhardt & Blumberg. Since the delivery of the Hamburg-Süd reefers POLAR HONDURAS and POLAR COSTA RICA by Lübecker Flender Werft in 1980, the construction of reefer vessels had been suspended at German shipyards. Two replicas of the HANSA BREMEN class, CHIQUITA BREMEN and CHIQUITA ROSTOCK, were built in the following 2 years at Schichau Seebeckwerft in Bremerhaven, then part of the Bremer Vulkan shipyard group, for employment by the Great White Fleet of the US group Chiquita Brands International.

At 156.9 m length and 23 m breadth the 21 knots fast ships offer a cargo hold volume of 16,725 m^3 (590,654 cubic feet). After delivery, HANSA BREMEN, HANSA VISBY, HANSA LÜBECK and HANSA STOCKHOLM were chartered by Stockholm-based Cool Carriers. In 2013, HANSA VISBY and HANSA STOCKHOLM changed name to BALTIC SUMMER and BALTIC SPRING, resp., under the flag of Ost-West-Handel und Schiffahrt in Bremen, which in 2022 still listed 22 reefer vessels in its fleet (Fig. 1.25). In early 2022, BALTIC SUMMER was en route from Puerto Bolivar, Ecuador, to Helsingborg, BALTIC SPRING from St. Petersburg to Dakar, Senegal. The former CHIQUITA BREMEN was at the same time underway from that African port to Guayaquil under the name BALTIC WINTER for Ost-West-Handel und Schiffahrt. ROSTOCK REEFER ex CHIQUITA ROSTOCK, HANSA BREMEN and HANSA LÜBECK, at last managed by Dutch company Seatrade, were scrapped in 2017.

The era of conventional refrigerated ships such as the HANSA BREMEN is coming to an end. On many routes, conventional transport on pallets has already been replaced by shipments in refrigerated containers. Loss making

Fig. 1.25 In October 2019, the former HANSA VISBY transported a cargo of bananas from Puerto Bolivar/Ecuador to St. Petersburg and still served on the same route in February 2022 under the management of Ost-West-Handel und Schiffahrt as BALTIC SUMMER (Ralf Witthohn)

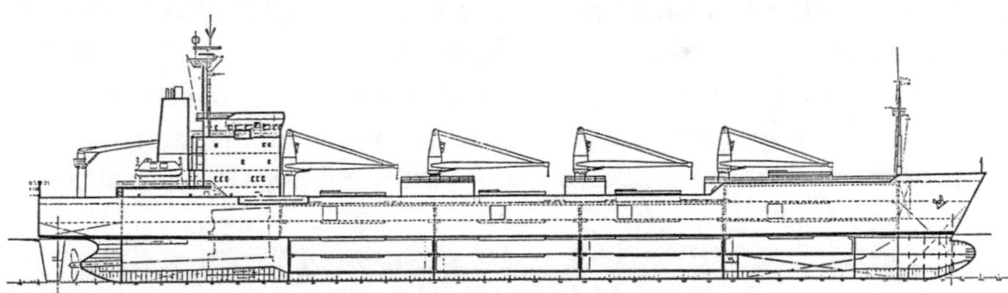

Fig. 1.26 Designs as of this 15,570 m^3 (550,000 cbf) reefer developed by Flender Werft in Lübeck remained in the shipyard's drawers due to the increasing switch of refrigerated cargoes to container ships after the 1990s (Flender)

charter rates during the container shipping market collapse in wake of the 2008 financial crisis let shipowners look for employment alternatives. They undercut the charter rates of conventional reefer vessels and consequently displaced them (Fig. 1.26).

1.2.2 Oranges from Durban

Unlike bananas, which are harvested unripe, some types of fruit can only be sent on their journey when ripe. These include oranges, of which South Africa exports around 1.2 million tonnes each year from May to November. The total export of fresh fruit reached a weight of 3.3 million tonnes in 2017. It is primarily transported by container ships, which have been deployed between South Africa and Europe since 1976 as part of the South Africa Europe Container Service (SAECS). Numerous mergers and acquisitions in container shipping reduced the number of participating shipping companies. In 2019, Maersk Line, its former South African subsidiary Safmarine, Japan's container alliance ONE and Deutsche Afrika-Linien from Hamburg were still involved in the weekly liner service offering an average capacity of 7300 teu. They provided a total of eight ships of between 6300 and 8800 TEU (Twenty-foot Equivalent Unit) capacity in a port rotation with partly double calls including Rotterdam, London Gateway, Bremerhaven, Rotterdam, Algeciras, Cape Town, Ngqura, Durban, Ngqura and Cape Town.

Due to the high proportion of fruit cargo to be cooled, the SAECS ships have up to 1700 electrical connections. In the beginning of 2022, five of the ships, SANTA RITA, SANTA TERESA, SANTA CLARA, SANTA ISABEL and SANTA CRUZ, wore the red Hamburg Süd colour (Fig. 1.27). They carry between 7150 and 8200 teu and have up to 1600 reefer plugs resulting in a 39% reefer share of the total cargo volume. The MOL PROFICIENCY contributed by Japan's consortia ONE can cool 2570 teu of its 6350 teu resulting in 40% of the intake. For this ability, yet, the freighter had to be equipped with additional generators in 2014. Since then, the total number of the gensets is being numbered nine with an aggregated output of

Fig. 1.27 In the SAECS service from Europe to South Africa Maersk Line's 8200 teu ship SANTA URSULA can carry nearly 40% of its cargo in reefer containers (Ralf Witthohn)

19,140 kVA. In 2020, ONE deployed the 9300/9100 teu charter ship AKADIMOS and MAIRA XL with 1500/1700 reefer plugs. Deutsche Afrika-Linie contributed the newly purchased 6589 teu ship DAL KALAHARI fitted with 1162 reefer plugs in 2021. A second SAECS liner service with smaller container ships was temporarily set up during the fruit season over several years, but last established in August 2009. German consortia member (Figs. 1.28 and 1.29).

1.2.3 Organic Coffee from Honduras

It is accomplished in April 2017. After a two-year reconstruction period in Elsfleth and an 8 months lasting trip across the Atlantic, the sailing vessel AVONTUUR discharges 200 bags each containing 70 kg of organic coffee from Honduras and a few barrels of rum and gin at the Schlachte quay in Bremen. In an extraordinary project, Captain Cornelius Bockermann and 160 volunteers have rebuilt the former two-masted gaff-sail schooner, built in the Netherlands in 1920 and used as a motor-driven coaster until the 1980s, at last for daily passenger trips, into a sailing ship fitted with sails of 612 m² area and an auxiliary engine of 224 kW output. Under the slogan Timbercoast, *ecologically thinking producers and*

Fig. 1.28 Reefer container vessels of the MAERSK LEON type are serving in liner trades with high shares of cooled cargo, such as the routes from South Africa, South America and Australia (Ralf Witthohn)

Fig. 1.29 Temperatures in reefer containers have to be steadily controlled as well aboard the ships as ashore (Ralf Witthohn)

traders bring organic products in a climate-friendly way to the consumers on the AVONTUUR, as an exemplary alternative to conventional ocean shipping. From 2016 on, the sailing vessel undertook a series of voyage to Central America, at last in March 2021 to Puerto Cortes (Fig. 1.30).

1.2.4 Cocoa from Ghana

A long line of trucks, highly loaded with sacks, are waiting to be discharged on the quay of Tema. On top of the bags two or three local workers are heaving the sacks on pallets, which are lifted into the holds by the ship's cranes. Not far from the Ghanaian port, the cocoa-growing areas extend far beyond the border to the Ivory Coast. One million farmers contribute a significant share to Ghana's economy.

By harvesting about 800,000 tonnes per year, Ghana is the most important supplier of cocoa next to Côte d'Ivoire. In West Africa, South and Central America in particular, a total of around 4 million tonnes are cultivated annually. Important receiving countries are the Netherlands, the United States and Germany, where the annual per capita consumption is 4 kg. But the majority of the beans are processed in the Ivory Coast itself. The Ghanaian cocoa industry is organised by the state Cocoa Marketing Board. Together with

Fig. 1.30 After extensive restoration and conversion back to a sailing ship, the twin-masted AVONTUUR set course for Central America for the first time in July 2016 (Ralf Witthohn)

internationally operating companies such as the US Cargill Group, the Noble Group in Hong Kong or the Danish Maersk Line, its sister company APM Terminals in cooperation with Bolloré Africa Logistics and the Ghanaian port company, it was decided to build a multi-purpose terminal in Tema with an investment volume of 1.5 billion US dollars. In June 2019, the first construction phase was inaugurated for cargo handling at two berths, while two more berths are to be opened during 2020. The terminal is to have an annual handling capacity of 3.5 million TEU.

Whereas cocoa was traditionally transported in sacks—nowadays mostly loaded into containers—it is meanwhile also shipped in bulk and discharged from cargo ships into weather-protected halls with the aid of cranes and conveyor belts. Such a facility is working in Amsterdam, which is regarded as the world's largest cocoa import port. By a market share of 26%, the Netherlands is the largest importing country, ahead of Germany and Belgium. The cocoa imported into the Netherlands in 2015, mainly from West Africa, amounted to 1.1 million tonnes, worth EUR 2.1 billion in 2018.

1.2.5 Antarctic Vegetables for the Mars Flight

On 8 October 2017, a special container showing the inscription EDEN ISS swings into the hatch of the multi-purpose cargo vessel GOLDEN KAROO at C. Steinweg's South-West Terminal in Hamburg. The liner of Hamburg-based shipping company MACS is bound for South Africa and, after having complemented its cargo at Antwerp and Leixoes, arrives at Cape Town via Walvis Bay/Namibia a month later. In Cape Town, the container is reloaded onto South Africa's research icebreaker S. A. AGULHAS, which delivers the box shortly before Christmas to the German Antarctic station Neumayr III of Bremerhaven-based Alfred Wegener Institut (AWI). At the end of December, the container becomes the scene of an extraordinary experiment. In the hostile environment of the eternal ice, a scientist from the German Aerospace Center (*Deutsches Zentrum für Luft- und Raumfahrt; DLR*) starts a plant breeding research project in Antarctica. As part of the EDEN-ISS project, he cultivates vegetables under largely self-sufficient conditions in order to test the food supply for manned missions to the moon and Mars. This is done using a technique known as aeroponics, in which plants are cultivated sterilely without soil and sprayed with a computer-controlled water-nutrient mixture (Fig. 1.31).

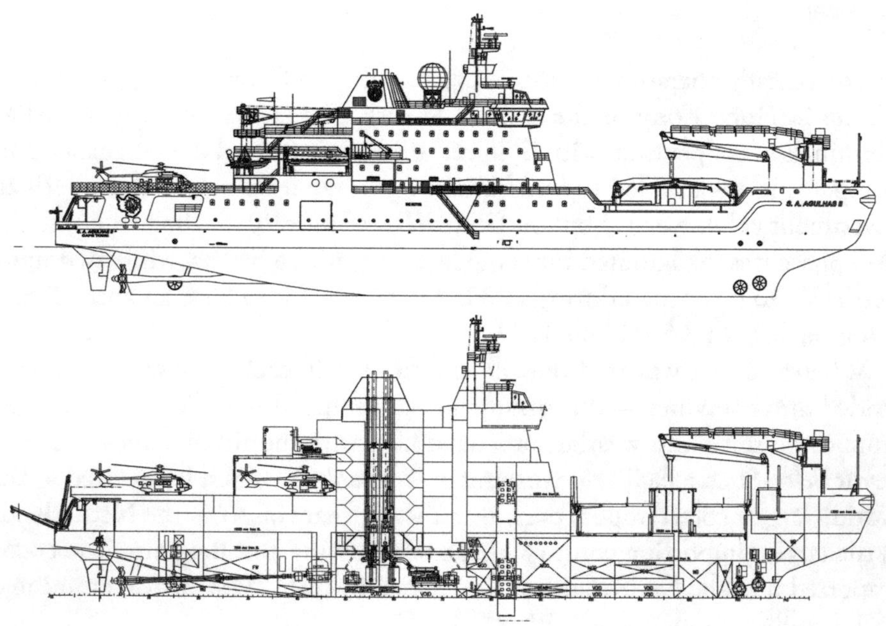

Fig. 1.31 South Africa's research icebreaker S. A. AGULHAS was built by STX Finland in Rauma (STX Finland)

Since 1982, the research icebreaker POLARSTERN has regularly been supplying the crew living at Neumayr Station. 60 tonnes of provisions, equipment and fuel in tank containers were every year lifted by the ship's crane to the about 12 m high ice shelf edge. Up to four of these containers are equally pulled to the station by snow cats on sledges. In line with the ship's dual function as a supply and research vessel, its small container capacity of only about 70 teu has to be distributed among provisions and research equipment. POLARSTERN is also equipped with helicopters, the use of which under Antarctic conditions has, however, twice prompted serious accidents. The ship regularly takes fresh provisions for the Antarctic station in Cape Town. They are supplemented by flights from South Africa, which are yet only possible during the Antarctic summer.

Built at costs of EUR 100 million, the 119 m length, 25 m breadth POLARSTERN is powered by a 14,100 kW generating propulsion system. It enabled the vessel to cover over 1.5 million nm on its sailings to the polar zones. In September 2019, POLARSTERN started a one year lasting expedition from Tromsøy, Norway, during which the vessel was willingly trapped in the Arctic ice to act as scientific base for 600 people from 19 countries. During this time, POLARSTERN was burning marine diesel oil (mdo) and by this stressing the otherwise relatively low burdened Arctic environment. Developing a more environmental propulsion concept is especially challenging for ships undertaking long lasting expeditions. Nevertheless experts ask, which sense it makes to further research a profound, obvious environmental disaster, as climate change is, instead of taking effective measures to fight it. In December 2019, 100 members of the expedition team were replaced by new personnel carried to the POLARSTERN by the research icebreaker KAPITAN DRANITSYN. Another crew change took place in January 2020 with help of the Russian research icebraker AKADEMIK FEDOROV. A personnel replacement planned in April by plane from Spitsbergen had to be cancelled due to a closure of the isle in the wake of the COVID-19 pandemic. Instead, the Swedish icebreaker ODEN was intended to carry out the team change (Fig. 1.32).

POLARSTERN is to be replaced by a newbuilding. In November 2010, the countrie's Science Council (*Wissenschaftsrat*) in Berlin had issued a recommendation for the project; in December 2015, shipyards were invited to submit bids. The operator of the first POLARSTERN, Rostock-based Reederei F. Laeisz, was awarded a contract for advising the construction and management of the newbuilding, for which Hamburgische Schiffbau-Versuchsanstalt (HSVA) carried out tests with a model based on innovative lines. In February 2020 Germany's Ministry of Research cancelled the tender proceedings,

Fig. 1.32 Germany's research icebreaker POLARSTERN will be nearly 50 years old, before it is to be replaced by a newbuilding (Ralf Witthohn)

because no acceptable offer would have had been made. Competing for the building contract at estimated costs of about EUR 650 million were Lloyd Werft at Bremerhaven and the Finnish design and engineering company Aker Arctic from Helsinki. The tender process was to be re-started under supervision of AWI with the aim of commissioning the newbuilding in 2027. In June 2022, Germany's parliament agreed to provide additional financial means.

In addition to the Neumayr III station, AWI operates the Dallmann laboratory at the Argentinean Carlini station in Antarctica, for which logistics tasks are handled by the partner, as well as the Kohnen station and the Drescher ice camp. The supply and exchange of Arctic research sites is much easier because they can be reached from the air all year round. AWI has a 4800 m^2 store in the Bremerhaven free port near the ship's regular berth for keeping goods to be shipped with POLARSTERN as well as equipment and spare parts of the ship, such as a 28 tonnes streamer winch, which is used for seismographic research on marine geology. AWI, which is open to international research with foreign scientists, also provides logistical support, for example for Russia, which supply ship AKADEMIK TRYOSHNIKOV took over a container of the institute with spare parts for a left construction vehicle

Fig. 1.33 To optimize the design of the new POLARSTERN, tests were carried out by using this model in an ice tank (Ralf Witthohn)

in Bremerhaven in February 2018. Bremerhaven has over decades also been used as logistic supply location for the Russian research icebrecaker AKADEMIK FEDOROV, which, for example, took over kerosene for the operation of the ship's helicopters, before leaving for Cape Town in the beginning of 2020. Joint German-Russian research activities, which included the participation of Russian scientists in the 2020/21 MOSAIC expedition, have been suspended after the Russian attack against Ukraine (Fig. 1.33).

Prior to the POLARSTERN II project plannings, a European project initiated by AWI to build one of the most expensive research ships ever planned had been abandoned. The investment in the AURORA BOREALIS, to be used as an icebreaker, research and drilling vessel, was estimated at EUR 450 million. The Hamburg engineering office Wärtsilä Ship Design Germany and AWI informed the public in 2007 about the widely elaborated but finally not realized design. At a length of just under 200 m and a breadth of 49 m, the newbuilding was to have a displacement of about 65,000 tonnes. Its diesel-electric propulsion system was designed for an output of 94,000 kW, the accommodations were to provide space for 120 persons. The research institute derived the need for the ship to be financed from state funds from Europe's interest in understanding the Arctic environment and clarifying unresolved climate issues. The ship should not search for raw materials but carry out basic research. In contrast to POLARSTERN, which is only operational in Antarctica and the Arctic during the respective summer seasons, the new ship should be able to undertake research voyages all year round.

In order to withstand temperatures down to -50 °C, the use of FH36 grade steel in thicknesses of up to 70 mm and an outer skin to be tripled in some areas were planned. In contrast to conventional icebreakers, which capabilities are essentially based on a reinforced foreship, AURORA BOREALIS was to be built with a correspondingly high lateral strength due to its planned use as a dynamically positioned drilling vessel. In this way, it should still be possible to maintain the position in drifting ice of 2.5 m thickness. In order to have sufficient mobility at the same time, the ship's breadth in the waterline was chosen to be a few metres smaller than the largest breadth of 49 m at the height of the deck above, namely 45 m at a maximum draught of 13 m.

AURORA BOREALIS was laid out to drill through a moon pool in the ship's bottom in a water depth of 4000 m up to 1000 m deep into the seabed under a closed ice cover. Six transversally acting thrusters under the keel should enable the dynamic positioning of the ship. In order to resist the pressure of the pack ice, it was planned that the ship could move with the help of a ballast system and thus free itself. The design proposed a propulsion system acting on three ice-reinforced fixed-pitch propellers with a diameter of 6.5 m rotating side by side under the stern. One of them should be sufficient to reach the service speed of 12 knots in open water, the two outer ones would have been needed above all to generate additional power during ice breaking. They should enable the ship to reach a speed of 2 to 3 knots when sailing through several years old ice of 2.5 m thickness, and a maximum speed of 15.5 knots in ice-free water. Of the total output of 94,000 kW generated by eight diesel engines a maximum of 81,000 kW was to be available for the three electrical propulsion motors (Fig. 1.34).

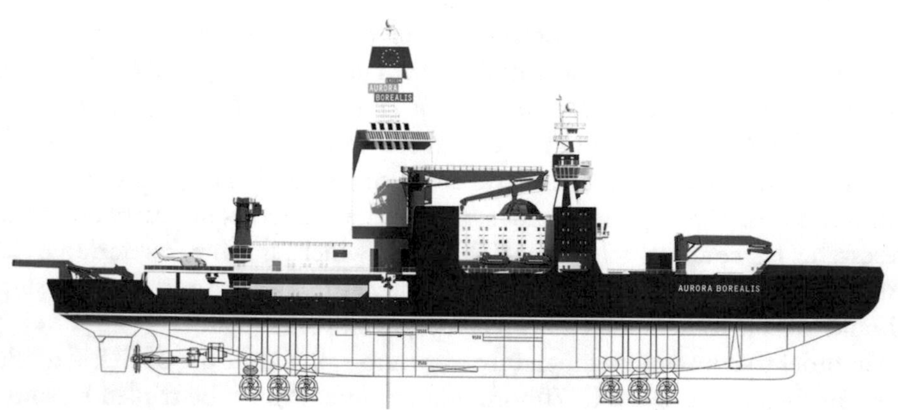

Fig. 1.34 One of the most sophisticated projects of a research vessel to be named AURORA BOREALIS did not go beyond the design stadium (AWI)

1.3 Salt and Sugar

In addition to being used as a basic chemical in the production of chlor-alkali, PVC and soda, as a fertilizer in the form of potash or as a spreading agent, salt is shipped as a food additive. The most important exporter is Australia, which sells more than 10 million tonnes per year to Japan, China, South Korea and Indonesia, ahead of China, which supplies over 2 million tonnes per year to the United States. The Kassel/Germany-based company K + S is the world's largest producer with production facilities in Europe, North and South America. Apart from mines, salt is obtained by evaporating sea water.

Like salt, sugar is transported in bulk, in bags as breakbulk or in containers. In 2018/2019, Brazil exported 19.6 million tonnes, followed by Thailand (11.5 million), Australia (3.8 million) and India (3.4 million). In 2016 Brazil had exported 43% of all sugar worth USD 10.4 billion, followed by Thailand with 9% and India with 6%.

1.3.1 Fish Salt from Zarzis

4000 tonnes of salt are loaded in the hold of the Ukrainian coaster SHUYA, when the ship grounds on Scharhörn reef in the Elbe estuary in strong northwesterly winds. The ship is on way from Zarzis in Tunisia. There, the high-quality salt has been extracted from the sea in salt works and was loaded for Fredericia in Denmark to be used as a preservative in the fish processing industry. For 5 days, the salvage attempts by Cuxhaven-based tugs and the authoritie's multi-purpose emergency vessel NEUWERK fail, as the notorious grinding sand holds the ship in its clutches, until it can at last be taken to Cuxhaven. Built in 1994 by Cassens-Werft in Emden for Russia's White Sea—Onega Shipping, the 4143 dwt coaster has since 2016 been owned by the Ukrainian shipping company Profy from Odessa, operating the ship under the flag of Antigua & Barbuda (Fig. 1.35).

Fig. 1.35 The Ukrainian coaster SHUYA carried a cargo of salt when grounding in the Elbe estuary in July 2017 (Ralf Witthohn)

SHUYA's salt came from the Sabkhet Lâadhibet saltworks, which extend over an area of 3200 ha at a distance of 70 km from the Zarzis port of loading. Constantly renewed by the underground contact of brine with seawater, the capacity of the deposit is estimated at several ten million tonnes. Pumps convey the high-percentage brine so that crystallization takes place quickly and without any further concentration process; evaporation being accelerated by the hot climate. The salt is exported to Europe, North America and African countries, where it is used in the food industry, but also as a gritting agent. Salt imports in Germany regularly take place via Wismar, which handled 690,000 tonnes in 2016. Of these, 533,000 tonnes were de-icing salt and 32,000 tonnes fishery salt (Fig. 1.36).

1.3.2 Sugar for Toronto

Carrying a cargo of raw sugar from Paranagua, the 30,000 dwt bulk carrier TORRENT has reached the Quay East branch basin at Toronto. The Redpath Sugar refinery has a modern, German-built discharge facility here, which lifts up to 600 tonnes of sugar out of the holds every hour. The 160 m long sting basin is too short to accommodate the entire length of the 185 m long bulk carrier of Greek's Navarone shipping company, so that the stern of the ship protrudes into the

Fig. 1.36 In the port of Wismar, several hundred thousand tons of salt are handled every year, as to be seen here from the Russian 3200 dwt cargo vessel IVAN SHCHEPETOV (Ralf Witthohn)

harbour. At a breadth of 24 m, however, the TORRENT, in 2010 built by Weihai shipyard in Shandong, has been designed to pass through the locks of the St. Lawrence seaway. Coming from the Atlantic, the vessel has passed Québec, Trois Rivières, Lac-Saint-Pierre and Montreal before reaching Lake Ontario and Toronto at its eastern shore. There the raw sugar is refined and distributed to Canadian and US ports on special "Lakers" designed for sailing the Great Lakes—St. Laurence seaway system.

1.3.3 Sugar for Bulgaria

To be classified at the lower end of the handysize scale is a 16,190 dwt bulk carrier delivered by the shipyard Shin Kochi Jyuko in Kochi in 2001 under the name CENTURY SEYMOUR to Japan's Dowa Line. After being sold to a Greek shipping company it was renamed BARU SATU in 2011. At only 137 m length and 23 m breadth, the ship features four cargo holds being served by four cranes, including a double crane. Shin Kochi built a long series of similar handysize bulker carriers. Over a period from 2000 to 2015 Dowa Line received the similar dimensioned CENTURY ELKHORN, CENTURY CYPRESS, CENTURY SEYMOUR, CENTURY QUEEN, CENTURY PEARL, CENTURY VENUS, CENTURY MELODY, CENTURY ROYAL, CENTURY DREAM, CENTURY BRIGHT, CENTURY EMERALD and CENTURY GOLD. While the ships of this type accept transport orders mainly in the Far East, the CENTURY SEYMOUR, after being sold and renamed BARU SATU, also sailed for European destinations. In July 2013 the ship had loaded a cargo of sugar for Bulgaria when it collided with the Greek bulk carrier KATHERINE off the isle of Andros/Greece and severely damaged it on the port side. Renamed LADY SEMA, the ship undertook a voyage from Istanbul to Ravenna in February 2022 on behalf of Turkish owners.

Dowa Line, founded in Tokyo in 1957 to transport coal from the port of Kyushu, extended its activities in 1961 to a liner service between Japan, Hong Kong and Singapore and set up a New York branch in 1980. In 2019, Dowa Line America operated 26 bulk carriers ranging between the 5700 dwt type BALSA 81 and the 24,300 dwt ship OREGON HARMONY originating from the shipyards Shin Kurushima, Shin Kochi and Toyohashi (Fig. 1.37).

Fig. 1.37 Built within a series of a dozen 16,000 dwt bulk carriers, BARU SATU came under Greek and, as LADY SEMA, under Turkish management and subsequently regularly sailed European waters (Ralf Witthohn)

1.4 Meat

Unlike fish, which is shipped deep-frozen if not chilled under ice as fresh fish, meat is often transported at temperatures of −1 °C to 0 °C. This atmosphere ensures maturation during the usually several weeks lasting sea voyage. In the past, the meat was transported in refrigerated rooms of general cargo freighters or in refrigerated ships, but the switch to container transport necessitated the development of refrigerated containers. Since the Australian trade with its meat exports was considered an important market for a long time, right at the beginning of containerisation in the early 1970s, a functioning refrigerated container technology was a prerequisite for the introduction of the new transport technology. Hamburg-Süd began to operate a cross-trade service from Australia/New Zealand to the east coast of the United States and Canada as early as in 1971 by putting three reefer container ships into service. They were able to cool almost half of their 1200 containers. For this purpose, the specially developed 'porthole' system with a central refrigeration system working via ventilation shafts was used, usually named Conair system according to the name of one of the developing firms. Over the years it was replaced by containers with their own refrigeration unit and power supply via the ship's sockets. With these, however, a high air exchange rate became necessary to dissipate the waste heat from the holds. Much later than in the Australian trade—from 1989 on—large new container ships were also put into service by Hamburg-Süd for employment between Europe and the East Coast of South America.

1.4.1 Container Ship for 2100 Reefer Units

After delivery by Korea's Hyundai Samho shipyard in November 2016, VALPARAISO EXPRESS of Germany's shipping company Hapag-Lloyd leaves

Busan and sails under ballast conditions to Qingdao, 480 nautical miles away, to pick up a full load of brand-new containers. Within five weeks, the newbuilding crosses the Pacific to take over an initial cargo for Northern Europe at Valparaiso. In addition to Hapag-Lloyd containers, this includes boxes from partners CMA CGM and Hamburg Süd. Chilean beef and beef products account for a significant proportion of the cargo showing high growth rates. By its 2100 reefer connections supplying 40% of the ship's total container capacity, the vessel is featuring one of the highest reefer shares. The number of containers corresponds to a volume of 110,000 m^3 of reefer cargo and is thus about ten times the capacity that is being achieved by the largest conventional reefer vessels. The 10,589 teu newbuilding and its four sister ships were built according to the same design as newbuildings for Hamburg Süd, but have higher lashing bridges for improved cargo securing. At 333 m length, 48.2 m breadth and 14 m draught, the ship is able to pass through the new Panama canal locks and therefore referred to as a neopanamax vessel. Via Callao, Manzanillo, Cartagena and Caucedo, the German-flagged VALPARAISO EXPRESS sets course for Rotterdam, its homeport Hamburg, London and Antwerp (Fig. 1.38).

1.4.2 Meat from South America

During the voyage from the East Coast of South America to Europe on the traditional route of the Maersk-owned Hamburg Süd company, the cargo shares of a container ship consist of about 25% poultry, pork, beef and fish, 22% coffee, tobacco and other foodstuffs such as cereals, peanuts, juices and sugar, 20% fruit, vegetables and plants, 18% chemicals, plastics, resins, paper and wood, 12% car parts, steel products and machine parts as well as 3% other cargo such as leather, shoes, textiles, ceramics and stones. From Asia to the east coast of South America, mainly finished products such as electronics

Fig. 1.38 Hapag-Lloyd's VALPARAISO EXPRESS was the the lead ship in a series of five such units for the South American West Coast route, the type of which had previously been delivered to Hamburg Süd (Ralf Witthohn)

and household goods, car parts, steel products, machine parts, chemicals, plastics and resins are transported.

Hamburg Süd, which had from its beginnings in 1871 concentrated its activities on the South America East Coast trade, expanded its business after the second World War to include numerous other trade lanes, several times by integrating acquisitioned shipping companies. Despite this expansion, the companie's special position in transport of refrigerated cargoes, originally stemming from its traditional shipping route served, was retained. Before being taken over itself by Denmark's A.P. Møller/Maersk Line in 2017, the Hamburg owner completed a ten-year newbuilding programme, which started in 2004 and meant the commissiong of a total of 41 large post-panamax reefer container ships. By this, Hamburg Süd made the leap to become one of the ten largest liner shipping companies in terms of capacity. The reefer container ships were ordered in several lots from Korea's Daewoo and more recently from Hyundai shipyard. The fleet expansion was the largest investment project of the company ever. As second largest German liner operator after Hapag-Lloyd it ranked ninth among international container ship operators by disposing 116 vessels including the chartered units with a total capacity of 600,000 teu. This corresponded to 2.6% of the world's container fleet. For the newbuilding series, the shipping company chose ship designs which took into account the respective type of cargo type. This included the choice of a large ship's depth in order to create more below deck room for the protected carriage of reefer containers accumulating in high proportions (Fig. 1.39).

Fig. 1.39 As a typical reefer container ship type Hamburg Süd's first postpanamax vessels of the MONTE SARMIENTO class feature a higher than usual depth (Ralf Witthohn)

Being involved in cross-trade shipping including Panama canal passages for decades, the newbuilding series meant the first acquisition of postpanamax container ships for Hamburg Süd. Since the expansion of the Panama canal they fulfil the criteria of neopanamax ships and are therefore no longer subject to restrictions in the Atlantic-Pacific relation. The 41 newbuildings for Hamburg Süd are divided into eight units of the MONTE CERVANTES class with a loading capacity of 5560 teu, six ships of the RIO DE LA PLATA class carrying 5900 teu, ten of the SANTA CLARA class with 7150 or 7800 teu, ten CAP SAN NICOLAS type carriers of 9800 teu and seven newbuildings of the SAN CLEMENTE/SAN FELIPE design achieving capacities of 9000 teu. Four CAP SAN NICOLAS vessels are long-term chartered from Greek's Enesel shipping company, four SAN FELIPE class units of 8700 teu from Ship Finance International, a ship financing company managed by Norwegians, listed in New York and registered in Bermuda. Unlike previous Hamburg Süd vessels, the latter two classes do not represent exclusive designs for the Hamburg owner, but are ship types realised for other shipping companies as well (Fig. 1.40).

Fig. 1.40 Four SAN FELIPE class units of 8700 teu capacity were chartered from a US-based ship financing firm by Hamburg Süd (Ralf Witthohn)

Three ships of the RIO class, RIO DE LA PLATA, RIO DE JANEIRO and RIO NEGRO, were in 2008 built by South Korea's Daewoo concern, three units, RIO BLANCO, RIO BRAVO and RIO MADEIRA, the following year by its Romanian shipyard site Daewoo Mangalia Heavy Industries (DMHI). The first two newbuildings were initially deployed between Asia and South America's East Coast, then switched to the Europe—South America route. The six newbuildings replaced the MONTE class vessels on the South Atlantic route, which were transferred to the Asia—East Coast service. Ten SANTA type vessels of 7150/8200 teu capacity with 1600 reefer plugs, SANTA CLARA, SANTA ISABEL, SANTA CATARINA, SANTA CRUZ, SANTA RITA, SANTA ROSA, SANTA TERESA, SANTA URSULA, SANTA BARBARA and SANTA INES, were from 2010 to 2012 also supplied by DMHI. Hamburg Süd continued its fleet expansion in 2013 by adding ten 10,500 TEU Hyundai-built vessels, CAP SAN NICOLAS, CAP SAN MARCO, CAP SAN LORENZO, CAP SAN AUGUSTIN, CAP SAN ANTONIA, CAP SAN RAPHAEL, CAP SAN ARTEMISSIO, CAP SAN MALEAS, CAP SAN SOUNIO and CAP SAN TAINARO, of which four were chartered from Greek's company Enesel.

The high reefer capacity, created by 1365 sockets on the MONTE and RIO classes, made this type the world's largest reefer cargo transporter at the time of delivery. The ships are fitted with 665 connections in the holds and 700 on deck. However, the number of connections is not a fixed indication of the reefer capacity, because in some 40-foot bays on the hatches there are both front and rear sockets. The class note RCP1085/25 issued by Germanischer Lloyd gives evidence, that the ship has 1085 stowage spaces for 40-ft reefer containers, which in turn have a fruit cargo share of 25%. The increase of the nominal container intake of the MONTE CERVANTES type by 6% from 5560 teu to 5905 teu of the RIO DE LA PLATA was achieved through the relatively simple design measure of lengthening the ship in by a 40-ft bay. This corresponds to an additional 14.4 m long section, resulting in a ship length of 286.5 m, the ship's breadth remaining unchanged at 40 m. In order to keep the scope of changes to a minimum, the number of cargo holds in front of the engine room was increased from six to seven. Compared with other postpanamax designs, the ship class has several special features. These include the relatively large depth of 24.2 m up to the main deck. It allows nine layers of containers to be stowed in the holds, thus creating a higher proportion of protected reefer containers below deck. The "tonnage depth" determining the ship's measurement is figured at 19.46 m (Fig. 1.41).

Due to the lengthening and increase in draught the deadweight capacity of the MONTE class, represented by the 2004 to 2007 built MONTE OLIVIA,

Fig. 1.41 Hamburg Süd's 2009 built RIO BLANCO represents the lengthened version of the companies's MONTE class (Ralf Witthohn)

MONTE PASCOAL, MONTE ROSA, MONTE SARMIENTO, MONTE VERDE, MONTE AZUL, MONTE ALEGRE, MONTE ACONCAGUA and MONTE TAMARA increased from 71,475 tonnes to 80,454 tonnes, the container capacity according to stability criteria from 4381 teu with a weight of 14 tonnes for the MONTE ships to 4816 teu for the further developed newbuildings. On the RIO DE LA PLATA, 2985 teu of 8.5 feet and 2745 teu of 9.5 feet height can be stowed side by side in 14 rows in the holds, on the hatch covers 2920 teu or 2770 teu in 16 rows, according to the line of sight restrictions. In front of the engine room there are 14 40-ft bays, behind the superstructure there is one 40-ft bay, behind it a cargo hold for another 40-ft bay and aft above the winch deck a last one, totalling to a number of 17. 17 pontoon hatch covers of the ship are designed for a stack load of 120 tonnes for 40-ft containers. However, the draught of 12.5 m was amended. Though initially also intended for the RIO class, it was finally increased to 13.5 m. This resulted in a freeboard of 5.974 m for the RIO class as specified by Germanischer Lloyd, compared to 6.274 m of the MONTE class. Up to seven layers of two 20-ft containers plus two layers of 40-ft containers can be loaded in the 40-ft bays. The hatch covers can accommodate 16 rows and a maximum of six or seven layers, depending on the container height and the line of sight regulations.

Another design characteristic unusual for this ship size is the relatively far back installation of the main engine, which has the advantage of a shorter propeller shaft, but worsens the line of sight conditions, thus reducing the deck container capacity. The cargo space behind the superstructure is limited to one hold. The forward 40-ft bay of this hold reaches one container layer below the main deck and is without hatch cover, the rear bay reaches four layers deep. In the holds, the containers are secured by cell guides for 40 ft. long containers. On deck are lashing bridges between the hatches, from which the lower container layers are secured by means of lashing rods in addition to twistlocks. The vessels are equipped with three MAN 9L32/40 diesel generators putting out 4100 kW each and one MAN 6L32/40 diesel generator of 2700 kW to satisfy the high energy demands originating from the large proportion of refrigerated containers. During handling operations, a heeling compensation system improves the ship's stability. The ballast system with a total capacity of 20,500 m^3 is served by two pumps each delivering 500 m^3/h. The ships' propulsion system is of conventional concept. The main engine is of type Wärtsilä 8RTA96C, which achieves a maximum continuous rating (MCR) of 45,760 kW at 102 rpm and a nominal continuous rating (NCR) of 38,900 kW at 96.6 revolutions. It is enabling a speed of 23 knots at 85% MCR plus 15% Sea Margin on a draught of 12.5 m. The engine acts without gearbox on the propeller shaft, which turns a six-bladed fixed pitch propeller. The daily consumption is stated to be 153.8 tonnes, while the range is figured at about 18,000 nm. The content of the heavy oil tanks was designed for 8050 m^3, that of the diesel oil tanks for 350 m^3.

By the introduction of the CAP SAN NICOLAS class, Hamburg Süd changed its design philosophy from the one-island type to the then more common two-island type, even though the realised cargo capacity of 9814 teu could have been covered by a one-island ship type. The ships are fitted with 2100 reefer plugs. Five auxiliary diesel engines, each with an output of 4900 kW, are installed to generate mainly the energy for cooling purposes. A seven-cylinder MAN B&W 7S90ME-C9.2 crosshead engine with an output of 40,670 kW provides a speed of 21 knots. The energy of the auxiliary engines thus reaches 60% of the power of the main engine, which consumes 148 tonnes of heavy oil (IFO 180) per day at 21 knots. The 333.2 m length, 48.2 m breadth CAP SAN NICOLAS class has a draught of 14 m, and a corresponding deadweight capacity of 124,458 tonnes. After delivery in 2013/14 CAP SAN NICOLAS, CAP SAN MARCO, CAP SAN LORENZO, CAP SAN ANTONIO, CAP SAN AUGUSTIN, CAP SAN RAPHAEL, CAP SAN ARTEMISSIO and CAP SAN MALEAS were employed in the weekly Europe—South America East Coast route operated jointly by Hamburg Süd and Maersk Line, with CMA CGM, COSCO, Hapag-Lloyd and MSC actually taking slots. CAP SAN SOUNIO and CAP SAN TAINARON operated in the Eurosal service to the west coast of South America with the Hamburg

Fig. 1.42 Hamburg Süd expedites the same reefer container ship type that Hapag-Lloyd employs in its West Coast service as the CAP SAN MARCO class to the East Coast of South America (Ralf Witthohn)

Süd charter vessel SAN CLEMENTE, Hapag-Lloyd's VALPARAISO EXPRESS, CALLAO EXPRESS, CARTAGENA EXPRESS and GUAYAQUIL EXPRESS and with CMA CGM NIAGARA and CMA CGM TANYA of French CMA CGM (Fig. 1.42).

In 2019, CAP SAN SOUNIO participated in the North Europe/Far East/US West Coast service with calls at Felixstowe, Rotterdam, Bremerhaven, Tangier, Salalah, Nansha, Hong Kong, Yantian, Xiamen, Los Angeles, Yokohama, Ningbo, Shanghai, Xiamen, Tanjung Pelepas and Colombo within Maersk/MSC's cooperation 2M, as the only ship of this class along with MSC TINA, MARGRETHE MAERSK, MSC VALERIA, MSC KALINA, MSC EVA, MAERSK EMDEN, MSC VEGA, MSC ALTAIR, MAERSK EUREKA, MAERSK ENSENADA, MAERSK ENPING, MSC AMBITION, MSC ARIANE, MSC RAPALLO and GUNHILDE MAERSK. In 2022 the Hamburg Süd ship was serving in a Maersk loop between the Mediterranean and the Middle East with the similar CAP SAN LAZARO, CAP SAN TAINARO and the 8059 teu carrying MAERSK GUATEMALA, MAERSK GENOA and MAERSK GIBRALTAR. CAP SAN ARTEMISSIO changed hands from Greek Enesel company to Japan's Kyowa Kisen in 2019, but remained on its initial route to South America (Fig. 1.43).

Fig. 1.43 In 2019, the high-reefer capacity container vessel CAP SAN SOUNIO sailed with 15 container ships of other types on one of the longest container routes including four continents, in 2022 it ran between the Mediterranean and the Middle East (Ralf Witthohn)

1.4.3 60,000 m³ Cold Store on EMMA MÆRSK

The buyer of Hamburg Süd, Denmark's A. P. Møller/Maersk group, has as well a long tradition in refrigerated transport dating back to 1936. Since the first use of a refrigerated container in 1964, Marsk Line has developed to the world's largest transporter of reefer boxes. While the operator distinguishes only four different sizes of standard containers of 20, 40 and 45 feet in length and a high cube 40-ft version with capacities of 33, 67, 85, and 76 m³ for uncooled cargo it disposes of not less than five types of containers for refrigerated cargo, one of which is 20 ft. long. All others measure 40 ft., which is the usual length for reefer containers. The integrated refrigeration unit and the insulation reduce the contents of the 20-ft container to 27 m³ compared with the non-tempered container, the 40-ft units vary between 60 m³ and 68 m³ contents, their maximum total weight is 34 tonnes, that of the payload 30 tonnes maximum. For the EMMA MÆRSK, which was commissioned in September 2006 as the largest container ship at that time, Maersk Line names a stowage capacity of 1000 rfeu of a total of a 17,816 teu intake, enhanced from 15,550 teu by raising the wheelhouse in 2016.

This is less than the 1109 rfeu of the L-class postpanamax refrigerated container ships built for Maersk Line's services to South America and Africa with a high proportion of reefer cargo, and the 1085 rfeu of the MONTE CERVANTES class of Hamburg-Süd specially designed for reefer voyages. After all, EMMA MÆRSK has cooling facilities for more than 60,000 m^3 of cargo and thus three times the capacity of a conventional reefer vessel, like the 20,900 m^3 DITLEV LAURITZEN as one of the largest such units and in service since 1990. The reefer has a total container loading capacity of 480 teu, of which 236 teu are stowed in the holds. On deck of the EMMA MÆRSK there are 356 electrical connections for containers on the hatches and 644 in the holds. The high temperature development caused by the refrigeration units of the containers led to a correspondingly large design of the hold ventilation systems. Depending on the size of the hold, their capacities are between 3600 m^3/h and 63,000 m^3/h. The number of fans ranges from two in rooms 1, 14, 21/22 and 23 via four fans in rooms 2/3, 15/16, 17/18 and 19/20 to seven in holds 6/7, 8/9, 10/11 and 12/13 and even eight in rooms 4/5. The cargo space ventilation system requires a high basic electrical energy supply, which is supplied by five auxiliary diesel engines of 4320 kW each coupled to Siemens generators (Fig. 1.44).

1.5 Juices, Alcohol and Water

Special wine and alcohol tankers are used to provide producers with spirits. After wine was first transported in bottles and after the Second World War increasingly also in separated built-in tanks of general cargo liners, especially in the Spain and Portugal trade, sophisticated tanker newbuildings or converted ships were employed from the 1950s onwards. As early as 1946, the Swiss merchant fleet included the 475 dwt wine tanker LEMAN of Geneva-based owner Marivins, which had let convert the ship from a British standard-type war tanker. In 1979, a second ship of this name was acquired, which had been built in 1965 as the general cargo vessel NORDENHAMERSAND, but converted into the wine tanker STAINLESS ANNE in 1974. The relatively small fleet of wine tanker newbuildings included the 2271 dwt SENEGAL built in 1960 at the Verolme shipyard in Heusden as SENEGAL for Rotterdam-based Nederlands-Franse Scheepvart and purchased by Rickmers Linie from Hamburg in 1965. Two years later, Rickmers commissioned the

Fig. 1.44 Between the hatches of the EMMA MÆRSK are 356 plugs for reefer containers installed, in the cargo holds 644 (Ralf Witthohn)

3929 tdw wine tanker newbuilding MONI from a French shipyard, but sold both ships in 1969.

1.5.1 Concentrate from Santos

For the transport of juice from its production sites to stainless steel storage tanks in the port of Santos Brazil's Citrosuco company uses insulated tank trucks or barrels stored in refrigerated containers. In order to increase efficiency the following sea transport of pasteurized orange juice is mainly carried out as a dearomatized concentrate, which only has a sixth of its original volume. The concentrate yet demands more cooling energy, as it is transported at a temperature of −10 °C, while juice is shipped at 0 °C.

The Citrosuco group operates own tankers to export the products of its plantations from Santos to Europe, the United States, Australia and Japan. The orange juice is discharged at the companie's own terminals in Ghent, Wilmington, Newcastle and Toyohashi. Citrosuco has a fleet of five special tankers, including the specially designed 205 m length and 32.2 m breadth

43,000 dwt tankers CARLOS FISCHER and PREMIUM DO BRASIL. Four cylindrical tanks for the transport of fresh juice or juice concentrate with a total capacity of 29,000 m³ are vertically installed in each of the four cargo holds of the ships. In order to ensure their complete emptying, the tank bottoms are inclined. Cooling is provided by three electrically operated compressors located in the deckhouse, the primary circuit of which uses ammonia, the secondary brine, which circulates from the evaporator through heat exchangers. Fans distribute the cold air into the insulated holds. The hulls of the juice carriers were in 2002/2003 built at Romania's Mangalia Shipyard and fitted out by Kleven Florø in Norway (Fig. 1.45).

1.5.2 Alcohol for Bacardi Production

Until 2018, regular imports of alcohol from Guatemala, Cristobal and Puerto Quetzal to Montreal, Barcelona and Cuxhaven were carried out by the tanker JO SPIRIT of the Norwegian shipping company Jo Shipping. On behalf of the Bacardi group, it undertook a rum collection tour in the Caribbean. By its deadweight capacity of 6285 tonnes, the ship was one of the smallest in the transatlantic trade. Its 20 stainless steel tanks have a volume of 5175 m³. In

Fig. 1.45 The 204 m length CARLOS FISCHER can load 29,000 m³ of juice in 16 cargo tanks (Ralf Witthohn)

2010, a separate tank storage facility with a capacity of 1800 m^3 had built in Cuxhaven for the import of alcohol. From there, the alcohol was transported by tank lorries to the Bacardi plant in Buxtehude, Germany, for processing. The Bacardi production was yet closed in 2018 and relocated to Pessione/Italy. The 1998-built JO SPIRIT was sold to Turkish owners and scrapped at Aliaga/Turkey in November of the following year (Fig. 1.46).

Transports of alcoholic beverages from the Caribbean to Europe have a long tradition. At Geesthacht/Germany, Ernst Menzer shipyard built the 4600 dwt tankers BONAIRE and ANTIGUA in 1982 as the largest ships ever constructed on the Upper Elbe river. These tankers were employed by Hamburg-based Kompass shipping company to import spirits on the Caribbean-Europe route. From 1986 the service was supported by CARUPANO, a 6679 dwt products tanker built as BOMIN EMDEN at Husum/Germany in 1980 (Fig. 1.47).

Fig. 1.46 The Norwegian tanker JO SPIRIT regularly supplied a distillery in Buxtehude with rum from the Caribbean (Ralf Witthohn)

Fig. 1.47 Built at Geesthacht in 1982, the alcohol tanker BONAIRE sailed as SABAHAT TELLI after 2007 and was still registered under this name for Gemiciler Denizcilik Sanayi at Istanbul in 2022 (Ralf Witthohn)

1.5.3 Aquavit Across the Line

With cargo rolled on its decks at Bremerhaven, Zeebrugge and Southampton, the vehicle carrier TORTUGAS sets course for South Africa. After a stopover at Santander, the ship calls at Durban and East London. The voyage is then continued to Australia with stops at Fremantle, Melbourne, Port Kembla and Brisbane. From the Fifth Continent the carrier heads for Singapore, Masan, Kobe, Nagoya and Yokohama, then via South Korea, Malaysia and Singapore back to Europe, where Turkish, Greek and Italian ports are visited, before the voyage around the world ends in Northern Europe. In addition to cars and vehicles of all kinds, there is also stowed a container with aquavit on the loading deck of the TORTUGAS during the entire voyage.

The transport of Norwegian aquavit to Australia and back is legendary. Due to the movements of the ship during the long sailing, including two equator crossings, the so called line aquavit matures and becomes milder. The positive taste effect occurred by chance in 1805 during the voyage of a sailing ship to Australia and back. This procedure was then continued on on car carriers of the Norwegian shipping company Wilh. Wilhelmsen and is nowadays still done on the ships of Wallenius-Wilhelmsen. While the shipping took place first in old oak barrels, later in tank containers, the Aquavit is actually also transported as concentrate. This circumstance led to a lawsuit about the designation line aquavit which was brought before Germany's Federal Court of Justice. The complaint against the use of the name, if the aquavit was not shipped in oak barrels, was dismissed in the last instance (Fig. 1.48).

1.5.4 Water for Kimolos

The Greek water supplier painstakingly pulls the hose pipes to the beach of his island. Shortly afterwards, the pumps are started on the near-by anchored tanker MARIA and the urgently needed water begins to pulsate through the hoses. The provisional water supply by tankers is a daily business on many Greek islands, carried out under the responsibility of the Athens Water Supply and Sewerage Company (EYDAP). Customers of the 2000 m³ capacity ship are the municipalities of Lavrion, Patmos, Amorgos, Nisiros or Kimolos. For a few hours at a time, MARIA anchors in the immediate vicinity of the beach or moors at the quays of the harbours to deliver the elixir of life. The 1974-built MARIA suspended its service and was broken up at Aliaga/Turkey in 2020.

In addition to many Greek islands, there are other regions that depend on the supply of drinking water by tankers. Until the 1960s, Lanzarote and

Fig. 1.48 Cargo that rolls on board the vehicle carrier TØNSBERG includes containers in which aquavit is transported to Australia and back (Ralf Witthohn)

Fuertaventura were supplied in this way from neighbouring islands, such as smaller islands of the Canaries are still today. Already in 1984, long-term, ultimately not realised plans became known to use the German 142,000 dwt tanker BADEN for the supply of drinking water to Gran Canaria. In 2010, the US company S2C developed plans to supply India with water from Alaska, which were yet also given up. In 2015, a similar project for the shipment of water from Alaska to California became public.

1.5.5 Bremen Malt for Izmir

In 2 days, 4500 tonnes of malt are loaded on to the Dominica-flagged cargo vessel CARINA at the quay of the Avangard maltworks in Bremen. As one of four plants in Germany producing 75,000 tonnes per annum the site is the only one that can be reached by ocean-going ships. The malt is destined for Izmir, where the ship is scheduled to arrive 13 days later at the end of January 2020. The malt works are part of the Russian-backed Avangard concern and produce Pilsen malt and wheat malt, which is dispatched by barge, truck, container, train or ocean-going vessels. In this case, the export is carried out by a former German-owned container feeder vessel. The 7562 dwt CARINA was built at Sietas yard at Hamburg in 1990 for

Fig. 1.49 After conversion to a bulk carrier, the 1990-built former German container feeder CARINA carried malt from Bremen to Izmir (Ralf Witthohn)

captain Hans-Peter Wegener, who chartered the ship out to Finnish operator Containerships as CONTAINERSHIPS III for almost a decade, later to the Scandinavian feeder line operators Unifeeder and Saimaa Lines. The 500 teu carrying CARINA was laid out to carry special Europallet containers, but was decelled at the beginning of the container shipping crisis in 2010 and converted for breakbulk and bulk transport for Swedich company Österströms. In 2013 ownership changed to Poland's Gdanska Zegluga, in 2018 to a Malta-registered company, the homeport to Portsmouth/Dominica. Located at a basin of the former AG „Weser"shipyard, the Bremen malt works are occasionally also called by larger bulk carriers including the 37,000 dwt Singapore-flagged MARITIME CHAMPION (Fig. 1.49).

1.6 Vegetable Oil

Palm oil is one of the most controversial raw products for the food industry, since primeval forests are often cleared to replace them with palm plantations. This happens primarily in Indonesia, Malaysia and Brazil. Since the oil obtained is processed in factories of multinational corporations worldwide, it is transported by sea in so-called product tankers, the deadweight capacity of which is often in the range of 50,000 tonnes. Shipping in new tankers is preferred, as their cargotanks are not yet contaminated by other cargoes. Palm oil is mainly used in a large number of foodstuffs and animal feeds, but also in cosmetics, candles, detergents and as biofuel. Other vegetable oils are also shipped in larger quantities. Turkey, for example, is both exporter and importer of sunflower oil.

1.6.1 Palm Oil for German Biscuits

On 17 May 2015, the tanker newbuilding STI BRONX calls for the first time at a German port to discharge a residual cargo of vegetable oil for Brake. In addition to the usual reference to a smoking ban on tankers, there are two slogans on the front of the superstructure that contradict the use and the operating conditions on the palm oil transporter, which, at this time, still burns mainly heavy fuel oil: "Conserve the Energy" and "Preserve the Environment". The first voyages took the 49,990 tdw ship of Scorpio Tankers, a corporation listed on the New York Stock Exchange and managed by Emanuele Lauro, around the globe. After delivery by Korean shipyard SPP Shipbuilding, the tanker took its cargo in Southeast Asia for several European ports, including Genoa, Liverpool, Rotterdam and, finally, Brake. Within a few months further Scorpio newbuildings of various sizes, named STI VIRTUS, STI HIGHLANDER, STI AQUA, STI ST CHARLES, STI REGINA, STI MANHATTAN, STI NOTTING HILL, STI SOHO, STI WESTMINSTER and STI ST ALBANS BAY, some of them named after London or New York districts, call at the Lower Weser ports of Brake and Nordenham to deliver residual cargoes (Fig. 1.50).

The customer for palm oil in Brake is a grease refinery founded in 1912, which had belonged to the globally active Wilmar Edible Oils group after 2006 and was by Wilmar in 2016 contributed to Olenex Edible Oils in a joint venture with the US Archer Daniels Midland Group (ADM). The tank capacity of the refinery amounts to 105,000 tonnes, its production capacity of raw materials for foodstuffs such as margarine and ice cream was increased sixfold following the expansion in 2009. At its own 166 metres long pier, 1000 tonnes of the raw material, most of which comes from Asia, can be handled per hour. The produced oils and fats are delivered to customers 80% by truck and 20% in ships. Brake supplies other locations in Scandinavia and Eastern Europe. As a result, two tankers occasionally are berthed at the pier at the same time, including the smaller vegetable oil tankers STAR ARUBA and STAR CURACAO being regular guests. Callers from the Far East in

Fig. 1.50 On the bridge front of the palm oil transporting STI BRONX, the demand for the preservation of the environment is emblazoned (Ralf Witthohn)

2020/2021 included the Japanese 50,000 dwt tanker newbuilding PACIFIC JADE, the similar sized Danish TORM TROILUS and the Norwegian ANDREA VICTORY.

The production of palm oil in deforested jungle areas is severely criticized and seen by environmental activists as a "tragedy" for Indonesia. In September 2010, at the request of a Green member of parliament, the state parliament of Lower Saxony/Germany dealt with the problem of palm oil production. It was explained that the Wilmar Group would use extensive areas of tropical primary and secondary forests. The group is said to treat local people ruthlessly. According to informations of the environmental organisation Robin Wood, 16 local residents had in August 2010 ended up in prison because they wanted to till their fields, which had been cultivated for generations. The parliamentary request related to public subsidies for the refinery, which were denied by the state's government to have been granted. Various application attempts by the Brake plant would not have been successful. The promotion of fuels from renewable raw materials would only be permitted in exceptional cases and limited to innovative plants for the production of second-generation biofuels. According to the Lower Saxon government, the high investment costs regularly result in too few job effects. In July 2017, the biscuit factory

Fig. 1.51 The construction of new tanks increased the storage capacity of the Brake refinery to 105,000 tonnes of vegetable oil from 2009 on (Ralf Witthohn)

Bahlsen in Hanover announced that it would no longer produce its butter biscuits intended for Eastern Europe with palm oil, but instead with butter (Fig. 1.51).

1.6.2 Ukrainian Sunflower Oil for Istanbul

On Wednesday evening, after 24 h sailing time, the freight ferry SEA PARTNER of Riga-based Sea Lines arrives from Chornomorsk/Ukraine in the port of Karasu/ Turkey on the south coast of the Black Sea. The weather deck of the ro-ro passenger ship is cowded with trailers bearing the labels of companies like UNO, Atasoy and Gülmen. Among the trailers are transporters of sunflower oil extracted in the Ukraine for Turkish customers. Eleven hours later, the ship departs from Karasu for Chornomorsk near Odessa, which is reached on Friday morning. In July 2019, the former Turkish destination of the ferry line, the ro ro terminal at legendary Haydarpasa on the Asian side of Istanbul, was replaced by the port of Karasu in the district of Sakanja. From there, it takes the ferry 31 h to reach the Ukraine. The freight ferry can carry up to 130 trucks and 166 passengers—mainly truck drivers—in 44 cabins across the Black Sea (Fig. 1.52).

Fig. 1.52 A Latvia-based company operates the passenger and cargo ferry SEA PARTNER between Turkey and Ukraine (Ralf Witthohn)

SEA PARTNER, which in 2021 was sold to Istanbul-based owners and still served the Turkey—Ukraine route in early 2022 as CENK T and in June 2022 the Thessaloniki - Limassol run, belongs to a series of 11 newbuildings which Korea's Hyundai shipyard in 1977/1978 initially constructed as pure freight ferries according to a design of company Knud E. Hansen at Copenhagen for Sweden's Stena Line. No other ro ro vessel type has been used on so many ferry routes and modified in so many different ways as the 17 knots achieving so called *Stena Searunner* class. This applies to SEA PARTNER, which came into service in May 1978 as the eighth ship of the series under the name ALPHA ENTERPRISE in charter of Piraeus-based Arghiris Line. In the following year, the ro ro vessel was rebuilt at Hapag-Lloyd Werft in Bremerhaven and supplying local shipbuilders according to the demands for the transport of lorries between Volos in Greece and Tartous in Syria. The ship was renamed SYRIA und sailed within the Hellas Syria Express Line of operator Nick Soutos in Piraeus. The ferry service allowed to circumvent the land route through Turkey, which was regarded as unsafe. The Soutos group, which had been active in shipping since 1972 and was later mainly engaged in container logistics, part-owned the ship. In order to create space for the accommodation of 144 drivers, the ship was fitted with additional superstructures aft. By this conversion it became a passenger ship with an enhanced safety level, especially regarding is leakage safety to be achieved by additional bulkheads. Above the weather deck a further cargo deck was installed, accessible from the inside via a seesaw ramp and an additional stern ramp on the same level and bringing the number of trailer decks to three. The parking facilities for passenger cars on the moveable decks and in most of the lower spaces were closed. In order to counteract the stability-reducing installations, 1.8 m wide

Fig. 1.53 The 2012-broken up STENA CARRIER represented the original type of the Stena Searunner series (Ralf Witthohn)

so called sponsons were added at 80 m of the waterline. They increased the ship's breadth from 21.7 to 25.3 m. During another rebuilding act at the Bremerhaven shipyard in 1981, SYRIA was lengthened from 151.9 to 184.6 m by the insertion of a midship section (Fig. 1.53).

The same conversion and later lengthening was carried out by Hapag-Lloyd Werft in 1979 and 1982, respectively, on the STENA RUNNER. This ro ro passenger ship was commissioned as ALPHA PROGRESS. It was operated under the name HELLAS in the same liner service as SYRIA. In 1980, the Bremerhaven shipyard similarly converted STENA TRANSPORTER to BALTIC FERRY and MERZARIO HISPANIA to NORDIC FERRY for a Channel service of UK-based Townsend Thoresen European Ferries. Initially fitted with seats for 166 passengers, the capacity of both ships was in 1986 increased to 258 persons in cabins and 424 on deck during another conversion at shipyard Wilton-Fijenoord in Rotterdam. The vehicle capacity was increased to 207 passenger cars and 1170 lane metres for trucks (Fig. 1.54).

Another shipyard involved in the conversions of the Stena ferry series was Blohm + Voss in Hamburg, where STENA SHIPPER was in 1987 rebuilt to the passenger ferry GRAIP on behalf of Sweden's shipping company Nordström & Thulin for their Gotlandslinjen service to Visby. Comprehensive additional superstructures to host 768 berths and 832 deck seats more than tripled the ship's tonnage from 5539 gt to 18,779 ft., the largest measurement that one of the Stena ships reached. Sold to China, the ship ran under the name CHONG MING DAO between Hong Kong and China and was built back to 14,471 gt at Shanghai in 2006. It returned to Europe as KILMORE for a charter of the Irish Celtic Lines and was used in the second half of 2007 under the Bulgarian flag for Finnlines' Transrussia Express between Lübeck and St. Petersburg. In 2008, Logitec Lines employed the ferry between Marina

Fig. 1.54 SYRIA before its second conversion by lengthening at Hapag-Lloyd Werft (Ralf Witthohn)

di Carrara and Castellón de la Plana. After lay-up at Varna, the ship was scrapped at Aliaga in 2010.

By far a much smaller scope of redesign referred to ATLANTIC PROSPER and ATLANTIC PROSPECT, chartered by UK's Cunard shipping company for employment in the Atlantic Container Line (ACL) service after delivery in 1978. Initially, the ships' stability was in 1979 improved by the adding of sponsons at Bremerhaven. In 1986 both ships, operating as STENA GOTHICA and STENA BRITANNICA at that time, received an additional cargo deck at Lloyd Werft. The first ship named sank in November 2006 under the name FINNBIRCH after capsizing off the Swedish coast. Two of 14 crew members lost their lives. By 2017, eight ships of the series had been scrapped. In early 2022, in addition to the CENK T ex SEA PARTNER, the PELAGITIS, which had come into service as TOR FELICIA, was still in service, owned by Greek Ainaftis Shipping and operated by Blue Star Ferries in Greek waters (Fig. 1.55).

1.7 Livestock and Live Fish

Apart from people travelling on passenger ships, cruise liners or ferries, there are only a few creatures that are regularly transported by ship. This includes livestock since the early steamship times. Today, the largest animal transports take place from Australia and South America to Islamic countries to provide

Fig. 1.55 As Gotland ferry GRAIP, the pictured KILMORE reached the most comprehensive conversion state of all Stena Searunner units (Ralf Witthohn)

the population there with living sheep and cattle. In Australia alone, it is handled through 17 ports. The largest livestock carrier newbuilding began operation in December 2016 under the name OCEAN SHEARER. The vessel is able to accommodate 20,000 cattle or 75,000 sheep. It was in 2020 sold by Australian owner Wellard Ships to Livestock Transport in Safat/Kuwait and renamed AL KUWAIT. Another transporter named AL SHUWAIKH, which had been converted from a tanker in 1981 and was scrapped in 2012, was even able to house 125,000 sheep, fed by so-called cowboys in open stables. Animal transports also take place on remote routes in specially designed combi ships, which simultaneously carry passengers and freight.

A unique kind of short-distance transport is carried out by the rare type of livefish carriers. The idea had already been realised over a century ago on fishing cutters, which landed their living catch this way in a fresh state. Involuntary and highly unwelcome transports of living creatures took place in the ballast tanks of ships for a long time, leading to the introduction of alien species that cause damage to ecosystems, harbour structures etc. This is actually to be prevented by regulations prescribing technical measures, as ballast water treatment systems are.

1.7.1 Australian Cattle for China

In July 2015, Australia and China agree on a first contract on the export of slaughter cattle. 16 months later, the livestock carrier GLOUCESTER EXPRESS loads the first 1200 cattle in Portland, Australia, for transport to Shidao in the northern province of Shandong. The each 500 kg weighing Angus breed animals have grown up in the states of Victoria and South Australia and are, under the supervision of Australian experts, to be processed into meat for restaurants and hotels within 14 days after arrival in the Chinese port (Fig. 1.56).

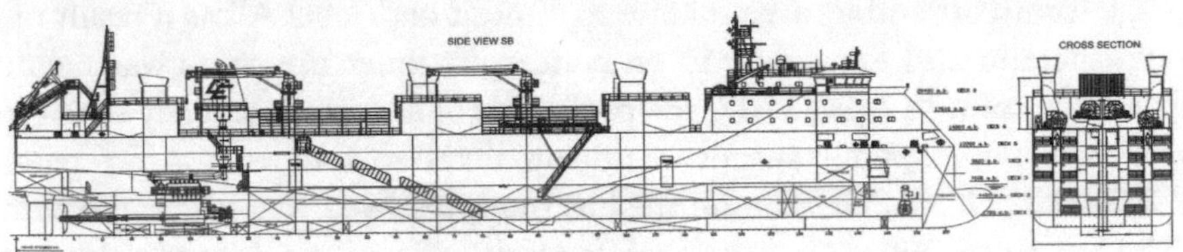

Fig. 1.56 The GLOUCESTER EXPRESS has stables on five decks, the cross-sectional drawing shows the supply, disposal and ventilation systems (Vroon)

The 18 knots achieving GLOUCESTER EXPRESS is one of seven livestock transporters of new architecture, which the Dutch shipping company Vroon received from the shipyard COSCO Guangdong in Dongguan/China from 2013 on. The ships have a vertical stem and a backward inclined, the forecastle deck arching superstructure front. Sister ships are named GALLOWAY EXPRESS, GANADO EXPRESS, GELBRAY EXPRESS, GIROLANDO EXPRESS, GREYMAN EXPRESS and GUDALI EXPRESS. The 134.8 m length, 19.8 m breadth ships can accommodate 4500 cattle on five cargo decks comprising an area of 4600 m^2. The animals are driven aboard over a side and distributed via internal ramps. On the upper decks there is longitudinal space for 17 boxes in rows of six. The feed reserves amount to 1200 m^3, the water reserves, of which 200 m^3 are produced daily, to 2500 m^3. The ventilation system is laid out for 90 air changes per hour. In 2022, cattle transports of the ships included sailings from Fremantle to Panjang/Vietnam, from Napier to Talcahuano/Chile and from Sines/Portugal to Haifa/Israel.

Vroon has been active in livestock transport for many decades. In 1982, the shipping company had the 14,750 dwt general cargo vessel HÖEGH PRIDE, built by the Wärtsilä shipyard in Turku in 1970, converted by J. L. Meyer shipyard at Papenburg into a 55,000 sheep carrying ship. It was renamed CORRIEDALE EXPRESS after a New Zealand sheep breed. In 20 previous projects, the German shipbuilding company had gained extensive experience in the development of automatic feed and drinking water supply as well as manure removal systems. This included important details such as non-slip deck coverings. Using the foldable ramps newly installed before the superstructure, 5000 sheep can per hour be driven on seven decks in the hold and five above the weather deck. The feed silo has a capacity of 1500 tonnes, the fresh water generator produces 80 tonnes per day. The vital ventilation system is laid out for 40 air changes per hour. The sailings from Australia and New Zealand to the Middle East meant great stress for the animals. In August 2002, more than 6000 of them died on CORRIEDALE EXPRESS during a voyage from Portland and Fremantle to Kuwait and Jebel Ali as a result of a brief generator and fan failure in great heat. At times the route was neither safe for the crew. In April 1987, the projectiles of a warship severely damaged the Philippine-flagged transporter during the Iran-Iraq war. After repair, CORRIEDALE EXPRESS continued to transport live cattle under the name KENOZ, since 2004 on behalf of Jordan's Hijazi & Ghosheh group at Amman. In recent years, the ship undertook transports from the Colombian loading ports of Mamonal and Cartagena to the Middle East with stopovers at Las Palmas, before it was scrapped in Bangladesh in May 2021 (Figs. 1.57 and 1.58).

1.7.2 Australian Sheep for Arabia

In the transport of animals by sea, high mortality rates have repeatedly been reported. In August 2003, 6000 out of 48,000 Australian sheep died on Vroon's CORMO EXPRESS, when the ship was unable to unload the animals on time due to an import stop imposed by Saudi Arabia because of the

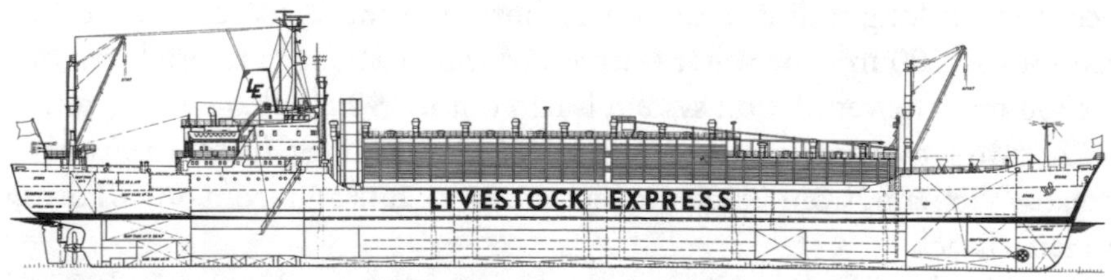

Fig. 1.57 Within 11 h up to 55,000 sheep are driven over the side ramp of the KENOZ ex CORRIEDALE EXPRESS (Meyer Werft)

Fig. 1.58 Nearly 40 years after conversion from a 1970-built general cargo liner the CORRIEDALE EXPRESS carried livestock (Ralf Witthohn)

danger of an epidemic. As consequence of the incident legal standards were introduced, including an Accredited Stockman acting as an experiened veterinarian on the vessels. Already in February 1975, 1300 cattle, 118 camels and 700 sheep had died on their way from Djibouti to Jeddah following the sinking of the animal transporter MOHAMEDIA converted from the German reefer PERIKLES (Fig. 1.59).

1.7.3 Sheep from Tierra del Fuego

Early in the morning the roll-on roll-off vessel TRINIDAD I leaves Puerto Montt on the northern shore of the Reloncaví Sound and heads south through the narrow fjords of the Chilean coast towards Puerto Nantales. On the ship's cargo deck sheeps are bleating, the species which is best adapted to the unspoilt region, ruffled by a constant cold wind. After 3 days sail through the navigationally most challenging area, TRINIDAD I reaches Puerto Natales. The Chilean ferry company Navimag bought the ship as its third at the end of 2015. The former paper transporter, built in 1993 under the name AMBER, can carry 100 trailers on 1278 lane metres, as well as people and animals. Even the construction required an extraordinary voyage, as the hull was built at the Serbian Sava shipyard in Macvanska Mitrovica and had to be towed downstream the Danube across the Black Sea to the Norwegian Fosen shipyard for completion. The ship arrived in Bremerhaven from Aqaba, Tartous and Augusta in March 2016 to handle rolling cargoes before its transfer to South America's West Coast.

The Chilean shipping company Navimag operates another, shorter line from Puerto Montt to Puerto Chacabuco. In addition to TRINIDAD I, it

Fig. 1.59 The death of 6000 sheep on the CORMO EXPRESS converted from the pictured car carrier MEDITERRANEAN HIGHWAY let enhance the efforts to reduce the high mortality rate during the sea transport of animals (Ralf Witthohn)

employed the ro ro vessel EVANGELISTAS, a ro ro vessel built in 1978 in Tokyo as ADMIRAL ATLANTIC by Ishikawajima shipyard, which was later named ROMIRA, DUKE OF FLANDRES, MAERSK FLANDRES and MAERSK FRIESLAND. After being converted into a passenger and cargo ferry in 2000, the ship initially operated as MAGELLANES and after 2007 as EVANGELISTAS. It offered rooms for 268 passengers and 1050 m of parking space for vehicles. EVANGELISTAS was demolished in India end of 2020. EDEN ex MAZATLAN STAR, a combined ferry built in 1983 in Le Havre as MONTE CINTO and employed since 2014, was scrapped in 2019. The ship had replaced AMADEO I, which sank in the Kirke Passage in August 2014 after haviong hit a rock. In March 2020, the operator took over from Guandong Bonny Fair Heavy Industry in Guangzhou/China the 18,600 gt/5500 dwt ferry newbuilding ESPERANZA offering accommodation for up to 274 passengers and 1800 lane metres for vehicles (Fig. 1.60).

1.7.4 Salmon from Faroe Islands Fjords

The fish transporter HANS Á BAKKA harvests live salmon from the fish farm located in a narrow fjord near the 200-strong village of Hvannasund. The fish are pumped from the salmon cage through a hose into the ship's cargo tank and then transported to the newly built fish factory in Glyvrar, 20 nm away. There, the

Fig. 1.60 Built in Serbia and Norway for paper transports, the ro ro vessel TRINIDAD I is actually employed along Chile's southern coast district (Ralf Witthohn)

salmon are processed. The three tanks of the HANS Á BAKKA, each 1000 m³ in size, can carry 400 tonnes of fish. In Glyvrar, the cargo is transferred ashore by using a suction hose. After processing in the factory, refrigerated trucks take the fish, chilled or frozen, to Torshavn on the only ferry connecting the Faroe Islands with mainland Europe. The NORRÖNA, operated by Smyril Line, arrives 27 ours later in Hirtshals, Denmark, from where the trucks roll on to the consumers (Fig. 1.61).

The outstanding importance of the countrie's fish industry became obvious, when the Faroese Prime Minister took part in the commissioning of the live fish transporter HANS Á BAKKA in July 2015. The live fish Carrier was built at Turkey's Tersan shipyard and is registered for Bakkafrost Farming in Glyvrar. At a length of 76 m, 16 m breadth and a draught of 7 m, the ship has a deadweight of 3700 tonnes. The diesel-electric propulsion system includes two Bergen diesel engines connected to generators delivering an output of 3000 kW to drive two Rolls Royce variable pitch propellers through a reduction gear. The characteristics of the controllable pitch propeller, its possibility to change fast speed and direction provide the high manoeuvrability required during operation at the fish farms. It is further increased by two transversally acting propellers, one installed forward and one aft. Since 2017, Bakkafrost has also been operating the former oil rig suppliers ROLAND and MARTIN for delousing measures in the salmon farms.

Built to a design different from the HANS Á BAKKA are the live fish transporters RO SERVER, RO NORTH, RO WEST and RO FORTUNE with an aft positioned wheelhouse. Based on hulls from Marine Projects at Gdansk they were delivered by Larsnes shipyard at Larsnes/Norway to the Norwegian fish farm Rostein in Harøy between 2016 and 2018. At 82.1 m length, 15.5 m breadth and 6.9 m draught, the vessels, designated well boats by the shipyard,

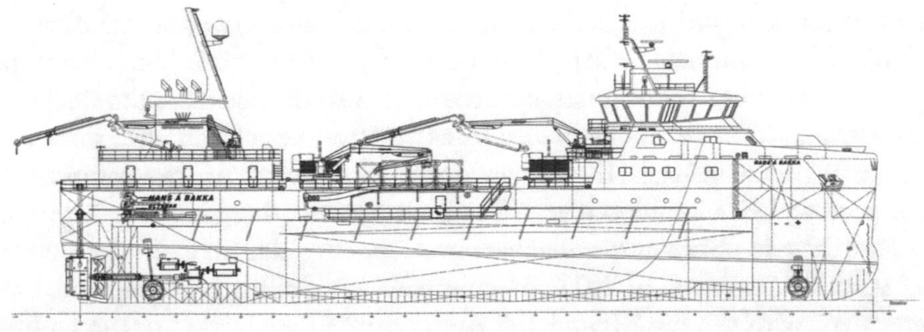

Fig. 1.61 The live fish transporter HANS Á BAKKA harvests salmon from Faroese fish farms (Tersan)

have a deadweight of 5500 tonnes. The ships are powered by a diesel-electric propulsion system comprising four diesel engines and two electric drive units, which act via a reduction gear on the shaft turning a variable pitch propeller. Two more units named RO VENTURE and RO VISION were delivered in 2020.

Further live fish carriers were ordered by the Norway's fish farming company Fjordlaks Aqua based in Alesund, a company of Hofseth International and Alliance Seafood. In November 2018, the 60 m live fish carrier LADY ANNE MARIE was delivered by Fincantieri's Norwegian shipyard Vard Aukra to Hofseth Aqua to serve the companie's five farm locations breeding salmon and rainbow trout. Another, competing concept is being realised on the factory and transport ship NORWEGIAN GANNET of Norway's Hav Line, on which the salmon is, after direct taking out of the farm, processed during the voyage to Hirtshals/Denmark. In 2018 delivered by Zumaia/Spain-based Balenciaga shipyard, the 94 m length, 18.5 knots achieving ship accommodates a crew of 53 and is able to deliver up to 2000 tonnes per week. Due to concerns about the handling of injured or bad fish, which is to be carried back to Norway, approval of this kind of combining the transport and processing has been limited in time.

1.8 Fish, Krill, Garnets, Whales

The industrialisation of agriculture with the consequence of increasingly polluted food made the consumption of fish appear to be a healthier alternative, even when more recent research showed that sea fish could also be polluted by the disposal of for example plastic parts absorbed by marine organisms. By this time, decades of overfishing had already caused a sharp decimation of fish stocks. The fishing activities of foreign nations off particularly the African coasts must be regarded as perfidious, because it means a twofold exploitation of animals and humans, when entire fleets of Asian countries, but also ships of European ones in continuation of colonial customs decimate the stocks so severely, that local coastal fishermen lost their livelihoods and were forced to engage in piracy instead. The damage inflicted on West African countries is estimated at USD 2 billion a year. The governments there are in a dilemma, because they do not want to endanger development aid from China by enforcing strict fishing bans. In 2017, 44 countries—excluding China—signed an agreement with the FAO (Food and Agriculture Organization of the United Nations) to prevent illegal fishing. Japanese whalers also write a dark chapter of exploiting the oceans, trying to justify the commercial and industrial hunting of whales by declaring it research work (Fig. 1.62).

The actual fishery research to identify stocks, as carried out by a number of countries at high costs, has in turn proved to be an often ineffective mean of conserving them. The catch quotas recommended by the scientists are regularly increased by the political bodies out of consideration for the fishing industry. The phenomenon of migration of entire fish species, which had already occurred earlier, for example when the North Sea herring disappeared in the mid-1960s, is increasingly considered as a result of climate change and the warming up of the sea. Facing the continuing exploitation of fish resources, the environmental organisation Greenpeace recommended at the beginning of 2016, that only sea fish from certain fishing areas should be consumed. This was described by representatives of the fishing industry as "unrealistic". The call to renounce also included saithe caught in large quantities off Alaska mainly by Russian fishing vessels, because the bottom trawls used destroy the seabed ecosystem.

In the wake of the Russian invasion of Ukraine, the US imposed a ban on the import of Russian fish, which is yet partly by-passed by transports via China. In Bremerhaven, Germany's leading fish processing location, company Frozen Fish International, a subsidiary of UK-based Nomad Foods and producing 48,000 tonnes of fish sticks and 20,000 tonnes of fillets mainly form Russian-caught Alaska pollock yearly, had to introduce short work for 440 employees due to disrupted supply chains, after Chinese factories were hit by COVID-19.

The replacement of side by stern trawlers at the end of the 1950s was accompanied by the introduction of factory ships on which, unlike the smaller fresh fish catching trawlers, the catch was mechanically processed and preserved by freezing. This concept extended the duration of the voyage from a few weeks to around 3 months, but also required larger vessels and crews. The

Fig. 1.62 CHANG HAI NO. 1 represents the type of large factory ships sent from China to the West African coast (Erich Müller)

Fig. 1.63 The 7419 gt factory ship ALASKA OCEAN is believed to be the largest US stern trawler for catching cod and white fish in the Bering Sea and processing them into fillets, surimi, fish oil and fish meal (Arne Münster)

number of 23,000 deep-sea fishing vessels recorded in 2017 far exceeds that of all other vessel types. According to the United Nations, the world fisheries by capture totalled at 90.9 million tonnes in 2016, down from 92.7 million tonnes in the pre-year (Fig. 1.63).

1.8.1 Saithe, Surimi, Fish Oil and Roe from the Bering Sea

The US trawler has sailed from Dutch Harbour on the Aleutian island of Unalaska to search for fish in the Bering Sea. As soon as the sensors indicates, that the net is filled with fish and the underwater camera confirms this, it is time to pull the 120 tonnes weighing net through the stern chute on deck. There, a deck-hand opens the cod end of the net and lets slide as much saithe into the hatch as the fish bunker can hold. 90 men and women are busy sorting and filleting the fish, then shock freezing the fillets at -40 °C. The saithe from the Bering Sea are exported all over the world by reefer vessels and processed into food like fish fingers or hamburgers. Most prized are the roes of saithe, which are highly valued by Asian gourmets. Surimi paste from the pollack is already being produced on board in order to be marketed later as a crab meat imitation. Omega-3 fish oil is also part of the production range, as is fertiliser, which is made from the heads and fins of the fish and accounts for 13% of the production.

The Bering Sea is also one of the main fishing grounds for Russia's giant trawler fleet. To extend their sea endurance, the catch is taken over by fish transporters and distributed by reefer vessels over the whole world. One of the reefers being employed for such transports is the 37,428 m^3 capacity

COLOMBIAN STAR of the Star Reefers company, which is operating from London and Oslo. In March 2014, the owner completed a conversion programme launched in 2012 on the basis of seven-year charter agreements. At costs of around USD 10 million each, CARIBBEAN STAR, COSTA RICAN STAR, COTE D'IVOIRIAN STAR and COLOMBIAN STAR, were extended from 154 to 184.5 m by the insertion of a container section including two side lifts of 12.5 tonnes lifting capacity on starboard. The ships, which had in 1997/1998 been built by the Shikoku shipyard at Takamatsu, were further equipped with an additional 45 tonnes revolving crane. The lengthening increased the number of the reefer compartments to 22. In April 2014, the COLOMBIAN STAR used its handling equipment at the Cuxhaven Steubenhöft to discharge frozen fish (Figs. 1.64 and 1.65).

1.8.2 Saithe from the North Atlantic

After a three-week fishing expedition, the Atlantic-capable deep-sea cutter SUSANNE circles the Kugelbake navigation mark at the Elbe estuary and enters the lock to the Cuxhaven fishing port to discharge its cargo of saithe. Compared to

Fig. 1.64 Kamchatka-based fishing company Akros supplies processing companies all over the world with saithe, halibut, salmon, herring, squid and rays from the Okhotsk and the Bering Sea (Ralf Witthohn)

Fig. 1.65 COLOMBIAN STAR uses its additionally fitted side loading arrangements and crane to handle frozen fish independently from landborne infrastructure (Ralf Witthohn)

the large factory trawlers, the smaller ship with a crew of only eight is part of the fleet of the Cuxhaven Kutterfisch cooperative, the largest German fresh fish supplier with shareholders from Finkenwerder and Bremerhaven.

After the expansion of the Icelandic fishing zone from 50 to 200 nm and the defeat in the so-called cod war in 1982, attempts by German fishing companies to use their factory trawlers in other regions such as off the American and African coasts were unsuccessful in the long term, as was krill fishery in Antarctica, which had initially been tested on an expedition. The catch of fresh fish, mostly saithe and redfish, which was still being undertaken by a smaller number of stern trawlers for several years, was finally discontinued as well and was in early 2022 carried out by ten large deep-sea cutters based in Cuxhaven, named ANTARES, BLAUWAL, CHRISTIN-BETTINA, JANNE KRISTIN, IRIS, BLEIBTREU, J. VON COELLN, VIKTORIA, WESTBANK and SEEWOLF. In June 2017, the members of the Kutterfisch-Zentrale cooperative signed a contract with Spain's shipyard Nodasa in Pontevedra about the construction of two newbuildings to replace four over 30 year old ships, to still be able to exploit the total annual quota of 12,000 tonnes. The first, EUR 8 million costing newbuilding christened JANNE KRISTIN was delivered in November 2018, the sister vessel IRIS in July

2019. The 35 m length, 10 m breadth ships of 680 gt are powered by a 750 kW main engine. Due to new work regulations the ships are manned by two crews of six, who alternate after the three weeks lasting voyages. To shorten the trips, the fish caught in the northern North Sea, off Norway and in the North Atlantic is regularly landed in Iceland or in Denmark. Catches by German-owned shipping companies are still only occasionally discharged in German ports.

As a large share of fish is caught in British waters, the Kutterfisch managers fear, to get lost of important fishing grounds after the departure of the United Kingdom from the EU. The withdrawal of an agreement on mutual fishing in coastal waters by the exit of the United Kingdom from the EU would have allowed the banning of foreign fishing vessels from a 200 nm zone, this for example forcing nearly all German herring and mackerel fishery to give up. Euro-Baltic Fischverarbeitung at Saßnitz/Rügen, founded in 2003 as part of the Parlevliet group, figured the possible loss of herring catches at about 40,000 tonnes per year. The Brexit deal finally allowed EU-based fishers to keep on for five and a half years under a progressing catch reduction scheme of up to 25% (Fig. 1.66).

Fig. 1.66 German deep-sea cutters like the Cuxhaven-registered newbuilding IRIS deliver their catch only still sometimes in the home port (Ralf Witthohn)

1.8.3 Fish from Iceland, the Faroes and Norway in Containers

Instead by trawlers, fish regularly reaches Germany on container ships operated in liner services and disposing of high refrigeration capacities. Once a week the container ships ARNARFELL and HELGAFELL of the Icelandic shipping company Samskip, operating from Rotterdam call at continental ports. End of 2019, the port rotation read as Reykjavik, Vestmannaeyjar, Kollafjordur in Faroe Islands, Rotterdam, Cuxhaven, Aarhus, Varberg, Kollajfordur, Reykjavik, Grundartangi and Reykjavik. The ships have 200 reefer connections, so that almost half of the container cargo of 421 40/45-ft containers (908 TEU) can be cooled. In Cuxhaven, the two Samskip ships were, until 2019, the only container ship customers. In 2007, the port operator had acquired a used container gantry crane from Bremerhaven for handling the boxes. 320 container movements during an approach of ARNARFELL in May 2013 were regarded as a handling record. For the return run, the containers are primarily filled with supply goods, chemicals, construction machinery or cars for Iceland (Fig. 1.67).

In August 2019, Cuxhaven was included in a second Samskip line, transporting mainly fish from Norway to the Baltic Sea region and consumer goods back. The line is served by SAMSKIP ENDURANCE, ATLANTIC COMET, SAMSKIP HOFFELL, SAMSKIP CHALLENGER and SAMSKIP COMMANDER by calls with varying rotations at Rotterdam, Hamburg, Cuxhaven, Bremerhaven, Gdansk, Klaipeda, Aarhus, Oslo, Tananger, Egersund, Husy, Maaloy, Bergen, Sauda, Kvinesdal, Holla and Aalesund. The service had initially been organized after the purchase of Euro Container Line by Samskip and the merger of Samskip's Norway line with the ECL service. Samskip as well transports refrigerated cargo from Iceland to Halifax, Boston and Norfolk. The shipping company moves a total of 600,000 containers

Fig. 1.67 Fish is transported from Iceland and the Faroes to European ports by regularly operating container ships such as HELGAFELL of the Icelandic shipping company Samskip (Ralf Witthohn)

(TEU) annually, while its subsidiary FrigoCare operates cold stores and terminals in Rotterdam and Aalesund (Fig. 1.68).

From the Cuxhaven ro ro terminal another regular ship connection to Norway was established in 2015 by the LNG-powered newbuildings KVITBJØRN and KVITNOS. They transport fish from ports on the Norwegian west coast between Bergen and Hammerfest, including many small harbours such as Bodø, Svolvaer, Tromsø, Trollebø, Agotnes, Tananger and Egersund, to Rotterdam and Eemshaven, while Cuxhaven was later cancelled from the regular timetable. On the return trip industrial goods, among other cargoes, are carried. The weekly liner service was being organized by Norway's shipping company Norlines, before being taken over by Samskip in 2017. The ships were in 2020 renamed SAMSKIP KVITBJØRN and SAMSKIP KVITNOS (Fig. 1.69).

In Cuxhaven, the Gooss cold store at Lentzkai has a capacity of 5000 tonnes, 30,000 tonnes reefer goods can be stored in the same companie's building at the eastern end of the New Fishing Harbour. There, ships chartered by the Dutch reefer shipping company Seatrade occasionally import up to 5000 tonnes of Alaska Pollock supplied by the US company Trident Sea. The frozen goods are transported by truck for processing throughout Europe. The import of fish in containers was granted to Cuxhaven, a former important fishing port by the Hamburg government, when it returned its port area

Fig. 1.68 The 657 teu feeder vessel SAMSKIP ENDURANCE, built by China's Zhoushan Shipyard according to a design of Germany's Peters shipyard, is chartered by Iceland's Samskip company to transport containerized fish cargo from Norway to North and Baltic Sea destinations from 2017 to 2020 (Ralf Witthohn)

Fig. 1.69 Mainly employed in the transport of fish, KVITNOS and KVITBJØRN, renamed SAMSKIP KVITNOS and SAMSKIP KVITBJØRN in 2020, were among the first ships using only LNG fuel (Ralf Witthohn)

there to Lower Saxony in 1991 under the condition of a waiver of competition in container traffic.

A second Icelandic shipping company, Eimsskip, exports fish from Iceland and the Faroe Islands to Europe and North America on six lines by employing container ships with a high proportion of reefer containers, like the chartered JONNI RITSCHER and VERA D. of 1856 teu, which replaced the DETTIFOSS and GODAFOSS disposing of a loading capacity of 1467 TEU and 400 reefer connections. In early 2022, they they served the so called Blue Line from Reykjavik and Grundartangi to Rotterdam and Bremerhaven every fortnight.

Special containers are available for the shipping of frozen fish. After discharge from the fishing vessel they are directly filled through hatches in the container roof and by this immediately exposed to the freezing process. According to a Maersk Line concept up to five refrigerated containers are connected to form a unit designated Sortie, which enables the cargo to be sorted. An airtight seal between the containers ensures complete refrigeration. Such transport solutions are primarily chosen for highly temperature-sensitive tuna frozen in brine. Tuna, swordfish and sea urchins destined for the Japanese sushi and sashimi market are transported either in their entirety or as fillets at extremely low temperatures down to −60 °C. The tuna is also transported at extremely low temperatures (Fig. 1.70).

Fig. 1.70 The Icelandic shipping company Eimskip exports fresh and frozen fish to Europe on container vessels such as JONNI RITSCHER ex CMA CGM CARIBBEAN (Ralf Witthohn)

1.8.4 New trawlers from Turkey and Norway

The last German-flagged factory ship from the pre-cod-war period, the 1972-built stern trawler KIEL, was in 2017 sold by Cuxhaven-based Deutsche Fischfang Union (DFFU) to Portugese owners. Since the companie's sale in 1995, DFFU is controlled by the Icelandic company Samherji. In January 2018, DFFU commissioned the 80 m length newbuildings CUXHAVEN and BERLIN built at Norway's Myklebust shipyard as replacements for operation in the North Atlantic. On the EUR 40 million costing floating fish factories a crew of 30 can daily produce 30 tonnes of fish fillet to be stored in a reefer room of 2000 m^3 capacity.

DFFU had in 2001 sold the wet fish trawler CUXHAVEN, built in 1990 by Mützelfeldtwerft for DFFU, to Dutch group Parlevliet & van der Plas, which converted the ship to a freezer named GERDA MARIA with a Rostock external identification sign (Fig. 1.71). Another Parlevliet ship, the 1988-built JAN MARIA, yet remained entered in the Bremerhaven registry. In 1996 the Rostock-registered Parlevliet subsidiary Oderbank Hochseefischerei had commissioned the 7278 gt newbuilding HELEN MARY as the largest factory vessel under the German flag to date. Warnemünde Hochseefischerei, another Parlevliet subsidiary, took over the 90 m long, 4270 gt stern trawler MARK for operation off Greenland, Norway and Iceland in May 2015. Turkey's Tersan shipyard built the ship according to a design supplied by the Norwegian

Fig. 1.71 In January 2018, the EUR 40 million costing new factory ship CUXHAVEN was equipped for the first voyage in its homeport which name it is carrying (Ralf Witthohn)

Fig. 1.72 The Turkish-built and Rostock-registered factory ship MARK helps the Dutch Parlevliet group to exploit the German fishing quota (Ralf Witthohn)

engineering company Skipsteknisk under supervision of the Norwegian-German classification society DNV GL (Fig. 1.72).

A sister ship of MARK, the Hull-registered KIRKELLA, was in 2015 delivered to UK Fisheries, a joint venture of Parlevliet and the Icelandic shipping company Samherji. On the 86.1 m length, 16 m breadth and 6.5 m depth trawler, 100 tonnes fish can be processed daily. Its hold capacity amounts to 2510 m^3. The cooling system comprises 12 vertical and three horizontal freezers with a total capacity of 761 kW. The high energy demand for freezing, cooling, processing and the net winch operation is generated by a 1530 kW auxiliary diesel engine. A shaft generator supplies 2875 kVA and can transfer 1000 kW from a Cummins diesel to the propeller shaft, if the main engine should fail. Main engine is a four-stroke Caterpillar engine of type MaK

8M32C able to burn heavy fuel oil und delivering 4000 kW at 600 rpm. A four-bladed controllable pitch propeller rotating in a nozzle as protection against the net accelerates the ship to a speed of 15 knots. The accommodation offers space for 34 crew members. Only a few months after delivery, KIRKELLA was transferred to the Spanish register, renamed LODAIRO and registered at Vigo. When at the year's end the fishing quotas are exhausted, trawlers are forced to assemble in port, as Parlevliet's MARK, LODAIRO ex KIRKELLA and GERDA MARIA were in Bremerhaven at the turn of the years 2018 and 2019. Parlevliet & van der Plaas entered the fish processing industry in February 2018, when the Dutch company took over Deutsche See, Germany's largest fish processor, with 22 branches and 1700 employees (Fig. 1.73).

1.8.5 Freezer SÓLBERG for Iceland

The distinctive Skipsteknisk signature is as well visible on the stern trawler SÓLBERG delivered by Tersan shipyard to the Rammi fishing company in Olafsfjördur/Iceland in May 2017. Carrying the fishing designation OF 1 for Olafsfjördur, the 79.8 m length, 15.4 m breadth and 6.2 m depth stern trawler

Fig. 1.73 Only a few months after delivery, the UK-flagged KIRKELLA was transferred to the Spanish register, renamed LODAIRO and homeported at Vigo (Ralf Witthohn)

is somewhat smaller than KIRKELLA and MARK, but different by its considerable rake of keel. Its operation in northern waters requires a high ice class which is met by the notation Ice(1C) of DNV GL. The ship's trawl deck is 9.1 m above the baseline. From there, the net is lowered through the stern chute into the water by help of the net winch. Through a hatch, the catch reaches the main deck at 6.2 m above the base, where it is processed by automated machines and frozen. The fish rooms are located on the two lower decks. The towing of the net requires a powerful propulsions system, consisting of 3000 kW four-stroke engine of type Wärtsilä 8L32, which acts through a reduction gear with Power Take-Off (PTO) on a variable pitch propeller turning in a nozzle (Fig. 1.74).

1.8.6 Seahake from Saldanha

In order to provide a sufficient catch quota for the operation of the Rostock trawler MARK, the trawler ATLANTIC PEACE of the last independent Bremerhaven deep-sea fishing company Ocean Food was in 2014 acquired by Parlevliet and sold to South Africa, where it is being operated as HARVEST ATLANTIC PEACE from Saldanha by the Sea Harvest company to catch so called cape hake along other species, as well as by-catch such as kingklip and monk. Sea Harvest was in 1964 founded by Spain's Pescanova Group and under control of Brimstome, a black owned investment consortium listed on the Johannesburg stock exchange. In 2019 the company operated a fleet of 42 single and twin fresh fish trawlers, factory freezer trawlers and freezer trawlers, 11 prawn vessels through its subsidiary Mareterram in Australia as well as fish factories in Saldanha and Mossel Bay/South Africa. The company produces

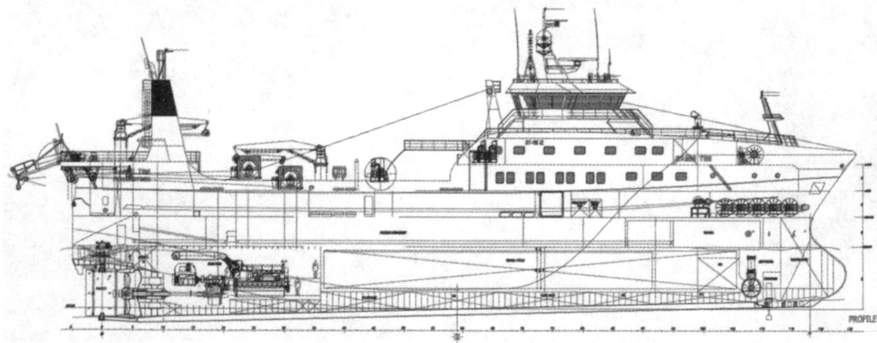

Fig. 1.74 Based on a Norwegian design, Turkey's Tersan shipyard delivered the freezer SOLBERG to an Icelandic shipping company (Tersan)

Fig. 1.75 ATLANTIC PEACE was in 2014 sold to South Africa, to catch seahake off the Cape of Good Hope (Ralf Witthohn)

100 different fish products for the South African market and 22 export countries, including discounters and supermarkets. The freezer trawler ATLANTIC PEACE had in 1987 been built at Sterkoder shipyard in Kristiansund/Norway under the name LONGVA III. The 57.1 m length, 13 m breadth ship was renamed KERMADEC in 1995 and hoisted the German flag as ATLANTIC PEACE in 1998 (Fig. 1.75).

1.8.7 Mackerel for Cameroon

After arrival from St. Petersburg, the Norwegian reefer vessel GREEN SELJE berths at the Labrador quay of the Bremerhaven fishing port in the evening. The once largest fishing port on the European mainland has not been a home of locally owned fishing vessels for years. Only a few stern trawlers of foreign shipping companies, which have registered their ships here to exploit the German catch quotas, occasionally still land their catch in the cold stores of the harbour at the mouth of the river Weser. In one of them are stored 1000 tonnes of deep-frozen wood mackerel in cartons, which are now to be loaded on the GREEN SELJE for transport to Cameroon. The cartons stowed on pallets are pulled along the ship on trolleys. Forklift trucks push them into baskets, which are lifted by the ship's own cranes into the cargo holds of the vessel. The mackerel have been caught by factory trawlers of the Dutch fishing group Parlevliet & van der Plaas. From Bremerhaven, GREEN SELJE sets course for Ijmuiden, the headquarters of the Dutch fishermen, to complete the Africa-destined cargo.

The four holds of GREEN SELJE are divided by bulkheads and tweendecks into 16 compartments, eight of which are independent of each other, so that a total of 2500 pallets can be stowed and cooled down to -25 °C. At a capacity of 7100 m³ (250,000 cubic feet), GREEN SELJE represents a smaller reefer vessel type suitable for all types of cooled cargo, frozen fish being among the regularly shipped cargoes. The ship was built by Kleven shipyard in Norway in 1989 for Rederiaktiebolaget Gustaf Erikson from Mariehamn under the name ERIKSON CRYSTAL. The sister ships GREEN AUSTEVOLL, GREEN FROST and GREEN KARMOY as well participate in the export of frozen fish to Africa (Fig. 1.76).

1.8.8 Frozen Fish from the Faroe Islands

Within 10 days, the reefer vessel BELOMORYE, owned by Karelian Navigating Co. in Murmansk, discharges a full cargo of frozen fish at Bremerhaven Columbuskaje. The fish has been loaded in Toftir on the Faroe Islands. After unloading from the ship's three holds with a total volume of 2410 m³ is completed, the 14.5 knots vessel sets course for the North Atlantic island again, this time destined to the port of Runavik. As one of only few reefers still equipped with conventional cargo handling gear the 77.8 m length and 12.7 m breadth vessel was in 1979 delivered by Germany's Büsumer Schiffswerft to a Cypriot company of

Fig. 1.76 By its own cranes, the reefer vessel GREEN SELJE loads 1000 tonnes of mackerel for Cameroon (Ralf Witthohn)

Fig. 1.77 The Russian-owned BELOMORYE took a cargo of frozen fish from the Faroe Islands to Bremerhaven, 36 years after being built at Büsum as YORKSAND (Ralf Witthohn)

Hamburg-based Reederei A. F. Harmstorf under the name YORKSAND (Fig. 1.77).

In 1978/1979, the shipping companies and shipyards owning Harmstorf group let build a total of five small reefer ships at its Büsumer shipyard on own account, the 2410 m^3 capacity WITTSAND, GROOTSAND, KNIEPSAND and YORKSAND as well as the RUNGHOLTSAND, the largest newbuilding of the shipyard with a reefer cargo intake of 3700 m^3. At a length of 95 m and 13.8 m breadth, the latter was designed with four hatches served by eight derricks. For the transport of fruit with a stowage factor of 117 cubic feet/tonne, the draught is 4.1 m, for cargo with a space requirement of 77 cubic feet/tonne 4.42 m, for 55 cubic feet/tonne 5.4 m and for heavy cargo a maximum draught of 5.8 m. As is customary on conventional refrigerated ships, the 4 × 5 m measurung hatch openings were kept as small as possible to optimise insulation of the holds, although this created unfavourable stowage space outside the hatchways. Powered by an Atlas MaK diesel engine of 2200 kW, the ship achieves a speed of 14 knots on a draught of 5.8 m. The range is figured at 10,000 nm. In early 2022 BELOMORYE made use of its traditional loading gear to take over frozen fish from trawlers on Russian fishing grounds in the Barentsz Sea and carry it to the Dutch port of Beverwijk.

The Harmstorf vessels, along with TURICIA and BASILEA built for Swiss owners, were employed by the International Reefers pool in the international fruit and fish trade. KNIEPSAND and YORKSAND were lengthended to 95.6 m in 1985. The former RUNGHOLTSAND and KNIEPSAND were in 2022 registered for the owner Vladkristall in Vladivostok and used in the Far East for the transport of frozen fish under the names ORION and PLUTONAS resp., while the SIRIUS ex GROOTSAND was owned by Arctic Shipping in St. Petersburg and navigated in Russian and North European waters. The Harmstorf shipyards in Büsum, Travemünde and Flensburg had in 1988 become insolvent. The fleet of a new Harmstorf shipping company founded under the old logo in 2000 became part of Schlüssel Reederei at Bremen in 2012 (Fig. 1.78).

1.8.9 Salmon Oil from Salthella

Two days after leaving Salthella on Norway's west coast, the tanker HORDAFOR V arrives at Bremerhaven Verbindungshafen. Since 2010, the quay has been used for discharging fish oil from small tankers. HORDAFOR V can carry 2550 m³ of the valuable stuff in ten tanks. The ship is named after its owner, a Norwegian company founded in 1983 in Austevoll and based in Salthella. Hordafor specialises in the production of protein additives and fish oil from salmon farming. The fish oil is used in food, as a feed additive in livestock farming and in numerous chemical products. In Bremerhaven it is filled into so-called flexi-bags and carried

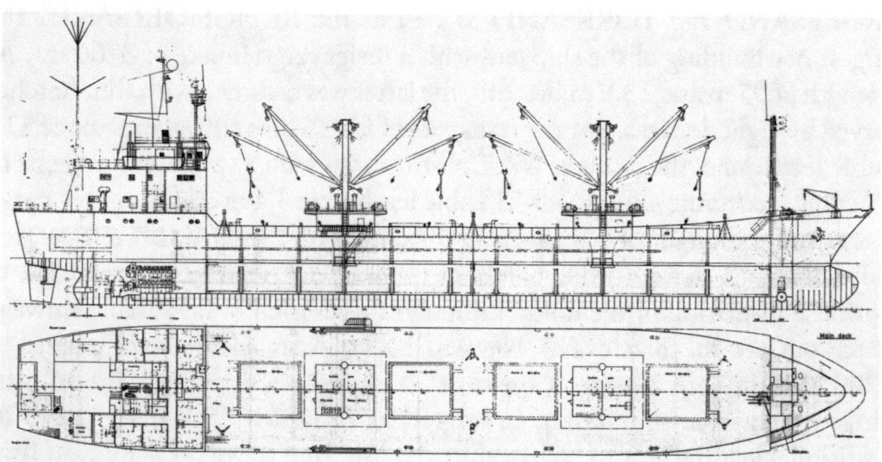

Fig. 1.78 The Büsumer Werft reefer design realized by the RUNGHOLTSAND shows four holds offering a capacity of 3700 m³ (Büsumer Werft)

Fig. 1.79 Until capsizing at Gdynia, the Norwegian tanker HORDAFOR V exported fish oil (Ralf Witthohn)

away in containers by truck and to overseas destinations on container ships. The recipients include chemical companies such as the US company GAC in Searsport/ Maine (Fig. 1.79).

HORDAFOR V is a product tanker built in 1997 under the name CRYSTALWATER by the Dutch shipyards Slob and Breko. In April 2017, the ship sank partly under water in a repair dock of Nauta yard in Gdynia, when the 1200 tonnes lifting floating dock lost its stability and overturned. The 80 m length, 11 m breadth tanker was salvaged only in April 2018 by the Dutch companies Smit and Multratug. Shortly after the accident of HORDAFOR V, which resumed service as CRYSTAL WATER, the owner purchased the 3230 dwt tanker AMETYSTH from Poland's Unibaltic company and renamed it HORDAFOR VI. In 2018 the 3550 dwt HORDAFOR VII ex YIGIT was purchased from Turkey. Hordafor also operated the waste transporters HORDAFOR and HORDAFOR II as well as the cargo vessel HORDAFOR IV. Further salmon oil transporters are managed by Sea Tank Chartering in Bergen disposing a fleet of 15 tankers of up to 6500 dwt in 2022 (Fig. 1.80).

1.8.10 Tuna from the Mediterranean Sea

While trawling for cod, redfish, plaice and other fish species is a catching method that damages the bottom fauna, herring, salmon and tuna are hunted in other ways. Using nets of up to 1.5 km length laid out by daughter boats, the drift-net fishery carried out for this purpose was banned from European waters in 2008. Tuna is caught in the western Indian Ocean, including the Red Sea, Arabian Sea and the Madagascar/Mozambique channel. Italy operated a fleet of tuna catching vessels in the Mediterranean, like the 43 m TENACE SECONDO, built in 1975 by Lucca shipyard according to a typical design in Viareggio and based in Catania (Fig. 1.81). The ship was in 2010

Fig. 1.80 The 5000 dwt tanker KEY WEST of Bergen-based Sea Tank Chartering transported fish oil to Bremerhaven, where it was refilled in bags, before being shipped world-wide in containers of Sorrento/Italy and Genêve/Switzerland based Mediterranean Shipping Co (Ralf Witthohn)

Fig. 1.81 The Catanian tuna catcher TENACE SECONDO came under the Libyan flag in 2010 as AL HARES 2 (Ralf Witthohn)

re-registered in Libya as AL HARES 2 and last reported in Malta in August 2021.

Somewhat smaller was the 41 m length DOMENICO PAPPALARDO, built and registered in Naples in 2004 and in 2017 rebuilt to the offshore supply vessel VIVRE-G on behalf of Turkish owners (Fig. 1.82). The Sicilian

Fig. 1.82 The Neapolitan DOMENICO PAPPALARDO had the typical appearance of a tuna vessel, before being rebuilt to an offshore supply ship (Ralf Witthohn)

FUTURA PRIMA is of different lay-out by a further forward positioned bridge. Built in its homeport Catania in 2005, the vessel has a length of 36 m.

1.8.11 Krill from the Southern Ocean

Facing the depletion of fish stocks, the commercial capture of krill is being considered as an alternative since about 1974, despite strong critics by researchers and environmentalists, because of the role of krill as staple food of whales. In the Soviet Union, krill was already processed to must at that time. The prospect of catches of up to 100 million tonnes per year, synonymous with world meat production, let krill appear as an attractive alternative for the gathering of proteins. In 1982 the catches superseded 500,000 tonnes. Companies from Norway, China, Korea, Chile and the Ukraine applied to the Australian-based Antarctic organization CCAMLR for a share of the catch of krill, which was limited to 620,000 tonnes in 2017. In 2014, 294,000 tonnes had been caught by these countries.

Catches of krill were carried out by the German fishing research vessel WALTHER HERWIG and the factory ship WESER as early as 1975 during an eight-month expedition to the Antarctic, which included, among other research work, the testing of a new deep-sea fish finder. Processing of the krill, i.e. the loosening of the meat from the shell, turned out to be problematic. Human consumption was also considered to be questionable due to the high fluorine content. On the other hand, krill contains high amounts of vitamins and omega-3 fatty acids. It is mainly used as animal feed.

In 2007, the Norwegian shipping company Emerald Fisheries at Alesund considered the conditions to be in place to deploy the fishing factory ship PAERANGI on a hunt for krill and to use the catch as food, animal feed or as an additive to cosmetic products. The ship had previously been active in the hoki fishery from New Zealand. Within 7 months, MWB shipyard in Bremerhaven converted the five-year-old stern trawler into one of the first krill catching vessels, named JUVEL (Fig. 1.83). At costs of EUR 65 million the ship was enlarged and processing equipment installed. The catching technology includes a gentle treatment of the krill by use of pumps which transports the krill from the trawl net still hanging in the water on to the ship's deck. In order to create the necessary space required for processing and storing the catch, the hull length was extended from 71.3 m to 99.5 m by insertion of a 28.2 m section. The cooling capacity increased from the original 1000 m^3 to 3300 m^3. In 2017 Emerald Fisheries declared its bankruptcy. In 2019 a Volda/Norway-based court decided in favour of Norwegian krill harvester Rimfrost, that the company was patent holder of the krill processing plant aboard the JUVEL and rejected a claim of Aker Biomarine Antarctic to use the patent. The ship was in 2020 sold to Oman and rebuilt to a pelagic trawler named JAWHART AL WUSTA and operating from Duqm/Oran.

In April 2017 the Aker BioMarine company of Stamsund/Norway ordered a NOK 1 billion costing krill fishing newbuilding from Norway's Vard Brattvag shipyard. The 130 m length ANTARCTIC ENDURANCE was delivered in January 2019 vessel to carry out fishing in Antarctic waters. In the following months the trawler operated from Montevideo/Uruguay and Walvis Bay/Namibia. Before, Aker BioMarine operated the trawlers ANTARCTIC SEA and SAGA in the South Atlantic. Refrigerated vessels were chartered to transport the captured krill. On behalf of Aker BioMarine, the krill is taken over in big bags by transport vessels in the South Atlantic and transported to Montevideo or other ports. One of the transporters engaged

Fig. 1.83 The hoki trawler PAERANGI was converted to one of the first krill catchers named JUVEL (Ralf Witthohn)

was the former Soviet fish transporter PROLIV DIANY employed under the name LA MANCHE. In October 2016 the vessel transported a cargo of krill to Bremen. The reefer operated from Montevideo, using the fender equipment already used in Soviet times to take over frozen fish. LA MANCHE was in 2021 replaced replaced by the newbuilding ANTARCTIC PROVIDER, delivered by China's Yantai Raffles Offshore shipyard to Aker Biomarine Antarctic in. Fitted with a 425 kWh energy storing system of type Corvus Orca the hybrid vessel undertook four supply and krill transport sailings to the Antarctic from Montevideo and subsequently carried a cargo of krill to Bremen/Germany in October 2021 via Las Palmas. In 5 days, the krill filled in big bags was unloaded from the four holds of the 159.9 m length, 27 m breadth reefer vessel with a total capacity of 40,000 m^3 by means of an automated crane system (Fig. 1.84).

1.8.12 Garnet from the North Sea

A glut of garnet (named Krabben *or* Granat *in German) has triggered a fall in prices. At an actual average annual income of EUR 134,000, this is threatening the existence of German, Danish and Dutch garnet fishers. Two Dutch wholesalers dictate the prices, until the fishermen, after coastal protests and a five-week fishing ban, in 2011 call for strike. The following year, a hundred fishermen from Lower Saxony and Schleswig-Holstein found the German Garnet Fishers' Association (Erzeugergemeinschaft der Deutschen Krabbenfischer) at Wardenburg near Oldenburg. From now on, the fishermen classify their products themselves and negotiate the prices by the producer association. The North Sea prawns are caught within the 12-mile zone by small cutters with low machine output and sold directly to the customer in the homeport or transported by refrigerated lorries to the*

Fig. 1.84 The Chinese-built supply and krill transport vessel ANTARCTIC PROVIDER regularly sails from Montevideo to the Antarctic, but carried a cargo of krill to Bremen in October 2021 (Ralf Witthohn)

screening points in Cuxhaven, Neuharlingersiel and Büsum, where they are sorted and delivered to the wholesalers.

In 2017, the German garnet cutter fleet comprised 95 ships with an average age of 40 years. The largest number of 16 cutters is registered in Friedrichskoog, nine of them sailed from Accumersiel, two from Brake, eight from Büsum, three from Cuxhaven, one from Dangast, four from Ditzum, three from Dorum, five from Fedderwardersiel, six from Greetsiel, one from Hooge, five from Husum, one from Kiel, eight from Neuharlingersiel, seven from Pellworm, one from Pogum, three from Spieka, six from Tönning, one from Varel, five from Wremen and two from Wyk on Föhr. In 2016 the average revenues doubled to EUR 272,000. Following a severe shortage of the world's unique North Sea shrimps, prices exploded in 2017 with wholesale prices of more than EUR 50 per kilo of peeled garnet. In 2019 were caught 7000 tonnes of shrimps, down from 15,200 tonnes the year before. In 2021, the number of fishing cutters had decreased to 87 units (Fig. 1.85).

Fig. 1.85 In the summer of 2009, a large demonstration of almost a hundred cutter owners against the price dictate paved the way to the establishment of a cooperative and to more stable prices (Ralf Witthohn)

1.8.13 Garnets to Tangier and Back

The garnet caught by hundreds of shrimp cutters off the German and Dutch coasts are still cooked on board and are therefore suitable for immediate consumption, but still have to be peeled. While the knowledgeable coastal dweller does not allow himself to be deprived of this work, the peeling poses unsolvable tasks for unsuccessfully developed crab-peeling machines as well as for the common consumer. As a result, resourceful Dutchmen developed a business model that was as unusual as it was profitable. In the beginning, the garnets were transported in refrigerated trucks to Poland and Belarus, and later to specially built factories in Morocco, where local women peeled them. The garnets only remain edible, if mixed with preservatives during the 5000 km journey. The land route leads via France and Spain to Algeciras, and from there by ferry to Tangier. In March 2020, this long way of getting the garnets peeled came to a stop, when in the wake of the COVID-19 pandemic restaurants were closed and Marocco was hit by the virus as well.

On several routes from Spain and Gibraltar to Morocco, the Spanish subsidiary FRS Iberia of Flensburg-based ferry operator FRS used a total of six ferries in 2017, including the high-speed ferry TARIFA JET. At a maximum speed of 42 knots the ferry managed the route from Tarifa to Tangier in 35 minutes. ALGECIRAS JET and CEUTA JET achieved 36 knots, while TANGER EXPRESS (Fig. 1.86), KATTEGAT and ANDALUS EXPRESS are conventional ferries running 20 to 23 knots. The routes established were Algeciras—Ceuta, Algeciras—Tanger-Med and Tarifa—Tanger-Ville. In 2018, the new lines Motril—Melilla and Huelva—Canary Islands were added. In March 2020, the passenger services between the Moroccon ports and Spain were suspended due to the COVID-19 pandemic. At this time, the schedule structure included six lines, besides the mentioned routes a connection from Tanger-Med to Motril. In May/June 2021 FRS started to employ TARIFA JET on a new route between Alcudia/Mallorca and Ciutadella/Menorca and the companie's fast catamaran SAN GWANN between Ibiza and Formentera. In August 2021, SAN GWANN hit a rock off Ibiza causing injuries to 35 of the 47 passengers. In the beginning of 2022, out of four routes between Spain mainland and the African continent the three lines to Tangier were temporarily suspended, only the connection from Algeciras to Ceuta was maintained by the 37 knots achieving LEVANTE JET, laid out for 675 passengers and 151 vehicles. The 5537 gt Incat design of 2015 had been operated by Japanese owners before and was purchased by FRS Iberia in August 2021.

Fig. 1.86 Germany's ferry operator FRS operates several ships across the Strait of Gibraltar, including the conventionally designed TANGER EXPRESS (Ralf Witthohn)

1.8.14 Fishery Protection Vessels

The Spanish Ministry of Labour maintains the fishery protection vessels ESPERANZA DEL MAR and JUAN DE LA COSA. They are classified as hospital ships, but medical support, which may as well be transmitted by radio contact, is only one of their duties carried out to support Spanish trawlers operating off the African coasts. Built in 2001 and 2006, the vessels as well render technical assistance to the about 5000 fishermen, in the same way as provided by other nations' ships without hospital ship status. Based in Las Palmas, the 4983 gt ESPERANZA DEL MAR was considered to be the world's largest hospital ship for civil purposes, before the 37,856 gt newbuilding GLOBAL MERCY of US-based Mercy Ships was commissioned in 2021. The twin-propelled ESPERANZA DEL MAR has room for 17 sick or injured persons or 30 castaways and can accommodate up to six containers of medical equipment in the forward hold. The ship can also carry out rescue, towing and salvage operations, including fire and oil fighting. A helicopter landing deck is fitted above the aft deck. The deck equipment includes several fast boats, including one for 17 persons. The smaller JUAN DE LA COSA can accommodate ten to maximum 22 patients (Fig. 1.87).

Although only a few deep-sea fishing vessels are still flying the German flag, the Federal Agency for Agriculture and Food still maintains a fleet of three protection vessels built between 2000 and 2009 and named SEEADLER,

Fig. 1.87 Spain's fishery protection vessel ESPERANZA DEL MAR is the largest hospital ship for civilian purposes (Erich Müller)

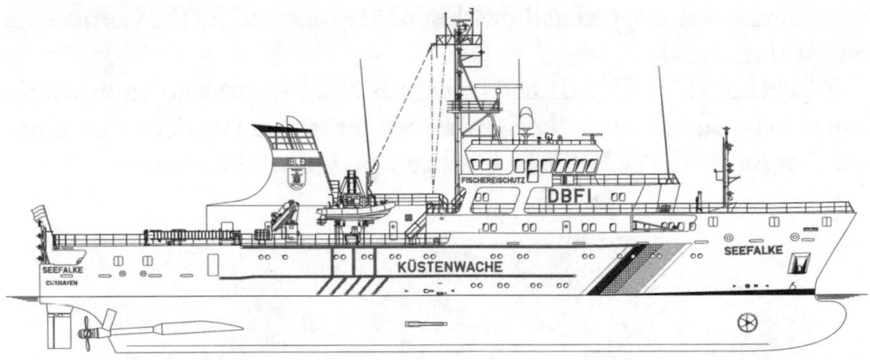

Fig. 1.88 The fishery protection vessel SEEFALKE was the second of three such newbuildings for Germany's Ministry of Agriculture in 2008 (Peene Werft)

SEEFALKE and MEERKATZE. In addition to medical and technical assistance, they carry out checks on catches, logbooks and mesh sizes of nets (Fig. 1.88).

1.8.15 Fishery Research Vessel from Romania

An important role in fishing is played by research vessels, on which scientists determine the size of fish stocks to make recommendations for catch quotas. These are to ensure the survival or recovery of the stocks, but are often adapted to the economic interests of the industry. For the Federal Republic of Germany, the EU-framework surveys as well as ecological, chemical and physical marine investigations in the North Sea, Baltic Sea and in the North Atlantic from the subtropics to Greenland have since 1993 been carried out by the fishery research vessel WALTHER HERWIG III of Bremerhaven-based Federal Institute for Agriculture and Food on behalf of the Ministry of Agriculture. Following an international tender, Dutch Damen shipyard was in March 2017 awarded the contract to build a new, EUR 85 million costing research vessel, which was to replace WALTHER HERWIG III by 2020. Its construction was to be based on a design of the Norwegian company Skipsteknisk. At a length of 84.7 m and a breadth of 16.2 m the ship was to be accelerated to a maximum speed of 15 knots by a diesel-electric propulsion system. It was to accommodate 26 scientists in addition to the 26-man crew and should be equipped with a DP1 class dynamic positioning system Due to disagreements about the design, the building was delayed, so that construction had not yet begun in 2019 and delivery was postponed to May 2023. In August 2021, the tender process was stopped and the ship newly tendered by the German government (Fig. 1.89).

WALTHER HERWIG III had in March 2020 to suspend an expedition planned to be carried out in the Sargosso Sea for research work on the eel until the end of April. The COVID-19 pandemic prevented the research team from

Fig. 1.89 The role of the Bremerhaven-based fishing research vessel WALTHER HERWIG III is to be taken over by a newbuilding, for which the first contract with Dutch Damen group was cancelled (Ralf Witthohn)

being exchanged, so that the expedition, which had started 6 March, was to be terminated already during the positioning voyage and the vessel returned to its homeport 25 March 2020.

In November 2017, a fishing research vessel similar to the German newbuilding WALTHER HERWIG was launched at Damen Shipyards Galati for the Ministry of Fisheries of Angola, off which coast foreign nations carry out fishing. The government of the African state had already received two fishing protection boats from Damen. The 74 m length and 16.4 m breadth BAIA FARTA provides accommodation for 29 crew members and 22 scientists and is to perform a variety of tasks in the fields of hydrography, acoustic measurements, pelagic and trawl fishing as well as plankton and environmental research. It is equipped with a special sonar and echo sounder system, but can also be used in emergencies to combat oil pollution or to undertake salvage action. The design of the BAIA FARTA, also drawn by Skipsteknisk, pays special attention to minimal noise emissions and vibration on and under water. In addition to Damen's Romanian company in Galati, the group's Schelde Naval Shipbuilding in Vlissingen was involved in the realisation (Fig. 1.90).

1.8.16 BRIGITTE BARDOT Against Whaling

In December 2014, the crews of a former Norwegian whalecatcher and a former Japanese research vessel, named BOB BARKER and SAM SIMON, launch the operation Icefish on behalf of the Sea Shepherd Conservation Society. Their intention is, to prevent six trawlers, including THUNDER, KUNLUN, YONGDING

Fig. 1.90 Like the new WALTHER HERWIG for Germany, the Angolan fishing research vessel BAIA FARTA was designed by a Norwegian engineering company for construction by the Dutch Damen group (Damen Shipyards)

and VIKING, from illegal fishing of cod and hake in Antarctic waters. After finding the THUNDER, a 110 days lasting pursuit over 11,500 nm takes place, at the end of which the crew of the THUNDER themselves sink their ship off the West African coast near Sao Tomé, because they have to fear arrest when calling at port. In April 2015, BOB BARKER and SAM SIMON come to Bremen to be equipped for the subsequent campaign. On their way up the Lower Weser they pass the trimaran BRIGITTE BARDOT, also a Sea Shepherd vessel, berthed at the Abeking & Rasmussen shipyard in Lemwerder. At the bow of BRIGITTE BARDOT a luscious blonde swings the pirate flag with a skull, a not really true symbol, because according to the environmental activists' understanding they are the guardians of nature, who try to keep illegally fishing and whale-hunting trawlers from their raids. To do this, however, they use pirate-like actions, cut nets and push the fishing vessels away. Two engines enable the 34.9 m long and 14.1 m wide BRIGITTE BARDOT, stabilized by two booms, to run 24 knots. Under the previous name OCEAN 7 ADVENTURER the trimaran in 1998 made a record-breaking circumnavigation of the world in 74 days. After visiting Bremen and Hamburg, BRIGITTE BARDOT expels pilot whales from the Faroe Islands to protect them from being killed by islanders (Fig. 1.91).

Fig. 1.91 The trimaran BRIGITTE BARDOT distributed pilot whales off the Faeroe Islands in 2015 to prevent their killing (Ralf Witthohn)

1.8.17 Stunned Fish on SPES NOVA

In November 2019, Damen Shipyards Maaskant at Stellendam/Netherlands delivered the newly developed trawler SPES NOVA to the La Hague-based fishing company Van der Zwan as the first of two such units. Realised according to a basic concept of the companies Vripack Naval Architects, Ekofish and the builder the 32 m trawler can carry out various fishing methods, including twin rigging, bottom trawl, seine fishing and flyshooting. The vessel is driven by a 1000 kW developing diesel-electric hybrid system supplying the power for a bollard pull of 20 tonnes, with an alternatively usable battery bank. The fishing equipment includes a netdrum, flyshooting winches, a central trawl winch and conveyor belts, the processing systems comprise the electrical stunning of the fish, an automatic heading and gutting machine, pre-cooling, automated weighing and flake ice machines. The hull lines of the ship ensure, that, in rough seas, the reserve buoyancy of the bow is reduced, so that pitching into the waves and stress on the hull is less strong (Figs. 1.92 and 1.93).

Fig. 1.92 The 2019-built trawler SPES NOVA embodies various design innovations like special hull lines, varying fishing methods and a hybrid propulsion system (Damen)

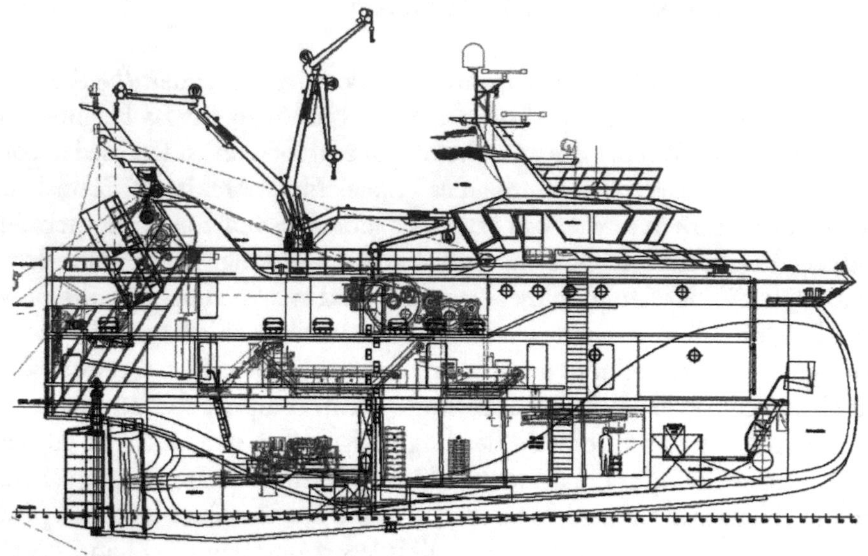

Fig. 1.93 Turning in a knozzle the large diameter propeller of the SPES NOVA is either driven by electrical motor or by battery (Damen)

1.9 Feedstuffs

The industrialisation of agriculture requires the provision of animal feed, which is partly guaranteed by imports. In Europe, the feed mainly comes from South American countries, from where soya and soya flour are shipped, from Argentina mainly soya flour. The largest consumer of soybeans is China, which, at an annual growth rate of 5% in 2017, imported more than 90 million tonnes from Brazil and the United States, each country contributing a share of over 40%. An almost all-year delivery period results from the season starting in September in the USA, while in Brazil harvesting takes place from February to June. The beans are processed into soy flour and soy oil. The latter is used for food, plastics, lubricants and biodiesel. Argentina's largest soya customer, importing 5 million tons annually, is Vietnam. A newcomer in soya production is Mocambique by exports through the port of Beira mainly to China. The export of US-produced soya to China was hit by the trade war between the two countries. In July 2018, the kamsarmax carrier PEAK PEGASUS loaden with 70,000 tonnes of soya beans for Dalian on behalf of the trading house Louis Dreyfus, arrived off the Chinese coast only shortly after China had imposed a 25% tariff on the cargo. Like the bulk carrier STAR JENNIFER, the PEAK PEGASUS was therefore ordered to wait off the coast for several weeks, before the ship entered the port of Dalian for discharge.

Other kinds of animal feeds like magnesium and urea are as well finding their ways into the stables by imports. Broker companies engaged in the shipment of animal feed include Banchero Costa in Genoa, Braemar ACM, Howe Robinson and Clarksons in London or Barry Rogliano Salles located in Neuilly-sur-Seine.

1.9.1 Feed for Oldenburg Pigs

For several hours the Panama-flagged 56,000 dwt bulk carrier SANTA URSULA inward bound from Paranagua anchors in the Outer Weser. Only then, the scheduled berth in Brake has become available and the supramax-sized ship can start its 5-h ride on the tidal wave. On early Sunday, the hatches are opened to prepare the unloading of the soy flour. A few days later, the mini-bulker APOLLO EAGLE of Stade-based L&L Shipping takes over its cargo from the same silo facility for further distribution in Europe. At this time, the next bulk carrier arriving from the South Atlantic, SANTA ISABELLA, is awaited. Often engaged in the transport of soya to Brake were bulk carriers of Hamburg-based shipping company Rudolf A. Oetker. In 2019, this included the 61,255 dwt SANTA JOHANNA, delivered to a Japanese owner by Shin Kurushima shipyard at Toyohashi/Japan in 2017 (Fig. 1.94).

Fig. 1.94 Before the companie's sale, Rudolf A. Oetker's supramax carriers including the SANTA JOHANNA often showed their striking Hamburg Süd red, when discharging soya from South America at Brake (Ralf Witthohn)

Rudolf A. Oetker was the name, under which the bulk shipping arm of Germany's Hamburg Süd group sailed over more than 60 years. When in December 2016 the conglomerate announced the sale of all shipping activites it disposed of a number of panamax and supramax bulk carriers. In addition to these ships mostly chartered on a long-term basis under SANTA names from Japanese owners, there were also bulk carriers employed without a change of name, such as the Kamsarmax carriers DIANA SCHULTE and DORA SCHULTE of company Thomas Schulte. In April 2019 the sale of the bulk shipping sector with the shipping companies Rudolf A. Oetker in Hamburg, Furness Withy in London and Melbourne as well as the Brazilian Alianca in Rio de Janeiro to China Navigation Co. in Hong Kong became effective. The sold companies operated a total fleet of 45 handysize, supramax and panamax class ships.

One of Europe's most intensively used agricultural area situated south of the city of Oldenburg/Germany has to rely on the import of animal feed. Offering large silo facilities the near-by port of Brake on the Lower Weser is the place for handling the fodder, primarily soy flour from South America, and Asia. Animal feed is also imported from ports like Chornomorsk/Ukraine, from where the 74,750 dwt bulk carrier FORTUNE TRADER transported a nearly full cargo to Brake in January 2020. The southern pier of the harbour extending along the Weser is reserved for the handling of animal feed, grain and other agricultural products. A 95 m high silo tower, visible far into the marsh landscape, is located adjacent to the quay. As one of the largest in Europe, the facilities of company J. Müller have a total capacity of 360,000 tonnes. They do not only serve to store the goods, but also to dry, grind, crush or mix them before being loaded onto trucks with the help of a self-service system. The agricultural products transported by bulk carriers up to panamax size are also re-exported by ships of the same size or distributed to European ports by coasters. Brake is one of the few ports where, after unloading, new cargo is occasionally available for bulk carriers, as sulphur from natural gas production is loaded in course of the same quay. As one of nine Lower Saxon ports Brake handled a total of 6.6 million tonnes in 2019, an increase of 5% compared to the year before. In 2020, 5.5 million tonnes, including 2.1 million tonnes of general and breakbulk cargoes, were handled, resulting in a decrease of 17% (Fig. 1.95).

Among the transporters of soya from Paranagua to Brake was in August 2019 one of the most innovative bulk carriers commissioned in recent years, the 95,263 dwt HARVEST FROST, delivered by Japan's Oshima Shipbuilding to the US company ADM Harvest Shipping as part of the Archer Daniels Midland Company. The post-panamax bulk carrier was the first vessel of its

Fig. 1.95 The storage facilities in Brake have a capacity of 360,000 tonnes for feed imported from South America or Asia (Ralf Witthohn)

size to use the Mitsubishi air lubrication system (MALS), which reduces the drag between the vessel hull and seawater by blowing air bubbles produced under the vessel's bottom. The effectiveness of the system benefits from the ship's large breadth of 40 m and its small draught of 12.5 m related to a length of 237 m. The supplier reported that it would help to achieve a 27% reduction of CO_2 emissions. The HARVEST FROST also features a new bow shape designed to reduce resistance, better propulsion from fins positioned forward of the propellers and special grooves placed in the propeller boss cap. Japanese classification society NK issued the first EEDI (Energy Efficiency Design Index) certification for a vessel fitted with the MALS system. The HARVEST FROST was commissioned in 2014, the sister vessels HARVEST TIME and HARVEST RAIN in 2015 (Fig. 1.96).

1.9.2 Soya for Sweden

The Finnish-flagged coaster NINA arrives on the Weser on 17 January 2015 on a voyage from Amsterdam, deep immerged to the loadline. The 2700 dwt ship is bound for Brake, where it is berthed at the port's breakbulk quay without a turning manoeuvre. After less than 20 h the discharge of the steel cargo is finished, and

Fig. 1.96 An innovative air bubble system helps the US-owned post-panamax bulk carrier HARVEST FROST to compensate the resistance imposed by the ship's small length to breadth ratio, which, in turn, eases to reach upriver ports like Brake called in August 2019 (Ralf Witthohn)

NINA proceeds to the nearby Agri terminal of company J. Müller to load soya, which has before been imported by bulk carriers from South America. On the following day Finland's Meriaura shipping company based in Turku and employing 16 coasters expedits the small freighter to Halmstad and Karlshamn.

For NINA, the trip to the Weser meant a return to its first home port of Brake, where it had been registered after construction in nearby Oldenburg. Captain Gottfried Serafin had his residence in the traditional shipping town of Brake, when he ordered a newbuilding at Heinrich Brand shipyard for delivery in 1987. To finance the new ship, he sold his 19-year-old 1400 dwt coaster CLAUS to Günther Braasch, who was also from Brake and who continued to operate the ship under the name ANTJE B. Serafin's newbuilding took over the name CLAUS from the old ship and in the same way painted in a striking green with a white bulwark. As Brand's shipyard had not taken part before in the construction of rhine-going coasters, of which hundreds of units had already been delivered by numerous German and Dutch builders, CLAUS was developed from a design supplied by Lauenburg/Germany-based shipyard J. G. Hitzler. A larger newbuilding series as planned in cooperation between the two shipyards was yet not realised. At a length of 82 m and a breadth of 12.6 m, the ship type was laid out to pass through the locks of the Finnish Saimaa canal. Driven by a throttled engine, it achieved a modest speed of not more than 11 knots. Special feature of the ship is its ice class E2 of Germanischer Lloyd for winter voyages in the Baltic Sea (Fig. 1.97).

But a low, two decks high superstructure and lowerable masts also allow the ship to sail the Rhine up to Duisburg. The double-hulled ship has only one, 51 m long cargo hold and an asymmetrical stern to improve the hydrodynamic conditions of the aft ship configuration. To minimise its tonnage measurement a deck opening between the forecastle and the forward hatch coaming has been arranged, which creates an ideal, lower tonnage deck. CLAUS was the second newbuilding in a series of three. First one was ATULA for Egon Blohm from Drochtersen, who, benefitting from the ship's high ice class, also found a buyer in Finland, when in 2006 Helmer Lundström in Nivelax took over the coaster, renamed it SABINA and let it manage by company Meriaura. In 2015 it was taken over by Albania-based Motega Shiptrade and renamed NS SABINA in 2017. The third newbuilding of the Brand series was named BETULA. After being sold, the ship broke apart in January 2004 when it grounded off Bilbao under the name DIANA UNO.

Brand's rhine-going coaster type was developed to a pure seagoing 3300 dwt newbuilding. To meet the requirements connected to an improved seaworthiness a higher forecastle was designed, and a four decks high superstructure was to enhance the sight conditions during container transport. The ship's depth was increased from 6 to 6.5 m, the length by 3 m, and a main engine of 2450 kW output chosen to enable a speed of 13 knots. After delivery as one of the last newbuildings in the builder's almost 150-year history, the ship started a charter under the name TAINUI (Polynesian for canoe) in New Zealand in 1993. Back in Europe, it was renamed AMICA and OSTSEE and

Fig. 1.97 Once built for a Brake-based shipowner, the rhine-going coaster CLAUS returned 28 years later to the Lower Weser as the Finnish NINA (Ralf Witthohn)

laid up at Bremerhaven for several months during the shipping crisis, interrupted by occasional employments. When AMICA was due to leave the shipyard after a stay at Bredo in May 2008, a propulsion defect occurred, which caused the ship's grounding in the vicinity of the Bremerhaven-based settlement of the managing Zirkel Transport und Schiffahrtsdienstleistungsges. At that time, the ship had already had a long history of detentions issued by port inspectors in Europe including sailing bans of up to 3 days. In 2013 registered for a Flensburg shipping company as ARSLAN 1 the ship was the following year classified by the Japanese company NKK. After a couple of ownership changes it was in 2019 converted by Yalova/Turkey-located Tersan shipyard to a self-discharging cement carrier with a pneumatically working handling system on behalf of Zagreb-based company Makar Navis. Under the name C3 MAGNAR it subsequently served mainly between Croatian ports (Fig. 1.98).

1.9.3 Fish Feed Supply Vessel AQUA FJELL

In 2014, Crist shipyard in Gdansk delivered the sophisticated fish feed supplier ARTIC FJELL to Artic Combiship in Florø/Norway. The 3200 dwt ship is used to transport fish feed in bulk or in big bags. The 69.7 m length, 15 m breadth vessel has 20 box-shaped holds with a capacity of 4800 m^3 for 2200 tonnes of feed, also being transported as pellets, in small silos. The ship's single hatch has a length of 36 m and a width of 12.6 m. The pneumatic unloading system, which handles 200 tonnes per hour, includes a crane with an outreach of 28 m placed in front of the superstructure. In combination with an Alstom dynamic positioning system it is used to directly supply the salmon basins. A gantry crane travelling along the hatch lifts 40 tonnes. The ARTIC FJELL, which is also equipped with two forklifts, is driven by a Wärtsilä 9L20 type main engine for a speed of 13 knots. The ice-strengthened ship is carrying out its tasks on behalf of Bergen-based feed supplier EWOS on the Norwegian coast together with three other ships of the shipping company. It is also employed to import the feed, for example from Brake/Germany, to Norway. In 2019, the ship was renamed AQUA FJELL (Fig. 1.99).

Norway's NSK Shipping placed an order at Tersan shipyard in Turkey about a similar 81.5 m length, 16 m breadth vessel for use by the globally active Biomar Group under the name NYKSUND since 2017. A special feature of this 3850 dwt newbuilding based on a 3125 dwt ship of 2012, is its gas-powered Rolls Royce Bergen propulsion system of 2160 kW output.

Fish fodder is also, besides grain, coal and scrap, one of the cargoes transported by the less sophisticated vessels of Norway's Halten company, founded

Fig. 1.98 The general cargo vessel AMICA, which had longer times of unemployment and was laid up at Bremerhaven in 2009, became a cement carrier 10 years later (Ralf Witthohn)

Fig. 1.99 The AQUA FJELL ex ARTIC FJELL is designed to transport only one type of cargo, namely fish feed (Ralf Witthohn)

at Trondheim in 2014. In December 2019, the 4933 dwt FEED RANA carried a cargo from Brake/Germany to fish farms near Averøy/Norway. The mini bulker purchased from Arklow Shipping Nederland as ARKLOW

Fig. 1.100 The coaster ARKLOW ROCK was fitted with a travelling crane to serve its new Norwegian owner for bulk transports including fish fodder under the name FEED RANA (Erich Müller)

ROCK has retroactively been fitted with a travelling gantry crane to independently handle the bulk cargoes carried. Along with the FEED RANA, Halten operated the FEED FISKAA, FEED HALTEN, FEED STAVANGER, FEED TRONDHEIM, FEED ROGALAND and FEED HELGELAND in 2022 (Fig. 1.100).

1.10 Wood and Wood Products

By a decrease of 4.7% compared to the previous year, worldwide overseas transports of forest products were down to 365 million Tonnes in 2020 from 383 million tonnes in 2019. The spectrum of wood-related goods includes roundwood, sawnwood, cellulose, pellets, panels, paper and fuels derived from wood are shipped. Important exporters of roundwood are Russia, New Zealand and the United States, while main importers are China, Austria and Germany. Canada, Russia and Sweden are leading exporters of sawnwood, China, the United States and the United Kingdom main importers. Wood pellets were exported by the United States, Canada and Latvia, imports to the United Kingdom, Denmark and South Korea. Brazil accounts for a large share of cellulose exports while China is the largest customer. In the second half of 2020, wood transports increased considerably, driven by demands from North America, where imports rose from 3.5 million cubic metres import in 2019 to 5 million in 2020, China and the overall building and furniture industry in the wake of the COVID-19 pandemic causing all-time high lumber prices. In 2020 the export of German raw lumber, for example, increased by 43% to 12.7 million cubic metres, of which 51% were shipped to China. e. g. from Nordenham by the bulk carriers SE KELLY, EIKE OLDENDORFF and ECKERT OLDENDORFF from October 2020 on. In April 2022, the EU, which had imported 3.75 million m^3 of sawn timber from Russia in 2021, imposed a ban on the trade including imports from Belarus.

1.10.1 Cross-Docking in Kotka

Eleven railway wagons roll to the landside entrance of the transshipment hall in the port of Kotka/Finland. Two wagons are loaded with paper for Immingham, Glasgow, Knowsley and Selby, four for Tilbury and Avonmouth and five for Zeebrugge. At the Hietanen X-dock terminal, as the facility is named by Finnish-Swedish operator Stora Enso, rail freight is reloaded into special containers to be shipped. Known as cross-docking this kind of cargo transfer needs large forklift trucks taking paper rolls out of the railway wagons and carrying them to the opposite port side gates of the hall. There they are loaded through the front end of large containers named SECUs (Stora Enso Cargo Units). The boxes have been especially designed for the transport of paper by the Swedish-Finnish forestry products group Stora Enso. At 13.8 m length, 3.6 m width and 3.6 m height, the SECU is larger than a 40-ft container and, at 80 tonnes, can carry three times its weight, though not being approved for road transport (Fig. 1.101).

On the base of a long-term agreement signed in 1999 Svenska Orient Lines (SOL) took responsibility for shipping Stora Enso products from northern Finland and Sweden to the European mainland and the UK. In 2017, the annual volume was stated to amount at 2.5 million tonnes. The largest vessels used for the distribution of paper products were the 2006/2007 built ro ro carriers TRANSPULP, TRANSTIMBER and TRANSPAPER, which were chartered from November 2016 on under the names THULELAND, TUNDRALAND and TAVASTLAND. SOL served two routes from Sweden to Belgium/Great Britain and from Finland to Lübeck, from where the service is continued to Antwerp, Zeebrugge and British ports. In September 2017, the timetable between Gothenburg and Zeebrugge included nine weekly departures of the ships FREYJA, SLINGEBORG, THULELAND, ELISABETH RUSS and SCHIEBORG. Between Oulu, Kemi and Lübeck

Fig. 1.101 Forklift trucks load paper rolls from railway wagons into SECU containers at Kotka's cross-dock terminal for shipment to European destinations (Ralf Witthohn)

TAVASTLAND sailed once a week from Oxelösund, Oulu, Kemi, Lübeck, Gothenburg, Zeebrugge to Tilbury the THULELAND, while the VASALAND served between Gothenburg, Kemi, Oulu, Antwerp and Zeebrugge. In cooperation with British P&O Ferries, SOL offered eleven weekly connections between Gothenburg and Tilbury, and seven from Teesport via Zeebrugge, with the participation of one own departure.

THULELAND, TUNDRALAND and TAVASTLAND are ro-ro carriers of 190.8 m length, 26 m breadth and 15.3 m depth to the weather deck. They can carry vehicles up to 5.2 m high loaded through the stern ramp on the main deck. During cargo handling, a heeling system from Intering compensates the ship's movements. By means of compressed air, water is transferred between two tanks arranged on the ship sides, thus generating an uprighting moment against the heeling moment. The most powerful of these systems are able to generate a moment of 5000 tm. From the main deck ramps create a connection to the upper and to the lower deck. The ice-strengthened vessels have a deadweight capacity of 16,000 tonnes and can load 155 SECUs over a lane length of 2950 m. Two diesel engines, delivering a total of 18,000 kW on one propeller shaft, allow a speed of 20 knots when both main engines are used and of 16 knots by one engine acting. Builder of the vessels under contract of AB Transatlantic in Skärhamn were Aker Finnyards at Rauma in 2006/07.

Twenty years after the shipment agreement between Stora Enso and SOL, a new shipping company named Wallenius SOL was founded in April 2019 by the Swedish Orient Lines and Wallenius to carry out sea transports of forest products in a network covering the same destinations in the Gulf of Bothnia, the Baltic and the North Sea. Initial customers signing long-term contracts were StoraEnso and Metsä Board. At the start, Wallenius SOL relied on five ro ro vessels, named BALTICA, TAVASTLAND, THULELAND, TUNDRALAND and VASALAND. The schedules comprised a service from Oulu, Kemi and Husum to Lübeck operated by three 2774 lane metres ships with two southbound departures and one northbound per week. A second one was connecting Oulu, Kemi and Pietarsaari with Lübeck, Antwerp, Zeebrugge, Tilbury, Zeebrugge and weekly maintained by 2774 lane meters vessels, THULELAND, TUNDRALAND and VASALAND. In November 2019, the port of Vaasa was included in the rotation. In February 2021, the chartered 3300 lane metres ro ro carriers FIONIA SEA and JUTLANDIA SEA established a weekly connection between Lübeck and Vaasa within their service from north Finnish ports to Zeebrugge/Antwerp (Fig. 1.102). After 1 year, the ships' charter switched to Sweden's Stena Line for employment between Rotterdam and Immingham. Instead, Wallenius Sol chartered the UK-operated freight ferry CADENA 3, until the company was to take delivery from China's CIMC Raffles shipyard of their own LNG burning, 5800 lane meters newbuildings BOTNIA ENABLER and BALTIC ENABLER, which completion was scheduled in autumn 2021, but had not taken place until January 2022. The Danish design office Knud E. Hansen developed the world's largest freight ferries of their kind, of which up to four units should be built. They feature 241.7 m length, 35.2 m breadth and are loaded through a 17 m wide stern ramp. The 27,000 dwt, 20 knots achieving ships will be able to run on green electricity from shore connections during port calls.

1.10.2 Eucalyptus Wood from Bahia

Stora Enso also procures raw materials for paper production by deep-sea transports. The Finnish-Swedish concern and its Brazilian partner Fibria each hold 50% shares in the pulp producer Veracel in Eunápolis, located in the Brazilian state of Bahia. The raw material is being produced there from eucalyptus trees since May 2005, with an annual production capacity of about 1.1

Fig. 1.102 Due to the delayed completion of the world's largest freight ferries Wallenius SOL had to charter a ro ro ship in place of the FIONIA SEA, which switched to a Stena employment on the Channel route (Ralf Witthohn)

million tonnes. Total investments in the plant amounted to USD 1.25 billion. The financing was provided by the Nordic Investment Bank (NIB), the Brazilian Banco Nacional de Desenvolvimento Economio e Social (BNDES) and the European Investment Bank (EIB). In 2001, the EIB granted a loan of USD 30 million for the planting of trees on an area of 26,200 ha, and in 2003 a further USD 80 million for the construction of the factory. The bleached eucalyptus pulp is produced for export and is the raw material for the production of tissue, printing and writing papers. The eucalyptus trees planted in the region since 1991 and harvested after about seven years are felled by automated harvesting machines that cut an average of 490 trees a day. In August 2005, Stora Enso's paper mill in Oulu received its first batch of 4000 tonnes of pulp from the 50% share of Brazilian production. The pulp is transported on barges across the Jequitinhonha river to the Atlantic port of Belmonte. In 2017, the Dutch Damen Group supplied the company with dredging and pumping equipment to ensure sufficient water depth at the loading berth (Fig. 1.103).

Fig. 1.103 Transfennica's ro ro vessel SEAGARD is a customer at the Finland centre in Lübeck (Ralf Witthohn)

1.10.3 Finnish Paper for Lübeck

The roll-on roll-off vessel SEAGARD of Finland's Bore-Reederei glides past a natural shore landscape with reed belts, flocks of birds and the thatched fishing houses of Gothmund before arrival at Nordlandkai almost in the city centre of Lübeck. The ship occupies the traveside of five berths of the Finland centre. Trucks, trailers and other vehicles roll over the ship's stern ramp. By its hall capacities of 150,000 m² the area is one of the main distribution sites for Finnish paper in mainland Europe. SEAGARD is operated on behalf of the Finnish shipping company Transfennica, a company of the Dutch Spliethoff group. Along the Lower Trave further ro ro vessels have opened their stern doors. At Seelandkai another Transfennica ship, the Finnish-flagged MIRANDA, is being discharged. Opposite in Schlutup is berthed the Dutch TRICA. At a length of 205 m, it represents the largest freight ferry type regularly sailing on the Trave. TRICA and its sister ships TIMCA, KRAFTCA, GENCA, PULPCA and PLYCA are equipped for the simultaneous transport of containers and rolling cargoes, and therefore designated con ro vessel. Their lane capacity is 2950 m, the container intake figured at 643 TEU. On behalf of the Finnish paper group UPM-Kymmene, daily departures are offered from Travemünde to Hanko and Paldiski. There are a total of two dozen ferry connections from Lübeck and Travemünde to the main Baltic Sea ports in Finland, Sweden, Poland, Russia, Lithuania and Latvia. Freight and passenger ferries pass the jetties of the Trave entrance every hour (Figs. 1.103 and 1.104).

In December 2019, test operations started at a new terminal for forestry products at Skandinavienkai in Lübeck-Travemünde, next to the existing passenger terminals. The terminal comprises a warehouse of 25,000 m² and a 5000 m² large ferry hall. In the 300 m long warehouse can be stored up to 35,000 tonnes of forestry products. They are received and distributed through

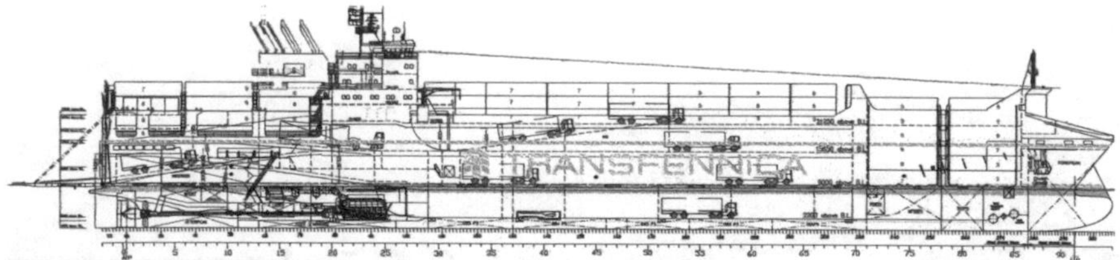

Fig. 1.104 The Dutch-owned TRICA stows trailers and containers on its cargo decks (Stocznia Szczecinska)

Fig. 1.105 End of 2019, the schedule of the KRAFTCA included calls at Hanko, Antwerp, Kotka and St. Petersburg (Ralf Witthohn)

six doors at every side of the warehouse. Simultaneously with the inauguration of the new Travemünde terminal the Seelandkai terminal was extended by 10,000 m^2. For 2019, the local city-owned port company LHG reported a 6% increase in the handling of paper, cellulose and sawn timber, followed by a 7.5% decrease in the next year (Fig. 1.105).

1.10.4 Sawn Timber from Wismar

After a few hours waiting time at the Mecklenburg Bay anchorage five nautical miles north of the island of Poel, the coaster FINJA takes the pilot from the transfer boat stationed at Timmendorf at the northern tip of the island. Within three quarters of an hour the ballasted ship owned by Jürgen Ohle from Drochtersen arrives at the port of Wismar to load 2000 tonnes of sawn timber.

At Wismar, the second largest seaport in Mecklenburg-Vorpommern next to Rostock, the handling of forest and wood products plays an important role.

A total of 1.3 million tonnes of industrial and sawn timber were imported via the port in 2016. Almost half of the processed coniferous wood is exported, primarily to North America, Asia and Australia, but also to UK, the Netherlands and Denmark. 1800 employees work in the Haffeld industrial estate in one of Europe's most modern centres for wood processing. The saw-mill and planing mill built in 1998 by the Austrian Klausner Group in the immediate vicinity of the seaport became part of the Russian company Ilim Timber in June 2010. As one of the largest plants of this kind in the world it produces an annual output of 120,000 tonnes by cutting 80 million boards. 60% of spruce and pine are imported over the Baltic Sea from Scandinavia, the Baltic States and Russia.

Another wood processing company, Egger Holzwerkstoffe, uses the advantages of the port for the production of chipboard, fibreboard and laminate. The plant also receives wood chips and sawdust. Every year, 900 employees process 2 million cubic metres of wood and handle 50 shiploads. Up to 800,000 m^3 of chipboard and 55 million square metres of laminate flooring can be produced from this raw material. 170,000 tonnes of glue and 41,500 tonnes of impregnating resin are required for this. The urea needed for production comes by ship from Russia, 90% of the wood from Germany, the remainder from the Baltic Sea region, Scotland and France. 90% of the production is exported to Europe, 10% to the rest of the world.

A third timber company, Hüttemann originating from Germany's Sauerland region, has in 1999 chosen Wismar as location of its laminated wood production with a capacity of 90,000 m^3 because of its proximity to the Baltic raw material and the export possibility by ship. Wismar is also the shipment place of wood pellets produced by German Pellets, a leading European manufacturer, which was, following insolvency in 2016, taken over by the US financial investor Metropolitan Equity Partners (MAP). In that year, 129,000 tonnes of planed sawn timber and 43,000 tonnes of fibreboard were exported via the Hanseatic city. The plant has an annual capacity of 256,000 tonnes, the port of Wismar indicated the export volume shipped as 3500 tonnes. In 2020, a total of 5.8 million tonnes of cargo was handled in the port of Wismar, a drop of 12% compared to the pre-year (Fig. 1.106).

1.10.5 Sawn Timber from Germany to Baltimore

Arriving from the Norwegian port of Herøya, the captain of the bulk carrier ANSAC WYOMING has ordered to open the hatch covers of his ship already during the voyage upriver. This helps to ventilate the holds of the partially loaded and

Fig. 1.106 Imports of fresh wood and exports of wood products dominate the Wismar handling activites (Ralf Witthohn)

slightly aft trimmed ship. Moved by a chain drive two each of the four covers of every of the five hatches begin to fold like a tent, thus continuing to protect the cargo from humidity in the damp December weather of 2017. In the next 2 days, the 33,100 dwt vessel takes over sawn timber from German production for Baltimore at the north pier of the port of Brake. The ship of Japanese owner MMSL has been chartered by Denmark's Ultrabulk shipping company, which is operating a liner service from Europe's north to the US east coast once to twice a month.

The so called parcelling service established by Ultrabulk represents a new kind of shipment procedures for mostly breakbulk cargo by trying to individually attract single cargo parties from different customers, instead of acquiring contracts for the transport of complete cargoes of a sole shipper. This compiling of parcels on bulk carriers copies the way in which general cargo vessels are traditionally filled. In December 2017, ANSAC WYOMING had started its round trip to Herøya and Brake from Baltimore, Wilmington and Tampa. The schedule is dependent on cargo volumes and may include other ports of call such as Antwerp, New London, Savannah or Port Canaveral. Open hatch bulk carriers with very wide hatches that do not require any stowage in the hold wings, are preferred, thus simplifying handling work and cargo securing. In the beginning of 2021, the bulk carriers serving the Ultrabulk line to the US East Coast/Gulf included WESTERN MIAMI, NORDMOSEL, CIELO DI VALPARAISO and TCZEW. A year later were engaged the 37,000 dwt OCEAN ECHO, the 50,000 dwt COUGA, the 45,500 dwt POLAND PEARL and the 39,000 dwt LEGIONY POLSKIE.

The Copenhagen-based shipping company, which together with its sister company Ultragas forms the Danish Ultranav group, recorded 31 panamax, 69 supramax and 63 handysize carriers in its fleet list at the end of 2017, the majority of which were chartered tonnage in addition to own and partly-owned ships. End of 2019, the number of supramax ships had increased to 70, the handysize units decreased to 60. At that time, Ultrabulk employed the 37/38,000 dwt ships GREAT PROFIT, ANSAC PRIDE, SUPER EMMA and SUPER NEELE from Brake to varying US East Coast ports including Baltimore, Wilmington, Port Canaveral, Lake Charles and Houston. Another Ultrabulk parcel service connected Black Sea and Mediterranean ports like Novorossiysk, Iskenderun and Istanbul with Caribbean, US Gulf and US East Coast destinations.

ANSAC WYOMING was in March 2015 delivered according to a frequently realized standard design of Japan's Kanda Shipbuilding. Japanese shipbuilders in particular have decades of experience in constructing bulk carriers of all sizes. While the largest units, the VLOCs (Very Large Ore Carriers) and Capesize ships, are built at the shipyards of the Mitsui, Mitsubishi or JMU groups, the numerous medium-sized builders are dedicated to the smaller bulk carrier types. These include Iwagi Zosen, Shimanami, Imabari, Minaminippon, Shikoku, I-S Shipyard, Saiki, Shin Kochi, Shin Kasado, Kanda and Naikai. In October 2010 the Imabari shipyard delivered the first newbuilding of its IS Bari-Star, developed from a 37,000 dwt handysize bulker, under the name NORD IMABARI. The shipyard had previously chosen similar designations for three more of its bulker classes: IS Nexter for 95,000 dwt, IS I-Star for 61,000 dwt and IS Brastar for 335,000 dwt vessels. The IS Bari-Star type was realised at the locations in Imabari and Muragame as well as at the former Watanabe Zosen site, now operating as Shimanami shipyard (Fig. 1.107).

At Imabari, NORD IMABARI was succeeded by 49 newbuildings of this type by 2017, including INA-LOTTE and DIANA BOLTEN for German shipping companies. IYO SEA, ULTRA FITZ ROY and CLIPPER BREEZE of the same handysize type were built in Muragame in 2015/2016. ULTRA FITZ ROY of Taiyo Sangyo Trading & Marine in Yokohama, delivered in January 2016, was as well contracted by Ultrabulk. The 180 m length, 29.8 m breadth and 15 m depth vessel achieves a deadweight of 37,918 tonnes on a draught of 10.5 m. The cargo holds have a capacity of 47,000 m^3 grain. A 6780 kW two-stroke engine drives the vessel to a speed of 16 knots. At the end of 2019, eleven ships were registered for Taiyo Sangyo. The Shimanami shipyard in Hakata participated in the construction programme from 2011 on with YOU & ISLAND and by 2021 had produced a further 72 units,

Fig. 1.107 Open hatch bulk carriers such as ANSAC WYOMING are suited to be employed in parcel services, as established by Denmark's Ultranav Group from the northern continent to US East Coast ports (Ralf Witthohn)

Fig. 1.108 CRIMSON MAJESTY from Shimanami Shipyard is one of more than 70 IS Bari-Star units delivered until 2021 (Ralf Witthohn)

including CRIMSON MAJESTY of MMSL Japan Ltd. in 2013, a company of the Marubeni Group in Tokyo, for which 33 general cargo vessels and bulk carriers of up to panamax size were registered in 2022 (Fig. 1.108). Marubeni took part in a joint development programme for the demonstration of unmanned ships, which was tested by a small passenger boat in Yokosuka.

1.10.6 Namura's Log Carriers

Together with its sister shipyard Hakodate Dock Co., Japan's Namura Shipbuilding developed a type of bulker named High Bulk 34E and specifically laid out for the carriage of logs. For the transport of tree trunks and timber are fixed steel supports arranged on the weather deck. The supports are foldable in way of the hatches. The 180 m length, 30 m breadth and 14 m depth ships carry 34,400 tonnes on a draught of 9.8 m. No hopper tanks as commonly found in bulk carriers have been arranged, so that the five holds are largely box-shaped. In order to still allow unloading in the corners of the holds, they have large hatch openings of 20.4 m width with relatively small wing tanks. The hatches are served by four 30 tonnes lifting cranes. The total volume of the cargo holds is figured at 42,900 m^3 for general cargo (bale) and 44,150 m^3 for grain. The construction of the ballast tanks is complying with the IMO guidelines against corrosion in order to improve ship safety. A two-stroke diesel engine of type MAN B&W 6S46ME complying with NOx reduction regulations gives the ships a speed of 14 knots. Fins are fitted at the rudder to improve hydrodynamic conditions for the reason of saving fuel.

From 2014 to 2017, Namura delivered 36 buildings of this type. Hakodate's contribution to the series in the same period included GLOBAL SYMPHONY, GLOBAL HERO, OCEAN RALLY, BERGE HAKODATE, BERGE DAISETSU, PEGASUS OCEAN, DREAM ISLAND, BERGE ASAHIDAKE, BERGE ANNUPURI, MARATHA PREMIER, BERGE RISHIRI, BENJAMIN CONFIDENCE, HILMA BULKER and HELGA BULKER. Some of the newbuildings were specially laid out for wood transport and therefore fitted with special supports at the bulwark. AFRICAN PIPER, which was delivered to MUR Shipping in Amsterdam in January 2015, transported a cargo of coal from Murmansk to power plants in Thamesport and Bremen in May of the following year. In 2016, the Dutch shipping company operated or commissioned almost 40 handysize carriers, most of which were built after 2014. In 2021, the fleet comprised 56 units (Fig. 1.109).

Among the few shipbuilders that are specialized in building multipurpose cargo vessels, particularly laid out for the transport of general and break bulk cargo, is Honda Heavy Industries in Saiki/Japan. First newbuilding in a series of eleven 16,954 dwt ships for Thorco Shipping was THORCO LIVA in 2012, succeeded by THORCO LEGEND, THORCO LEGACY, THORCO LINEAGE, THORCO LUNA, THORCO LEGION (Fig. 1.110), THORCO LANNER, THORCO LOGOS, THORCO LOGIC and THORCO LOHAS until 2016. Their equipment includes two 50-tonne cranes positioned starboard, by combination able to carry units of up to 100 tonnes weight. The 131.7 m length, 23 m breadth vessels stand out by their 16.4 m depth, which, at 9.6 m draught, results in a significant freeboard of 6.4 m. The large depth creates a cargo hold capacity of 26,000 m^3, the container capacity is yet only figured at 200 teu. Main engine is a 5200 kW

Fig. 1.109 Thanks to a smaller draught, but at the cost of less deadweight capacity AFRICAN PIPER is able to call at shallow ports (Ralf Witthohn)

Fig. 1.110 The 17,000 dwt tween decker THORCO LEGION, sold to Shanghai based Fairwind International and renamed FAIRWIND LEGION in 2020, seeks its cargo in world-wide trades (Ralf Witthohn)

two-stroke engine enabling a speed of 16 knots. Somewhat smaller then the THORCO L class is the 2010 Honda built 13,800 dwt THORCO GLORY, which ran aground in the Tateyama Bay/Japan in October 2013 during a typhoon.

Copenhagen-headquartered Thorco Shipping has world-wide located settlements for cargo acquisition. When the Philippine-flagged THORCO LUNA reached Bremen on one of its first voyages in May 2015 from Central America, the vessel was nearly showing a ballast draught. The fleet list of Thorco, then co-operating with the Clipper group, comprised 70 ships with deadweights of 8000 tonnes to 20,000 tonnes at that time, many of them time-chartered. In September 2016 Thorco teamed up with United Heavy Lift to found the new company Thorco Projects. In 2020, the Thorco fleet list comprised 20 multipurpose tween deck carriers of up to 20,000 dwt. Among nine units listed in May 2021 were five 16,900 dwt ships of the 2012- to 2015-built THORCO LIVA class.

1.10.7 Woodchip Carriers

Japan's old-established Hachiuma Steamship Co. in Kobe still operated, according to companie's data, four woodchip carriers in 2022, down from five ships in 2019 and seven ships in 2014. Typically, the GREEN PEGASUS, OCEAN PEGASUS and CRIMSON SATURN are fitted with sophisticated deck equipment comprising cranes, conveyors and hoppers to allow the self-sustained handling of cargo. Only the STELLAR STREAM II, which was built by Shin Kurushima Dockyard for Pegasus Shipholding in 2005, lacks of an own handling equipment. The 200 m length, 32.2 m breadth vessel has an extremely large depth of 22.8 m, a cargo volume of 102,276 m^3 and a deadweight of 50,471 tonnes related to a draught of 11.5 m. The freeboard dimension of 15.7 m thus exceeds the draught by more than one third. A 7545 kW generating two-stroke engine accelerates the ship to a speed of 15.6 knots.

Of the same dimensions and tonnage are Hachiuma's woodchip carriers CRIMSON SATURN and OCEAN PEGASUS, delivered by Shin Kurushima in 2001 and 2008, resp. The handling equipment fitted is yet reducing the deadweight to 49,997 tonnes of CRIMSON SATURN and 49,724 tonnes of OCEAN PEGASUS. Backed by Japan's largest shipping company NYK as main shareholder, Hachiuma primarily employs the ships in woodchip transports to Japan, South Korea and Taiwan as well as increasingly to China and India as a new market. The shipping company also manages bulk carriers up to the capesize class, car carriers, multipurpose cargo vessels and semi-submersible module carriers (Fig. 1.111).

STELLAR STREAM II took over the name from the companie's STELLAR STREAM, which was scrapped in 2014 after 24 years of service. The old STELLAR STREAM had been built by Sumitomo Heavy Industries in Oppama featuring a similar length and breadth, a depth of 21.9 m and a draught of 10.8 m. Its deadweight was figured at 42,744 tonnes. A two-stroke engine of 5360 kW output allowed the six-hatch vessel to achieve a speed of 13.8 knots (Fig. 1.112).

One of the largest woodchip carriers is the GLORIOUS SAKURA of Japanese shipping company NYK. At a length of 210 and 36.5 m breadth, the postpanamax ship carries 60,600 tonnes on a draught of 11.5 m. Oshima Shipbuilding delivered the vessel with a cargo capacity of 122,150 m^3 to Kitaura Kaiun in December 2008 (Fig. 1.113). Another NYK woodchip

Fig. 1.111 OCEAN PEGASUS is one of three Hachiuma woodchip carriers that are equipped with a special discharge system (Ralf Witthohn)

Fig. 1.112 The 1990-built woodchip carrier STELLAR STREAM came along without own handling equipment (Ralf Witthohn)

Fig. 1.113 As one of the world's largest chip carriers GLORIOUS SAKURA features special equipment for the self-sustanied discharge of cargo (Ralf Witthohn)

carrier is the Oshima-built SILVER PEGASUS, which in December 2007 loaded a record of 46,000 tonnes for Muroran in Japan at Eden, New South Wales, on its first voyage in charter of South East Fibre Export (SEFE). Later, the ship was as well engaged in the feed trade between South America and Europe. At 210 m length and 32.3 m breadth, the depth of the SILVER PEGASUS is figured at 23 m, the ship's draught is half of that, resulting in a freeboard of 11.5 m. No other cargo vessel type has such a high freeboard/draught ratio of 1 to 1. The ship's tonnage of 43,621 gt is correspondingly large.

Due to the unusually large freeboard, woodchip carriers are constructed without a raised forecastle, while the aft deck is arranged lower than the main deck for a better handling of the mooring equipment. The six holds of the 54,347 dwt SILVER PEGASUS have a capacity of 109,000 m^3. Its extensive loading facilities include three revolving cranes and four hoppers for loading by the conveyor system. The deep cargo holds of woodchip carriers harbour special dangers at the hold accesses. In September 2014, a Vietnamese officer of the SILVER PEGASUS plunged six metres into the empty hold during the discharge of soy flour in the port of Brake and fatally injured himself. The Federal Bureau for Marine Accident Investigation in Hamburg criticised the design of the access by assessing it as "inadequate", because the last part was 6 m high without technical necessity (Fig. 1.114).

Fig. 1.114 In 2014 safety deficits of the cargo hold accesses of the SILVER PEGASUS caused the death of a crew member, but the typical high freeboard architecture of the woodchip carriers as well requires a combined ascent of Jacob's ladder and gangway for pilots

Fig. 1.115 The sub-panamax carrier DAISHIN MARU had a cargo hold capacity of nearly 80,000 m^3 (Ralf Witthohn)

In 1996, Shin Kurushima shipyard launched the woodchip carrier DAISHIN MARU, which was after 2013 operated by a Singapore-based company under the name JUPITER and broken up in 2019. Its dimensions of 195 m length and 29 m breadth remained below the old panamax dimensions. At a depth of 17.8 m, the cargo hold volume amounted to 79,356 m^3, the deadweight was figured at 38,692 tonnes (Fig. 1.115).

1.10.8 American Pellets

A mobile crane discharges 2650 tonnes of wood pellets from the Faroe Islands-flagged coaster ATLANTIC in the Handelshafen of Bremerhaven. The pellets are a small party of 600 tonnes from the cargo of a bulk carrier, which had shipped it from North America to Fredericia/Denmark. A share of the pellets are temporarily stored in the halls of the former Bremerhaven timber trading company Chr. Kühlken, another part is transported by truck to the warehouse of the importer, Raiffeisen Bio-Brennstoffe at Münster. At the same time, a large consignment of feed imported from Russia is stored in sacks in the same area behind the double lock of the fishing port. The site is operated by HVG Service, which has a 120 m long berth for ships with a draught of up to 6.9 m on the harbour canal and one for 150 m length vessels with a draught of up to 6.2 m on the Silopier. After discharge is finished, ATLANTIC proceeds to Bremen to load fish meal for Denmark in the Holz- und Fabrikenhafen. The 3040 dwt ATLANTIC of Thorshavn-registered EP Shipping is a series type coaster built at the Sietas shipyard in Hamburg under the name MELTON CHALLENGER in 1980. In 2020 ownership switched to Pilin Fleet Management Florida (Fig. 1.116).

1.10.9 Russian Timber for Oldenburg, Swedish Cellulose for Minden

Departing the port of Brake, the Russian coaster LETNIY BEREG sails upriver on the Weser to the entrance of the Hunte river at Elsfleth, from where the ship proceeds to Oldenburg, in order to discharge its cargo of wood there. The ship's maximum draught of 3 m has been reduced by about 60 cm so that the 1750 dwt ship can navigate the meandering small Hunte under pilot guidance. A fortnight

Fig. 1.116 The coastal freighter ATLANTIC transported a part cargo of pellets from Fredericia to Bremerhaven (Ralf Witthohn)

later the ship leaves Oldenburg and sets course for Arkhangelsk, where operating Belfreight shipping company is based (Fig. 1.117).

Belfreight employs the cargo vessel primarily for transporting timber. Ten years after foundation in 2000 the company operated the cargo vessels LETNIY BEREG and PUR-NAVOLOK in the Kara Sea, Barents Sea, White Sea, Black Sea, Caspian Sea and Mediterranean Sea as well as on Russian inland waterways. For navigating rivers and canals, the overall height as well as the draught of the ships is limited. Their manoeuvrability is increased by a German-manufactured twin-engine propulsion system with two propellers, the visibility from the wheelhouse eased by a forward arranged superstructure (Fig. 1.118).

Built in 1991 by Volodarskiy shipyard in Rybinsk under the name SELIGER, the cargo vessel later sailed as AKMENA, UNDINE, DORADO I, EUROSPIRIT and UROS, before it was named LETNIY BEREG in 2009. The builder, located 280 km north of Moscow at the mouth of the river Sheksna into the Volga, delivered a longer series of this seagoing vessel type from 1968 to 1999. Previously, the yard completed 35 SORMOVSKIY type newbuildings. Renamed Slip, the builder subcontracted the hulls for the 4500

Fig. 1.117 Russian cargo vessels like the VALENTIN PIKUL export sawn timber from their homeland to the German ports of Brake, as pictured, and Oldenburg (Ralf Witthohn)

Fig. 1.118 The seagoing inland cargo vessel LETNIY BEREG, built at Rybinsk in 1991, is homeported in Arkhangelsk (Ralf Witthohn)

dwt newbuildings APIS, FRIEDA, BORNEO and CLAUDIA-ISABELL, which were completed by shipyard Peters in Germany from 1995 on. The last delivery took place in 1999, when the 3290 dwt freighter RYBINSK of 116 m length was handed over as the second largest ship in the shipyard's history to the Russian state enterprise Morinraschet at St. Petersburg.

Via the harbour of Brake is also shipped sawn wood on smaller cargo vessels. In November 2018, the Panama-flagged general cargo ship NEHIR left port bound for the Mediterranean with a full deck load of timber. Afterward the ship mainly traded in the Black Sea and Mediterranean. During a port state control at Novorossiysk in November 2019 eleven deficiencies were found by the inspectors, at Rotterdam six deficiciencies in the following month. The 4742 dwt ship is a building of the Weihai Donghai shipyard in Weihai and was delivered in 2010 as HERMANN W. to the German shipping company Zweite MLB Bulktransport under management of Marlink Schiffahrtskontor and Malta registry. HERMANN W was the third in a series of four such units, the sister vessels being named TIM B., JOERG N. and RENE A. Though by a length of 90 m and a cargo hold intake of 5780 m^3 hardly exceeding the dimensions of coasters, the ships are suited for worldwide trading. At a maximum speed of 11.5 knots achieved by a 1960 kW four-stroke engine manufactured in China they will mainly be competitive in bulk or breakbulk trades. HERMANN W, launched as MARLENE, was renamed CHRISTOPH M in 2010 and sold to Turkish owners for management by Gamma Denizcilik Nakliyat at Ankara as NEHIR. Classification switched from DNV GL to Japan's NK. The sister vessels TIM B., JOERG B. and RENE A were as well purchased by Turkish owners and sailing Mediterranean and Black Sea waters as SYRENA, ERGE and CELINE, resp., end of 2019 (Fig. 1.119).

Fig. 1.119 Built in China for German investors as HERMANN W and operated by Turkish owners since 2014 NEHIR was employed for export of sawn wood from Brake (Ralf Witthohn)

Wood product cargoes handled through the port of Brake were the main factor in the boost of general cargo volumes of 2 million tonnes in 2018. In 2020, mainly wood exports helped to reduce the port's negative result. Handling includes humidity and damage sensitive pulp (cellulose), which is regularly imported from Sweden by the 1998/1999 built coastal cargo vessels CELLUS, launched as HEIKE BRAREN, and TIMBUS ex BRAR BRAREN of 6350/6398 dwt capacity. Operated by the Rörd Braren company from Kollmar/Germany in a fleet of eleven vessels in 2020, the German-flagged CELLUS and TIMBUS have since their commissioning been chartered out to the Södra group, Sweden's largest forest-owner association by 52,000 members and the world's leading pulp producer with mills at Väro, Mörrum and Mönsteras. On their way from northern European ports to the Södra terminals, CELLUS and TIMBUS are mostly sailing in ballast, but also transport for example raw wood for processing by Södra. On the ships, Södra has, among other products, from 1994 to 2014 delivered 150,000 tonnes of chlorinefree cellulose from Växjo to Brake, from where it is further on transported by inland cargo ships to the Minden/Germany coffee and tea filter producer Melitta, reached within 8 days from Sweden. The exhaust gas system of the CELLUS has been fitted with a Selective Catalytic Reduction (SCR) catalysator from Siemens to reduce NOx and HC emissions. TIMBUS is fitted with

Fig. 1.120 For over 20 years after delivery by Peterswerft at Wewelsfleth/Germany, the 6350 dwt CELLUS carried pulp from Sweden's Södra production to Brake (Ralf Witthohn)

a cleaning system reducing sulphur by a dry process from the exhaust emissions (Fig. 1.120).

In January 2020, the Marshall Island-flagged, 57,536 dwt general cargo vessel BRASSIANA, managed by Korea's Pan Ocean carried a small cargo of cellulose from Brake to Punta Pereira/Uruguay.

1.10.10 Garage Type Forest Product Carrier SWIFT ARROW

The deficiency of ships fitted with a gantry crane to handle humidity sensitive forest products independently from weather was in 1991/92 tried to avoid by designing the *Totally Enclosed Forestry Carrier* (TEFC), as designated by Norway's Gearbulk company. This unusual solution for a forest product carrier was realized by Japan's Mitsui shipyard. By their high garage-like superstructure fitted to protect the cargo during handling operations the realised ships resemble vehicle carriers, but in contrast to those, forestry carriers have four 40 tonnes gantry cranes working at six cargo holds. The cargo is discharged through side doors, which fold up to protect the cargo during operation. The 185 m length, 30.4 m breadth, 18.2 m depth GROUSE ARROW, SWIFT ARROW and MOZU ARROW have a deadweight of 42,276 tonnes at a maximum draught of 12.2 m. Another of Gearbulk's TEFCs is the Hyundai-built JAEGER ARROW, delivered in 2001 for the combined transport of dry and liquid cargo, of which bitumen is carried in tanks of 2500 m^3 at a temperature of 260 °C. While the first three vessels are classified as general cargo vessels by Det Norske Veritas, Japan's NKK has classed the 23,529 dwt JAEGER ARROW as bulk carrier (Fig. 1.121).

1.10.11 Danish Pallet Carrier LYSVIK SEAWAYS

In December 2014, the Danish shipping company DFDS sent its coastal cargo vessel LYSVIK SEAWAYS for the first time on a route from Oslo and Halden to Bremerhaven and Hamburg. Under the name DFDS LysLine the short sea service was organized in a way, that the industrial goods could be disloaded at Oslo over the weekend and regionally distributed already on Monday. Built in 1998 and carrying 5.175 tonnes, the special vessel is able to combine the transport of containers with other types of cargo. These include paper rolls, project cargo and palletized breakbulk, which are handled by a lift in the side loading facility and distributed on the cargo decks with help of forklifts (Fig. 1.122).

1.10.12 Christmas Trees for the Antarctic Station

In order to supply its Antarctic bases Mawson and Davis, the Australian National Antarctic Research Expedition (ANARE) long-term chartered a German high ice-class newbuilding, which was designed especially for this purpose by a shipyard in Germany. Within 1 year after being awarded the

Fig. 1.121 The forestry carrier SWIFT ARROW is designated a Totally Enclosed Forestry Carrier and classified as general cargo ship (Ralf Witthohn)

Fig. 1.122 A side loading device provides the pallet transporter LYSVIK SEAWAYS with a high cargo handling flexibility (Ralf Witthohn)

contract by shipowner Günther Schulz from Wedel, Heinrich Brand Schiffswerft in Oldenburg delivered the icebreaking cargo vessel ICEBIRD. At 109.6 m length and 18.9 m breadth the type of the newbuilding corresponded in its basic concept to at that time realised single-propelled multi-purpose general cargo vessels with suitability for the handling and transport of containers and other cargoes by two revolving cranes positioned at the sides. Powered by a relatively moderate MaK-main engine output of 4400 kW, ICEBIRD was yet designed for continuous sailing through 1 m thick ice at 3 knots, a feature to be achieved by a series of design features. This included the development of a special hull configuration with a pronounced ice bow, cast stem and an 8° inclined shell plating against the ice pressure. Furthermore, the ship was constructed with a double hull in the two cargo holds and the engine room, so that a two-compartment status was achieved. Rudder spur, rudder hoe and ice fins in front of the special ice propeller also enhanced the ship's abilities to navigate in heavy ice. Other equipment details included a helicopter landing pad on the hatches with refuelling station, a tender suitable for sailing in ice and a removable three-storey living module for 100 scientists and station personnel positioned in front of the superstructure.

The vessel was classified according to the ice rules E4 of Germanischer Lloyd, while the foreship was designed in accordance with ARC1, but was given the notation ARC2 due to a higher than supposed steel quality. In order to be able to use the first Antarctic summer season as long as possible, final shipyard work was carried out during the first days of the transfer voyage. The first Antarctic trip with supplies, including Christmas trees for the upcoming festival, took place in October 1984 from Cape Town under the command of Ewald Brune. In 1981 Brune had lost the converted coastal cargo vessel GOTLAND II of the Schulz company at the shelf edge due to ice pressure. In December 1987 the ICEBIRD rescued 100 people from the research vessel NELLA DAN, which sank after fire off Macquarie Island.

The duties of the ICEBIRD were from 1994 onwards taken over by the newbuilding AURORA AUSTRALIS, built in 1990 in Australia by Carrington shipyard at Tomago. After that ship failed, the ICEBIRD was employed again in 1998, then owned by Norway's Riber Shipping as POLAR BIRD. In 2003 the ship was acquired by an Israeli company and sighted in 2017 under the Honduran flag as ALMOG in Haifa, still with the Hamburg city sign originally attached to the bow. The Australian Ministry of Environment and Energy ordered a 160 m length newbuilding, which concept was developed by the Danish design office Knud E. Hansen. Its keel was laid at Damen Shipyards Galati in August 2017 and it it was delivered in August 2021. Named NUYINA the research and supply icebreaker can accommodate 116 scientists and load over 100 TEU containers (Fig. 1.123).

1.11 Cotton on the Buriganga

The main cotton exporting countries are China, which accounts for about 30% of exports in value terms, India, the United States, Pakistan and Vietnam. The transport of cotton, which is for this purpose pressed into bales of about 218 kg weight, requires great care, because it can spontaneously ignite over temperatures of 120 °C due to its cellulose and oil content. A fire in the cotton cargo is very difficult to fight.

More than 100 bulk carriers are anchored at the mouth of the Karnaphuli, and a further 25 moored at the lower reaches. The ships are awaiting permission for entry to Chittagong port, one of the most important harbours on the Gulf of Bengal, along with Mumbai and Colombo. Chittagong is used for export of cotton

Fig. 1.123 The German Antarctic supply vessel ICEBIRD after launch at Oldenburg, still without its two cranes and the accommodation block (Ralf Witthohn)

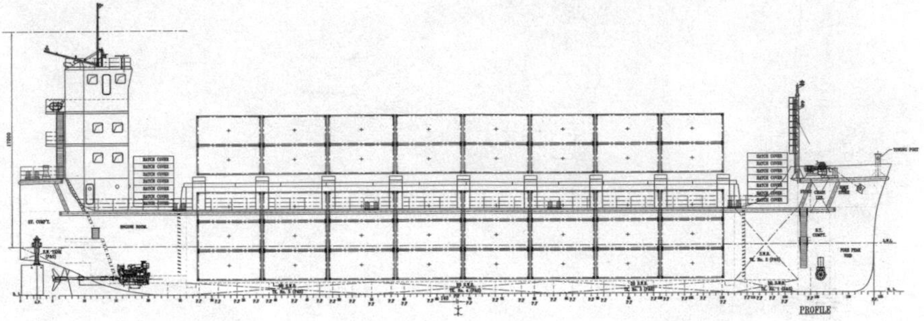

Fig. 1.124 Operating between Pangaon near Dhaka and Chittagong, the container ship HARBOUR-1 is designed for navigation on river and sea (Western Marine)

from Bangladesh, but also for exports from India, Nepal and Bhutan. One of the ships anchored on the Chittagong roads is the small container ship HARBOUR-1, which a community of Bangladeshi companies contracted at Western Marine Shipyard in Chittagong for delivery to the firm Neepa Paribahan at Dhaka in November 2015. Operator of the vessel is the Pangaon Inland Container Terminal company on the Buriganga River. At 82.5 m length and 15 m breadth the ship can carry a maximum of 176 TEU containers between the terminal in Pangaon and Chittagong. The largest part of the 130 nm distance runs on the river Meghna, which limits the draught of the HARBOUR-1 to 3.5 m. Accordingly, the ship is twin-propelled and fitted with two steering gears. Two 550 kW main engines enable a speed of 10 knots (Fig. 1.124).

Opened in 2013, the Pangaon terminal is located on the Buriganga River in Keraniganj, 20 km south of the capital Dhaka, to relieve the motorway between Dhaka and Chittagong. In order to promote the activities of the terminal, the Bangladeshi government made its use compulsory for 20% of the transports of cotton and machinery to the Dhaka region in January 2016. Foreign operators are offered the terminal as an alternative to shipping goods via Singapore and Colombo. The small port handles over 100,000 TEU annually and is used also by Indian ships.

2

Raw Materials

In 2018, the dry bulk volume transported by sea was figured at a total of 5.2 billion tonnes, an increase of 2.6% compared to the pre-year, when shipments had reached 5.1 billion tonnes. They grew again marginally to 5.25 billion tonnes in 2019, but decreased by 1.5% to 5.17 million tonnes in 2020. The largest share—1.5 billion tonnes—was accounted for by iron ore, which gained 3.2% growth in 2020. China received 76% (from 71% in 2019), Japan 7% and Europe 6% of the ore imports. Australia and Brazil were the most important exporters with shares of 58% and 23%, resp. The second most important bulk cargo is coal, of which 1.17 billion tonnes were transported by sea, 9.3% down from 1.28 billion tonnes year on year. The increase of smaller bulk trades had accelerated at a rate of 3.7% in 2018, compared to 2.8% in 2017, but shrinked by 1.5% to 5.2 million tonnes in 2020. The expansion had resulted from metals and minerals, like nickel ore, manganese ore, cement and bauxite, of which the latter still grew by 8.2% in 2020. In 2018, Guinea had consolidated its rang as leading bauxite exporter, particularly by shipments to China, which world market import share grew from about 20% to 77% in 2020, while Guinea's supply share expanded to 46% in the same year (Fig. 2.1).

The peculiarity of dry bulk cargoes means that they are shipped in large-volume ships which are defined as bulk carriers. Due to the high gravity of iron ore its transport, in particular, requires special ship constructions with strengthened hulls. Nevertheless, total losses of ore carriers are among the most frequent causes of serious shipping accidents. Other bulk goods, such as bauxite or coal, can also pose a major hazard to the ship and its crew, if the cargo shifts or if its moisture content is too high. Such catastrophic events

R. Witthohn, *International Shipping*, https://doi.org/10.1007/978-3-658-34273-9_2

Fig. 2.1 The 45,680 dwt product tanker SEYCHELLES PRELUDE, specially designed under safety aspects, is owned by the Seychelles Petroleum Company and operated by German Tanker Shipping (Ralf Witthohn)

happened over decades and meant the worst series of accidents in recent shipping history. They occurred mainly in the 1980s and 1990s, but were continued until recent times.

The vulnerability of ore carriers became evident, when in December 2011 the South Korean 404,398 dwt newbuilding VALE BEIJING of the Pan Ocean company, in 2014 renamed SEA BEIJING, suffered structural damage in a ballast tank during initial cargo operations at Ponta da Madeira/Brazil and had to be repaired and strengthened at its South Korean builder STX Shipbuilding in Jinhae. In March 2017, the 266,000 dwt ore carrier STELLAR

DAISY of Korea's Polaris Shipping was lost in the South Atlantic on a voyage from Ilha Guaiba in Brazil to China after water ingress occurred into the cargo hold. Only two of the 24 Filipino and Korean crew members were rescued from a life raft. A completely surprising casualty of tragic dimensions happened to the crew of the large, modern bulk carrier BULK JUPITER of the Norwegian shipping company Gearbulk in January 2015. The ship suddenly developed a strong list off the Vietnamese coast and sank within minutes. Of the 19-strong Philippine crew, only the cook was rescued, the first officer recovered dead, while the captain died after rescue. The 56,000 dwt ship had loaded 46,400 tonnes of bauxite for Qingdao in Kuantan, Malaysia. Investigations by the Bahamas flag state revealed that the loading operations had taken place during heavy rains and had been delayed for days. This resulted in a moisture content of the cargo of 20%. The investigating commission therefore assumed that the cargo had shifted or that free surfaces had been created by water on the cargo. The shipping company announced that due to the accident it would stop transporting bauxite on its ships until further notice. After the sinking, the International Maritime Organization (IMO) issued a circular warning against the transport of bauxite, if its moisture content should exceed 10%, thereby threatening liquefaction and spillage. The bulk carrier classified by the Japanese company NKK and delivered in 2006 represented a design, which has more than 170 times realised by Japan's Mitsui shipyard.

Bulk carrier losses of this kind were long time tolerated by shipping companies and supervisory authorities, although they caused the deaths of countless seafarers and revealed serious shortcomings in the management, classification and survey rules of the ships, which were often weakened by corrosion. According to an investigation carried out by Japan on behalf of the IMO, 2916 accidents occurred on bulk carriers between 1975 and 1996, killing 1890 seafarers. From 1978 to August 2000, the number of human lives lost was 1126. In 2013, despite numerous commissions of newbuildings and the scrapping of elder units, the number of losses was still 15 ships. It was not until June 2016 that IMO regulations on the safety of bulk carriers and tankers over 150 m in length came into force.

In 2021 the total number of active bulk carriers was figured at about 11,000 ships, of which were 2300 capesize, 2370 panamax, 3820 supramax and 2530 handymax carriers. Their aggregated dwt tonnes achieved 913 millions, up from 880 million tonnes in 2020, and represented 43% of the world's total fleet. The transport of raw materials is probably that shipping sector to be most affected by the decarbonisation processes to be initiated in the economy. This mainly refers to the energy sources coal, coke, oil and gas and thus to about one third of the total global sea transport volume carried out by bulk carriers, oil and gas tankers. The use of hydrogen instead of coal to produce "green" steel will mean a further reduction of coal shipments in bulk carriers. In spring 2021, yet, before the background of coal shortages and power outages, the daily capesize rates reached a 10-years high by over USD 40,000 per day, and thus three times as much as the average of the preceding 5 years period. During 2021, freight rates continued to increase rapidly. In October daily spot rates for capesize vessels topped USD80,000, but declined sharply afterwards.

The great dependence of the global economy on the still irreplaceable energy sources oil and gas is expressed in the associated high transport volumes, which had reached a total volume of 3.19 billion tonnes in 2018, an increase of 1.5% compared to the pre-year. The total comprised 1.89 billion tonnes of crude oil and 1.31 billion tonnes of other oil and gas cargoes. The latter included 0.29 billion tonnes LNG and 0.09 billion tonnes LPG. In 2019, the corresponding figures amounted to a total of 3.20 billion tonnes comprising 1.88 billion tonnes crude oil, 1.32 billion oil and gas, the latter with a share of 0.42 billion tonnes. In 2020 crude oil shipments were reduced by 7.8% to 1.7 billion tonnes, the other oil trades by 7.7% to 1.2 billion tonnes, reflecting less fuel demands for aircraft and automobiles. Only gas volumes expanded slightly by 0.4% to 0.48 billion tonnes. Oil thus still accounts for about one third of global energy consumption. India and China in particular recorded high growth rates. In 2020, oil was globally transported by tankers with a total deadweight of 619 million tonnes, equivalent to 29% of the world fleet.

After an increase of 12%, China overtook the United States as the largest crude oil importing country for the first time in 2017. In 2020, the country purchased 541 million tonnes, succeeded by India (201 million tonnes), South Korea (133), United States 124), Japan (123) and Germany, which imported 83 million tonnes (Fig. 2.2).

2.1 Ores and Metals

2.1.1 Iron Ore

For the transport of iron ore are employed the largest dry cargo ships ever designed. At deadweights of up to 400,000 tonnes, the Very Large Ore Carrier (VLOC) reaches the capacities and dimensions of the largest tankers and

Fig. 2.2 The Japanese-owned 181,700 dwt bulk carrier BULK SWITZERLAND, built in 2010 by Japan's Imabari shipyard and strengthened for heavy cargoes, is, at a length of 292 m and 45 m breadth, a typical example of a capesize vessel. The ship was managed by Monaco-based C Transport Maritime (CTM) in a fleet of 113 bulk carriers of sizes up to the newcastlemax class and with an aggregated 1.15 million dwt in 2020. In 2021, the ship was sold to the listed US Capesize operator Seanergy disposing 17 such vessels with an aggregated capacity of 3 million dwt. Renamed PATRIOTSHIP, the ship carried a cargo of iron ore from Brazil to China in February 2022 (Ralf Witthohn)

doubles the deadweight of the largest container ships. The ship size is a result of the weight of the cargo, which, at a stowage rate of 0.4 to 0.62 m^3 for 1 tonne, only fills a small share of the cargo hold volume. Out of this reason, the large modern ore carriers, which exclusively transport ore, are constructed with two longitudinal bulkheads. Only the middle space is reserved for cargo, the outer compartments are ballast tanks. Smaller-sized bulk carriers, on the other hand, which carry various types of bulk cargo such as ore, grain or coal, are constructed without longitudinal bulkheads. When employed in the ore trade, only every second cargo hold is loaded on these ships. This method is also used for bulk carriers of medium size, which structures must be strongly reinforced, if the transport of ore is intended.

2.1.1.1 Australian Iron Ore for China

The loading of the ore carrier OCEAN WORLD at Port Hedland requires a sophisticated scheme. The conveyor belt hoisted by a crane over the loading hatch first fills the last hold before the engine room with brown iron ore incessantly. The crane then slowly moves forward and loads the spaces so that the ship lies on a level keel when reaching the 18.2 m load line. The Hong Kong-flagged carrier has a total of nine hatches, the covers of which are being moved sidewards to enable charging operations. The loading of 220,000 tonnes of iron ore is completed within 1 week. Five tugboats are hitching up to pull OCEAN WORLD out of port. At a speed of 12 knots the ship arrives at Qingdao 13 days later. The next ship to be loaded at Port Hedland is the Turkish bulk carrier YASA DREAM. In total, the port on the north-west coast of Australia offers berths for up to a dozen large bulk carriers, with 15 or more ships often waiting on the road for receiving permit to enter.

The ore carrier OCEAN WORLD, delivered in 2011 by Guangzhou shipyard as its ninth newbuilding, represents the SDARI 230OC design of the Shanghai Design and Reseach Institute. At 325 m length, 52.5 m breadth and 24.3 m depth, the ship reaches a deadweight of 228,800 tonnes. Its nine cargo holds have a total capacity of 153,315 m^3, the water ballast capacity is figured at 134,758 m^3. Main drive is a MAN B&W 6S80MC-C diesel engine generating 22,500 kW at 76 rpm. OCEAN WORLD is the fifth newbuilding of the 230OC type from the Guangzhou shipyard in a series of 12 ships. It was commissioned by Ocean World Navigation in Hong Kong, which also took over the OCEAN CHINA, OCEAN HONGKONG and OCEAN LORD of the same type. The carriers YI DA, XIN YIA YANG, LI DA and ZHI DA were as well handed over to Hong Kong customers (Fig. 2.3).

Fig. 2.3 OCEAN WORLD regularly transports cargoes of 220,000 tonnes iron ore from Port Hedland to Chinese ports (Steffen Urbschat)

2.1.1.2 Capesize Carriers at Port Hedland

One of the numerous bulk carriers involved in supplying China with iron ore from Port Hedland is the capesize carrier GIANT SLOTTA. In autumn 2017 the ship made several trips to Rizhao, one of the steel production centres in the north of the country, which had stopped iron ore imports from North Korea in August 2017 due to UN sanctions against the dictatorship's nuclear armament policy. By a deadweight of 174,093 tonnes, the ship represents one of the most common bulker types, the capesize carrier. Delivered to Turkey's Geden Lines (Genel Denizcilik Nakliyate) in 2006 as AVORE, the ship was built at Shanghai Waigaoqiao Shipbuilding. In the following year, the ship was taken over by the Hamburg investment company Nordcapital, which financed it with support of private investors and gave it into the management of the tradition-rich Hamburg shipping company Vogemann under the name VOGEMASTER.

The collection of EUR 62.9 million, of which EUR 20.8 million was limited partnership capital, was undertaken by Nordcapital, after the financier had carried out numerous container ship financings. In the investor's first bulk carrier fund VOGEMASTER private investors were offered shares of a minimum amount of EUR 15,000. The capital collectors' forecast that "high demand can be expected in the long term" for this type of ship yet proved to be much too optimistic, when soon a market imbalance occurred due to high order numbers for bulk carrier newbuildings. The planned term, based on a charter agreement with Geneva-based Cargill International and a tonnage tax assessment, with annual payouts rising from 7% to 22% and a total planned payout of 232% until 2024 remained illusory. VOGEMASTER was sold to Kassian Maritime Navigation in Athens in February 2017 and renamed GIANT SLOTTA. In March 2021, ownership switched to Singapore-based Pioneer Bulk, the name to HIGHLAND PARK, while employment was found on the Australia— China route.

In contrast to pure ore carriers, the capesize GIANT SLOTTA is also suitable for the transport of other bulk goods, especially coal. The cargo area cross-section does not show a longitudinal division by bulkheads but a single hull construction with hopper, top and double bottom tanks. Of the 11 holds with a total capacity of 193,250 m^3, only the holds 1, 3, 5, 7 and 9 are loaded when iron ore is carried. Four of the holds can serve as ballast compartments in addition to the wing and bottom tanks. The total ballast capacity is figured at 140,000 m^3, of which 88,800 m^3 are arranged in the holds. A length of 289 m allows the 45 m breadth, 24.5 m depth ship to pass through the

Bosporus. On a draught of 18.1 m the deadweight amounts to 174,000 tonnes. A diesel engine of type MAN B&W 6S70 puts out 16,860 kW giving the ship a speed of 14 knots. Initially classified by the American Bureau of Shipping, the ship was temporarily supervised by DNV GL, subsequently by the Japanese classification society Nippon Kaiji Kyokai (NK) (Fig. 2.4).

2.1.1.3 Newcastlemax MADEIRA

At 292 m slightly longer than the GIANT SLOTTA is the 178,200 dwt bulk carrier MADEIRA, which was commissioned by the Athens-based Dryships group of majority owner George Economou after delivery by Shanghai Waigaoqiao shipyard in 2007. The structure of the ship, classified by American Bureau of Shipping, allows to keep the holds 2, 4, 6 and 8 empty. Thus the ship type is suitable for carrying ore, but also for lighter cargoes. In October 2017 the ship sailed from Norfolk/US to Juzhne/Ukraine. In February 2015, New York-listed Dryships, which also operates large tankers and offshore vessels, had 13 capesize bulkers built between 2001 and 2013 in their fleet, 24 panamax and two supramax carriers built between 1999 and 2012. At the end of 2017, were named four 180,000 dwt ships, designated as newcastlemax ships, the number of panmax/kamsarmax ships was 18. In October 2019, Dryships was to be de-listed from Nasdaq and to undergo a buyout transaction to Economou. At this time, the fleet comprised a total of 41 bulk carriers.

Shanghai Waigaoqiao has a long history of building capesize bulkers dating back to 2003. 37 newbuildings were delivered by 2007. More recent orders included 14 180,000 dwt carriers completed from 2014 to 2016 for Knightsbridge Shipping, a company of the Norwegian shipping tycoon John Fredriksen. At the beginning of 2015, the shipping company succeeded in signing 6 year charter contracts for 14 capesize bulkers with the German energy group RWE. The charter was agreed on the basis of the spot market index plus a surcharge based on the performance of the newbuildings (Fig. 2.5).

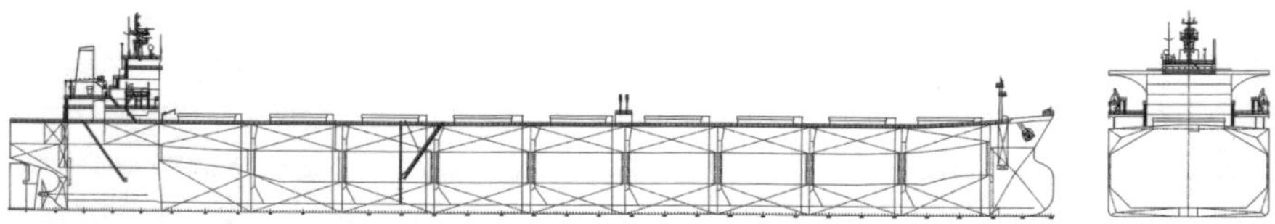

Fig. 2.4 The cross-sectional drawing of the GIANT SLOTTA shows the configuration of the cargo holds with hopper and wing tanks (Vogemann)

Fig. 2.5 A Shanghai-based shipyard delivered the capesize carrier MADEIRA to a Greek shipping company (Ralf Witthohn)

2.1.1.4 Iron Ore from Ponta da Madeira

Two large ore carriers, SEA MARANHAO and CAPE SAKURA, are being moored at the two jetties of Ponta da Madeira that extend for kilometres into the sea. In front of the Brazilian port, 11 more bulk carriers are anchored waiting for a free berth, another bulk carrier is approaching the anchorage. A few days before the loading of the capesize carrier GENCO CLAUDIUS had been completed. After leaving Ponta da Madeiras at the end of September 2017, the 169,000 dwt ship sets course for the Cape of Good Hope, the headland that gave its name to this type of ship. It must be circumnavigated by bulk carriers that are too large to pass the Suez Canal in a loaded condition. The arrival of the GENCO CLAUDIUS in Singapore is expected 34 days later. But the Southeast Asian port metropolis is not the final destination of the bulk carrier. Its cargo is destined for China. After its delivery in January 2010, the GENCO CLAUDIUS took up a time charter with Cargill International. For an initial period of 10.5 to 13.5 months, Genco Shipping, listed in New York, agreed to pay a daily rate of 36,000 US dollars.

The Ponta da Madeira ore terminal, located near the port of Itaqui, is owned by the Brazilian mining company Vale, the world's largest supplier of iron ore with an annual output of up to 380 million tonnes. Ponta da Madeira can accommodate ore carriers with a maximum draught of 26 m. Thus, the terminal is one of the countrie's ports, which can be reached by the ships of the valemax class featuring a deadweight of 400,000 tonnes. The iron ore comes from the Gerais mines and of the newly developed Carajàs deposit, from which a first batch of 26,500 tonnes was shipped in January 2017. It served to complete the cargo of three ore carriers of 73,000 dwt to 38,0000 dwt, which had received the largest share of their cargo from other north Brazilian mines.

In July 2009, New York-listed Genco Shipping took over the seventh of nine Capesize newbuildings as GENCO COMMODUS on the basis of an agreement concluded 2 years earlier with the Metrostar Management group. In the same month, the 170,500 dwt carrier took up an about 2-year time charter for the Morgan Stanley Capital group at a daily rate of 36,000 US dollars. The ship is a product of Sungdong Shipbuilding founded in 2007 in Tongyoung/South Korea. The shipyard also delivered the sister ships GENCO HADRIAN, GENCO MAXIMUS and GENCO CLAUDIUS. In 2007, Universal Shipbuilding in Tsu, Japan, supplied the GENCO TIBERIUS. In 2016, the Genco fleet furthermore comprised the 2007/08 Imabari-built capesizes GENCO AUGUSTUS and GENCO CONSTANTINE as well as the GENCO TITUS and GENCO LONDON from Shanghai Waigaoquiao. The total Genco fleet comprised 70 ships, 13 capesizes, eight panamaxes, four ultramaxes 21 supramaxes, six hanydmaxes and 18 handysize carriers. In 2017, Genco still operated 60 bulk carriers with capacities ranging from 28,000 tonnes to 1800,000 tonnes. For an extensive newbuilding programme, the shipping company took out loans amounting to 1.4 billion US dollars. End of 2019, the company, describing itself as the largest US-headquartered bulk shipping owner, listed 56 bulk carriers, including 17 capesizes, in March 2021 41 units, and in February 2022 44 units (Fig. 2.6).

Fig. 2.6 Capesize carriers like the GENCO COMMODUS are regular guests in Brazil's ore ports (Ralf Witthohn)

2.1.1.5 Brazilian Ore for Asia's Steelmakers

China's Bohai shipyard at Huludao realised a bulk carrier type of 210,000 tonnes deadweight by delivery of BERGE ZUGSPITZE (Fig. 2.7) and BERGE GROSSGLOCKNER to BW Bulk in Singapore in 2016/2017. In November 2017, BERGE ZUGSPITZE transported a cargo of iron ore from Tubarao/Brazil to Kisarazu/Japan, BERGE GROSSGLOCKNER was simultaneously sailing from Itaguai/Brazil to Gwangyang/Korea. Of the same type are the Bohai newbuildings GOLDEN SCAPE, GOLDEN SWIFT, PSU FIRST and STELLA TESS. The nine-hatch Berge vessels have a length of 300 m, a breadth of 49 m and a depth of 25.3 m, resulting in a cargo volume of 225,443 m^3. The total ballast capacity amounts to 122,870 m^3, the addition of 68,059 m^3 in the ballast tanks plus the filling of hold 5. The deadweight capacity achieves 211,170 tonnes at a draught of 18.6 m. It is increased by the use of high-tensile steel accounting for about 80% of the ship's steel weight. The two-stroke main engine of type Wärtsilä W6X72, built under licence by China's Dalian Marine Diesel Co., consumes 49 tonnes of diesel oil per day at an output of 13,060 kW, corresponding to 75% of the maximum continuous power and resulting in a range of 22,000 nm. Previously, Berge Bulk 2014/15 had taken over BERGE TAI SHAN and BERGE HENG SHAN of 216,461 dwt from Bohai shipyard. These 306 m length, 52 m breadth ships are being described as the world's largest bulk carriers by excluding VLOC.

2.1.1.6 400,000 dwt VLOC for Ore Transport

The Bohai ore carrier buildings BERGE ACONCAGUA, BERGE EVEREST, BERGE JAYA and BERGE NEBLINA, which were delivered between 2011 and 2013, are even larger by carrying 388,139 tonnes and measuring 195,199 gt. At 360.9 m length, 65 m breadth, 30.5 m depth and 23 m draught, they

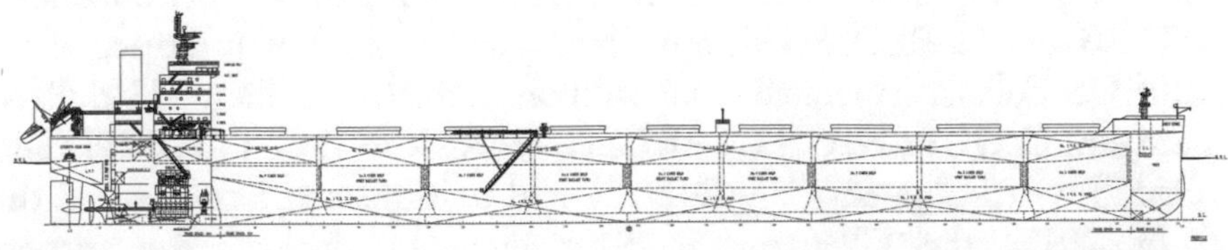

Fig. 2.7 An 80% share of high-tensile steel built in the hull of the 210,000 dwt BERGE ZUGSPITZE reduces the bulk carrier's weight and is thus suited to increase its deadweight (Bohai)

represent one of the largest dry cargo types ever developed. Designated valemax type their carrying capacity is only exceeded by the 400,000 dwt ships ORE BRASIL, ORE CHINA, ORE DONGJIAKOU, ORE HEBEI, ORE ITALIA, ORE KOREA and ORE SHANDONG built from 2011 on for the Brazilian mining group Vale. In February 2022, BERGE ACONCAGUA was on its way from Ponta da Madeira to Qingdao, BERGE EVEREST on way back to Tubarao/Brazil, BERGE JAYA from their to China, and BERGE NEBLINA moored in Shekou/China.

Out of a fleet of 83 bulk carriers with a total deadweight of 14.2 million tonnes in February 2022, the capesize segment in the Berge fleet includes four 181,500 dwt carriers BERGE ISHIZUCHI, BERGE TSURUGI, BERGE KOSCIUSZKO and BERGE MAWSON produced between 2011 and 2015 by Koyo Dockyard in Mihara and Imabari Zosen in Saijo. The 292 m length, 45 m breadth and 24.7 m depth ships have nine hatches, of which the second, fourth, sixth and eighth remain empty during ore transport. The speed of the ships propelled by a 18,660 kW two-stroke engine is given as 14 knots. BERGE ISHIZUCHI was delivered as CATHARINA BULKER, BERGE TSURUGI as SHOEI PROSPERITY. They were taken over by Berge in 2013 and 2015, resp. In April 2014, BERGE KOSZIUSZKO was handed over to the Japanese shipyard's financing company Shoei Kisen Kaisha and chartered from Berge under one of four contracts with the global mining group Rio Tinto. The contract included a purchase option after 3 years, which Berge realised.

On its way to becoming a producer of the world's largest bulk carriers, China's Bohai shipyard delivered a series of capesize carriers from 2003 onwards, including the 173,759 dwt C. UTOPIA and C. VISION of 292 m length, 45 m breadth and 24.8 m depth contracted by Changmyung Shiping. Founded in 1989, the Korean shipping company operated a fleet of 27 bulk carriers comprising 15 capesize, two kamsarmax, one panamax, four supramax, one handymax and three handysize carriers at the beginning of 2015. In April 2016, Changmyung applied for debt relief and attempted to consolidate the fleet. Eight capesize vessels were sold for scrapping or to other operators. C. ATLAS and C. BLOSSOM, built by Hyundai in 2009 at a price of 88 million US dollars, generated 18.6 million and 18.9 million US dollars, respectively. C. QUEEN, C. POLARIS, C. HARMONY, C. TRIUMPH and C. OASIS were scrapped, C. WINNER sold to Transmed Shipping. At the end of 2017, the Changmyung fleet included the capesize carriers C. ETERNITY and C. DISCOVERY, two panamax and three handysize units. End of 2019 were left C. UTOPIA and C. VISION along with two panamax and three handysize bulkers, and in 2022 the owner still had the C. UTOPIA registered (Figs. 2.8 and 2.9).

2.1.1.7 106,000 dwt Carrier from STX Dalian

The development of bulk carriers does not always appear to result in an improvement of economy. For example, the postpanamax bulk carrier ABML EVA delivered by the STX shipyard in Dalian in 2011 to Augustea shipping company of the Cafiero-Zagari families in Naples and in 2019 renamed STAR EVA by Greek owner Starbulk has a deadweight capacity of 106,660 tonnes, while the empty ship weighs 18,540 tonnes, i.e. it can carry 5.8 times its own weight of cargo and fuel. Augustea's panamax carrier ABY DIVA, built by Japan's Imabari yard in 2007 and sold in 2017, yet, was listed with a light ship weight of 10,228 tonnes and a deadweight of 76,600 tonnes, which means 7.5 times the ship's mass. By the same main engine, a 10,320 kW developing MAN B&W 6S60 engine, ABML EVA reaches 13 knots, ABY DIVA 14 knots. An advantage of the larger vessel, however, is its smaller draught of 13.6 m thanks to the larger ships breadth. The draft of ABY DIVA figures at 14.1 m.

At the same carrying capacity as the ABML EVA, another Augustea postpanamax carrier built by Oshima in 2010 as AOM ELENA and sailing as weighs only 14,680 tonnes corresponding to a deadweight/ship weight ratio of 7.3. The capesize ABML GRACE, also formerly operated by Augustea and since 2020 in US hands as RACE, features a weight of 22,040 tonnes and a deadweight of 172,316 tonnes resulting in a 7.8 to 1 ratio. The 253.5 m length, 43 m breadth and 19.8 m depth ABML EVA has a summer freeboard of 6.2 m. In winter the freeboard increases to 6.5 m, the draught is reduced to 13.3 m with a corresponding 103,780 dwt. The smaller tropical freeboard of

Fig. 2.8 The capesize C. UTOPIA was built by China's Bohai shipyard for a Korean shipping company that was unable to pay its debts in 2016, but still had the ship in its fleet in 2022 as the last unit (Ralf Witthohn)

Fig. 2.9 After the emergency sale of the Hyundai-built C. ATLAS and C. BLOSSOM the sister ships C. ETERNITY and C. DISCOVERY (photo) remained in the Changmyung fleet, but were later sold as well (Ralf Witthohn)

Fig. 2.10 The 2011 STX Dalian-built postpanamax carrier ABML EVA has a comparatively high light ship weight (Ralf Witthohn)

5.9 m means a draught of 13.9 m and the deadweight increasing to 109,540 tonnes. The ship's seven holds have a total capacity of 134,000 m^3 (4.73 million cubic feet) (Fig. 2.10).

2.1.1.8 81,400 dwt Panamaxes from Universal Shipbuilding

By an unusual register information in addition to the class notation, the classification society Det Norske Veritas certified, that the longitudinal frame construction of the panamax vessels MARIELENA and OKINAWA, built by Japan's Universal Shipbuilding in Maizuru in 2008 and 2009 and managed by Athens-based Sea Traders, has proven itself in worldwide service over more than three decades. The cargo hold area of the ships is exposed to strongly alternating loads from water pressure and cargo weight, in addition to the already constantly changing buoyancy forces by the waves' impact. This is even enforced during voyages in the ore trade, when the holds 2, 4 and 6 remain empty. End of 2019, OKINAWA repeatedly headed for Hay Point, Queensland/Australia, to load a cargoes of coal for China. End of 2021 the ship crossed the Pacific from Kaohsiung to Vancouver.

The 225 m length, 32.3 m breadth, 20 m depth and 14.4 m draught ships have a carrying capacity of 81,400 dwt. They have been developed from a panamax type, the building of which Universal Shipbuilding had concentrated on by delivery of high numbers since the 1990s. The OKINAWA type was introduced by TRITON OSPREY in 2007; further deliveries were KM MT. JADE, NCS BEIJING, SEA OF HARVEST, STAR OF ABU DHABI, MARCO, VATHY, RBD THINK POSITIVE, MIGHTY SKY, MIDLAND SKY, KM SYDNEY, NORD VENUS, NAVIOS MARCO POLO, KEY FRONTIER, KEY JOURNEY, AZALEA SKY, KEY SPRING, KM SINGAPORE, MELODIA, ARPEGGIO and KM SHANGHAI. Out of this panamax type arose a kamsarmax series, which started in 2015 by NAVIOS AMBER, ELSA S., MACHERAS, MARATHASSA, ABY VIRGINIA and BTG EIGER (Fig. 2.11).

2.1.1.9 76,000 dwt Panamaxes from Hudong Zhonghua

China's Hudong Zhonghua shipyard built more than 160 panamax carriers since 1987, including the 76,000 dwt NINGBO SEAL, NINGBO DOLPHIN, NINGBO WHALE and NINGBO SEA LION for Hong Kong-based Longsheng Holding from 2011 to 2013. The 225 m length, 19.7 m depth and 13.4 m draught ships are powered by a 10,000 kW two-stroke

Fig. 2.11 The construction of the Greek OKINAWA on longitudinal frames is based on 30 years of experience (Ralf Witthohn)

engine of type MAN B&W 5S60MC-C8 from the shipyard's production. In 2015, SBI CONGA, SBI BOLERO, SBI CAPOEIRA, SBI CARIOLA, SBI SOUSTA, SBI REGGAE and SBI LAMBADA came into service for the US shipping company Scorpio Bulkers, but have since then be sold. Extended to 229 m length they achieve a deadweight of 81,253 tonnes and are thus categorized as kamsarmax ships (Fig. 2.12).

2.1.1.10 LNG Burning VIIKKI in Quadrilateral Traffic

After the grabs of their own shipboard cranes have retrieved the last heave of iron ore from the holds, the bulk carrier VIIKKI of Finland's ESL Shipping leaves the steel mill's own harbour in Bremen with the course set for the Baltic port of Ust-Luga. Five days later the ship reaches the Russian port via Skagen to load coal for Naantali. Located about 100 km west of St. Petersburg and not far from the Estonian border, the port activities include, besides the coal handling facilities, a container terminal operated by the Hamburg/Bremen Eurogate group since 2011 and an oil terminal. In 2016, Russia exported 18 million tonnes of coal mined in the Kuzbass and other regions from Ust-Luga to European, African, Latin American and Asian ports. In just 3 days, VIIKKI takes over the coal for the Fortum power plant in Naantali/Finland located located at distnace of about 200 nm. Shortly afterwards, the bulk carriers ULTRA COUGAR, KAI OLDENDORFF and CEMTEX INNOVATION are being handled at the

Fig. 2.12 NINGBO SEA LION is one of more than 160 panamax bulkers built at Hudong Zhonghua shipyard (Ralf Witthohn)

terminal. During this time VIIKKI has already completed discharge at Naantali. There is generated electricity and district heating for the Turku region. Afterwards, VIIKKI undertakes another short trip of 300 nm over the Gulf of Bothnia to Lulea in Sweden, to load iron ore for the Arcelor-Mittal plant in Bremen again. The German port is reached within 3 weeks after the VIIKKI's previous call there. The following voyage from Lulea takes the Finnish ship to the Hansaport ore terminal at Hamburg, where the relatively small ore carrier is a rare customer at a site normally used by capesizes and VLOCs.

Shortly afterwards, a most unusual handling operation takes place for the first time, when, in mid-October 2019, VIIKKI proceeds in ballast to the Bremerhaven Eurogate container terminal, where the ship is berthed alongside the ESL bulk carrier KUMPULA, which has earlier the day arrived from the Canadian ore shipment site Milne Inlet. At 13 m draught, KUMPULA, another regular caller at the Bremen steelworks of ArcelorMittal, would not have been able to sail the Weser upwards. Therefore, part of the ore is being transshipped to VIIKKI at Bremerhaven. Ore transports from Milne Inlet, from where the first shipment had taken place in 2015, is restricted to the period from June to October. An attempt of the Baffinland Iron Mines company to extend it to 10 months from June to March was given up due to concerns expressed by the Nunavut Impact Review Board by citing hunters, who were dependent on a closed ice cover.

ArcelorMittal is together with Nunavut Iron Ore jointly owning the Canadian company Baffinland Iron Mines, which excavates high-grade iron ore at the Mary River operation in the Qikiqtani region of North Baffin, Nunavut. From the 2018 summer season on, Baffinland Iron Mines chartered the multifunctional Estonian icebreaker BOTNICA from Tallinn-based TS Shipping, with options for charters until 2022. The BOTNICA provides ice escorts for the iron ore carriers as well as oil spill and emergency response services from June to the beginning of November, when the icebreaker returns to Tallinn.

Upon delivery by the Jinling shipyard of the China Changjiang National Shipping Group in September 2018, the VIIKKI was the ninth ship in the fleet of Finnland's ESL Shipping. The sister ship HAAGA had been completed the proceeding month. The newbuildings were the owner's first ones to be equipped with LNG-burning MAN 5G45ME-C-GI main engines. The LNG fuel is supplied from a cylindrical tank positioned athwartships behind the superstructure. The 160 m long, 26 m wide two-hatch freighters can carry 25,600 tonnes on a draught of 10 m. They have been built according the high ice class 1A of DNV-GL.

Helsinki-based ESL Shipping, which mainly ships ore, coal and limestone, also operates the 56,000 dwt double-hull bulk carriers ARKADIA and

KUMPULA as well as three single-deckers of around 20,000 dwt. Like VIIKKI and HAAGA, ALPPILA and KALLIO are yet classified by DNV GL as general cargo ships, EIRA by Lloyd's Register as bulk carrier. At 159.2 m length, 24.6 m breadth and 9.4 m draught KALLIO achieves a capacity of 21,353 tonnes. The ship was built in 2006 on the Shanghai Edward Shipyard with only two hatches, in contrast to the EIRA and ALPPILA of the same size, which were constructed with three holds in 2001 and 2011 resp. ALPPILA was built by India's ABG shipyard in Surat. At 156 m length and 25.2 m breadth it features a dwt capacity of 20,500 tonnes, the 10 years older Tsuneishi-built EIRA of 19,625 tonnes. The latter carried a full cargo of Swedish iron ore from Lulea to the ArcelorMittal steelworks at Bremen in January 2020. In April 2020 a rare transport of iron took place, when the Liberian-flagged 56,000 dwt bulk carrier ORION, in 2007 completed by Mitsui shipyard and Greek-owned, delivered a cargo from the Finnish port of Kokkola. The ore in the form of pellets comes from one of Russia's leading iron ore mines at Karelsky Okatysh, part of OAO Severstal, on the base of long-term contracts signed with the Finnish, Kokkola-based company M. Rauanheimo as port operator for Russian transit cargo and with state-owned railway company Valtion Rautatiet (Fig. 2.13).

Fig. 2.13 At the quay of the Bremerhaven container terminal a cargo of Canadian ore was transferred from the LNG-powered bulk carrier KUMPULA of ESL to the companie's VIIKKI (Ralf Witthohn)

2.1.1.11 Iron Ore from Buchanan

After a week at anchor off the Weser estuary in October 2019, the supramax bulk carrier TERESA OETKER proceeds to the ArcelorMittal steel works at Bremen, to discharge its cargo of iron ore laden a month before at Buchanan/Liberia. From time to time, the German steel mill has to complement its iron ore supplies from Sweden by cargoes from other sources. The import of directly shipped iron ore from Africa is based on a 25-year concession signed by Luxembourg-based ArcelorMittal concern with the Liberian government in 2005. The contract allows the exploitation of mines at Gangra, Yuelliton and Tokadeh and included the renovation of the discharge facilities in Buchanan port. The high-quality ore has a ferrum share of 60% in its natural state and is transported by a 240 km long railway track from the Yekepa mine to Buchanan for shipment to Europe and Asia. An investment plan to build up an ore concentration facility with a capacity of about 15 million tonnes was reported to have been delayed due to the Ebola virus outbreak in West Africa. In 2018 ArcelorMittal commissioned a feasibility study to find out an optimal concentration solution for utilizing the resources at Tokadeh.

The supramax carrier TERESA OETKER bulker was in 2017 purchased by a single ship company named after the vessel and related to the Oetker family. It is operated by Orion Reederei at Hamburg, which carried out management services since 2016/17 for a number of bulk carriers belonging to the Oetker group, including LEON OETKER, FLORENTINE OETKER, BLUMENAU and BELO HORIZONTE (Fig. 2.14).

Delivered in 2010 under the name MASSALIA, the TERESA OETKER represents the standard design Crown 58, of which Yangzhou Dayang

Fig. 2.14 Destinations of the supramax bulk carrier TERESA OETKER in 2021 included the Australien ports of Kembla and Newcastle, Shikokuchuo/Japan and Vostochnyy in Russia (Ralf Witthohn)

shipyard in Yangzhou/China delivered more than 50 units. At 190 m length and 32.2 m breadth the vessel achieves a deadweight of 57,970 tonnes. The Crown 58 type was enhanced by cooperaton of French owner Setaf Saget with the Sinopacific Shipbuilding Group in Shanghai. According to the owner, the 63,200 dwt supramax type supersedes the earlier version by 3200 dwt (5.2%) due to a reduction of the light ship weight on the base of optimized strength calculations. Until 2014 the 199 m length, 11.3 m design and 13.3 m scantling draught panamax type was contracted over 100 times. At 14 knots the 8300 kW developing MAN B&W main engine of type 5S60ME-C8-T2 burns about 27 tonnes of heavy oil per day, resulting in a range of 20,000 nm. For sailing Emission Control Areas (ECAs), the specific tank capacities amounted to 450 m^3. The fuel savings compared to the forerunning type are said to reach 18%, achieved by extensive model tests at Hamburgische Schiffbau-Versuchsanstalt (HSVA) and the fitting of a larger, slow turning Nakashima propeller, the savings in the total transport economy are figured at 20%.

Laid out to carry all kinds of bulk cargo, including iron ore, grain, cement, the ships can also transport steel products and stow cargo on the hatches. The first two newbuildings for Setaf Saget were delivered in February and March 2012 as JS AMAZON and JS COLORADO by Yangzhou Dayang Shipbuilding, until 2015 followed by JS DANUBE, JS GARONNE, JS LOIRE, JS MEUSE, JS RHIN, JS RHONE, JS CONGO, JS MEKONG, JS MISSISSIPPI, JS MISSOURI, JS NARMADA, JS SANAGA, JS VOLGA, JS YANGTSE and JS COLUMBIA. Voula/Greece based Allseas Marine ordered four newbuildings of the same type. GENTLE SEAS and PEACEFUL SEAS were delivered in 2014 and enlarged the Greek owner's bulker fleet to eight panamax, five supramax and five handysize ships. The fleet was already reduced to only 12 units 1 year later in favour of extending the box carrier fleet to 12 ships. MAGIC SEAS was completed in 2016. Participant of the building programme was Zhejiang Shipbuilding at Fenghua, where from 2015 on the keels of the JS POTOMAC, SYROS ISLAND, DELSA, ARAGONA, ANDROS ISLAND and ALBERTA were laid (Fig. 2.15).

Other bulk carriers of the supramax class that supplied the ArcelorMittal steelworks at Bremen in January 2020 with Liberian-mined ore were GRANDE ISLAND and the 60,063 dwt LOWLANDS HOPE, in March end April followed by the 2017 Iwagi-built Taiwanese 63,400 dwt KM WEIPA and the 2019 Oshima-built 62,000 dwt ULTRA INITIATOR from the Philippines. The 58,110 dwt GRANDE ISLAND, delivered as GL CORONILLA by Tsuneishi's Chinese shipyard location in Daishan in 2009, was retroactively fitted with a scrubber system as were about 800 bulk carriers

Fig. 2.15 The French-owned JS MEUSE represents the enhanced Chinese standard type Crown 63 (Ralf Witthohn)

until 2019. The vessel belongs to a design family named TESS (Tsuneishi Economical Standard Ship) by Japan's Tsuneishi shipyard group, which is traditionally involved in bulk carrier construction. From1984 on the builder developed bulk carriers of various sizes, designated TESS35, TESS40, TESS45, TESS52 and TESS58. Of the latter type, the shipbuilding group's Chinese site at Zhoushan delivered the GL PRIMERA, GL LA PAZ and GL CORONILLA as the first units of this type in 2009. Other carriers of the TESS58 took as well part in ore shipments to Bremen, including MAINE DREAM, which in March 2012 delivered a party of iron ore from Narvik. In April 2020, the 61,300 dwt PRT ACE arrived from Narvik. The ship was in 2014 delivered on own account by the Japanese builder Iwagi Zosen. In May 2021, the supramax carriers JORITA and PUFFIN ISLAND became part of the supply chain of Liberian iron ore for the Bremen steel works. Im February 2022, the 56,141 dwt BULK BAHAMAS delivered a cargo of Swedish ore to the AM steelworks.

The Tsuneishi Cebu shipyard, founded in 1997, had initially laid the keels of smaller bulk carrier types, until it delivered the first supramax carrier, named CENTURY SEA, in 2001 and completed around 100 newbuildings of the TESS58 type between 2006 and 2014. In 2009, the state-owned Portuguese Portline received PORT SHANGHAI from Zhoushan. In 2014, MC Shipping in Tokyo took over the newbuilding TEAL BULKER built in Zhoushan and chartered it out on a long-term basis to the traditional Danish shipping company J. Lauritzen, which in 2017 also managed TERN BULKER and EVER ALLIANCE of the same type built on Cebu (Fig. 2.16).

2.1.1.12 Canadian Iron Ore for Salzgitter Via Panama

In January 2019, the Japanese capesize carrier CAPE AMAL leaves the ore terminal at Port of Sept-Iles near Vancouver bound for the Hamburg handling facility Hansaport, from where the ore is transported to the Salzgitter steelworks by rail. The Panama canal extension has created a new ore supply route connecting Canada's west coast with Europe. Since then the canal can accept capesize cariers featuring a designed breadth of 45 m. As the maximum Panama canal draught is yet restricted to 13.1 m, the maximum 14.7 m draught of CAPE AMAL cannot be exhausted. 18 days after departure from Vancouver the capesize carrier passes Balboa, 3 weeks later Hamburg is reached (Fig. 2.17).

The Vancouver ore terminal at Port of Sept-Iles was opened in March 2018, when the 206,030 dwt MAGNUS OLDENDORFF loaded a first cargo of ore containing 66% iron for Asian steelworks. The ore was delivered by Canada's mining company Champion from the Bloom Lake mine in Québec by railway transport. In December 2019, a fleet of capesize and newcastlemax

Fig. 2.16 After being fitted with a scrubber system as visible by the widened funnel casing, the former HUDSON TRADER II transported iron ore from Buchanan to Bremen in January 2020 under the name GRANDE ISLAND (Ralf Witthohn)

Fig. 2.17 The 182,930 dwt capesize carrier CAPE AMAL, delivered by Tadotsu shipyard to Kawasaki Kisen Kaisha in June 2018, half a year later carried out one of the first ore transports from North America's west coast to Hamburg through the Panama canal. In April 2020 the ship took armed guards on board to protect crew, ship and the cargo of iron ore against pirates on a voyage from Richards Bay along the east African coast to the Persian Gulf (Ralf Witthohn)

carriers, named ASIAN BLOSSOM, CAPE MATHILDE, KSL SINGAPORE and NSU TRUST, were, among several smaller bulk carriers, anchoring in the bay of Sept Iles and awaiting a free berth.

2.1.2 Canadian Copper Concentrate for Wilhelmsburg

In September 2017, the 25,000 dwt Polar Sea cargo vessel NUNAVIK leaves Deception Bay on Canada's east coast loaden with a cargo of copper ore destined for Pori, Finland. In the following month the voyage is continued to Antwerp and Brunsbüttel, where residual cargoes are discharged. Since January 2007, the Elbe port of Brunsbüttel operates as transhipment facility for the Hamburg-based Aurubis group on the base of a contract signed for an initial period of 20 years and including an investment of over EUR 38 million. About 1.3 million tonnes of concentrate are annually imported through Brunsbüttel in ocean-going vessels from copper mines all over the world. The copper is discharged by crane and transported by a conveyor system to be weighed and sampled. An automatic storage system carries the ore, sorted by type, into 23 storage boxes in a hall of 246 × 75 m (Fig. 2.18).

Fig. 2.18 In Brunsbüttel, copper ore is unloaded from seagoing vessels, mixed and transported on to the production site at Hamburg-Wilhelmsburg in barges (Ralf Witthohn)

After mixing, the ore is transported onward in the barges KAJA JOSEPHINE and SOPHIA SORAYA to the Aurubis works located in Hamburg-Wilhelmsburg on the Müggenburg canal, which is not accessible by seagoing vessels. There, high-purity copper is smelted from the copper concentrates as well as from recycling materials, reaching a production of about 400,000 tonnes of marketable copper cathodes every year. These are processed in the plant into various copper or copper alloy wire and profile products. In 2020/2021 the Hamburg site had a concentrate throughput of 2.25 million tonnes, 5% less than in the period before. The concentrates are supplied from from Chile, Peru, Bulgaria, Brazil, Argentina, Canada and Georgia with shares varying from 19% to 6%. As byproduct the group generated 51 tonnes of gold, 949 tonnes of silver, 41,000 tonnes of lead, 3900 tonnes of nickel, 10,000 tonnes of tin and 8800 tonnes of zinc in 2020/2021.

The supplier of the copper concentrate imported on the bulk carrier NUNAVIK is Canadian Royalties, a company which began exports via the port of Nunavik in November 2013. The necessary infrastructure had previously been built up for 735 million Canadian dollars by China's Jien Canada Mining company and transferred to the bank Forbs and Manhattan, when debts of 54 million dollars were incurred by the companies Nunavut Eastern Arctic Shipping, Desgagnés Transarctik and Nuvumiut Developments, among others. In the year after its opening, NUNAVIK was the first Arctic cargo vessel to sail the Northwest Passage in a loaden state and without icebreaker escort, transporting 23,000 tonnes of nickel ore from Deception Bay to the Chinese port of Bayuquan. NUNAVIK, which is operated by the Canadian shipping company Fednav, thus shortened the voyage by about 40% compared to the Panama route. Fednav is active in Canada's Arctic shipping operations for 60 years and describes the annual transport volume from the mines there as to exceed two million tonnes (Fig. 2.19).

The icebreaking bulk carrer NUNAVIK delivered in January 2014 by the Tsu location of the Japan Marine United shipyard group is powered by a 22,100 kW MAN main engine and able to sail through 1.5 m thick ice at a speed of 3 knots. The vessel was built according to the ice class ICE-15 of Det Norske Veritas. It is designed for operation in outside temperatures of -30 °C. The 188.8 m length, 26.6 m breadth and 15.7 m depth ship of the handysize class has a draught of 10.2 m and carries 25,000 tonnes in five holds, which are served by three cranes.

2.1.3 Nickel Ore

2.1.3.1 Nickel Ore from New Caledonia

Bulk carriers of supramax and handysize types are berthed at jettys around the French island of New Caledonia to undertake loading operations of nickle ore from nearby mines. On the South Sea island, located east of Australia, 10% of all the world's nickel deposits are suspected. Their mining has been going on since the beginning of the colonial era. On the west coast near Poya there is moored the 56,000 dwt bulker BALTIC K, further south near Doniambo a handysize bulker

Fig. 2.19 The ice-going bulk carrier NUNAVIK transports concentrate from the port of Nunavik to copper producers all over the world (Ralf Witthohn)

of 28,000 dwt, named after the town, in the same bayt the similar VALENTE VENUS. Not far away, the cruise liner VOYAGER OF THE SEAS has arrived in Nouméa, coming from Port Kembla near Sydney. After departure the next day, the ship will pass the narrow between the southern tip of New Caledonia and the island of Ouen. In a bay there, the tug CAP NDOUA pushes the 51,900 dwt bulker ASHIYA STAR to the loading berth. At the eastern coast of the island three more bulkers are being charged, named OCEAN RALLY, SM EMERALD and KIRISHIMA SKY. Immediately after the latter has left for Hososhima, its berth is occupied by the ultramax carrier SM EMERALD, under assistance of the tug MARCEL VIRATELLE.

The nickel ore carrying bulk carrier BALTIC K of Singapore-based New Ocean Shipmanagement is a 2012-delivered unit of the series-built type Future 56 from the shipyard IHI Marine United at Yokohama. At 190 m length and 32.2 m breadth the supramax ships achieve a deadweight of around 56,000 tonnes. The long series started in 2007 by the delivery of SANKO KING, followed by 29 newbuildings until 2012. The IHI sister shipyard in Kure, later renamed Japan Marine United, realised newbuildngs of the same bulk carrier type from 2006 onwards, the JEWEL OF SHINAS, TTM HARMONY, JEWEL OF DUBAI, JEWEL OF KURE, DST QUEEN, KURE HARBOUR, UNITED HALO, VENUS HALO, WESTERN TOKYO, WHITE HALO, SANSHO, ETERNAL TRIUMPH, GLORIOUS HOPE, CAPE HENRY, VSC CASTOR and in 2014 as the last new building ALKYONI SB (Fig. 2.20).

Fig. 2.20 Supramax carriers like the ALKYONI SB are suited to ship nickel ore, as is done from New Caledonia (Ralf Witthohn)

2.1.3.2 Nickel and Palladium from Siberia

In 2006, the Russian mining group Norilsk Nickel, in 2016 renamed Nornickel, took delivery of the Polar Sea cargo vessel NORILSKIY NICKEL from Aker Finnyards in Helsinki at a price of EUR 70 million. The prototype was in 2008 followed by four more ships of this class, named MONCHEGORSK, ZAPOLYARNYY, TALNAKH and NADEZHDA. These were yet built at Wadan shipyards in Wismar and Warnemünde. Wadan yards had belonged to the Aker shipyard group, when the contract was signed. As one of the largest producers of palladium, nickel, cobalt, platinum, rhodium and other materials such as silver, gold, iridium, ruthenium, selenium and tellurium, Norilsk Nickel employs almost 57,000 people on the Taimyr peninsula, located in the polar zone. Credited by an international bank consortia the employees are occupied with the exploration and exploitation of such raw materials. The group estimates the deposits to reach 22.3 million tonnes of copper, 12 million tonnes of nickel and 1700 million tonnes of ores. In the first quarter of 2022, Nornickel, which accounts for about 40% of the global, partly sanctioned palladium supply, reported an up to 12% decline year-on-year of the material. From Norilsk and Krasnoyarsk the raw materials are shipped to Murmansk via the Siberian port of Dudinka.

In order to facilitate the transport of the raw materials, the vessels have specially been designed for the transport of ore and containers in an ice-covered sea. They have been constructed with a pronounced ice stem and a large deadwood for a better course stability. Furthermore, the vessels are designed according to the double acting principle, invented and patented in Finland. It enables the ships to break ice in both directions. When manoeuvring astern, the water inflow between the hull and ice caused by the propeller reduces the resistance created from the ice cover. In this direction, ice up to 1.5 m thick can be sailed through at a speed of 2 knots. The wheelhouse windows and equipment are therefore positioned in a way, which provides good visibility aft. The diesel-electric propulsion system from Wärtsila and ABB consists of three main generators, each with an output of 8314 kVA, providing a speed of 15.3 kn in free water by means of a 13,000 kW azipod propulsion system with a non-variable pitch propeller. The four-hatch ships, classified according to ice class ARC7 at a maximum draught of 9 m, can operate at temperatures of down to −50 °C. At 169 m a length, 23.1 m breadth and 14.2 m depth, the ships can carry 18,486 tonnes on a maximum draught of 10 m, 389 containers in the holds and 259 units on the hatches.

In 2011, Nordic Yards in Wismar, then in Russian hands, built the 18,900 dwt ice-going tanker ENISEY according to the same basic design for Norilsk Nickel. In place of the dry cargo holds tanks with a total capacity of 20,650 m^3 were arranged. The double acting principle had already been implemented in 2010 on the 70,000 dwt tankers MIKHAIL ULYANOV and KIRILL LAVROV delivered to Sovcomflot by Admiralty Shipyards in St. Petersburg. Russia's largest shipping company uses the tankers to transport oil from the Prirazlomnaya field in the Barents Sea. The 257 m length, 34 m breadth tankers can operate at temperatures as low as -40 °C and pass through ice up to 1.5 m thick without the assistance of icebreakers. Like the NORILSKIY NICKEL, they are equipped with a diesel-electric propulsion system, in this case comprising four Wärtsilä diesel engines and two 8500 kW azipods from ABB (Fig. 2.21).

In March 2018, CSSC Shanghai Shipyard began the construction of the first 105,990 dwt bulk carrier to be delivered to the Estonian shipping company Platano Eesti for use under the special requirements of Arctic voyages with and without icebreaker assistance in termperatures as low as -25 °C. In April 2019 the ship was launched under the name ADMIRAL SCHMIDT showing a special stem trying to combine icebreaking capabilities with the relatively large beam of a bulk carrier. Delivery took place in September, the sister vessel VITUS BERING was completed the following month. Due to the lack of port infrastructure in the Arctic, the ship, which was classified by DNV GL according to ice class IACS PC (6), Polar Code B, is equipped with four 40 tonnes lifting cranes and grabs, a novelty on a cargo vessel of this size,

Fig. 2.21 Like Canadian-operated high ice-class cargo vessels, Nornickel's Polar Sea carriers are outside the navigation period employed on other routes, including TALNAKH, which called at Brake/Germany in May 2010 (Ralf Witthohn)

also known as mini caper. The 250 m length, 43 m breadth, 21.8 m depth and 14.5 m draught ship type has a cargo capacity of 130,000 m^3. ADMIRAL SCHMIDT is conventionally driven by a two-stroke 7 × 62 Winterthur main engine directly acting on a controllable-pitch propeller. The owner, which until then had been active in the operation of reefers, contracted two newbuildings and agreed on a third as an option, which was not realized. The extraordinary capabilities of ADMIRAL SCHMIDT were not needed, when the bulk carrier undertook a voyage from Eregli/Turkey to Baltimore/US in April 2020, while VITUS BERING was on way from Richards Bay/South Africa to Klaipeda/Lithuania (Fig. 2.22).

2.1.4 Zinc Ore from Townsville

At the end of August 2019 the loading operations into the bulk carrier SEA FALCON in Townsville/Australia are finished. Located in the province of Queensland in the continent's northeast, the port is Australia's number one shipment place of copper, lead and zinc ore. The cargo of the 37,150 dwt vessel is destined for Europe, the greater part for England and a residual part for Germany. In the first week of October, the SEA FALCON reaches Northfleet on the Thames, 40 km east of London. The cargo is intended for the firm Britannia Refined Metal, part of the international Glencore group. One week later, the ship is moored at Nordenham-Einswarden/Germany. The remaining cargo is to be processed in the

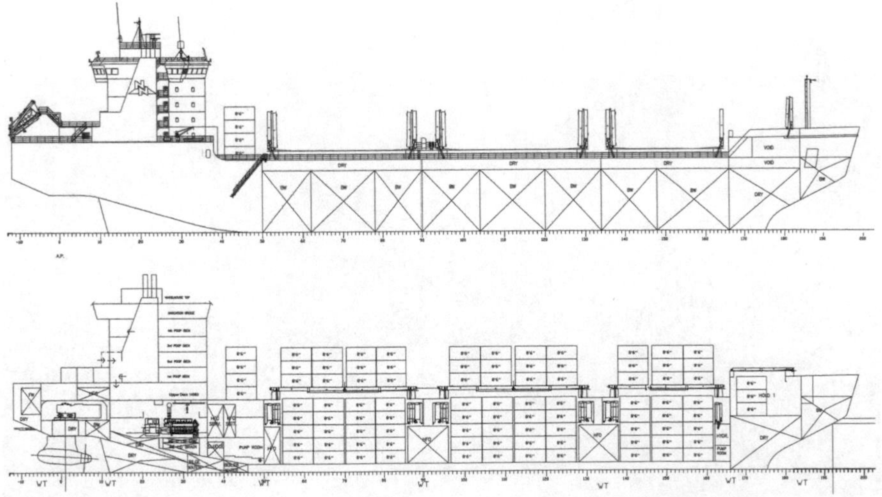

Fig. 2.22 During its voyages in the Polar Sea, the NORILSKIY NICKEL breaks hard ice by sailing aftwards (Aker Finnyards)

group's Nordenham works. As one of the world's largest commodity traders, Glencore transports, processes and markets metals and minerals, in addition to consumer goods and agricultural products. In the Nordenham plant, approximately 160,000 tonnes of fine zinc and its alloys as well as 14,000 tonnes of lead/ silver concentrate, 1700 tonnes of copper concentrate and 500 tonnes of cadmium are annually extracted by an electrolysis process. Zinc ore as raw material is imported from Alaska, the Ivory Coast and other countries via the works' own quay (Fig. 2.23).

The zinc smelter is located next to the lead-producing company Weser-Metall, which belongs to the French Recylex group and obtains its raw material primarily from old batteries and only part of it by sea, for example from Peru. Oxide shares not to be recycled by the smelter until 2017 were exported to China by Chinese ships. For larger loads of 60,000 tonnes, the ships have to be lightered at the berth of the Rhenus Midgard company a few hundred meters downriver. There is a 21 tonnes lifting crane available, which discharges the ore out of the holds and transports it via a conveyor belt into a hall complex with a total floor area of 25,000 m^2. Weser-Metall was due for sale by Recylex in 2020 because of financial reasons, and after insolvency refounded by the Glencore group under the name Nordenham Metall in April 2021. In August of the same year the two firms were re-united under the name Glencore Nordenham. 400 workers achieve a yearly output of 164,700 tonnes zinc and 230,000 tonnes of sulphuric acid.

The two zinc and lead producing companies had already before concluded a cooperation agreement for the use of the handling quay. Both smelters are

Fig. 2.23 In October 2019, a part cargo of ore from Australia was discharged from the German-owned bulk carrier SEA FALCON in Nordenham/Germany (Ralf Witthohn)

successors to Friedrich-August-Hütte, founded in 1906. The plant then worked in synergy with a fertilizer plant constructed at the same time. It used sulphuric acid processed from the roasting of ores to decompose phosphate. Co-founder was the Bremen shipping company E. C. Schramm (later F. A. Vinnen). By importing raw phosphate from Tampa, Florida, a better utilisation of their ships mostly engaged in the carriage of emigrants was achieved on the return voyage. In this respect, the founding of the company in Nordenham was one of the rare examples in which the usual causality for the demand of sea transport was reversed. Fertilizer production was maintained as a guano plant until closure by BASF in 1988.

The Nordenham caller SEA FALCON was delivered by AVIC Weihai shipyard at Weihai/China in May 2017 in a bulk carrier newbuilding series of a 181 m length and 30 m breadth type. The owner of the ship is registered as BNB 1067 GmbH & Co. KG, managers are the AVB Ahrenkiel Vogemann Bolten shipping company in Hamburg. The builder, which started production in 2006, concentrated from 2015 on the construction of this bulk carrier class, of which the Hamburg shipping company received SEA ANGEL, SEA BREEZE, SEA HAWK and as last unit SEA FALCON in 2016/17. They were, in 2017/18, succeeded by LILY OLDENDORFF and BULK TRADER for Germany'a Oldendorff Carriers. Another unit was GOLDEN GRAINS for Minship in Schnaittenbach/Germany in January 2018 (Fig. 2.24). The owner had taken over the newbuilding HARVESTER from Weihai the previous year. In May 2019 China Merchants Offshore Engineering acquired loss-making AVIC Weihai shipyard.

Sailings from Townsville via London to Nordenham have become a regular route for the supply of zinc ore. In February 2018 it was taken by the 60,300

Fig. 2.24 Bavarian Minship company designates the Weihai-built handysize bulk carrier GOLDEN GRAINS as a self-trimming single decker (Ralf Witthohn)

dwt supramax carrier STH SYDNEY, built in 2016. A bulk carrier of similar size to the SEA FALCON employed in the same supply chain was the ANSAC KATHRYN operated by the Sincere Industrial Corp. in Taipei/Taiwan. Japan's shipyard Kanda in Kawajiri developed a bulk and a general cargo version of this ship type and delivered a total of more than 100 units. The general cargo type has wider hatches and is designated Kanda 32 Open, by this contrasting to the bulk carrier version designated Kanda 32 BC, including the ANSAC KATHRYN. The ship's five holds, served by four cranes, have a capacity of 40,896 m^3 bale or 42,857 m^3 grain. Sincere Shipping had a total of 17 bulk and general cargo vessels of the handysize class in its fleet in 2018. The owner already took over the first newbuilding of the 177.1 m length, 28.4 m breadth and 14.3 m depth Kanda type under the name ANSAC ASIA in 1998. The first general cargo version of the type were in 2003 C.S. GREEN and C. S. VICTORY. By widening to 28.6 m, the carrying capacity of the general cargo ships IVS KNOT and IVS KINGLET delivered to Island View Shipping (IVS) of South Africa's Grindrod Shipping in 2010/11 was increased to 33,140 tonnes. The ship class was then enlarged again by extending it to 179.9 m length and 30 m breadth. IVS TEMBE, delivered in 2016 with special equipment for log transport, thus achieved a loading capacity of 37,735 tonnes. Sister ships are IVS KESTREL, IVS PHINDA, IVS THANDA, ANGELIC ZEPHYR, KANDA LOGGER, KULTUS COVE, KEY WEST and KODIAK ISLAND (Figs. 2.25 and 2.26).

The shipment of Zinc ore holds, like some other cargoes including coal or cotton, the risc of self-hating. In December 2019, the Szczecin fire-brigade

Fig. 2.25 As the bulk carrier version of a 32,000 dwt standard type of Kanda shipyard the ANSAC KATHRYN transported a cargo of Australian cinc ore to Europe in September 2017 (Ralf Witthohn)

Fig. 2.26 The open hatch general cargo version from Kanda is represented by the IVS KINGLET (Ralf Witthohn)

was called in to assist in the discharge of the 39,641 dwt bulk carrier LINTAN of the Swire group's China Navigation Co., based at Hong Kong. The LINTAN had loaded zinc sulphide in a Far Eastern port. As liquefaction of the cargo was discovered and fears grew about the development of hydrogen, special safety measure were taken.

2.1.5 Bauxite

The yearly quantities of bauxite shipments worldwide vary greatly and were described to have reached 3.1% in total dry bulk trade in 2019. In 2020, the total volume amounted to 138 million tonnes. Main exporters are Australia, Brazil and Guinea, the latter having exported about 64 million tonnes or 14% of the world's bauxite shipments in 2020.

The port of Kamsar/Guinea thus lent its name to an entire class of bulk carriers. Bulk carriers had been the first ship type that reached the 32.2 m width limit of the Panama canal locks in large numbers from the mid-1960s onwards. Designated as panamax carrier their usual length came out as about 225 m. From 2002 onwards, the special navigational conditions of the bauxite terminal at Kamsar/Guinea were taken into consideration when designing this class of ship. They resulted in the development of special so called kamsarmax ships, which length was extended to 229 m. The longer ship regularly resulted in a deadweight of over 80,000 tonnes. When leaving the port of Kamsar, their draught must not exceed 13.8 m. Originally developed by Japan's Tsuneishi Shipbuilding, the number of deliveries of kamsarmax bulkers exceeded those of conventional panamax ships for the first time in 2010 and reached around 140 units each in 2013 and 2014, while the panamax ships of the old length were reduced to a few. The success of the Kamsarmax type is attributed to its high economic efficiency, while the number of bulkers per year actually calling at Kamsar being far below the total number of existing units.

Of much larger size than the Kamsarmax type bulk carriers are ships that are loaded on the Kamsar anchorage. In February 2022, this included the Liberian-flagged capesize carrier BEI JU STAR destined for Qingdao at 18.4 m draught, the capesizes CAPE OASIS, XIN MAY, GOLDEN CROWN, SPRING BRIGHT, WINNING BROTHER, DELOS, WINNING ZEPHYR and the newcastlemaxes JIN NIU STAR, NEWMAX and SUNNY MOREBAYA and SAMJOHN VISION. Largest ship was the 268,000 dwt BERGE FUJI.

2.1.5.1 Bauxite from Kamsar

With warning sirens sounding, the gantry crane starts moving and, within 5 days, retrieves 63,000 tonnes of bauxite from the holds of the bulk carrier NORDMOSEL of the German-owned, Cyprus-based shipping company "Nord" Klaus E. Oldendorff. As soon as the mineral from Kamsar has been processed at the AOS plant in Stade, it will be re-exported by an ocean-going vessel, such as the bulk carrier CORNELIA of Germany's Upper Palatinate shipping company Mineralien Seetransport departing after a 2-days loading time of 10,500 tonnes of aluminium hydroxide (Fig. 2.27).

The AOS (Aluminium Oxid Stade) plant, which was established in 1973 on 55 hectares, is the only large-scale plant of its kind in Germany. It has a

Fig. 2.27 A special handling facility at Bützfleth can discharge 650 tonnes of bauxite per hour from bulk carriers like the NORDMOSEL (Ralf Witthohn)

production capacity of more than 1 million tonnes of aluminium oxide. This is sold for the extraction of aluminium by electrolysis process. The aluminium hydroxide also produced is exported worldwide in bulk carriers for various applications. The bauxite provided for the production process by the parent company Dadco Alumina & Chemicals reaches the plant from Australia, West Africa, South America or Asia in bulk carriers up to panamax or kamsarmax size. Every hour, a grab retrieves a maximum of 650 tonnes of the raw material from the bulk carriers, which carry up to 80,000 tonnes, and transports it via a closed belt system for processing.

2.1.5.2 Bauxite for Grundartangi

The 31,900 dwt bulk carrier MAESTRO LION arrives at an inopportune time off the Weser estuary. The anchor has to be dropped, because the berth in Bremen is not yet available. Hurricane Herwart has announced its arrival over the North Sea. At around 11 pm on 29 October 2017, the ship, which shows a ballast draught of 6.2 m, has to pick up anchor to prevent the chain from breaking in the sea piling up higher and higher to wave heights of 7 m. Over the next 24 h, the ship is circling endlessly in the German Bight. Simultaneously the bulk carrier GLORY AMSTERDAM grounds with dragging anchors off the isle of Langeoog. The MAESTRO LION has come from the Icelandic port of Grundartangi, where the ship has discharged bauxite from Vila do Conde, Brazil (Fig. 2.28).

The Nordural Grandartangi plant produces aluminium with a weight of over 300,000 tonnes yearly. The Icelandic site was chosen because the high energy consumption necessary for production can be covered by geothermal sources and by hydropower from the Sultartangi dam. This is fed by glacial rivers for a prospected period of 50 years, limited due to the consequences of climate change. The plant is one of the most important industrial sites in

Fig. 2.28 The 32,000 dwt bulk carrier MAESTRO LION supplied the Icelandic aluminium plant at Grundartangi with Brazilian bauxite (Ralf Witthohn)

Iceland and owned by the US company Century Aluminium Company, which has three more sites for aluminium production in the United States. The construction of a second smelting facility planned since 2004 in Iceland with an annual capacity of up 360,000 tonnes in Helguvik was put on hold after the outbreak of the economic crisis in 2008, but is still being pursued. The Grandartangi plant is supplied with 150,000 tonnes of carbon anodes required for aluminium production from the group's own plant in Vlissingen. Century Aluminium is also involved in anode production in the autonomous Chinese region of Guangxi Zhuang, which as well supplies the Icelandic smelter.

Operated by Switzerland's Maestro Shipping the Marshall Islands-flagged bulk carrier MAESTRO LION was in 1999 launched at Japan's Saiki Jukogyo shipyard under the name OCEAN SPRING, commissioned as SWIFTNES, later named UBC SALVADOR and renamed H SEALION in 2019. It was employed by the Glencore trading group to supply the Icelandic aluminium plant. The ship represents a 171.8 m length, 27 m breadth standard type with a carrying capacity of 31,850 tonnes. The development of the design has continuously been kept on by Saiki. Five units, SWIFTNES, SIRNES, SANDNES, STORNES and SPRAYNES, were built between 1999 and 2001 and chartered by Norway's Jebsen company Jebsen. A number of newbuildings were built for Cyprus-based Athena Shipping, a company of the German Hartmann group. UBC SACRAMENTO, UBC SINGAPORE, UBC SAIKI, UBC STAVANGER, UBC SANTOS, UBC SANTA MARIA and the 1 m wider UBC SAGUNTO, UBC STOCKHOLM and UBC SALAVERRY strengthened the fleet of United Bulk Carriers (UBC), a US-based company as well belonging to the Hartmann Group. Maestro Shipping also took over UBC SVEA ex SIRNES as MAESTRO TIGER (Fig. 2.28).

A similarly remote port as Grundartangi in Iceland is Ardalstangen, located at the end of the Norwegian Sognefjord. Near-by, the international group Norsk Hydro, headquartered in Oslo, also uses hydropower to produce aluminium from bauxite mined in its own deposits in northern Brazil. The bauxite is transported in bulk carriers such as the 37,000 dwt ANSAC PRIDE, which delivered a cargo to the Ardal plant in early 2018 or the similar-sized, Marshall Island-flagged GABRIELLA in February 2022. The finished products, which include 204,000 tonnes primary aluminium, 279,000 casthouse products and 172,000 tonnes of anodes annually are exported worldwide by multipurpose cargo vessels.

2.1.5.3 Korean-Built Kamsarmax Carriers

The Samonas Brothers shipping company, founded in 2007 by the London-based Greeks Leon D Samonas and John D Samonas, operated five Panmax bulkers in 2015, of which OINOUSSIAN LION and OINOUSSIAN LADY were built at the Japanese shipyards Hitachi and Sasebo, MAVERICK GENESIS, MAVERICK GUNNER and MAVERICK GUARDIAN 2010/12 at the Korean STX shipyard. The latter three kamsarmax ships, flying the Greek flag, have a length of 229 m, a breadth of 32.2 m and carry 80,717 tonnes on a draught of 14.45 m. At the once maximum Panama canal draught of 12 m, meanwhile being insignificant after the waterway's extension, they still achieve 61,500 dwt. The ships' seven cargo holds are closed by sidewards movable hatch covers, the fourth hold can serve as ballast tank, which, together with the wing, double bottom and peak tanks, totals to 36,175 m^3. They can be emptied within 12 h. At the request of Crédit Suisse, the shipping company's Korean buildings were arrested for non-payment of the loan and from November 2016 put up for forced auction. MAVERICK GUNNER sailed from 2017 as PANAGIOTIS for Empire Bulkers in Athens. Independent from the dissolved Samonas Brothers company, the shipping firm John Samonas & Sons in early 2022 listed two capesize and newcastlemax units each in their fleet (Fig. 2.29).

Korean-built kamsarmax carriers also hoisted the flag of Athens-based Thenamaris shipping company. In 2016 the operator acquired the 81,640 dwt ships LIBERTY DESTINY and LIBERTY DAWN, built in 2012/13 by Daewoo shipyard, from US shipping company Liberty Maritime and renamed them SEATRIBUTE and SEATRUST. In 2019, the ships were resold and

Fig. 2.29 Four years after commissioning by London-based Samonas Brothers the kamsarmax carrier MAVERICK GUNNER was sold by auction (Ralf Witthohn)

renamed MSXT OCEANUS and ASTARTE, resp. Thenamaris als took over the equally sized newbuilding SEASTRENGTH from Hyundai in 2011, while the 82,600 dwt SEAJOURNEY and ELLINA from Japan's Tsuneishi shipyard were commissioned in 2008/09. The bulk carrier segment of the Thenamaris fleet at the end of 2017 included five capesize and kamsarmax carriers, one panamax carrier and four units each of the ultramax and supramax classes. In the beginning of 2020 were named five capesizes and kamsarmaxes each, one panamax unit and four supramaxes, 1 year later seven capsizes, six kamsarmaxes, one panamax, four ultramaxes and two supramaxes. After only 1 year of operation SEATRUST was sold to Greek Diana Shipping and renamed ASTARTE, while SEATRIBUTE was in 2019 purchased by Chinese owners and operated as MSXT OCEANUS (Fig. 2.30).

2.1.5.4 Kamsarmax Carriers from Sanoyas

Among other Japanese builders which developed kamsarmax carriers is the shipyard Sanoyas Hishino Meisho in Mizushima. In 2010, the builders completed the 83,369 dwt ore carrier KEY BOUNDARY ordered by Nippon Yusen Kaisha (NYK). After being sold to the BW Group in 2016, the vessel operated under the name BW BARLEY as one of six bulk carriers in the tanker-dominated fleet. In 2021, the ship switched into Greek hands and was renamed ALLIANCE. The 2008 built TORM SALTHOLM of 83,685 dwt was employed by the Danish shipping company Torm, before it was renamed PEDHOULAS COMMANDER in 2013 by New York Stock Exchange-listed owners Safe Bulkers.

Originating from Greek Alassia Steamship, Safe Bulkers managed eight further kamsarmax ships at the beginning of 2018, the younger ones of which were built in China since 2012 as PEDHOULAS BUILDER, PEDHOULAS

Fig. 2.30 In 2013, Korea's Daewoo shipyard was the supplier of the 82,000 dwt LIBERTY DAWN, which was transferred to Athens Thenamaris shipping in 2016 and renamed SEATRUST (Erich Müller)

FIGHTER, PEDHOULAS FARMER, PEDHOULAS CHERRY and PEDHOULAS ROSE. The ships were sold and leased back on a 10-year bareboat basis for a daily charter rate of 6500 US dollars, with the option of buying them back at a fixed price. 14 panamax carriers of 74,000 dwt to 78,000 dwt built in Japan between 2003 and 2016 formed the largest share of the Safe Bulkers fleet. In addition, the shipping company had 12 postpanamax bulkers as well as three capesize carriers of Japanese and Chinese production delivered from 2010 onwards. The shipping company tried to delay the delivery of further newbuildings during the shipping crisis. In early 2022, the company operated 39 bulk carriers comprising 12 panamax, seven kamsarmax, 15 postpanamax and five capsize class ships with a total cargo capacity of 3.7 million dwt and had nine bulk carriers on order, six kamsarmaxes and three postpanamax ships (Fig. 2.31).

2.1.5.5 Hundreds of Standard Bulkers from China

In order to equalize the lead of Japanese and Korean shipbuilders in bulk carrier construction, Chinese shipyards made use of the transfer of know-how from both foreign and domestic design companies. The former were brought into play by the shipping companies that placed the orders and expected technical results which met their requirements. Following this concept, China created their own centrally managed offices. Their designs were realized in large numbers by several builders, thus ensuring a high degree of amortization of the development costs. Even newly established shipyards without great technical experience were thus able to make international offers for competitive ships, which the customers could not ignore regarding the low price level. The concept also allowed a high output of hundreds of mid-size bulk carriers from around 2010 onwards.

Fig. 2.31 The hull of the bulk carrier KEY BOUNDARY delivered to NYK in 2008 is reinforced, so that the holds 2, 4 and 6 may remain empty, when ore is transported (Ralf Witthohn)

Against the background of the economic crisis, this contributed to considerable tonnage oversupply. As a result, there were numerous cancellations of newbuilding contracts, and some of the Chinese builders, which had sprung up out of thin air, soon stopped activities again. Although a significant proportion of the newbuildings were needed to supply China with raw materials, they basically became participants in the international market and increased its imbalance. Furthermore, from 2006 onwards, hedge funds and private equity firms played a similarly disastrous function in international bulk shipping as German investment companies had done in association with German state banks in container shipping. The crisis triggered by the high tonnage growth caused numerous insolvencies, especially of listed shipping companies in the United States. During 2021, yet, before the background of coal shortages and power outages, freight rates increased rapidly and reached a 12 years high, when in October daily spot rates for capesize vessels topped USD80,000, but declined sharply afterwards (Fig. 2.32).

A main provider of standard designs is the Shanghai Development and Research Institute SDARI. The office created a family of larger bulk carrier types designated Dolphin. Of the Dolphin 57 type, Germany's Conti group received the 56,969 dwt supramax carriers POS ACHAT, POS ALEXANDRIT, POS ALMANDIN, POS AMAZONIT, POS AMETHYST, POS AMETRIN, POS AQUAMARIN, POS ARAGONIT, POS AVENTURIN and POS AZURIT. They were built at Jiangsu Hantong shipyard in Tongzhou between 2008 and 2014. Another supplier of the 190 m length, Panamax breadth type was Taizhou Sanfu shipyard, which delivered 14 units, starting by the FREDERIKE SELMER for Hamburg owner Wehr and including CONTI PERIDOT, CONTI PYRIT, CONTI LAPISZLAZULI und CONTI LARIMAR for the Conti group. COSCO Guangdong completed 30

Fig. 2.32 As one of hundreds of Chinese-built bulk carriers the five-hatch 63,400 dwt BALTIC WASP, built in 2015 by Yangfan shipyard in Zhoushan within a series of four units for the US shipping company Genco, represents one of the largest supramax designs (Ralf Witthohn)

newbuildings of the type until 2013, including HUA HENG 166 and HUA HENG 167, delivered to the Grand China Logistics Group at Shanghai in 2011. From 2009 on Taizhou Kouan Shipbuilding participated in the building programme by launching AS VICTORIA, AS VALENTIA, AS VIRGINIA, AS VENETIA, AS VINCENTIA, AS VALERIA, AS VALDIVIA and AS VARESIA for Hamburg owner Ahrenkiel and further ships named VEGA AQUARIUS, VEGA ARIES, VEGA TAURUS, ATLANTIC MEXICO, IKAN PARANG, ATLANTIC MERIDA and DATO SUCCESS. Nine newbuildings of type Dolphin 57 came from shipyard Nanjing Wujiazui, among them in 2010/11 KIRAN EUROPE, KIRAN AMERICA, KIRAN AFRICA and KIRAN EURASIA for Istanbul based Kiran Holding. COSCO (Zhoushan) Shipyard was the builder of 38 units, including OCEAN BEAUTY, OCEAN MASTER, OCEAN LOVE and OCEAN LADY for Athens based Oceanstar (Fig. 2.33).

The same principal particulars as the Dolphin 57 type, but a smaller deadweight of 53,492 tonnes on a draught of 11.1 m is featured by the supramax carrier type Diamond 53. The delivery list included eight newbuildings for China's Chengxi Shipyard named RUI NING with consecutive numbers. The 190 m lenth, panamax breadth design, of which Chengxi contributed a total of 70 ships, is designed with a double hull und a strengthened tank top for cargoes weighing up to 1.35 tonnes per cubic metre. In May 2015, the Tian Jing registered RUI NING 3 transported rotor blades for wind generators from Shanghai to Brake on its hatches. Chengxi delivered its first units in 2005/2006 as SPAR LYRA, SPAR LYNX, SPAR VIRGO, SPAR TAURUS, SPAR CANIS and SPAR SCORPIO to Norway's Spar Shipping. The type was also produced by Vietnamese shipyard Nam Trieu at Haiphong, which launched VELA, SOPHIA, MARIETTA BOLTEN, LUCIA BOLTEN,

Fig. 2.33 Taizhou Kouan started an eight ships series by delivering the Dolphin 57 type vessel AS VICTORIA to Hamburg based Ahrenkiel company in 2009 (Ralf Witthohn)

THAI BINH SEA, LUCIA BOLTEN and OCEAN QUEEN. Another 26 Diamond 53/56 class vessels were launched by Yangzhou Guoyu-Werft at Yizheng. Built by several shipyards at a three-digit number, the Diamond and Dolphin designs from SDARI helped Chinese builders to win a leading role and to supersede the bulk carrier delivery figures of competing Japanese shipyards (Fig. 2.34).

SDARI also developed a larger version named Dolphin 64 from the Dolphin 53 and Dolphin 56 type bulk carriers. This was built over 300 times by Chineses shipyards The supramax carriers achieve a deadweight performance similar to the one of the first panamax bulk carriers, though those were 25 m shorter. A six ship comprising series of 63,800 dwt ultramax vessels was in 2013 contracted by Norway's Spar Shipping with Jiangsu Hantong shipyard. The newbuildings from Jiangsu Hantong were delivered from January 2015 on under the names SPAR ARIES, SPAR APUS, SPAR PYXIS, SPAR OCTANS and SPAR INDUS. The double hull ships are of 199.9 m length, 32.3 m breadth, 18.5 m depth and go 13.3 m deep. The cargo intake of five holds is figured at 78,930 m^3. On its first voyage to Europe, SPAR APUS unusually carried project cargo consisting of China made tower sections for wind generators to Bremerhaven. Jiangsu Hantong delivered the same type for domestic owners. Founded in 1990 by Helge Eide Knudsen as an offshoot of a textile company Spar then operated one panamax, 24 supramax and four handymax class ships. In 2021, the fleet management concentrated on the 24 supramaxes still operated.

Fig. 2.34 The supramax bulk carrier WU YI HAI of type Diamond 53 was delivered by Chengxi in 2008 (Ralf Witthohn)

2005 founded HTM Shipping from Shanghai took over JIA YUE, BAO FORTUNE and BAO RUN in 2016, and BAO TONG and BAO LUCKY from Chengxi Shipyard in 2014. The first Dolphin 64 unit was AMBER CHAMPION completed at Chengxi Shipyard in March 2013 after only 10 months of construction time. 2008 founded shipyard COSCO Zhoushan delivered DARYA TIANA and DARYA CHAND to Hong Kong-based KC Maritime in 2015. On one of its first voyages DARYA CHAND transported a cargo of zinc ore from Australia to European ports, of which the last part was discharged at Nordenham. The Dolphin 64 type was also built by Sainty Shipbuilding (Yangzhou) in Yizheng. In 2014 and 2015 the yard completed INTHIRA NAREE and ISSARA NAREE on behalf of Thailand's Precious Shipping. The owner cites a light ship weight of 11,678 tonnes, and a deadweight of 63,519 tonnes. Four 30 tonnes lifting cranes still have a safe working load of 24 tonnes when using the crabs of 12.5 m^3 capacity. Of the same type are SARITA NAREE, SARIKA NAREE, SAVITREE NAREE, SAVITA NAREE and SUNISA NAREE, in 2015/16 delivered by Taizhou Sanfu shipyard at a price of about USD 25 million each (Figs. 2.35 and 2.36).

2.1.6 Uranium Ore, Plutonium

2.1.6.1 Yellow Cake Via Walvis Bay

The two blue painted tugs of Namibia's national port company Namport push the bulk carrier to the Walvis Bay quay. The ship is to load over several hundred tonnes

Fig. 2.35 Sainty Shipbuilding equipped the Thai-owned ISSARA NAREE of Dolphin type 64 with crabs (Ralf Witthohn)

Fig. 2.36 Jiangsu Hantong delivered the Dolphin 64 type BAO LUCKY to a Shanghai-based owner (Ralf Witthohn)

of a rare commodity that has been extracted in the open-cast mining of the Rössing mine 70 km inland from Swakopmund. There, large tipper vehicles move the material broken up by blasting in the 3 × 1.5 km large, 390 m deep pit. Depending on the level of radioactivity, the uranium ore is sent through crushers until it has a maximum size of 19 mm or is stored as second-class material. After water is added, the raw material is turned into slurry, which is ground and transformed into a sulphuric acid solution in tanks by using ferrous sulphate. By separation, thickening and contact with resin, the uranium ions are separated until a concentrated uranium solution is produced, which in turn is mixed with organic substances and being added gaseous ammonia. The ammonium diuranate produced in this way is filtered in drums to form a yellow paste known as yellow cake. The ammonia is removed by roasting and the uranium oxide produced is made ready for shipment in metal drums.

The Rössing mine, which was discovered in the Namibian desert in 1928 and covers an area of 25 km^2, was not examined in detail until the late 1950s. In 1966, the British Rio Tinto group secured the mining rights to the country's first uranium mine and began exploiting it in 1976, accounting for about 2.5% of global production by an annual capacity of 4500 tonnes. Namibia had a 5.2% share of uranium oxide production in 2016, surpassed by Kazakhstan with 39%, Canada with 22%, Australia with 10% and Niger with 6.5%. Uranium is purchased at a share of 57% from Europe, Asia and Africa, 30% from North America and 13% from Japan. Demand yet experienced a slump due to the Fukushima catastrophe.

2.1.6.2 German Uranium Hexafluoride to Russia

In November 2019, the Panama-flagged general cargo vessel MIKHAIL DUDIN loads a cargo of 600 tonnes of Uranhexafluorid at Amsterdam. The radioactive material comes from the German uranium enrichment plant at Gronau and is to be transported via Skagen to the port of St. Petersburg, then further on to Novouralsk. Russian environmentalists criticize, that the isolated town would be misused as nuclear waste deposit being an illegal action. Official Russian sources yet describe the material as "strategic reserve". It could be enriched and used as fuel in fast breeder reactors. Ten years before, the transports of nucelar waste from Germany had been stopped.

The 3030 dwt MIKHAIL DUDIN is a coastal and inland cargo vessel delivered by Volgogradskiy shipyard at Volgograd/Russia in a series of nine newbuildings to NWS Nine Balt Shipping at St. Petersburg in 1996. The ship is operated by Aspol-Baltic Corp. at St. Petersburg. Nuclear material has already been transported from Bremerhaven to Russia in 2000 in containers on deck of the 1960 dwt Russian coaster BUGULMA. The former hopper dredger, built by Far East-Levingstone shipyard at Singapore in 1988, had only in the same year been converted to a cargo vessel (Fig. 2.37).

Fig. 2.37 Already two decades before the MIKHAIL DUDIN, the Russian coaster BUGULMA transported containers with labels warning of radioactive material from Bremerhaven to Russia (Ralf Witthohn)

2.1.6.3 Uranium Hexafluoride on the Largest Con Ro Ships

On 8 August 2017, three of the world's largest con-ro freighters make a rare simultaneous stopover in German waters. The newbuilding ATLANTIC SUN is on its maiden voyage heading from Dunkirk towards the Elbe estuary. At Hamburg, in the ship's port of destination, are already assembled two sister ships. The ATLANTIC SAIL, which has arrived from Antwerp the night before, is being unloaded at the Unikai terminal in the Hansa port. The ATLANTIC SEA is under repair at Blohm + Voss shipyard for a month now. As the Unikai berth is occupied, the ATLANTIC SUN has to anchor in the Elbe estuary.

A total of five Korean built con ro vessels of an innovative design, named ATLANTIC SUN, ATLANTIC SAIL, ATLANTIC STAR and ATLANTIC SKY, replaced five elder con ro ships from October 2015 on. Those passed on their title as the world's largest con ro carriers to the newbuildings. Orderer of the ships at Hudong-Zhonghua Shipbuilding in Shanghai was Atlantic Container Line (ACL), a subsidiary of the Italian Grimaldi group and based in New Jersey/US. The 55,649 dwt ships maintain a liner service across the North Atlantic. By a tonnage of 100,430 gt they are the largest ships ever built for the combined transport of rolling cargoes and containers cargo. At the same time, they supersede all pure ro ro ships (Fig. 2.38).

The principle of a combined transport was first introduced by Atlantic Container Line in the 1960s and has been retained in the third, constantly enlarged generation of vessels. The mixture of vertically and horizontally handled cargo indirectly facilitates cargo securing, which has remained inadequate to this day in conventional, hatch cover-fitted container ship that lack of the geometrically impossible arrangement of securing guides on deck. Since the greater part of the hold on the ACL ships is used for rolling cargo, cell guides can be installed in the hatchless area above deck. ACL points out that

Fig. 2.38 ATLANTIC STAR and four sisters feature a specially designed con ro architecture (Ralf Witthohn)

since its foundation no container has been lost overboard, which is otherwise not uncommon, especially in stormy weather. In the ship's forward compartment, in which containers are stowed under deck, no hatch covers have been arranged, so that the cell guides could be extended beyond the main deck and thus secure the deck containers.

The 296 m length, 37.6 m breadth and 10.25 m depth ACL newbuildings were designed by International Maritime Advisors (IMA) in Dragør/Denmark. The Chinese builders commissioned the design office Knud E. Hansen in Helsingør to adapt the IMA design to the shipping companie's requirements, which resulted in a modified version of the initially published lay-out. Like the ships of the predecessing type G3 built in Sweden, France and the UK, the newbuildings were equipped with an angled stern ramp starboard. However, the garage for the rolling cargo is not anymore located aft but at half the ship's length with two accommodation decks for the crew and the wheelhouse above. Moving the superstructure with the wheelhouse forward is in line with the trend in the design of large container ships. As in the case of mega container ships, the new ACL concept resulted in a separation of the superstructure from the engine casing, which is located aft on port side. The chosen position of the superstructure improves the visibility conditions over the forward rows of containers, of which eight 40-foot bays and one 20-foot bay are located forward and seven 40-foot bays plus two 20-foot bays aft of the superstructure. In the cell guides, the containers above the weather deck are stowed six layers high in front of the bridge and behind the bridge in seven layers and 13 transverse rows. A total of 200 sockets are provided for remotely monitored refrigerated containers.

The relatively large freeboard of 12.7 m from the loadline to deck 4 at 22.95 m, chosen out of safety reasons related to the ship's open hold configuration, encouraged the designers to do without a raised forecastle and to arrange the first row of containers only about one container length behind the almost vertical stem. Instead of a forecastle, the deck cargo is protected by a high breakwater. The G4 ships can carry 3800 teu containers, while the deck area for rolling cargo of 28,900 m^2 results in a car capacity of 1307 units. The ships are accelerated by a 22,000 kW two-stroke engine to 18 knots.

The newbuilding quintet took over the tasks of five con ro vessels of the ATLANTIC CARTIER type built in France and Sweden in 1985 and lengthened in South Korea in 1987. In May 2013 on one of the vehicle decks of the ATLANTIC CARTIER a fire broke out at Hamburg. It turned out to be a serious accident, which proved to be particularly problematic, because the vessel was carrying hazardous materials such as ammunition, rocket fuel and radioactive material in the form of uranium hexafluoride. According to the investing Bundesstelle für Seeunfalluntersuchung (BSU) in Hamburg, its gross cargo weight of 8.9 tonnes included 10.3 kg of pure, extremely dangerous material, 7 kg of which was uranium. The extensive fire, which was fuelled by the petrol of the parked cars, could only be extinguished by use of the ship's own CO2 fire-fighting system and after dangerous goods loaded in containers or on flats had been taken ashore from the ship's hold as a precaution. BSU was unable to determine the cause of the fire, but found that recurring welding work was required on the ship after its lengthening, in order to repair cracks in the hull. Such work was also carried out during the stay in Hamburg, before the fire broke out (Fig. 2.39).

2.1.6.4 Plutonium to the United States

On 26 January 2016, the automated identification system (AIS) of the British nuclear transporter OCEANIC PINTAIL reports Barrow-in-Furness on the west coast of England as its location. This is misleading, as the special ship is at this time berthed in the port of Nordenham/Germany, to take over nuclear material in secrecy and under strict security measures. The nuclear fuel brought in by truck comes from European plants and is to be transported to the United States. Police boats from all German coastal countries have gathered around the ship for escort to the Weser estuary (Fig. 2.40).

Designated irradiated nuclear fuel material carrier, OCEANIC PINTAIL was built by Mitsubishi shipyard in Kobe in 1987 on behalf of UK-based Pacific Nuclear Transport Ltd (PNTL) under the name PACIFIC PINTAIL. The ship was to carry out the transport of highly dangerous radiation material on behalf of Japanese nuclear power plants. In 2012 the ship was renamed OCEANIC PINTAIL. The basic safety principles of its design include twin propellers and a double hull. Its equipment comprises artillery

Fig. 2.39 The transport of uranium hexafluoride became particularly explosive by a major fire on ATLANTIC CARTIER (Ralf Witthohn)

Fig. 2.40 The specially designed OCEANIC PINTAIL transporting plutonium from Nordenham (Ralf Witthohn)

cannons and other defence systems, such as water cannons against environmental activists. In 2020, the ship was sold for scrap.

The special safety aspects demanded a large accommodation area to offer room for the additional crew. At 104 m length and 16.6 m breadth OCEANIC PINTAIL had a deadweight of only 3865 tonnes, while its gross tonnage was figured at 5087 gt. The five holds of the OCEANIC PINTAIL were fitted with guides on the side tanks for containers to be stowed crosswise. A few days before the arrival of the OCEANIC PINTAIL in German waters, a local English environmental organisation reported the departure of two further PNTL transporters, the 4.408 dwt PACIFIC EGRET, built in 2010, and the 4916 dwt PACIFIC HERON of 2008 from Barrow-in-Furness for the purpose of transporting plutonium from Japan to the United States. The transports to Charleston were being carried out on behalf of the US-led Global Threat Reduction Initiative (GTRI), under which plutonium and highly enriched uranium are shipped to the United States for safekeeping. Generally, two ships of the company are sailing together for mutual security.

2.1.7 Black Ilmenite for White Colour

It takes the grab crane only 10 h to fetch 3000 tonnes of ilmenite from the hold of the German coaster WILSON ALMERIA at the pier of the Kronos Titan works in Nordenham-Blexen at the Weser estuary. The black shimmering slate rock is destined for the German plant of the Dallas-based US concern Kronos International. Using sulfuric acid the raw material is converted into titanium dioxide, a white pigment for use in the paint and chemical industry. The sulfuric acid is delivered by the inland tanker ACIDUM, which annually transports about 150,000 tonnes from the Upper Weser. The ilmenite originates from Hauge i Dalane on the Norwegian southwest coast. There, the Kronos subsidiary Titania operates its own mine. Every year, 50 to 60 cargoes of ilmenite reach the quay of the plant in Blexen in small bulk carriers of up to 3500 dwt. The Malta-flagged WILSON AMERIA is owned by Schifffahrtsbetrieb Kapitän Siegfried Bojen from Moormerland/Germany and chartered by Bergen/Norway-based Wilson Ship Management, which operated 113 bulk carriers of between 1500 and 9000 dwt in the beginning of 2020 and 128 1 year later (Fig. 2.41).

The Norwegian mine, founded in 1902, has been working since 1916. Of the US concern's five production facilities, two are located in Germany. In addition to a Leverkusen site, the younger plant at Nordenham was founded in 1969. The facility is regularly supplied by ships owned or chartered by Wilson. The by-product iron salt is used in wastewater treatment to precipitate phosphate and as a chromate reducer in cement. The sulphuric acid is also being recycled following the protest of Greenpeace environmentalists against its dumping in the North Sea carried out until 1989.

Fig. 2.41 Coastal ships like WILSON ALMERIA transport a total of 150,000 tonnes of ilmenite from Hauge i Dalane to the Nordenham-Blexen production site every year (Ralf Witthohn)

2.1.8 Lithium from the Uyuni Salt Desert

In September 2016 Bolivia exports a first symbolic cargo of 10 tonnes of lithium to China. The material worth USD 70,000 has been extracted from the salt lake of Uyuni. The lightest of all metals is the basic material for the production of batteries, which importance is increasing, since the decarbonization of the economy has become a must. The state-owned company Comibol is investing in a production plant for the manufacture of 10,000 tonnes of lithium per year and later of lithium cathodes from the reservoir, which represents, by a guessed volume of 20 million tonnes, an estimated 70% share of the world's lithium deposits.

Efforts to develop the export business are hindered by Bolivia's inland location on the South American continent. Since the Saltpetre War (1878 to 1884) and the annexation of a 400 km long coastal strip by Chile, the country has no access to the Pacific Ocean. Bolivia uses the river ports Central Aguirre, Gravetal and Jennefer on the Paraguay-Paraná waterway that connects the landlocked nation with the Atlantic Ocean and the countries of Argentina, Brazil, Paraguay and Uruguay, some of which offer Bolivia free port privileges. About 500,000 tonnes of cargo, mainly soybean flour and oil, are moved through Port Jennefer, while some high transit costs affording 110,000 containers are annually imported through the Chilean port of Arica, another freeport used by Bolivia. From 1979 to 1986, the modern 12,544 dwt general cargo vessel ALEMANNIA of Germany's Hapag-Lloyd company sailed under the flag of Bolivia after being sold to Lineas Navieras Bolivianas (LINABOL) and renamed BOLIVIA.

2.2 Coal

2.2.1 Borneo Coal for Ishikawa

On the last December day of 2014, the postpanamax bulk carrier ALAM PERMAI leaves the roads of Taboneo in South Borneo, where a floating crane has filled the seven holds of the anchored vessel with coal. Port of destination is Ishikawa, where the Okinawa Electric Power Company operates one of its power plants. Delivered by Japan's IHI shipyard to Singapore-based Pacific Carriers Ltd (PCL) in 2005, the 87,050 dwt ALAM PERMAI is, by its breadth of 36.5 m, exceeding the panamax dimension by over 4 m (Fig. 2.42).

In 2016, the ALAM PERMAI was reregistered without a change of name for Malaysian Bulk Carriers (MBC), a company established by PCL together

Fig. 2.42 The postpanamax bulk carrier ALAM PERMAI participated in coal transport to Japan (Ralf Witthohn)

with Malayan Sugar Manufacturing Co. As the largest Malaysian bulk carrier operator, MBC operated three sister ships of ALAM PERMAI, named ALAM PADU, ALAM PENTING and ALAM PINTAR, as well as seven supramax and six handysize bulkers at the end of 2019. Shortly afterwards, the ALAM PERMAI ownership switched to Jakarta-based Energi Global Sejati, the ship's name to URMILA. In 2017, the much larger PCL fleet included twelve 38 m wide postpanamax bulkers built by the Japanese shipyards Imabari and Sasebo and featuring deadweight capacities of between 85,000 and 96,000 tonnes. The fleet as well comprises three capesizes, four mini-capers, seven panamax, 18 supramax and 22 handysize bulk carriers. Until the end of 2019, the number of ships in the smaller classes slightly changed to five panamaxes, 20 supramaxes and 15 handysize class ships. In 2021 the fleet had shrunk to eight vessels, three kamsarmax and supramax ships each and two handysize bulkers.

2.2.2 Russian Coal from Ust-Luga and Murmansk

Following delivery by the Japanese shipyard group Japan Marine United (JMU) in April 2017, the first order deploys the ANDROMEDA OCEAN from Beilun, a city district of the eastern Chinese port of Ningbo, to Dunkirk/France. From there, the kamsarmax vessel proceeds to Ust-Luga/Russia to load coal for Nordenham/Germany. The maximum deadweight of the vessel is 80,979 tonnes at a largest draught of 14.4 m, too much for the port of Nordenham. Therefore, the ANDROMEDA OCEAN has only been loaded with Russian coal up to 12.3 m draft. After Russia's invasion of Ukraine and a EU ban to come in force from

August 2022 on, coal imports through Nordenham initially kept on, like by the 75,000 dwt ANTHOS from Ust-Luga in May 2022, but increasingly originated from other sources. Already in April 2022, the 57,000 dwt supramax carrier TONY SMITH discharged a cargo from New Orleans, in June the 82,000 dwt MARVELOUS STAR imported coal from Tubarao/Brazil. Although Nordenham even accepts 180,000 dwt capesize carriers, they can berth only in a part-loaded state (Fig. 2.43).

ANDROMEDA OCEAN, which was renamed HAMPTON OCEAN in 2021, represents a panamax series type which the Japanese shipyard group has been building in its 1906 founded Maizuru shipyard, later named Hitachi shipyard, since 1973 and continuously developed over a period of 42 years from 61,250 dwt to almost 81,000 dwt. From 2015 onwards, the kamsarmax version was created by extending the length from 225 to 229 m. HAMPTON OCEAN is managed by Singapore-based Diamond Bulk Carriers. Powered by an 8880 kW two-stroke diesel engine the ship achieves a speed of 15.7 knots. An even larger modern designed bulk carrier that was engaged in the supply of coal to Nordenham was in January 2020 the 2015-built postpanamax ship HARVEST RAIN of 95,253 dwt reaching the port at 12.5 m draught, a sister vessel of the HARVEST FROST.

A draught-reducing solution similar to that for ANDROMEDA OCEAN is being developed for the bulk carrier newbuilding WESTERN MANDAI for the transport of coal to the Bremen steelworks. After delivery at Japan's Imabari

Fig. 2.43 Following its maiden voyage to Europe, the kamsarmax vessel ANDROMEDA OCEAN transported a cargo of coal loded through hatches which are opened by sideward travelling covers, from Ust-Luga to Nordenham (Ralf Witthohn)

shipyard in February 2015, the first voyage of the supramax class vessel leads across the Pacific to Vancouver for loading Canadian grain at the Richardson International Terminal and from there through the Panama canal into the US Gulf. At Murmansk, the 61,285 dwt five-hatch vessel takes on a consignment of coal for Bremen, this only partly using the ship's capacity. Instead of the largest draught of 13 m, WESTERN MANDAL reaches the iron and steel port at an immersion of just 8.9 m at the end of March 2016, after a waiting period of several days on the Outer Weser anchorage. Two months later the long-term charter by Oslo-based shipping company Western Bulk ends, and the ship is renamed AFRICAN ARROW by its Japanese owner (Fig. 2.44).

Coal handling in the Bremen Industriehafen area is carried out by the Weserport company, a joint venture of the Rhenus group and the steelworks ArcelorMittal, which at the same time is the greatest customer. At four terminals are, in addition to coal, handled ore, steel, other kinds of bulk cargoes as well as components for wind power plants and industrial plants.

2.2.3 Canadian Coal Through the Northwest Passage

The course set for Pori in Finland, the bulk carrier NORDIC ORION leaves Vancouver on 6 September 2013 with a cargo of 73,500 tonnes of coal. For the first time, a coal carrier is not heading south, but tries to reach Europe through the Northwest Passage north of the American continent. Global warming enables a large loaden cargo vessel for the first time to sail on this shipping route under ice-breaker escort. The ship of the Danish shipping company Nordic Bulk Carriers arrives at its Finnish port of destination on 9 October 2013 after a stopover in Nuuk on Greenland. The chosen route is 10,000 nm shorter than through the

Fig. 2.44 The supramax newbuilding WESTERN MANDAL took a cargo of coal from Murmansk to Bremen in the first year of service (Ralf Witthohn)

Panama canal and allows an additional cargo of 15,000 tonnes to be loaded, because the ship is not bound to the draught restrictions of the canal (Fig. 2.45).

The NORDIC ORION is a panamax carrier delivered in 2011 to Japan's Sanko Steamship by Oshima Shipbuilding under the name SANKO ORION. At 225 m length, 32.3 m breadth, 19.4 m depth and 14.1 m draught the ship achieves 75,600 dwt. SANKO ORION and the sister ships SANKO ODYSSEY, later named NORDIC ODYSSEY, are classified by DNV GL according to the regulations of the ice class 1A. In 2012 the ships were transferred to Nordic Bulk Carriers operating from Hellerup and Singpore. The company belongs to the Newport/US-based Pangaea group, which also includes Phoenix Bulk Carriers, Americas Bulk and Seamar Management in Athens. Sovcomflot took over the management of the ice-going bulk carriers in 2015. The Danish shipping company had in summer 2010 already sent the 43,700 dwt bulk carrier NORDIC BARENTS from Narvik through the Northeast Passage to China. In 2022, Nordic Bulk Carriers disposed, in addition to the aforementioned ships, of the panamax carriers NORDIC OSHIMA, NORDIC OLYMPIC, NORDIC ODIN and NORDIC OASIS, built in Oshima from 2014 to 2016 in accordance with the ice class 1A, and the handymax ships NORDIC BARENTS and NORDIC BOTHNIA, built in 1995 as FEDERAL BAFFIN and FEDERAL FRANKLIN by Daewoo shipyard for ice operations in Canada's Fednav fleet (Fig. 2.46).

2.2.4 Coal from Tanjung Pemanciangan

Among the many bulk carriers series-built by Asian shipyards, there are only a few individual designs, like the two five-hatch vessels UNISON STAR and

Fig. 2.45 Despite its high ice class affording a stronger hull structure, NORDIC ORION achieves, by 75,600 tonnes, the usual deadweight of a panamax bulk carrier (Ralf Witthohn)

Fig. 2.46 NORDIC OASIS, last of four panamax carriers delivered by Oshima Shipbuilding for Arctic operations in January 2016, transported coal from Wyssozk to Bremen after being lightered in the port of Nordenham on one of its first voyages (Ralf Witthohn)

UNISON POWER, delivered in 2011 and 2012 by STX Shipbuilding to Unison shipping company in Taipei/Taiwan. The two 189 m length, 30 m breadth and 15 m depth sister ships have a deadweight of 38,190 tonnes at 10.4 m draught. The 13.5 knots fast ships are the only products of a Korean builder in Unison's handysize fleet. In 2017, almost all the other bulk carriers of the 1980-founded shipping company came from Japanese shipyards, SAPAI from Shin Kurushima, HALUS from Mitsubishi, UNISON SPARK and MERIT from Kitanihon, UNISON LEADER from Kanasashi and UNISON MEDAL from Oshima. Latest newbuldings are the Hong Kong-flagged 37,296 dwt single deckers UNISON JASPER, 2019 built at Oshima, and the sister vessel UNISON SAGE delivered in January 2020. They are mainly employed in the Far East. In October 2017, UNISON STAR loaded a cargo of coal in the port of Tanjung Pemanciangan/Indonesia, UNISON POWER took a cargo in the Far Eastern Russian port of Vanino, the country's second largest coal port on the Pacific coast (Fig. 2.47). In early 2022 UNISON STAR sailed from Port Klang to Iskenderun and Constanta, the UNISON POWER from Pyeongtaek and Anegasaki to Shuaiba.

Fig. 2.47 UNISON STAR and its sister ship UNISON POWER of Taiwan's Unison shipping company are occasionally employed in the transport of coal (Ralf Witthohn)

2.3 Stones, Sand, Cement

2.3.1 Granite from Eide

At the characteristically named road Steinhuggervegen directly leading along the fjord, the mini bulk carrier PREGOL HAV loads 2200 tonnes of granite at Eide/Norway. After a 4 days lasting voyage, during which the ship, fully loaded on 4.2 m draught, sails at a moderate average speed of 7.5 knots, it enters the Weser estuary and swings into the tributary Geeste. The 82.5 m length ship overcomes the water level difference to the fishing port in the double lock within 40 minutes and turns to port into the old Handelshafen of Geestemünde, where the Russian crew members moor their ship on the west side. The granite cargo processed into chippings is retrieved from the hold by help of a mobile port crane and loaded on trucks to be used for construction purposes on North German roads. Within only 2 days, discharge of the stones from the PREGOL HAV is completed, and the ship leaves for the Blexen anchorage at a draught reduced to only 2.8 m to wait for new orders from the shipping company.

The Antiguan-flagged PREGOL HAV has been sailing since 2013 for the HAV shipping company at Bergen/Norway, which also has an office in Oslo and carries out ship management activities from Kaliningrad/Russia. In 2017, the company operated a total of 16 second-hand coastal cargo vessels built between 1982 and 1999 and featuring deadweight capacities of 2150 tonnes to 2850 tonnes. Two years later, the number of ships was increased to 17, which still was the fleet's size in 2022. Built in 1985 under the name KURT JENSEN at Peters Werft in Wewelsfleth/Germany, the PREGOL HAV sailed as EXPLORER for the German owner Ernst Strahlmann from Brunsbüttel/Germany before being sold to Norway. Ships of the same Peters yard type in the Hav fleet at this time were ARCTICA HAV, ATLANTICA HAV, BALTICA HAV, BRITANNICA HAV, CELTICA HAV, DANICA HAV,

GERMANICA HAV, ICELANDICA HAV, NORDICA HAV, SWEDICA HAV and IBERICA HAV, mostly employed in bulk trades along the Baltic and North Sea coasts (Fig. 2.48).

2.3.2 Building Materials for Hamburg

In the early morning of Monday, the British-owned bulk carrier YEOMAN BANK arrives from Glensanda on the Sottish west coast at the Elbe port of Brunsbüttel. Harbour tugs push the ship to the river quay. Shortly afterwards the ship's own discharge boom swings over the quay edge, and in only 20 h 36,000 tonnes of mineral building materials are taken from the holds. Already the following morning YEOMAN BANK leaves for Garrucha in Spain. Its discharged cargo will be fractionated into three sizes in a screening machine and then forwarded to the regional construction industry and by barge to Hamburg.

Built within 8 years from 1974 to 1982 by Eleusis Shipyards in Greece, the 205 m length, 27.2 m breadth YEOMAN BANK has a capacity of 43,728 dwt. The vessel is operated along with the YEOMAN BONTRUP, YEOMAN BRIDGE, BONTRUP AMSTERDAM and BONTRUP PEARL by the company Netherlands-based Bontrup Aggregates to transport mineral products from Glensanda quarry, Europe's largest granite quarry, Bremanger Quarry in Norway, Stevin Rock in Ras Al Khaimah/United Arab Emirates and other rock armour quarries in Europe. The material is used for ready-mixed concrete, asphalt, track ballast, road construction, protective material for offshore pipelines and concrete structures. In December 2019, Port State Control inspectors acting on behalf of The Paris Memorandum of Understanding (Paris MoU) found 13 deficiencies on the YEOMAN BANK

Fig. 2.48 The Norwegian-owned coaster PREGOL HAV carried 2200 tonnes of granite chippings from Eide to Bremerhaven (Ralf Witthohn)

at Royal Portbury/UK, including corroded bulkheads and hatchways, a cracking in a deck and defects of the fire fighting equipment (Fig. 2.49).

At a capacity of 96,700 dwt, YEOMAN BONTRUP and YEOMAN BRIDGE are the world's largest self-discharging bulk carriers. Built in 1991 by Japan's Hashihama shipyard in Tadotsu, the 250 m length, 38.1 m breadth ships can unload up to 6000 tonnes per hour by an automated discharging system of Sweden's Consilium company. Originally ordered by British Steel, the ships were primarily intended to transport iron ore with a density of 3 tonnes per cubic metre and of heavy coal weighing 0.8 tonnes per cubic metre. For this purpose, the five cargo holds were provided with 181 hydraulically operated floor openings through which the cargo slides onto three conveyor belts running parallel at a speed of 3 m per second. At the end of the loading space there are two belts running athwartships, which transport the cargo to a 35 m vertical pocket lift. Via a chute, the cargo is transported to an 84 m long, 180° swivelling boom. The cargo is loaded by shore side facilities into the cargo holds, which are each closed by two hatch covers. YEOMAN BONTRUP fulfilled a special task when it delivered 90,000 tonnes of foundation material at hourly rates of 4500 tonnes for the construction of the new London Gateway container terminal, opened in 2013.

In 1990/91, Daewoo shipyard in South Korea completed the 77,548 dwt self-unloaders YEOMAN BROOK and YEOMAN BURN on behalf of the Norwegian shipping company Fearnley & Eger on the basis of a 20-year charter by the Yeoman company. They were taken over by Oldendorff Carriers in Lübeck shortly after being commissioned. The ships were able to discharge

Fig. 2.49 The 1982-built bulk carrier YEOMAN BANK has been fitted with self-unloading equipment in 1991 and is since then supplying construction companies in northern Europe with building materials (Ralf Witthohn)

cargoes with a specific weight of up to 2 tonnes per cubic metre by using a discharge system supplied by Consilium. In the case of this system, two conveyor belts fed by 124 hydraulically operated bottom flaps run under the floors of seven cross-sectionally W-shaped cargo hold floors and transport the material to the conveyor tower in front of the bridge house. From there, the cargo was carried ashore via a 76 m long boom. During repair work at Bremerhaven in 1994, the discharge system of the YEOMAN BROOK caught fire and had to be replaced in 4 months lasting work. A similar accident occurred in 2013 on the YEOMAN BONTRUP at Glensanda, when a fire broke out in the conveyor tower and spread to the swung out conveyor arm. YEOMAN BURN was chartered on to Canada Steamship Lines, with which Oldendorff entered into an employment cooperation agreement. The ship ended service for Yeoman in 1994, was renamed BERNHARD OLDENDORFF and came in 2018 over Abu Dhabi interests into Chinese hands, like its sister vessel YEOMAN BROOK. Both ships were broken up in 2021 (Figs. 2.50 and 2.51).

Fig. 2.50 A fire in the conveyor system of the YEOMAN BROOK was difficult to extinguish and required 4 months of repair in 1994 (Ralf Witthohn)

Fig. 2.51 A new conveyor belt for the YEOMAN BROOK was installed with the help of a floating crane (Ralf Witthohn)

2.3.3 Irish Coastal Vessel ARKLOW BRAVE

Stones from Glensanda are also transported in smaller bulk carriers, so-called mini bulkers, which are, however, usually categorized by the classification society as general cargo carriers. In January 2018, the 8660 dwt ship ARKLOW BRAVE discharged such a cargo at Nordenham. The ship originates from a sextet of newbuildings built in 2014/15 at Dutch shipyard Ferus Smit for Irish-based Arklow Shipping. The sister ships are named ARKLOW BANK, ARKLOW BAY, ARKLOW BEACH, ARKLOW BEACON and ARKLOW BREEZE. The 119.5 m length, 15 m breadth, 7.2 m depth, 12.5 knots achieving ships have two cargo holds with a total capacity of 9900 m^3. Following the current shipbuilding trend, their design shows a vertical stem and, to create the largest possible hatch opening, a special prominent lines configuration around the forward hatch corner.

Of similar architecture as the ARKLOW BRAVE class, but smaller are the 2998 gt ships ARKLOW VALE, ARKLOW VALIANT, ARKLOW VALLEY and ARKLOW VIEW, delivered by the Dutch shipyard Royal Bodewes. At a length of 86.9 m and a breadth of 15 m, they carry 5160 tonnes at a draught of 7.2 m. Comparable designs had already been realised between 2002 and 2011 on 15 ships of the ARKLOW R- and ARKLOW F- classes from shipyards in Spain and the Netherlands.

The ship dimensions into which coastal shipping has advanced in recent decades became even more obvious, when KfW IPEX-Bank announced in January 2018 that it would provide EUR 51 million together with DekaBank to finance four 16,500 dwt bulk carriers of Arklow Shipping for employment in the North Sea and Baltic Sea area. The credit was covered by the Hermes insurance. The Irish family business is having the vessels built at Leer, the German location of the Dutch Ferus Smit group. The ships are constructed in accordance with the ice class 1A for Baltic Sea operations in the transport of grain, coal, ores and stones. Deliveries were scheduled between September 2018 and February 2020, but the first completion took place only in September 2019 by the 8511 dwt ARKLOW ABBEY, in 2020 succeeded by ARKLOW ACCORD, ARKLOW ACE, ARKLOW ARCHER, ARKLOW ARROW and in 2021 by ARKLOW ARTIST. From 52 bulk carriers in the beginning of 2018 the Arklow fleet thus increased to 61 units 2 years later. In 2022, the fleet comprised 56 ships (Fig. 2.52).

2.3.4 Building Sand from the Atlantic

While fairway deepening and maintenance is often combined with the reclamation of land for a new port, industrial or residential areas by using the dredged sand, the pure extraction of the raw material from the sea for use as

Fig. 2.52 Six newbuildings of the ARKLOW BRAVE type from Dutch shipyard Ferus Smit are operated by Irish Arklow Shipping in European waters (Ralf Witthohn)

building material is gaining importance. The Spanish shipyard Astilleros de Murueta, named after its location, built a suction dredger for such a purpose. Accordingly, the dredger therefore does not feature typical dredger lines with a pronounced tonne-like bulbous bow, but those of a relatively fast, cargo carrying ship. STELLAMARIS was in December 2012 delivered to Dragage-Transport Ettravaux Maritimes (DTM) of La Rochelle/France, a company of the Libaud and Italcementi group. The dredger extracted sand from the Bay of Biscay and transported it to ports on the French Atlantic coast for use in the construction industry. In 2022, the ship was engaged in the supply of sand to the Brest, Lorient and La Pallice. The 102.7 m length and 15.5 m breadth ship can dredge up to a water depth of 45 m and store in a hold of 2600 m^3 capactiy. For this purpose, the dredger is equipped with an electrically operated 600 kW submersible pump at the suction head and a 1800 kW pump for transferring the material ashore. The main engine of the 14 knots achieving ship turns a 3000 kW generator to supply the pumps. At a draft of 6.9 m, the cargo capacity of the hopper dredger is figured at 4720 tonnes (Fig. 2.53).

2.3.5 Chinese-Built Cement Carriers CEMCOASTER and CEMCLIPPER

The transport of loose cement is problematic under two aspects. Dust generation and sensitivity to moisture complicate the cargo transfer by conventional grabs, so that cement is regularly handled by the ship's own pneumatically working equipment. These demands required the development of a special ship type, known as cement carrier, which is, due to its sophisticated equipment, often realized in newbuildings. There were, however, also carried out a number of conversions from general cargo vessels or coasters. The often light

Fig. 2.53 The French suction dredger STELLAMARIS fetches building sand from the Bay of Biscay (Murueta)

grey painted cement carriers can also be employed in the shipment of granulated blastfurnace slag and fly ash. The greater problem of transporting these types of cargo results from their high risk of shifting. After loss of stability, a series of cement carriers sank with their crews.

In May 2019, Germany's Frankfurt-based KfW IPEX bank announced, that it would finance two cement carrier newbuildings of the Hamburg-based Baltrader Capital, a company of the Brise group. The Chinese shipyard Fujian Southeast Shipbuilding was contracted to supply them at a price of EUR 15.7 million. Financing is covered by the Chinese export credit insurance company Sinosure. Delivered in 2021, the cement carriers CEMCOASTER and CEMCLIPPER are laid out for later refit to dual fuel operation by the main engine either burning liquefied natural gas (LNG) or diesel oil (MGO). The ships were developed by Hamburg-based SDC Ship Design & Consult in cooperation with Brise for the European shortsea trade and are classified by DNV as gas ready. The ships' pneumatically working handling device can be operated by the shaft generator or by auxiliary engine power.

In 2022, Brise operated a fleet of 12 cement ships. One of the cement carrier conversions referred to the series type general cargo ship GDYNIA, built in 2000 by Polnocna shipyard at Gdansk for Poland's Euroafrica Shipping Line. Baltrader let the vessel rebuild into the cement transporter CEMISLE 9 years later. The 100.6 m length, 5183 dwt vessel was extended to 120 m to accommodate the cement handling equipment. As additional consequence of the lengthening, the ship's deadweight increased to 6700 tonnes. The loading rate is figured at 500 tonnes, the discharge rate at 250 to 300 tonnes per hour. Handling is either carried out pneumatically via pipes of 6 to 14 inches diameter in up to 12 cement trucks or mechanically by the ship's crane. In 2017, CEMISLE was the largest ship in the Brise fleet. At that time, the shipping company operated eight more ships of 3000 to 4950 dwt and converted into cement freighters between 2001 and 2016 for the European trade. Of CEMSTAR, CEMLUNA, CEMBAY, CEMSKY, CEMVALE, CEMSOL, CEMSEA III and CEMGULF several ships were originally built as coasters at Hugo Peters shipyard in Wewelsfleth/Germany.

In January 2015, the CEMFJORD of the Brise group, rebuilt from the coastal freighter MARGARETHA in Swinouscie in 1998, was lost with its entire crew of eight during a voyage from Aalborg/Denmark to Runcorn/UK in heavy weather off the British west coast without an emergency call being received. The ship was discovered floating 1 day after the last sighting of the ship off the Pentland Firth, but sank the following day. The eight-man crew from Poland and the Philippines remained missing despite a large-scale search operation. A British investigative commission criticised the master's route planning, safety deficiencies bypassed by exceptions of the flag state Cyprus and the presumed insufficient stability of the ship. The Port State Control in Runcorn had found 11 deficiencies 14 months earlier and imposed a three-day sailing ban due to a defective lifeboat device. The accident was not the first catastrophic accident of a German cement carrier. The repeatedly overloaded cement carrier SCANTRADER of Lübeck-based shipping company Beutler sank in February 1990 with a bulk cargo of 2200 tonnes on the voyage from Bilbao to Sheerness in the Bay of Biscay without a trace of the 12-man crew (Fig. 2.54).

2.4 Diamonds from Namibian Waters

In 1997, Rickmers Lloyd Dockbetrieb in Bremerhaven/Germany converted the diving support vessel SPRUT into the diamond mining vessel KOVAMBO. The original ship had in 1979 been built at Wärtsilä shipyard in Turku/Finland under the name SWAN OCEAN and registered for the company Arktikmorneftegazrazvedka in Murmansk/Russia. During the conversion at Bremerhaven, the 17.5 m wide ship was lengthened from 77 to 92 m by inserting an additional section, in order to accommodate the conveyor and sorting equipment for diamond mining. This included a crawler which was lowered into the water with the help of the stern frame during the search off

Fig. 2.54 Converted from the Polish general cargo vessel GDYNIA, CEMISLE can load 500 tonnes of cement per hour with the help of a pneumatic handling system (Ralf Witthohn)

Fig. 2.55 The diamond mining vessel KOVAMBO was deployed off Lüderitz to search and excavate diamonds from the seabed by using a crawler (Ralf Witthohn)

the Namibian coast. The excavated seabed was conveyed by a large dredging pump via a hose to the aft deck of the ship for sorting. The diesel-electrically powered KOVAMBO was laid up at Cape Town in March 2013 and reported to sail the Mediterranean in 2019. Off Namibia, the company Diamond Fields International alsp employed the DF DISCOVERER to mine diamonds, which is reported to have ceased operations in 2012. The underwater extraction of diamonds is also carried out by means of drilling rigs or suction pipes (Fig. 2.55).

2.5 Fertilizer

The main agricultural fertilizers used are potash, phosphate and urea. Their largest suppliers are the companies Mosaic, Uralkali, Yara, PotashCorp, Belaruskali, OCP, Israel Chemicals and the German company K + S Kali as the world's largest potash producer with mining sites in Europe as well as in North and South America.

2.5.1 German Potash for India, Urea from Egypt

Six days after leaving Port Said, the bulk carrier FURNESS PORTLAND reaches the island of Socotra off the Horn of Africa, a dangerous pirate territory. The military personnel on board are keeping a close eye on nearby ships, especially fishing boats that start their forays from the Somali coast onto merchant vessels. In 11 days, the bulk carrier will enter the Indian port of Kakinada to discharge its cargo of 35,320 tonnes of potash salt. The FURNESS PORTLAND sails under the funnel mark of Furness Withy, a shipping company founded in England in 1891 and taken over by Germany's Oetker group in 1990. Oetker, later sold to Singapore-based Swire group, has chartered the ship from the Singapore shipping company Grace Ocean. The vessel, which in 2021 changed hands to Greek owners and was renamed ELPINIKI, took over its cargo at the quay of K + S Kali in Hamburg 1 month earlier. Ships with a draught of up to 12.3 m can berth there, but the handling operation, which usually takes 3 days, must be completed, before the ebb tide begins, in order to avoid touching the ground. Potash is exported at the potash quay in the first instance, imports of urea from Egypt, for example, are rarer (Fig. 2.56).

K + S Transport, which is located at Wilhelmsburg since 1927, handles about 380 ocean-going ships with a total capacity of 3 million tonnes

Fig. 2.56 The 37,045 dwt bulk carrier FURNESS PORTLAND, in 2021 sold and renamed ELPINIKI, transported a cargo of 35,000 tonnes of potash from Hamburg to Kakinada (Ralf Witthohn)

annually at its 500 m long quay. The company uses two grabbers and three ship loaders. More than 90% of the export volume, of which 26,000 tonnes can be stored in deep bunkers every day, is delivered by rail. One million tonnes is filled into containers and transported by truck to the container terminals. The total storage capacity of the terminal is 405,000 tonnes, its conveyor system is 12 km long. As the world's leading potash producer, Kassel/Germany headquartered K + S has numerous production sites in Europe, North and South America. In August 2017, the company and its partner Pacific Coast Terminals opened a new storage shed for 1,600,000 tonnes of potash in Port Moody, a port area of Vancouver, Canada. There, the product from up to 3 km long freight trains with a capacity of 18,000 tonnes is transhipped into ocean-going vessels of up to 70,000 dwt. The wagons take the fertilizer on a three-day journey from the Bethune production site in Saskatchewan through an 1800 km long stretch of the Canadian Pacific through the Rocky Mountains to the Canadian west coast for shipment to South America and Asia.

In addition to nitrogen and phosphate, potassium chloride, which has a nutrient content of 60%, is one of the fertilizers produced in Germany. K + S Kali extracts about 10 million tonnes annually at four sites. The largest consumer countries in the EU are France, Poland, Germany, Great Britain and Spain (Fig. 2.57).

Fig. 2.57 KTG Kali-Transport Gesellschaft exports fertilizers by bulk carriers from Hamburg-Wilhelmsburg (Ralf Witthohn)

2.5.2 Potassium Chloride for New Holland

On 5 December 2005, the rhine-going coaster MARITIME LADY of a Norwegian shipping company is on its way down the Elbe from Hamburg to New Holland/UK with a cargo of 1800 tonnes of potassium chloride. Just as the ship reaches Brunsbüttel, the German container ship ARCTIC OCEAN, on its way from Muuga to Hamburg, leaves the lock of the Kiel canal. The feeder attempts to cross the Elbe fairway, but after an incorrect estimate of the distance and unclear arrangements with the right of way vessel MARITIME LADY, the two ships collide. The bulbous bow of ARCTIC OCEAN tears a 5 m hole into the starboard double hull of the coaster, which capsizes within minutes. The crew is rescued by the Brunsbüttel pilot boats. One hour later, the tanker SUNNY BLOSSOM, leaving the lock, collides with the drifting wreck of MARITIME LADY, suffers severe propeller damage and drifts across the Elbe until it runs aground on its south bank. After a few days, MARITIME LADY is towed keel up to Cuxhaven Amerika-Hafen, where it is turned and pumped empty with help of the Danish floating crane SAMSON. The coaster, built in 1984 by Krupp Ruhrorter shipyard in Duisburg/Germany, is later dismantled at Cuxhaven (Fig. 2.58).

Fig. 2.58 When the coaster MARITIME LADY was brought keel up from Brunsbüttel to Cuxhaven by a special crane catamaran and a floating crane 1 week after collision and capsize, the cargo of potassium chloride had totally been lost into the river Elbe (Ralf Witthohn)

2.5.3 Magnesium from Kymassi

At the entrance to the idyllic and old Cretan port of Souda, the silos of the flour mill tower up into the blue Mediterranean sky. Founded in 1928, the plant can grind 450 tonnes of wheat and produce 500 tonnes of animal feed per day. In January 2017, the Turkish coaster SEABEE discharges 3000 tonnes of rapeseed in the natural harbour. The next port of destination for SEABEE is Kymassi on the east coast of mainland Greece. Here the 3500 dwt vessel, in 1990 built at Dutch Bodewes' shipyard, loads 3.000 tonnes of loose magnesium. The mining company Terna Mag has in 2013 begun to quarry the mineral again after an interruption of 14 years. The cargo is ordered for Germany, where it is to be added to animal feed. At the end of February, the Cook Islands-flagged SEABEE of Istanbul-based Nismar Shipping reaches the Handelshafen of Bremerhaven. A grab dredger transfers the dusty cargo from the cargo hold of the ship directly into trucks (Fig. 2.59).

2.5.4 Russian Calcium Dihydrogen Phosphate for Saint Brieuc

In the port of St. Petersburg, a crane lifts bag by bag into the two cargo holds of the 5500 dwt coastal cargo vessel KETLIN. Each of the more than 5000 white bags

Fig. 2.59 The Turkish-owned cargo vessel SEABEE discharged 3000 tonnes of magnesium from Kymassi in Bremerhaven (Ralf Witthohn)

weighs 1000 kilograms. They are labeled Monocalcium phosphate, the trivial name for calcium dihydrogen phosphate. One of the world's largest artificial fertilizer producers, the Russian company Apatit in Cherepovets, supplied the material to be used as fertilizer, feed or in the food industry, for example as baking powder. The phosphate comes from a deposit in the village of Bikov Otrog in the Saratov region of the Balakovo district north of the Caspian Sea. At the end of February 2018, the Malta-flagged KETLIN leaves St. Petersburg and 5 days later, after transite through the Kiel canal, reaches its berth in the Handelshafen of Bremerhaven. Most of the cargo is discharged by a truck-mounted crane within 3 days, but 1500 bags remain in the holds. They are destined for Saint Brieuc. The managers of the Brittany port are relieved, that the freighter arrives there at a draught of only 3.5 m, and, by this, goes considerably less deep than the ship's maximum of 5.3 m. Initially, a draught of 4.8 m had been announced, which would have hardly allowed access to the small port in northern France usually accepting only up to 4.1 m draughts (Fig. 2.60).

With strongly changing tidal conditions, Saint Brieuc can only accommodate vessels of the KETLIN size 2 weeks a month during the spring tide period. During the fortnight of nipp tides this is not possible. Due to the difficult navigational conditions, the otherwise ideal natural harbour, protected by a long mole, was threatened by closure. In 2017, 350,000 tonnes of cargo were handled, two thirds of which were imported. In addition to fertilisers, animal feed and wheat, the imports handled also include timber from Finland, Sweden and Estonia. Among other cargoes, porcelain earth is exported to Egypt, Sweden and Belgium, while minerals from nearby deposits are exported

Fig. 2.60 One tonne weighing big bags filled with phosphate were transported from Russia to Germany and France by the coaster KETLIN (Ralf Witthohn)

to Norway. The port was also involved in logistical tasks for the laying of an Atlantic fibre-optic cable from the US East Coast to Saint Brieuc and Cornwall, which was completed in 2001. The construction of an offshore wind farm with 62 Siemens 8 MW wind turbines by Ailes Marines in a bay near the port, shall yet be carried out from Saint-Quay-Portrieux as base port due to the tidal dependence at Saint Brieuc. The construction of another seaward quay there has been delayed since 2016.

As one of the largest vessels that can reach Saint Brieuc, KETLIN is operated from Tallinn by the Estonian branch of the Hamburg shipping company Hansa Shipping. The 107 m length, 15.3 m breadth hull of the ship was built by Marine Projects in Gdansk, which supplied it to the main contractor Bodewes Shipyards in Hoogezand/Netherland. In total, the Polish shipyard produced 22 such hulls, primarily for Dutch builders. KETLIN, delivered in 2006 under the name EMSRUNNER, sailed from 2014 to 2016 as ELISE before being renamed EMSRUNNER again by 2018. Hansa Shipping took over the ship in January 2018 together with the sister ship KLARIKA ex EMSCARRIER. KERSTI ex BOTHNIADIEP of the same type had already been integrated into the fleet in September 2015. In March 2018, the shipping company listed a fleet of 29 coasters, end of 2019 of 31 vessels, and in early 2022 33 units. After shipowners retracted suitable tonnage for the export of fertilizer, Russia in April 2022 announced temporary restrictions for its export. The sanctions of the EU, on the other hand, included certain types of fertilizer, but only its export to EU countries (Fig. 2.61).

Fig. 2.61 Nearly 1 year after KETLIN, in January 2020, another small coaster, the 4210 dwt KAJA, as well operated by Hansa Shipping, carried another cargo of calcium dihydrogen phosphate from St. Petersburg to Bremerhaven (Ralf Witthohn)

2.6 Crude Oil

The COVID-19 pandemic and the related production cuts caused a fall of about 30% in the consumption of oil in the beginning of 2020 and let expect a decrease in transport demands. In fact, between 2019 and 2020 tanker trade decreased by 7.7% from 3.2 billion to 2.9 billion tonnes. Crude oil transport was hit most by a 7.8% drop, as imports declined in most of the important markets, except China with an 8% growth. In 2019, Saudi Arabia had the lead in the seaborne export of oil by reaching a daily average of 6.94 million barrels, followed by Iraque achieving 3.52, Russia 3.43, the US 2.73 and the UAE 2.19 million barrels. Yet, after the break-up of the OPEC and amidst an oil price war between Saudi Arabia and Russia and the prospect, that the market could be flooded with oil, the rates for very large crude carriers (VLCC) surged up to tenfold, from USD 30,000 in the beginning of March 2020 to USD 300,000 in the maximum only a week later. The price run was catalyzed by the use of such tankers as floating depots for the storage of cheap oil. In April, however, Saudi Arabaia, Russia and the OPEC countries agreed to cut oil production by 9.7 million barrels per day in May and June 2020, a rate that was considered to be hardly large enough to fully compensate the drop in demand caused by the pandemic. In February 2022, the rate for very large crude carriers (VLCC) were reported at US21,000, for suezmax carriers at USD15,500 and for aframax ships at USD18,000.

2.6.1 Nigerian Oil for Come by Chance

The 150,000 dwt tanker AIAS passes through Placentia Bay on the south coast of Newfoundland, takes the pilot at Red Island, leaves Long Island on port and anchors southwest of Sound Island in the deepwater of Come By Chance. A fleet of tankers is gathered in the bay. The ballasted 115,000 dwt TOFTEVIKEN leaves the terminal of Come By Chance on the same day to take a waiting position, the likewise empty 105,000 dwt SIGMA INTEGRITY arrives at New Brunswick after an 850 nm voyage from Canaport, the 50,000 dwt ARCTIC BREEZE comes from Carteret near New York and the 110,000 dwt tanker NS CHALLENGER with cargo from Corpus Christi. From the anchorage of the AIAS to the destination berth at Whiffen Head it is still 4.4 nm to go. The tanker has left the Forcados terminal in Nigeria 3 weeks earlier with a cargo of crude oil, which is now to be processed at the nearby refinery of Come By Chance. There is at this time discharged the cargo of the 126,000 dwt tanker NAVION HISPANIA. Under a 15-year charter agreed on in 2015, the tanker provides a shuttle service together with two to three more ships between the oil fields off the Canadian coast and Come By Chance.

Inaugurated in 1998, the ice-free deepwater terminal at Whiffen Head serves as a storage and transfer station for oil, which is produced 325 km east of the Newfoundland coast in the northwest Atlantic. The site has two jetties and six tanks with a capacity of 3.3 million barrels. In October 1998, the first cargo was received from the Hibernia field, in February 2002 from the Terry Nova field and in June 2007 from the White Rose deposit. From November 2017 on, oil from the Hebron field was accepted as well, although it is too heavy to be stored in the existing tanks and has to be directly transhipped at the Newfoundland Transhipment Terminal from three purpose-built Canadian flagged shuttle tankers to exporting tankers. Among the shuttle tankers, which were used in 2018 handling about half of the cargo, were HEATHER KNUTSEN and JASMINE KNUTSEN. The neighbouring Come by Chance refinery, which was commissioned in 1973, also has a deep-water terminal. The refinery with a daily capacity of 2000 m^3 also receives crude oil from deposits in the North Sea, West Africa and the Persian Gulf. Its products are mainly sold to customers on the North American east coast and in Europe.

The carrier of the Nigerian crude oil, the tanker AIAS, was then owned by Greek shipping company Capital Ship Management. Its fleet of six VLCC, five suezmax, two aframax and 27 MR/handysize product tankers included a number of ice-going ships in 2017, i. e. one suezmax, two aframax and 14 Medium Range (MR) product tankers. In 2022, the Greek owner operated 12 VLCC, nine aframax, ten MR and two smaller tankers. AIAS, which was sold to Singapore-based company Navig8 in 2021 and renamed, is one of a long series of newbuildings which keels were laid since the 1990s at Japan's Universal Shipbuilding in Tsu. The 274.2 m length, 48 m breadth tanker was built in 2008 under the name WALTZ together with its sister ship TANGO for Frisia Schiffahrt and was initially operated by the Hartmann shipping company at Leer/Germany (Fig. 2.62).

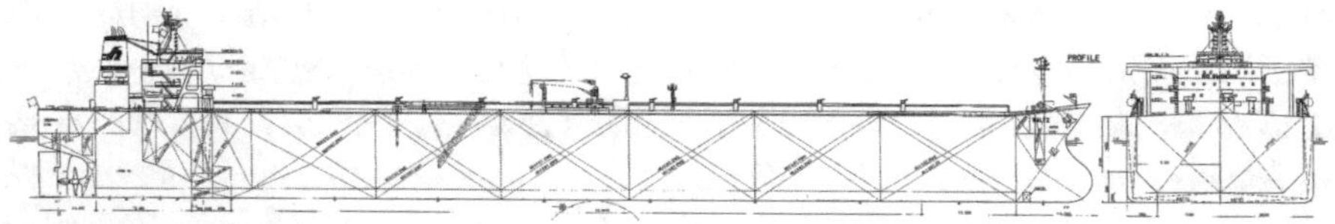

Fig. 2.62 The 150,000 dwt tanker AIAS, originally named WALTZ and from 2021 on RIDGEBURY JUDITH, transported a cargo of crude oil from Nigeria to Canada in early 2018 (Universal SB)

2.6.2 Venezuelan Oil for China

The Venezuelan-Chinese joint venture CV Shipping, a foundation of state-owned PDV Marina from Venezuela and Petrochina International (Singapore), had built four very large crude carriers (VLCC) by China's Bohai Shipbuilding Heavy Industry. Named after battlefields where Peru, Colombia and Venezuela fought for their independence from Spain, AYACUCHO, BOYACA, CARABOBO and JUNIN were delivered between 2013 and 2015. The tankers subsequently transported Venezuelan crude oil from the Puerto José Terminal to the People's Republic of China. The tankers were registered for single-ship companies in Singapore. In May 2020, AYACUCHO was registered on behalf of the Venezuelan government and renamed MAKSIM GORKIY. After insurance for the three other tankers had been withdrawn due to US sanctions, the Venezuelan-Chinese partnership came to an end in August 2021. JUNIN, BOYACA and CARABOBO were transferred to Chinese ownership and renamed THOUSAND SUNNY, XING YE and YONG LE, resp.

At 332 m length, 60 m breadth and 30.5 m depth, the tankers achieve a deadweight of 320,000 tonnes on a draught of 22.6 m. Their 15 cargo tanks and one slop tank have a capacity of 341,000 m^3, two ballast tanks in front of the engine room and the forepeak of 95,000 m^3. The cargo is discharged by three pumps with a capacity of 5500 m^3/h each. Two oil-fired boilers generate 45 tonnes of steam per hour for cleaning the tanks. The main engine of type Wärtsilä 7RT-flex84T-D puts out 33,250 kW at 76 rpm, accelerating the tankers to a speed of 15 knots at a daily consumption of 69 tonnes. High-strength steel accounts for 70% of the total steel weight of the vessels, which were designed with a double hull. Two newbuildings of the type named SVET and SCF SHANGHAI were in 2013/2014 delivered to Liberian companies of Russia's company Sovcomflot, which describes them as the largest ships in the history of Russian merchant shipping.

From the José Terminal, operated by state-owned Venezuelan PDVSA Petróleo y Gas in close proximity to the Mochima National Park, oil of the grades Ameriven-Hamaca, Cerro Negro, Sincor and Zuata Sweet varieties is exported. Ships of up to 320,000 dwt can be loaded at a mooring buoy. The port has four finger piers with eight berths where tankers with a maximum draught of 11.8 m can dock, as well as a berth for transhipment of petroleum coke into ships up to 100,000 dwt (Fig. 2.63).

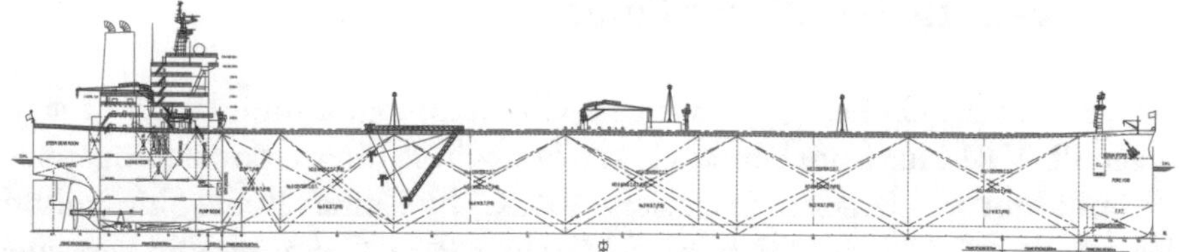

Fig. 2.63 Four 320,000 dwt tankers of the AYACUCHO type were built in China for the import of Venezuelan crude oil (Bohai)

Fig. 2.64 The panamax tanker JILL JACOB, which was in 2017 sold to Greek owners and renamed PRESTIGIOUS, seen discharging Siberian crude oil in Wilhelmshaven (Ralf Witthohn)

2.6.3 Siberian Light for Wilhelmshaven

20 nautical miles west of Helgoland the 15 knots running oil tanker JILL JACOB, coming from Tuapse on the Black Sea, changes course from 90° to 150°. The pilot, who has already boarded by helicopter at the German Bight buoy, advises to sail the ship in a large arc towards the mouth of the Jade. At a speed reduced to 7 knots, the Flensburg tanker is 3 h later positioned two nautical miles north of Minsener Oog. Between Schillig and the island of Mellum, the vessel continues its voyage off the coast of Wangerland for another 2 h southwards. Finally, four tugs push the 73,000 dwt vessel to the southern discharge head of the 1200 m long NWO bridge. Already the following day, 57,000 tonnes of Siberian Light crude oil are discharged into the storage tanks of the terminal. From there, the oil is pumped via pipelines to refineries in western Germany or Hamburg (Fig. 2.64).

Over the past 60 years, the company Nord-West Oelleitung (NWO) in Wilhelmshaven has unloaded more than 1 billion tonnes of crude oil from more than 20,000 tankers. The tanker terminal is comparable to similar facilities in Trieste, Rotterdam or Marseille. The oil is pumped ashore by the tankers' pumps via three discharge heads with transfer capacities of $2 \times 12{,}000\ m^3/h$ and $16{,}000\ m^3$. It is then stored in 35 floating roof tanks with a total capacity of 1.6 million cubic metres. At the two smaller jetties, tankers of up to 130,000 dwt can be berthed, the larger one can accept 260,000 dwt ships of up to 20 m draught, but also larger units, if they are part-loaded. The pump capacities mean, that the berthing time is hardly exceeding 24 h. Caverns in Rüstringen and Friedeburg with a capacity of 15 million cubic metres are connected. From there the oil is transported via a 391 km long pipeline with an annual capacity of 16.3 million tonnes to refineries in Lingen, Gelsenkirchen-Scholven, Gelsenkirchen-Horst and Wesseling. A second 142 km long pipeline can transport 8 million tonnes to Hamburg every year. In 2016, the NWO terminal handled 242 tankers with 18.1 million tonnes corresponding to a supply rate for Germany of almost 20%. Of this amount, 13.5 million tonnes were pumped to West Germany and 4.3 million tonnes to Hamburg. The average quantity imported per ship then was 74,700 tonnes. In 2018, a total of 15.5 million tonnes were discharged from 197 tankers, meaning a reduction from 19 million tonnes in the year before. In 2019, the crude oil handling volumes at Wilhelmshaven reportedly increased, and in 2022, NWO reported a total number of over 18,000 tankers having discharged more than 1 billion tonnes in the terminal's history. Ultra-large tankers with a deadweight of 300,000 tonnes or more reach the port less frequently, but bring, for example, partial loads of the Forties type oil from the British sector of the North Sea for the Shell group. NWO was founded in 1956, with Ruhr Oel, BP Europa, Shell Deutschland and Holborn Europa Raffinerie having purchased stakes, the latter by a 20% stake.

North of the NWO facility, Wilhelmshavener Raffineriegesellschaft, once belonging to ConocoPhilips, operated two jetties. The eastern jetty lies a few hundred metres further in deeper water and is accessible by fire-fighters on a tugboat. The refinery, which was opened by Mobil Oil in 1976, processed low-sulphur North Sea oil and had an annual capacity of 13 million tonnes, from which 170 different oil products were processed, including petrol, diesel oil, propane and butane. The refinery was decommissioned by ConocoPhilipps in 2011 and has since been operated by a local subsidiary of Rotterdam-headquartered HES International as Germany's largest independent liquid bulk terminal, offering 1.3 million cubic metres storage capacity.

2.6.4 Russian Crude Oil for Rotterdam

Oil tankers pass the headland of Skagen almost every hour. The Greek 115,724 dwt crude oil carrier NISSOS SANTORINI of Kyklades Maritime, bound from Primorsk to Bilbao, is followed by the slightly smaller Liberian-flagged 106,500 dwt CE-NIRIIS sailing from Ust-Luga to the oil refinery of Milazzo on the north coast of Sicily and by the 110,077 dwt ATLANTIC EXPLORER en route from Gdansk to Rotterdam. As the tankers circumnavigate the Danish cape on a westerly course, the tankers GIJON KNUTSEN, HILDA KNUTSEN, SIRI KNUTSEN and GAS GALAXY are anchored there, protected behind the peninsula, and await their next assignment. Only 24 h before, ATLANTIC EXPLORER has passed the offshore facilities of the Baltic 1 wind farm off the island of Rügen on port. However, renewable energies have not yet become established enough to replace the export of Russian oil from the Baltic Sea. ATLANTIC EXPLORER arrives 2 days later off the largest Dutch port. The aframax tanker of 244.5 m length and 42 m breadth was built in 2008 by the Mitsui shipyard for Mitsui O.S.K. Lines (Fig. 2.65).

The oil terminal of Primorsk, a small town of a few thousand inhabitants located northwest of St. Petersburg near the Finnish border, is considered the largest in the Baltic Sea. There, a pipeline system operated by Russia's Transneft ends, through which about 65 million tonnes of crude oil have since 2005 been pumped yearly from the Timan-Petschora Basin, Western Siberia, the Urals, the Volga region and Kazakhstan. The terminal on the north coast of the Gulf of Finland replaced the function of the Ventspils oil port, which was lost by Latvia's independence. Primorsk has six jetties with a total length of 1.9 km and a maximum fairway depth of 17.8 m for tankers up to 150,000 dwt. In 2015 the terminal handled 60 million tonnes of crude oil and oil products equivalent to 30% of the Russian exports. The port is operated by Russia's NCSP Group, which also manages the Black Sea port of Novorossiysk

Fig. 2.65 The Japanese 110,000 dwt ATLANTIC EXPLORER transported crude oil from the Baltic Sea to Rotterdam (Ralf Witthohn)

and in 2014 handled a total of 131 million tonnes. As part of its sixth package of sanctions against Russia, the EU, which had imported EUR48 billion worth of crude oil and EUR23 billion of refined oil in 2021, in June 2022 banned all the countrie's seaborne crude oil and petroleum products after phasing out periods of six and eight months, resp., for existing contracts (Fig. 2.64).

2.6.5 North Sea Oil for Brunsbüttel

During night, the Bahamas-flagged shuttle tanker PEARY SPIRIT reaches the Elbe estuary several hours too early on its voyage from Mongstad to Brunsbüttel. The 109,000 dwt ship, which has previously discharged a part load in Gothenburg, completes one more round west of Helgoland before heading for port at a draught of 11 m and a speed of 13 knots. Before reaching Scharhörn reef, PEARY SPIRIT meets its sister ship MATTERHORN SPIRIT, which has just vacated its berth in the Elbe port of Brunsbüttel (Fig. 2.66).

In the vicinity of Bergen, the Mongstad terminal is located sheltered in the Fensfjord. The terminal receives all the oil produced from the North Sea fields Troll B, Troll C, Kvittebjørn and Fram and transported to the coast in pipelines. In addition, a substantial part of the oil from the Heidrun field is shipped to Mongstad. From the terminal operated by Statoil, the temporarily stored crude oil is exported to North America, Europe and Asia. Two jetties are available for this purpose, on which tankers with a carrying capacity of up to 380,000 tonnes can be moored. There is also undertaken direct

Fig. 2.66 After the import of North Sea oil for German refineries the shuttle tanker MATTERHORN SPIRIT is seen departing Brunsbüttel bound for the loading port Mongstadfjord (Ralf Witthohn)

ship-to-ship transfer (STS operation) by mooring the tankers next to each other at a jetty able to berth 440,000 dwt tankers.

One of the recipients, the Elbe port of Brunsbüttel, supplies the Shell refinery in Heide with crude oil via seven pipelines. In addition to domestic, relatively heavy oil from the Mittelplate and Dieksand locations off the near-by coast, the refinery there receives 60% of its annual processing of 4 million tonnes from abroad. Every year, some 2.4 million tons are transshipped in tankers of up to 80,000 dwt, mainly from the North Sea fields and Russia. About half of the refinery products, including benzene, tuluene and xylene, are again shipped by sea or inland waterway vessels via the Brunsbüttel oil port on the Kiel canal. At a capacity of 550,000 m^3, this port equally serves as storage and handling site. The tide-independent port has five loading stations for ships of up to 235 m length and 10.4 m draught. 1000 m^3 of crude oil can be transshipped per hour, a maximum of 300 m^3 of oil products. A number of German tanker shipping companies are involved in the distribution of finished and semi-finished products, including Essberger, Büttner, German Tankers, Gefo, Peters, Glüsing and Dettmer.

2.6.6 Crude Oil from the Shetlands

It takes only 2 days to discharge the 100,000 dwt crude oil tanker, which has arrived from the Sullom Voe terminal on the Shetland Islands in the Kattwyk port of Wilhelmsburg. There, US-based Zenith Energy operates the former terminal of Royal Dutch Shell, which is still a major customer. To the east, on the Grasbrook, Shell produces lubricating oils from base oils in two oil blending plants. The oil is, among other places, imported from Qatar, the United States, South Korea and the Netherlands. It is received by either inland vessels or tankers of up to 6000 dwt at a terminal located at the Reiherstieg. Shell also buys the raw material from the neighbouring plant of the Nynas group. The lubricating oil produced is transported by up to 60 trucks a day or deposited in an external storage facility on the Süderelbe river.

At Harburg, the Nynas Group produces bitumen and naphthenic speciality oils. Nynas had taken over the Shell site in 2014 and 2 years later the Shell refinery on the Hohe Schaar, which it converted for an annual capacity of 3.3 million tonnes. Nynas then was minority-owned by Neste Oil of Finland and PDVSA of Venezuela. PDVSA sold 35% of its majority share to a Swedish foundation, thus reducing its stake to 15% in 2020. Before, Nynas had been hit by US sanctions. Neste, on its turn, subsequently sold its 49.99% share to Dubai's Bitumina Industries.

Zenith Energy had in 2018 extended its terminal net in Amsterdam, Bantry/Ireland and Palermo/Colombia by purchasing Shell's tank farm in Harburg. The last producer of fuels in Hamburg thus is the Holborn Europa Raffinerie located in Harburg and part of Oilinvest (Netherlands), owned by Libya's state fund, the Libyan Investment Authority. The refinery is supplied with crude oil via pipeline from Wilhelmshaven, of which up to 5 million tonnes are processed annually. In 2022, Holborn Europa held 100% of Norddeutsche Ölleitung (NDO) and 20% of Nord-West Oelleitung (Fig. 2.67).

2.6.7 Oil Across the Caspian Sea

While rail ferry transports are declining worldwide due to the construction of new bridge connections and due to a growing share of truck traffic, an unusual form of oil transport on special rail ferries is still carried out across the Caspian Sea. Due to the shallow water depths and strong water level fluctuations of the inland sea, special demands are placed on the design of the ro ro ferries. They were built at Pula/Croatia by the Uljanik shipyard. The carriers primarily load tank wagons transporting crude oil, oil products and propane gas produced in the region, but can also carry other liquids, gases, dry goods and hazardous cargo as well as trucks, cars and other vehicles.

Back in the 1960s, the Soviet shipyard Krasnoje Sormowo at Gorki (Nizhny Novgorod) had already delivered five ships of the smaller SOVIETSKIY AZERBAIDZHAN type of 2520 dwt, to the Soviet Union. A first series of eight 3960 dwt ferries of the SOVIETSKIY DAGESTAN design was built at Uljanik shipyard/Croata between 1984 and 1986. From 2005 onwards, the 5985 dwt MAKHACHKALA-1 to MAKHACHKALA-4, later renamed SHAHDAG, KARABAKH, AGDAM and AKADEMIK ZARIFA ALIYEVA,

Fig. 2.67 Until demolition in 2017, the Finnish 100,000 dwt tankers ALFA BRITANNIA and ALFA GERMANIA supplied Hamburg refineries with crude oil (Ralf Witthohn)

were delivered to the Russian Federation. Two more newbuildings of 5398 dwt were put into service in 2012 under the names BARDA and BALAKEN for operation between Baku, Kurik, Turkmenbashi and Alat by Azerbaijan Caspian Shipping. In contrast to the Soviet single-deck constructions, the Croatian-built ships were designed to accommodate railway wagons on two levels. This increased the capacity from 28 to 52 wagons, but also made it necessary to equip them with a sophisticated wagon lift supplied by the Finnish company MacGregor. The passenger capacity was reduced from 290 to 12 passengers. Through a stern door the rail wagons roll on the main deck on two tracks of 1520 mm Russian gauge in the rear and four tracks in the front area. Switches enable the wagons to be moved to the outer tracks. The equipment includes two wagon pushers, which develop a force of 680 tonnes. An electro-hydraulically operated lift lowers two wagons at a time onto the double-track lower deck. In order to compensate the high weights during handling, the ships are equipped with an antiheeling system which pump has a capacity of 1000 m³/h. Due to the small maximum draught of 4.6 m, the ships' design is characterised by a low silhouette. Driven by two main engines of 4000 kW total output the twin-propelled rail transporters achieve a speed of 14 knots (Fig. 2.68).

Two more wagon carriers from the Uljanik shipyard group were in 2016 contracted by state-owned railway company Kazakhstan Temir Zholy (KTZ) from Astana, named Nur-Sultana since 2019. At 154.8 m length and 17.5 m breadth, the newbuildings largely correspond to the MAKHACHKALA type,

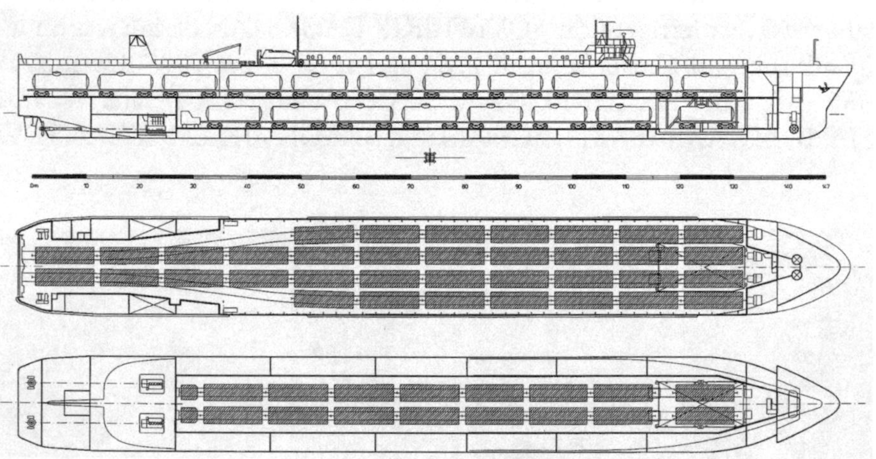

Fig. 2.68 Four tracks on the upper deck and two on the lower deck, connected by a wagon lift forward, create a capacity of 52 tank wagons on the MAKHACHKALA type (Uljanik)

yet with a forward-shifted bridge. The capacity was increased once again to 54 wagons. Their deadweight capacity of 5250 tonnes and the main engine power of 4080 kW for a speed of 14.5 knots hardly deviate from the forerunners. The delivery of the first unit was scheduled in 2017, but until early 2020 no more news had emerged about the contract. The railway company is planning invested in a ferry terminal in Kurik with an annual capacity of 4 million tonnes. Together with partners in Georgia, Turkey and China, the trans-Caspian route is to be expanded as part of the New Silk Road project, which is yet requiring axle changes due to different gauge widths.

2.7 Gas

Sea transport of methane is only economically viable by liquefaction to LNG, the abbreviation for liquefied natural gas. LNG accounts for about 10% of gas supplies, which in turn covers about a quarter of global energy needs. Unlike natural gas distributed via pipelines in its natural state, LNG is being liquefied by cooling to −163 °C when to be transported by tankers. In 2018, 318 million tonnes of LNG, equivalent to an increase of 8.9 per cent year-on-year, were shipped. In 2020, gas transports amounted to 480 million tonnes, equivalent to a 0.4% growth year-on-year.

The use of LNG as marine fuel has been discussed for a couple of years, but so far only materialized in a few cases on mostly smaller ships, like the 2015-built Norwegian KVITBJØRN and KVITNOS (see 1.8.3 Fish transport from Iceland, the Faroes and Norway in containers), the German coastal passenger ships HELGOLAND (10.5.2) and OSTFRIESLAND or the container feeder WES AMELIE (3.1.4). At costs of USD35 million the first large container vessel was converted in 2020 at a Shanghai shipyard to use LNG as fuel, the 15,000 teu capacity carrier SAJIR of Germany's operator Hapag-Lloyd. After being renamed BRUSSELS EXPRESS, the ship was to take LNG bunker two times per round trip in Singapore and Rotterdam (Fig. 2.69).

Though the burning of lng is suited to reduce the emission of toxic gases like those with sulphur contents from using diesel oil, the release of climate damaging CO2 is decreased at only low rates. A study of the Intergovernmental Panel on Climate Change (PCC) published in the beginning of 2020 pointed to greater anthropogenic contribution from incomplete combustion of LNG and leakage towards atmospheric methane levels, which would have had so far been underestimated by 25% to 40%. Before CH4 degrades to CO2, methane would increase the greenhouse effect considerably.

Fig. 2.69 After being fitted with an LNG fuel system, Hapag-Lloyd's container vessel SAJIR was renamed BRUSSELS EXPRESS (Arne Münster)

Like LNG, Liquefied Petrol Gas (LPG) is shipped in a liquid state in gas tankers. Being produced in industrial processes, LPG shipments, in addition to propane and butane, often refer to ethylene, which becomes liquid at -104 °C. Other cargo types are ammonia and VCM. The shipment of gas requires high safety standards. At LNG terminals, for example, the use of mobile phones is prohibited. Communication between the ship and the loading station is carried out there via fibre optic link.

2.7.1 LNG

While overall gas exports increased only slightly by 0.4% in 2020, LNG trade was reported to have grown by 1.1%, although gas shipment projects in the United States and in Australia had to be postponed due to weak prices. In February 2020 the COVID-19 pandemic started to affect the volumes of LNG shipments. Australian cargoes for China had already decreased to 26 deliveries from 40 in the month before. Particularly a reduction of China's gas consumption was estimated to cut shipments to the country by up to over 6 million tonnes, this being enforced by low oil prices. At the same time, yet, particularly Asian demands let recover US LNG exports by 15%. LNG is shipped from about 20 countries. Until 2015, Middle East countries were leading the way, but they were replaced by the Asia-Pacific region, partly because Yemen lost its supplier state as a result of the on-lasting war there. In 2018, Qatar held 25% of the world market share. Australia accounted for 22%, the United States, Malaysia and Nigeria for 7%, Russia for 6% and Indonesia for 5%.

In late 2019, Vietnam announced the construction of its first LNG import terminal with shipments planned from 2022 on. In Australia are followed at least five LNG import projects. By a total of 153 million tonnes, Asia-Pacific countries with Japan and Korea leading had by 48% the largest share of imports in 2018, despite a declining market there over a 5 year period. Next was the Asia region (China, India, Pakistan and Bangladesh) with an aggregated amount of 86 million tonnes. LNG is purchased by more than 35 countries, including the newer customers Egypt, Pakistan, Jordan and Poland, where the Swinoujscie LNG terminal handles imports since 2016. More than 400 special tankers are available for this purpose. In 2019, 43 LNG tanker newbuildings were commissioned.

New transport requirements arose from the extraction of ethane by the fracturing of shale rock, in particular through US exports of this kind of gas, also known as shale gas, to Europe and Asia. For this purpose, special ethane tankers with trilobe pressure tanks cooled down to −89 °C were for the first time delivered by Nantong CIMC Sinopacific Offshore & Engineering Co. from 2015 to 2017. Designed for the transport of either LNG or LPG the Danish-flagged tankers JS INEOS INSIGHT, JS INEOS INGENUITY, JS INEOS INTREPID, JS INEOS INSPIRATION, JS INEOS INNOVATION, JS INEOS INVENTION and JS INEOS INTUITION are operated by Copenhagen-based company Evergas and chartered out to Ineos, a group of privately owned, London-headquartered chemistry companies. The 180.2 m length, 26.5 m breadth combi ships cool down their butane, ethylene, LPG, vinyl-chloride or chemicals cargoes in three tanks of 35,000 m^3 to −104 °C minimum. The 6488 kW dual-fuel ships are propelled by either diesel engine or gas turbine to a speed of 16 knots.

In March 2016, JS INEOS INTREPID carried a first load of shale gas from the US to Europe, by loading at Marcus Hook near Philadelphia and discharging at Rafnes/Norway, to compensate reduced gas volumes from North Sea sources. In the beginning of 2020, INEOS INSIGHT undertook a voyage from the US port Marcus Hook to Frierfjorden/Norway, JS INEOS INGENUITY was on its way back from Rafnes/Norway to Marcus Hook as was JS INEOS INTREPID. JS INEOS INSPIRATION had left Marcus Hook for Grangemouth, JS INEOS INVENTION sailed on the opposite course, JS INEOS INNOVATION crossed the Pacific from Taixing to Panama, while JS INEOS INTUITION was bound for Houston from Grangemouth.

Another gas tanker operator, Japan's Mitsui OSK Lines, in 2016/17 commissioned the Samsung-built 85,000 m^3 capacity ETHANE CRYSTAL fitted with membrane tanks and designated Very Large Ethane Carrier (VLEC). The 227.9 m length, 36.5 m breadth, 22.4 m depth and 12.3 m draught ship is laid out for a minimum cargo temperature of -94 °C and a minimum temperature of the secondary barrier material of -163 °C, according to the ship's classifier American Bureau of Shipping. In the beginning of 2020, the Marshall Islands-flagged ETHANE CRYSTAL was on a voyage from Dahej/India to Houston in ballast. Sister vessels are ETHANE EMERALD, ETHANE OPAL, ETHANE PEARL, ETHANE SAPPHIRE and ETHANE TOPAZ delivered from the Korean builder.

Following the commissioning of the first LNG-powered cargo ships, the need arose to operate bunkering tankers for their supply. In February 2017, a consortium of Mitsubishi, NYK, the French energy group ENGIE and the Belgian natural gas supplier Fluxys put into service the 5000 m^3 tanker ENGIE ZEEBRUGGE. The bunkering tanker, built by Korea's Hanjin Group and operated by London-based NYK LNG Shipmanagement, is itself powered by gas oil, marine oil or LNG. In June 2017, ENGIE ZEEBRUGGE for the first time supplied the car carriers AUTO ECO and AUTO ENERGY with LNG fuel in Zeebrugge/Belgium, a traditional port for handling LNG and cars. The ship was renamed GREEN ZEEBRUGGE in 2020 (Fig. 2.70).

Subsidised by the European Union as part of the Blue Baltics project was the lng bunkering tanker KAIROS, delivered by Korea's Hyundai Mipo shipyard to Uranos Vermögensverwaltungsges. On behalf of Gasum AS in Tananger/Norway in 2018 and managed by Bernhard Schulte Shipmanagement. In April 2019, the 117 m length, 20 m breadth ship supplied LNG to the container ship WES AMELIE, later several times to the Meyer Werft cruise liner newbuildings IONA and AIDACOSMA during their fitting out and trial runs stage at Bremerhaven. Smaller LNG-burning coastal ships are usually supplied by trucks (Fig. 2.71).

Gas tankers are offered the unique propulsion solution of using their own cargo as fuel. Although in principle this is an old idea, the realisation of which was attempted on LPG tankers as early as the 1970s, the prerequisites for its technologically reliable implementation were only created a few years ago. Copenhagen-based engine manufacturer MAN B&W designed a drive system that can be operated with both heavy fuel oil and more environmentally friendly gas. By gas injection of the so-called boil-off rate, the vaporizing

Fig. 2.70 Operating from the LNG port of Zeebrugge/Belgium, GREEN ZEEBRUGGE ex ENGIE ZEEBRUGGE was one of the first tankers supplying LNG bunker (Ralf Witthohn)

portion of the liquid cargo is used for propulsion. This requires the installation of compressors. In double-walled high-pressure pipes, the gas is fed to the engine's combustion chambers via special gas valves. The first LNG tankers to be powered in this way were nine newbuildings of the 174,000 m^3 type CREOLE SPIRIT for Teekay shipping company. They were chartered out to the US company Cheniere after delivery by Daewoo shipyard from February 2016 on. The vessel was employed to export LNG from Sabine Pass in Louisiana.

2.7.1.1 Brunei Gas for Japan and Korea

After the Fukushima catastrophe and the shutdown of the country's nuclear power plants, Japan became even more dependent on natural gas supplies to cover its energy needs. In addition to deposits in the Middle East, those in Southeast Asia also play an important role. Brunei covers about 15% of Japan's requirements, figured at 4.3 million tonnes in 2018. The 130-hectare Brunei LNG terminal in the port of Lumut on the northwest coast of Borneo handles

Fig. 2.71 LNG-driven cruise liner newbuildings of Meyer Werft were supplied by the bunkering tanker KAIROS (Ralf Witthohn)

methane from the South-West Ampa, Fairley, Gannet and Egret gas fields and from the Total fields Jamalul Alam and Maharaja Lela. The government of Brunei Darussalam has a 50% stake, while Shell Overseas and Diamond Tenaga Investment have 25% shares each in the terminal. The customers in Japan are Osaka Gas, Tokyo Gas and the energy company Jera in Tokyo, in South Korea Kogas and in Malaysia Petronas. In Lumut there is a large liquefaction plant including three storage tanks with a total capacity of 195,000 m^3, the annual turnover reaches 6.7 million tonnes. A pipeline leads into the open sea on a 4.5 km long jetty enabling the berthing of LNG tankers from 75,000 m^3 to 155,000 m^3 with a maximum draught of 11.6 m. The 75,000 m^3 to 78,000 m^3 tankers BEBATIK, BUBUK, BILIS and BELANAK of Brunei Shell Tankers, built in France in 1972 and 1975, were regularly involved in LNG exports from Lumut. BILIS and BUBUK were scrapped in China in 2014/15, followed by BEBATIK and BELANAK in 2018 (Fig. 2.72).

The latest LNG newbuildings in the fleet of Brunei Gas Carriers are the 147,000 m^3 to 155,000 m^3 capacity ships ARKAT, AMALI, AMANI and AMADI, built between 2011 and 2015 by two South Korean shipyards, Daewoo Shipbuilding & Engineering (DSM) in Okpo and Hyundai Heavy

Fig. 2.72 For more than 40 years the tankers BEBATIK, BUBUK (photo), BILIS and BELANAK were engaged in LNG transport, most recently from the port of Lumut in Brunei (Steffen Urbschat)

Fig. 2.73 The twin propulsion system of the Korean-built LNG carrier ARKAT facilitates the berthing manoeuvre supported by two to three tugs at the jetty of Lumut/Brunei, which is exposed to the open sea (Steffen Urbschat)

Industries in Ulsan. The operator of the tankers is a joint venture of the government of Brunei, Shell Gas and Diamond Gas Carriers founded in 1998 in Brunei Darussalam. The 284.2 m length, 43.4 m breadth ARKAT has four membrane tanks. The ship is driven by a diesel-electric dual fuel twin-engine system. Two ABB pods with a total output of 23,340 kW from ABB enable a speed of 19.5 knots, equally ensuring high manoeuvrability. The 137,000 m^3 ABADI was built in 2002 by Mitsubishi Heavy Industries in Nagasaki, Japan. At 290 m length and 46 m breadth the tanker is equipped with five spherical tanks manufactured in accordance with the Moss-Rosenberg patent (Fig. 2.73).

In 2020, Japan's Advanced Hydrogen Energy Chain Association for Technology Development (AHEAD) launched a piloting project, in which hydrogen from the Brunei methane gas plant is, by using an organic chemical hydride method, utilized to produce methylcyclohexane (MCH) by hydrogenation from toluene (CH3). The MCH is transported in tankers at an ambient temperature and pressure to Japan, where in a dehydrogenation process hydrogen is extracted from the MCH and used in a thermal power plant to produce energy by driving gas turbines. The toluene is then shipped back to Brunei and processed there again.

2.7.1.2 Siberian Gas from Sabetta

In March 2017, a large LNG tanker reaches the port of Sabetta on the Yamal peninsula for the first time. The Sovcomflot-operated CHRISTOPHE DE MARGERIE only carries out a test run, methane gas from the terminal is not loaded on the 172,600 m^3 capacity newbuilding. As one of 15 carriers from Korea's Daewoo shipyard for employment in the Yamal project, the ship can sail through 2.1 m thick ice. Participants in the project, which started shipments in December 2017, are the Russian company Novatek, the French Total group, China's CNPC and the Silk Road Fund. A sailing time of 18 days is assumed for exports to China via the Northern sea route, 14 days less than the way via the Suez canal. In July 2017, CHRISTOPHE DE MARGERIE transported a first LNG cargo in 19 days from Melkøya/Norway to Boryeong in South Korea via the Northeast Passage. In early 2022, the vessel supplied LNG cargoes from Sabetta to Milford Haven, Zeebrugge and Montoir (Fig. 2.74).

The starting signal for one of the largest and most difficult port construction projects in history was given in July 2012. EUR 15 billion were available to build the port of Sabetta in the north-east of the Yamal peninsula. For this purpose, a 49 km long, 295 m wide and 14 m deep channel had to be dredged. In the first construction phase between 2 August and 9 October 2013, about 10 million cubic metres of a total of 70 million cubic metres were excavated. The work could only be carried out during the 12-week ice-free period. The Belgian shipping company de Nul deployed an armada of 19 ships, including the suction dredger AMERIGO VESPUCCI and the hopper barge ASTROLABE. Since its completion in 2017, Sabetta can accommodate LNG tankers up to the 170,000 m^3 class (Fig. 2.75).

Fig. 2.74 In March 2017, the ice-breaking newbuilding CHRISTOPHE DE MARGERIE was the first large LNG carrier in the port of Sabetta (Sovcomflot)

Fig. 2.75 The Belgian dredger AMERIGO VESPUCCI and the hopper barge ASTROLABE carried out work on the fairway to the Siberian port of Sabetta (Ralf Witthohn)

2.7.1.3 Methane for Yokohama

The methane cargo taken over by the tanker MUBARAZ at the offshore terminal Das Island 160 km off the Abu Dhabi coast is being discharged 2 weeks later at the Futtsu power station of the Tokyo Electric Power Co. (TEPCO) near Yokohama. By its capacity of 5 GW, the world's second largest gas-fired power plant helps to meet the increased demand for imported energy following the shutdown of Japan's nuclear power plants. The cargo of 135,000 m³ LNG transported by the MUBARAZ is sufficient to supply the power plant for 1 week.

MUBARAZ was the first of four tankers delivered in 1996/1997 by Kvaerner Masa shipyard in Turku to Abu Dhabi National Oil Company (ADNOC), followed by the sister ships MRAWEH, AL HAMRA and UMM AL ASHTAN. Fulfilling a long-term supply contract between Tokyo Electricity Production Co. (TEPCO) and the national Abu Dhabi Gas Liquefication Co. (ADGAS), the 290 m length ships are equipped with the Moss-Rosenberg and Kvaerner-Moss tank systems. Developed in Norway this system can be easily identified by its visible spherical halves made of aluminium alloy. The tankers were the first of their size to be fitted with only four instead of the usual five tanks, in order to simplify cargo handling, which is carried out by eight pumps of an hourly output of 11,000 tonnes. The MUBARAZ design approach resulted in a larger ship width of 48.2 m compared to four AL KHAZNAH class buildings equipped with five tanks and delivered to the same owner by Japan's Mitsui shipyard simultaneously. At almost the same length of 293 m, those are only 45.8 m wide. The larger tank diameter also resulted in a larger depth of 27 m for the Finnish-built ships against 25.5 m, while an identical draught of 11.3 m resulted in an only slightly larger deadweight of 72,590 tonnes for the MUBARAZ compared to 71,540 tonnes of the AL KHAZNAH type. Identical is also the ship's speed of 19.5 knots achieved by a steam turbine system. Shortly after the contract was awarded in 1993, the state shipping company started a training programme, which 250 cadets had completed when, in July 2011, one of the graduates took over the command of the MUBARAZ from a European captain for the first time (Fig. 2.76).

2.7.1.4 New Containment Technology on SAGA DAWN

In December 2019, the first LNG carrier fitted with a newly developed cargo containment system named LNT A-Box was delivered to Saga LNG Shipping by China Merchants Heavy Industries (CMHI) at its Jiangsu shipyard in Haimen. The 45,000 m³ SAGA DAWN had completed 3 weeks of gas trials in June. Its design was carried out by Singapore-headquartered LNT A-Box and FKAB from Gothenburg. The cargo tanks of type IMO independent prismatic type A are, as primary containment, held by laminated wooden supports, while a liquid-tight thermal insulation is attached to the hull compartment as an independent secondary barrier. The cargo tanks have a volume of

Fig. 2.76 The LNG tanker MUBARAZ, delivered to Abu Dhabi by Kvaerner Masa in 1996, was the first large LNG tanker to have only four instead of a number of five spherical tanks as was common at that time (Ralf Witthohn)

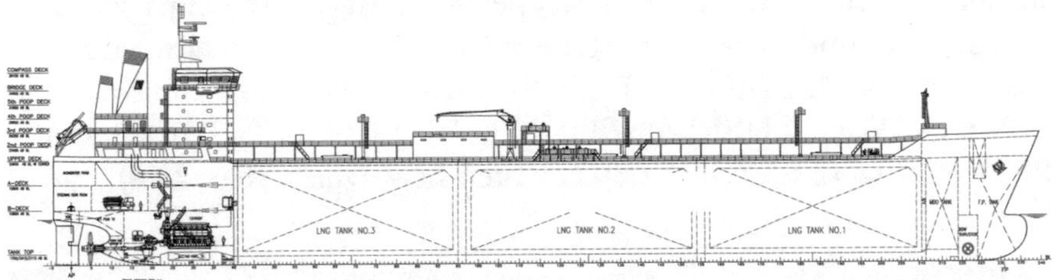

Fig. 2.77 The 2019-built SAGA DAWN features a new LNG containment system (FKAB)

about 45,000 m^3. At 193.4 m length, 30 m breadth and 20 m depth, SAGA DAWN, which had been launched in December 2017 and was delivered in November 2019, achieves a deadweight of 31,712 tonnes. In 2021, Singapore-based Farenco Shipping took over the management of the vessel, which was subsequently renamed LNG JIA XING (Fig. 2.77).

2.7.2 LPG

Being pushed by strong demands in India and Europe, the worldwide shipments of Liquefied Petrol Gas (LPG) experienced a 7.8% growth in 2018, up from 2.2% in 2017. While despite decades of planning the construction of a first German LNG terminal in Wilhelmshaven did not materialize and was only revived, together with plants in Stade and Brunsbüttel, in the wake of the energy crisis caused by the Russian attack on Ukraine, there are two jetties for LPG ships of up to 137 m length in use at the northernmost point of the tanker piers on the river Jade. From time to time, smaller tankers carry liquefied petroleum gas to Wilhelmshaven, where it is processed by the Vynova plant on Voslapper Groden. The company, which belongs to the International Chemical Investors Group (ICIG), has a production site there for the manufacture of vinyl chloride and polyvinyl chloride (PVC) with an annual capacity of 380,000 tonnes. The required substances ethylene dicloride and ethylene are delivered by ship and stored in two large tanks.

Due to the lack of profitibality of their ethylene carrier operations, Copenhagen/Denmark-based Evergas announced in January 2019, that it had sold eight smaller, 12,000 m^3 capacity ethylene carriers, named JS GREENSTAR, JS GREENSKY, JS GREENSEA, JS GREENSUN, JS GREENSAIL, JS GREENSPEED, JS GREENSTONE and JS GREENSAND, to the Unigas consortia formed by the German companies Bernhard Schulte and Sloman Neptun with the Danish operator Ultragas. The ships, which had been employed under a pool arrangement with Piraeus-headquartered Eletson, were renamed DELTAGAS, THALEA SCHULTE, HAPPY KESTREL, HAPPY OSPREY, THEKLA SCHULTE, THERESA SCHULTE. HAPPY PEREGRINE and ETAGAS, resp., by their new owners (Fig. 2.78).

Fig. 2.78 Unigas consortia members Sloman Neptun, Bernhard Schulte and Ultragas took over eight Evergas ethylene carriers, including the former JS GREENSAND, which was renamed ETAGAS by Sloman Neptun shortly after arrival in Bremerhaven in July 2019 (Ralf Witthohn)

2.7.2.1 LPG Tankers from Turnu-Severin

From 1989 to 1992, Brand Werft in Oldenburg, which had already been active in gas tanker construction since 1960, delivered a series of five 8250 m^3 LPG tankers, each costing 60 million Deutsche Mark. First ship was TEVIOT for Geo Gibson in Edinburgh for employment in the Unigas consortium, succeeded by NORGAS CHRISTIAN delivered to Chemikalien Seetransport Hamburg. The latter represented a version shortened from 132.2 m to 126.2 m by alteration of the stern and bow configuration and the reduction of the cargo space. The sister vessel NORGAS PATRICIA was handed over to Oslo-based company Skaugen, IGLOO BERGEN to Harpain Shipping and IGLOO TANA to ship investment company Norddeutsche Vermögen, both in Hamburg.

The ships were equipped with three insulated bilobe tanks supplied by Meyer Werft, the gas system was manufactured by LGA Gastechnik from Remagen. The compressors were designed to keep ethylene at −110 °C and 5 bar pressure as well as other gases such as propane, butane, ammonia and VCM in liquid state. But also the transport of chemicals with a maximum weight of 0.972 tonnes/m^3 was allowed. TEVIOT was the only ship to be equipped with a two-stroke Sulzer engine, the four following ships with MAN B&W four-stroke engines, which ensured smoother operation. The propeller of the TEVIOT detached from the shaft on a Mediterranean voyage, which caused a claim by the shipowner against the shipyard. It was recognized by a British court and resulted in insolvency proceedings, at last causing the end of the shipbuilding activities at Oldenburg in 1997. TEVIOT was renamed NORGAS TEVIOT in 1996 and NORGAS CARINE in 1998 and in 2018 broken up along with NORGAS PATRICIA, IGLOO BERGEN and IGLOO TANE, NORGAS CHRISTIAN 2 years later (Figs. 2.79 and 2.80).

The gas carrier type experienced a renaissance, when Hartmann Reederei from Leer ordered an 8500 m^3 newbuilding based on the same design from Romania's Severnav shipyard in Drobeta-Turnu Severin. GASCHEM BALTIC was delivered in 2004 as the largest ship built on the upper Danube until then. Its fitting out proved to be complicated. Due to draught restrictions on the Danube, the gas tanks could only be equipped in a dock of Santierul Naval in Constanta. For this purpose, the entire main deck with deckhouse and equipment, which had already been set up at the shipyard, had to be removed again in Constanta. Due to a low water level in summer 2004, the transfer voyage from Turnu-Severin to Constanta could only be carried out with a delay of several months. Further newbuildings of this type were

Fig. 2.79 TEVIOT was the first of eight ethylene tankers built in Oldenburg and Turnu Severin (Ralf Witthohn)

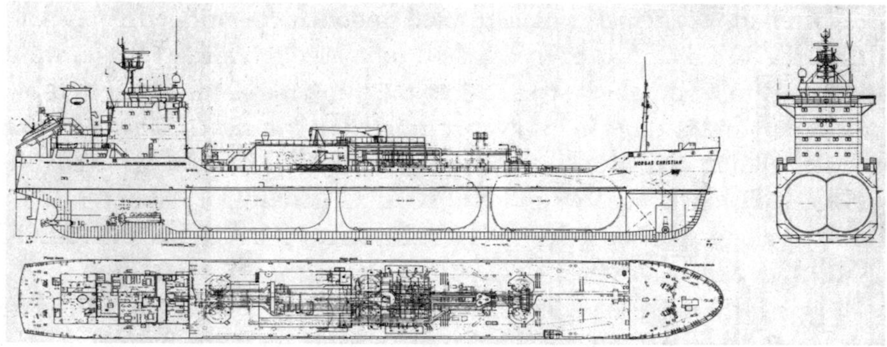

Fig. 2.80 The construction of LPG tankers of the NORGAS CHRISTIAN type in Oldenburg was continued more than 10 years later in Turnu-Severin (Ralf Witthohn)

delivered as GASCHEM ATLANTIC and GASCHEM ARCTIC in 2009 and 2010 (Fig. 2.81).

The ships have a double bottom, no side tanks, but wing and hopper tanks, which thanks to their triangular geometry surround the cargo tanks, so that the void space is kept small in favour of the ship's leak stability conditions. Because the criterion for the loading capacity of gas tankers is the cargo volume instead of the usually considered cargo weight, the design dispensed with

Fig. 2.81 The ethylene tanker GASCHEM BALTIC, built in 2004 on the upper reaches of the Danube, is based on a type designed 15 years earlier in Oldenburg (Ralf Witthohn)

a bulbous bow that would have increased the deadweight capacity. Instead, the design relied on slim lines, which were enabled by converging the 6.5 m apart symmetry axes of the two forward cylindrical cargo tanks of 9.5 m diameter. This forward double tank has a capacity of 2380 m^3, the second and third of 3020 m^3 each, the deck tank for preparatory gassing of 100 m^3. The reliquefaction plant located in the deckhouse comprises five compressors, two of which are screw compressors.

In 1998/1999 the Romanian shipyard had already built three 4360 m^3 tankers of the DANUBEGAS class, and in 2001 the 3500 m^3 ships SAARGAS for Hartmann Reederei. The owner took over further newbuildngs from Santierul Naval in 2007/2008 by commissioning the 6500 m^3 capacity GASCHEM SHINANO, GASCHEM MOSEL, GASCHEM LEDA and GASCHEM RHONE. The use of the former Brand design for the GASCHEM BALTIC type resulted in further design cooperations for the type of a 35,800 m^3 tanker of 188.3 m length, 29 m breadth and 17.5 m depth, of which GASCHEM BELUGA, GASCHEM ORCA and GASCHEM NARWHAL entered service in between 2016 and 2019. The tankers, built by Yangzhou Dayang shipyard and operated by Hartmann Gas Carriers, have three trilobe tanks for the carriage of LPG, CNG (compressed natural gas),

LNG or liquefied CO2. They are powered by the first MAN 7G50ME-GIE (Gas Injection Ethane) two-stroke engine, which burns ethylene, thus enabling them to take advantage of the boil-off rate of the own cargo, as well as HFO, MDO or gas oil. The LEG tankers are on long-term charter to the Saudi company SABIC Petrochemicals and are to transport ethylene between Houston and SABIC's Teeside production facility. The GASCHEM BELUGA class was in 2020 succeeded by the ethylene carrier GASCHEM DOLLART from Hyundai Mipo with a cargo capacity of 22,000 m^3 and in December 2021 the pressurized LPG tanker GASCHEM HOMER started a trio of three 5000 m^3 LPG tankers from Chinese builder Nantong CIMC Sinopacific Offshore & Engineering, which are able to burn either MGO or LPG in their MAN main engine.

2.7.2.2 Ethylene from Le Havre

Within 18 h after arrival from Tees, the gas tanker GASCHEM ISAR of Leer-based shipping company Hartmann discharges 2300 tonnes of ethylene in the port of Bützfleth near Stade/Germany. The next morning, the gas tanker TANJA KOSAN pumps 4700 tonnes of propylene ashore from its cargo tanks. Immediately thereafter, BETAGAS of Sloman Neptun arrives from Terneuzen with 3100 tonnes of propylene, followed in the evening by HAPPY HARRIER of the Bernhard Schulte shipping company with 2000 tonnes of ethylene from Le Havre and STOLT TERN, loading 1350 tonnes of the same gas for Bamble in southern Norway (Fig. 2.82).

Fig. 2.82 HAPPY HARRIER discharging ethylene from Le Havre at the Stade-Bützfleth LPG terminal of Dow Chemical (Ralf Witthohn)

The short loading and unloading times of the tankers ensure that there is a constant coming and going in the port located a few kilometres north of Stade. The timetable follows the production processes at the Stade plant of Dow Deutschland, which went into operation in 1972. Sixteen units produce epoxy resins, glycerine, polycarbonates, propylene oxide, propylene glycol and other substances primarily for the manufacture of plastics, cosmetics and pharmaceuticals. For this purpose, brine is supplied from the Harsefeld salt dome as basic material via an almost 30 km long pipeline. About half of the annual production of 2.7 million tonnes (2018) is transported from the Bützfleth quays by sea-going and inland waterway vessels. The gas tanker DANUBEGAS of the Hartmann shipping company, sailing under the German flag, is a frequent visitor. It carries propylene oxide to Terneuzen, for example, while Sloman Neptun's KAPPAGAS transports ethylene to the Scheldt. Other German tankers regularly bring raw materials, such as formalin from Rotterdam or propylene from Rafnes in Norway.

2.7.2.3 Ammonia for Antwerp

It takes the LPG tanker GAS GROUPER 5 days to carry the cargo loaded in Sillamäe to Antwerp. Opened in 2005, the Baltic Chemical Terminal of the Estonian port has loading facilities for urea and ammonia. Four of the storage tanks for urea each hold 20,000 tonnes, two for ammonia 30,000 tonnes each. In 2017, GAS GROUPER is one of six 35,000 m^3 LPG tankers in the fleet of Greece's Prime Marine, which also operates 27 product tankers and one crude oil tanker. Sister ships of GAS GROUPER built by Hyundai in South Korea in 2009 are GAS SNAPPER, GAS MANTA and GAS COBRA (Fig. 2.83).

Fig. 2.83 The ammonia tanker GAS GROUPER on its way from Estonia to Belgium (Ralf Witthohn)

2.7.3 Hydrogen

In December 2021, the world's first tanker for the transport of hydrogen was delivered by the Kobe shipyard of Japan's Kawasaki Heavy Industries. By cooling down to −253 °C, SUISO FRONTIER is able to carry the gas at 1/800 of its gas-state volume in a vacuum-insulated double-hull tank of 1250 m^3. After completion in late 2020 the SUISO FRONTIER is to test a new international supply chain by shipping liquefied hydrogen produced in Australia to Japan. In early Kawasaki is a member of the Hydrogen Energy Supply-chain Technology Research Association (HySTRA), together with US company Iwatani, Shell Japan and Japan's J-Power. The companies were subsidized by the government of Australia and the countrie's State of Victoria to build gas refining, hydrogen liquefaction and shipment facilities. The supply chain includes the mining of brown coal, gasification and gas refining at the Latrobe Valley mining site, the 150 km long land transport of hydrogen by trucks, liquefaction and loading at the port of Hastings/Victoria and the sea transport over a distance of about 4900 nm. The 116 m length, 19 m breadth SUISO FRONTIER is diesel-electrically driven to achieve a speed of 13 knots and operated by HySTRA participant shipping company "K" Line, part of the Kawasaki group. In January 2022, the tanker arrived in the port of Hastings, to take a first cargo of hydrogen, which is being processed by burning brown coal, for Kobe.

2.8 Oil Products and Chemicals

Tankers are subject to special safety standards, as their cargo poses considerable risks to the environment in the event of an accident, as was demonstrated by numerous spectacular oil spills. Although crude oil tankers can have particularly catastrophic consequences due to their large cargo capacity, heavy fuel oil transported in so-called product tankers means an even greater risk to the environment due to its high viscosity. In Germany, Lindenau shipyard at Kiel acquired a reputation for designing tankers of up to 45,000 dwt to high safety standards, contracted by domestic and foreign shipping companies. Specialisation in this type of tankers could not prevent the yard from discontinuing newbuilding activities in favour of repair work in 2012. In principle, the construction of a mere tanker hull is a work rather simply to do and thus easily to copy in low-cost countries like China.

2.8.1 Product Tankers from Kiel

Lindenau supplied the state-owned Seychelles Petroleum Co. (SEPEC) in Victoria on the Seychelles island of Mahé with the product tankers SEYCHELLES PRIDE in 2002, SEYCHELLES PIONEER and SEYCHELLES PROGRESS in 2005, SEYCHELLES PRELUDE and SEYCHELLES PATRIOT in 2007/08 and finally the much smaller SEYCHELLES PARADISE in 2009. The company had been founded in 1985 to supply the country with oil, after the Shell group had terminated this service. However, the tankers ordered by the shipping company are also employed in other trades under management of the Bremen-based shipping company German Tanker Shipping. Their deadweights, as crucial criteria of oil tankers, increased from 32,580 dwt to 37,558 dwt and 45,350 dwt, while the last newbuilding achieves 1786 dwt. The ships are capable of transporting chemicals, oil and oil products with a specific weight of up to 1.1 tonnes/m^3 in world-wide services. The SEYCHELLES PRIDE is since January 2022 managed by India's shipping company TruGro Ship Management under the name ELLES PRIDE and reported trading in the Persian Gulf area.

At an overall length of 189 m, 32.2 m breadth and 17.05 m depth, the deadweight of SEYCHELLES PRELUDE and SEYCHELLES PATRIOT is figured at 37,820 tonnes on the design draught of 10.5 m and at 45,350 tonnes on the freeboard draught of 11.8 m. The main dimensions have been chosen to reach an optimum ratio of deadweight to cargo volume. This was calculated on a maximum specific cargo weight of 0.67 tonnes/m^3 at the design draught and of 0.84 tonnes/m^3 at the maximum draught, based on homogeneous cargo. The design of the ship's lines was developed according to the Computational Fluid Dynamics (CFD) method, the lines were examined for ballast, design and freeboard draught. This included research work under aspects of resistance and propulsion, but also took into account sea-keepability and suitability for navigation in ice. Special attention was paid to the design of the foreship shape, to limit the loss of speed in heavy seas and to reduce the load on the bow structure. Compared to the previous foreship forms, the occurring forces were thus reduced by 40%. During the model tanks tests it was also proven that the chosen lines would reduce the flooding of the deck. At the same time, ice tank tests showed very good ice-breaking properties, allowing to maintain speed in ice. Compared to the first two twins for SEPEC, the ice classification was increased from E2 to E3.

Longitudinal and transverse bulkheads divide the cargo section of the ship into ten cargo and two slop tanks with a total capacity of 51,770 m^3 at 100%

filling. The double hull design ensures the protection of all cargo and oil storage tanks according to MARPOL regulations. Its construction has been designed under both strength and vibration aspects. The class designation of Germanischer Lloyd is 100 A5 E3 Chemical Tanker Type-2, Oil Tanker COLL-3 ESP VEC ERS BMW-S Environmental Passport, "Suitable for Carriage of various Oil Products" MC E3 AUT INERT. The notation sign COLL-3 indicates the construcitional reinforcements against collision damages.

For the first time, the design feature of a „predetermined breaking point "in the double hull was introduced, the shipyard having applied for a patent of this novelty. The predetermined breaking point is designed in such a way, that it can absorb the energy generated by a collision and thus prevent damage to the inner hull of the cargo tanks. This would significantly reduce the risk of environmental pollution compared to conventional double hull designs in the event of a cargo leakage. In a long-term measurement, the classification society Germanischer Lloyd evaluated the occuring loads. The swedged bulkheads of the cargo tanks result in smooth inner walls, which lead to shorter discharge and tank cleaning times as well as less cargo residues. At the same time, a special epoxy coating of the cargo and ballast tanks increases their service life.

The 2 × 5 cargo tanks, evenly distributed to both sides of the central longitudinal bulkhead, have different volumes from 3400 to 5500 m^3, this creating flexibility in the loading variants. On deck there are two additional cargo tanks of 240 m^3 volume. The ship's classification as an IMO Type 2 chemical tanker contributes to a larger variety of different cargo types. Twelve hydraulically operated submersible pumps with a capacity of 500 m^3/h are used to discharge the cargo. The maximum discharge rate achieved is 3000 m^3/h. The cargo pumps are operated from a central, electro-hydraulically operated remote control system, which is installed in the cargo control room and which also operates the ballast pumps. The cargo is heated up to 65 °C by stainless steel heat exchangers installed on deck, one for each cargo tank. The two deck tanks are brought to the same temperature by heating rods which are as well of stainless steel. A special residual draining system (super stripping) ensures that the cargo tanks are emptied almost completely.

The crossover has five connections, so that it is possible to separate five different types of cargo. The computerised cargo monitoring system allows a series of measurements to be watched off the monitors in the cargo control room. In this way the level of the cargo and slop tanks content is determined by radar, the temperature of the cargo by three sensors mounted at different heights and the tank pressure at preset alarm points. As well asessed is the pressure on the manifolds at equally fixed alarm points, the temperatures of the cargo heating system, the draught by four sensors and the temperature in the heat exchangers of the cargo heating system. A cargo computer is connected online to the measuring systems for the cargo tank levels and the tank contents of the ballast tanks as well as the most important storage and consumption tanks in the engine room. It enables permanent control of the longitudinal strength and stability of the ship during loading and discharge operation. Certain loading cases can be calculated in advance, as loading and unloading processes can be simulated. The cargo computer can simulate any leakage, so that the survivability of the ship can be ensured for each loading case. The cargo equipment also includes two tank cleaning heaters, a tank cleaning pump and two permanently installed tank cleaning devices in each tank. A crane with an SWL of 10 tonnes at 21 m outreach is installed at the manifold.

The tanker is propelled by a four-stroke engine of type MAN 8 L58/64 achieving an output of 11,200 kW at 428 rpm and acccelerating the ship to a maximum speed of 16.9 knots at 100% MCR on the design draught of 10.5 m. The reduction gear reduces the speed of the controllable pitch propeller of 5.9 m diameter to 115 rpm. The gearbox has a power take-off, which drives a shaft generator of 1700 kVA output at 1500 rpm. Simple and safe engine operation is ensured by a computerised monitoring system and a power control for the shaft generator and auxiliary engines. In addition to the manual control of the electrical power generation from the engine control room, a computer is installed in the wheelhouse. By five operating states programmed, the nautical personnel can select the most economical power generation based on the respective operating conditions. Steering gear room, diesel generators and the separator stations are located in separate rooms.

The design and equipment of SEYCHELLES PRELUDE and SEYCHELLES PATRIOT already complied with regulations that came into force later, such as the IMO regulations on ballast water exchange. The main engine and auxiliary diesel engine complied with the MARPOL Annex VI Convention on NOx and SOx emissions. The fuel system allowed for separate bunker storage of low-sulphur fuels. The installation of two settling tanks and two daily tanks for the main engine reduced the time needed to switch between normal and low-sulphur fuel. Other environmental features include a vapour recovery system for the cargo tank ventilation, environmentally friendly fire extinguishing agents and coolants that do not harm ozone. Further measures to protect the environment include a bilge water treatment with only a minimal residual oil content and TBT-free painting of the shell plating. The newbuildings meet both the environmental and safety standards of EXXON Marine (Figs. 2.84, 2.85, and 2.86).

Fig. 2.84 In order to limit the loss of speed in heavy seas and to reduce the load on the bow structure, special research efforts were undertaken in the design of the fore-ship shape of the 45,350 dwt SEYCHELLES PROGRESS, which is representing the largest of four tanker class sizes from Lindenau shipyard for Seychelles Petroleum Co. (Ralf Witthohn)

Fig. 2.85 Predetermined breaking points in the double hull of SEYCHELLES PROGRESS shall reduce the risc of oil pollution in case of an accident (Ralf Witthohn)

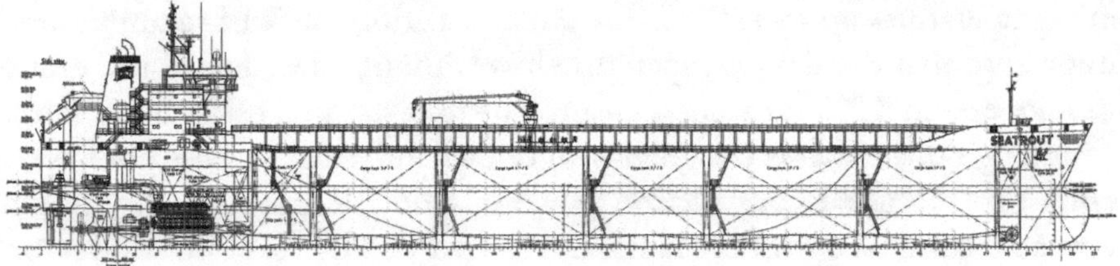

Fig. 2.86 Built in 2006 for German Tankers, SEATROUT is similar to SEYCHELLES PRELUDE delivered one year later (Lindenau)

2.8.2 Naphtenic Oil for Singapore

Inbound from Malta, the tanker ASPHALT SPLENDOR reaches Agioi Theodori on the Peleponnes in the beginning of June 2017. The Corinth refinery is operated by Motor Oil Hellas and the largest industrial complex in Greece, where LPG, naphta, kerosene, diesel and petrol are produced from crude oil. The next leg of the voyage will take the ASPHALT SPLENDOR from the Middle East via Suez, Gibraltar and Skagen to Nynäshamn, where the tanker's cargo is discharged at the Nynas refinery within 4 days at the end of September. From Sweden the next order sends the tanker, with a few thousand tonnes of cargo, via the Kiel canal to Antwerp, which is called 4 days later. ASPHALT SPLENDOR passes through the Berendrecht lock, is guided by two tugs to the Vierde Havendok and moors at the southern quay. Already the following day, the tanker leaves Belgium in ballast with the course set for Gibraltar.

ASPHALT SPLENDOR was delivered to the US shipping company Sargeant Marine from Boca Raton in December 2015 by China's AVIC Dingheng Shipbuilding in Yangzhou as the world's largest tanker designed for the transport of asphalt. In 2018, management of the ship switched to Rotterdam-based Valt BV. The vessel is also suitable for carrying oil products. Its construction was a challenge for the shipyard and could only be completed after a few setbacks. A number of special design solutions had to be found to make the tough cargo pumpable at high temperatures. Unlike conventional tankers, in which the ship's own construction elements serve as tank walls, asphalt tankers—similar to gas tankers—are equipped with tanks that are installed independently of the hull. ASPHALT SPLENDOUR has four blocks of four tanks each, resulting in a total of 16 cargo tanks. While the blocks 2 and 3 have a cubic shape, the forward and rear blocks had to be adapted to the shape of the ship's lines. At a block coefficient of 0.83, the underwater hull corresponds to the degrees of fullness usual for handymax tankers. The insulated tank blocks are supported on 560 special foundation components, which are also intended to prevent a thermal bridge between the hot tanks and the ship hull. ASPHALT SPLENDOR is designed for cargoes with temperatures up to 170 °C. The tank insulation made of ceramic wool is designed so that the daily temperature loss without operation of the heating does not exceed 3 °C. Heating is ensured by four thermal oil coils per tank.

Although the tanker is designed with double hull sides, it does not have, in contrast to the usual construction practise, a double bottom. The distance between the lower edge of the tanks and the flat keel of the vessel corresponds to the usual height of a double bottom. The same distance exists between the main deck and the roof of the tanks, which was constructed parallel to the camber of the deck beam. At 179 m length, 30.6 m breadth and 16.8 m depth, the 11,423 tonnes weighing tanker carries 36,962 tonnes on 10,4 m draught. The capacity of the 16 cargo tanks amounts to 35,666 m³. Three cargo pumps have a capacity of 500 m³/h each. ASPHALT SPLENDOR is powered by a two-stroke diesel engine of type Wärtsilä 5RT-flex50D built under licence by Yichang/China and developing 6400 kW output at 99 rpm. Directly acting on the shaft the main engine enables a speed of 14 knots. The tanker is part of the largest asphalt tanker operating company, which was founded in 2016 by the international Vitol Group and Sargeant Marine under the name VALT and in the beginning of 2020 had a dozen 11 special tankers with an annual transport volume of about 1.3 million tonnes in their fleet. The capacity of ASPHALT SPLENDOUR was superseded by the 46,000 dwt ASPHALT STAR and ASPHALT EAGLE, which had in 2003 and 2010 been converted (Figs. 2.87 and 2.88).

The Nynas refinery in Nynäshamn specialises in the production of naphthenic oils and bitumen for use in road construction, corrosion protection and roofing. In addition to the Nynäshamn plant, Nynas owns facilities in Gothenburg and a plant in Harburg, acquired in 2016, where up to 350,000 tonnes of base oil is produced, corresponding to 40% of Nynas' total production of naphthenic base oils. Furthermore, a refinery in Eastham/UK is operated jointly with Shell, as there are co-operations at other sites. In 2017, Nynas' network comprised eight plants and 23 depots, the most important of

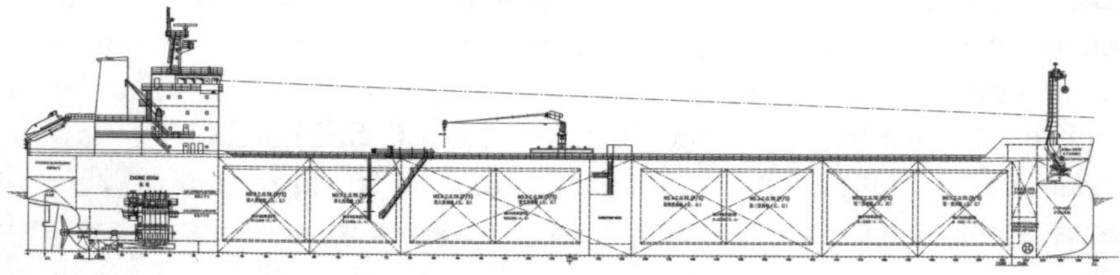

Fig. 2.87 By a cargo volume of 35,700 m³, ASPHALT SPLENDOR was considered to be the world's largest asphalt tanker newbuilding during construction in 2015 (Avic Dingheng)

Fig. 2.88 In October 2017, the world's largest asphalt tanker newbuilding ASPHALT SPLENDOR undertook a voyage from Nynäshamn to Antwerp (Ralf Witthohn)

which are located in Antwerp, Houston and Singapore. In 2016 the company transported 700,000 tonnes of oil by sea. For the supply of bitumen, 15 depots in Europe are maintained with mixing plants included.

A typical supply chain for Nynas products starts in the Lake Maracaibo region of Venezuela, where heavy, viscous oil is produced. The group's refineries are designed to process this type of oil, which is transported from Puerto Miranda to Europe by about 35 voyages of crude oil tankers of the 65,000 dwt class. The Gothenburg site has a pure bitumen refinery with a distillation tower in which the crude oil is broken down into various components suitable for further processing. In the upper section of the tower, light components, so-called distillates, are extracted and processed into naphthenic base oils. The product is then shipped to Nynäshamn, where unwanted components are removed during a hydrotechnical process and refined by chemical processing. One of the products produced in this process forms a component for transformer oil. It is then shipped by sea to a port of destination, where Nynas has a distribution centre, for example Singapore. From there the special oil is transported in tankers to a transformer factory. Each transformer requires a maximum of only 22 tonnes, so shipments can also be made in tank containers. In June 2017 the ultra-large container ship MADRID MAERSK loaded Nynas products for Shanghai, Qingdao, Bangkok, Singapore, Jakarta and Melbourne in Antwerp.

The company's second important production segment, bitumen, is particularly suited to be processed from Venezuelan oil, but also from some North Sea wells, especially from a field north of Aberdeen. Shuttle tankers, an 80,000 dwt ship every 3 months, bring this oil to the refinery in Nynäshamn. There the crude oil is stored in caverns until it is pumped into tanks from which the distillation plant is fed. The distillates are used for the production of special oils as well as components for bitumen and fuel. A large number of bitumen transports are carried by sea to North European ports, but the last leg of the voyage is often carried out by tank trucks on roads, for which construction bitumen is also being used.

2.8.3 Bunker for the GLOVIS COURAGE

Arriving from Hamburg, the Greek coastal tanker SIKINOS leaves the Bremerhaven Nordschleuse and slowly manoeuvres astern into the Nordhafen dock, where the Norwegian car transporter TORONTO is moored at the west quay, opposite of the Korean car carrier GLOVIS COURAGE berthed on the east side of the dock. Between the two ships there is only 100 m room for the tanker to manoeuvre along the port side of GLOVIS COURAGE. With the help of the tanker's deck crane the SIKINOS crew hoists two large rubber fenders into the water. Shortly afterwards the pumps of SIKINOS start to fill fuel into the bunker tanks of the car carrier. After 12 h the bunkering operation is completed and SIKINOS leaves bound for Rotterdam to load new bunker fuel (Fig. 2.89).

After two and a half years of construction the 4600 dwt SIKINOS was in August 2011 delivered by Fujian Southeast Shipbuilding to Piraeus-based Aegean Bunkering for operation under the Maltese flag. The Chinese shipyard, which has been active in the construction of smaller ships since 1984, built the double-hull tankers MILOS, SERIFOS, KITHNOS, AMORGOS, KIMOLOS, SYROS, MYKONOS, SANTORINI, PAROS, NAXOS, ANDROS, DILOS, IOS and ANAFI from 2007 on, the SIKINOS as the last unit of the series. In 2017, Aegean Bunkering employed a total of almost 60 bunker tankers worldwide, four of them in Port Elizabeth, one in Algeciras, 14 inland tankers in the ports of Antwerp, Rotterdam and Amsterdam, four tankers in Fujairah, five in Gibraltar, EVA SCHULTE chartered from the Schulte group in the Gulf of Mexico, an inland tanker in Hamburg, two seagoing vessels each in Jamaica, Singapore, Trinidad and Las Palmas, two barges in Vancouver, one in the North Sea, eight in Piraeus and nine others worldwide. In August 2016, company founder Dimitris Melissanidis sold his 22% share in the shipping company, which was founded in 1995 and is listed on

Fig. 2.89 Two large rubber fenders are being positioned at the port side of the Greek bunker tanker SIKINOS before berthing alongside the Korean car transporter GLOVIS COURAGE (Ralf Witthohn)

the New York Stock Exchange. In April 2020, the fleet comprised 59 bunker tankers in world-wide operation. Renamed MM Marine as a subsidiary of Mercuria Energy Group, the company disposed of 31 sea-going tankers in early 2022.

2.8.4 Mineral Oil in Kattwyk Harbour

Opposite the Reethe silo facilities in Hamburg, the 23,000 dwt Danish tanker TORM FOX is moored at the dolphins of the Vopak terminal to handle petroleum products. At the same time, the Swedish 4700 dwt tanker BRO GOLIATH is being moored in a branch basin off the Reethem, while the TERNHAV, fully loaded with 12,000 tonnes, arrives in Kattwyk harbour south of the Reethe. A mooring boat assists the berthing manoeuvre of the Swedish tanker at the jetty (Fig. 2.90).

Fig. 2.90 The area around the Kattwyk harbour is the Hamburg centre of liquid cargo handling (Ralf Witthohn)

2.8.5 Styrene for Wismar

In the morning, the tanker PUCCINI of Hamburg-based operator GEFO leaves the industrial port of Moerdijk on Hollands Diep and reaches the North Sea via Dordrecht, the Oude Maas and Rotterdam. From there, the course is set along the Frisian Islands to the mouth of the Elbe and the Kiel canal. After circumnavigating Fehmarn, the 3600 dwt tanker reaches 2 days later Wismar on Germany's Baltic coast. Carrying a cargo of styrene for the Jackon plant, the ship turns over starboard and berths at the tanker pier, its bow in the direction of the sea. The Norwegian Jackon company processes the styrene, of which about 60,000 tonnes are yearly shipped to Wismar, as basic material for the production of polystyrene beads to be used as insulation material. The Wismar pier for liquid goods also serves the company of Egger Holzwerkstoffe, which gets one of its raw materials, methanol, in tankers. This is pumped to the plant via a dedicated pipeline (Fig. 2.91).

Tanker operator GEFO developed a comprehensive programme to extend its fleet to a number of 13 small tankers in 2020. Apart from the 1999/2000-delivered VERDI, BELLINI, MOZART and DONIZETTI the company disposed of the 2009/10-built BRAHMS, BARTOK, BEETHOVEN and BERNSTEIN, in 2018/ 2019 purchased from Turkish owners. The last unit of a 6600 dwt newbuilding quartet, started in 2014 by FIDELIO, OTELLO and TRAVIATA, was in 2016 completed as NABUCCO at Turkey's Tersan shipyard. The tanker was equipped with an exhaust gas cleaning

Fig. 2.91 The tanker PUCCINI and its sister ships ROSSINI und BELLINI occasionally supplied the Wismar polysterene plant with styren, until their sale to foreign buyers in 2019/21 (Ralf Witthohn)

system in 2019. Largest ship in the GEFO fleet is the 7750 dwt GIOCONDA, in 2018 as well built in Turkey. The owner opted to equip a newbuilding series of 7000 dwt tankers of the TOSCA type contracted at AVIC Dingheng Shipbuilding in China for delivery from 2021 on with main engines ready to burn LNG. The newbuildings CORELLI, CAVALLI, PAGANINI, RAVEL, DIABELLI and SCARLATTI of 3900 dwt, delivered from China's Jinling Dingheng shipyard, along with the TOSCA type ships NORMA and ARABELLA will bring the GEFO fleet to 25 units in 2022 (Fig. 2.92).

2.8.6 Asphalt from Canadian Oil Sands

In March 2017, the Canadian shipping company Transport Desgagnés commissioned the first asphalt tanker equipped with a dual fuel main engine. The employment of DAMIA DESGAGNÉS, for which the Royal Bank of Canada was registered as owner, is based on a contract with Canada's Calgary-based Suncor group. The group specialises in the extraction of synthetic crude oil from oil sands and is considered the world's largest bitumen producer. It operates refineries in Edmonton, Sarnia, Montréal, Quebec and Commerce City. DAMIA DESGAGNÉS, which is mainly sailing the St. Laurence Seaway, is a newbuilding of the Turkish shipyard Besiktas Tersane in Altinova. The 135 m length, 23.5 m breadth tanker has 12 cargo tanks and two slop tanks with a total capacity of 15,280 m^3. They are heated by means of two thermal oil boilers with a capacity of 3000 kW each through stainless steel heating coils and emptied by four cargo pumps with a capacity of 400 m^3/h each. The cargo

Fig. 2.92 TRAVIATA belongs to a products tanker quartet built in Turkey for the German operator GEFO (Ralf Witthohn)

tanks are equipped with a portable washing system and kept permanently gas-free. At a draught of 7.9 m the deadweight reaches 14,719 tonnes.

Equipped with a 5450 kW dual fuel engine for a speed of 13 knots, DAMIA DESGAGNÉS is able to meet, by using LNG (Liquefied Natural Gas) as fuel, the requirements of the Emission Control Area (ECA) applicable to the Great Lakes. The Wärtsilä 5RT-flex 50DF two-stroke engine can be operated alternatively with heavy fuel oil or diesel oil. The French classification company Bureau Veritas has awarded the ship, classified as an oil/bitumen tanker, the smallest ice class Polar 7. According to this class, the hull must be reinforced and the engine power strong enough to pass through ice of 0.6 m thickness. In 2022 Transport Desgagnés operated a fleet of 23 vessels, comprising 12 tankers, 9 general cargo vessels and one combined passenger and cargo ship (Fig. 2.93).

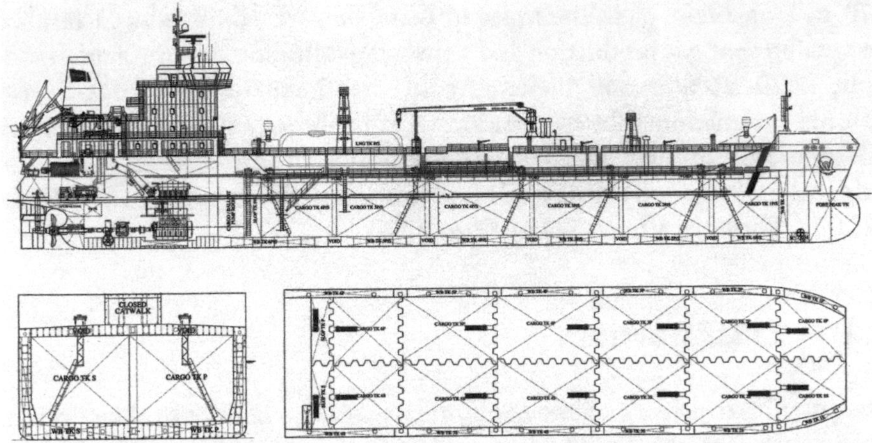

Fig. 2.93 Thanks to an LNG burning main engine, the Canadian asphalt tanker DAMIA DESGAGNÉS met the conditions of the American ECA zone (Desgagnés)

2.9 Pure Sulphur for Jorf Lasfar

It takes the 4252 dwt tanker SULPHUR GENESIS of Tokyo-based Taiichi Tanker Co. 1 week to sail from Brake on the Lower Weser to Jorf Lasfar in Marocco. There, its cargo of liquid sulphur is delivered to the Bunge Maroc Posphore factory, where the stuff is used for the production of fertiliser. The factory belongs to the state-owned, Casablance-headquartered OCP group (Office Chérifien des Phosphates), which has been operating since 1920.

The SULPHUR GENESIS, was built in 2008 especially for the transport of liquid sulphur at the Onishigumi shipyard in Mihara. The cargo which the tanker loaded at Brake, is elemental sulphur and a by-product from the purification of German natural gas. It was processed in Großenkneten near Oldenburg and Voigtei near Diepholz under the direction of BEB Erdgas und Erdöl. Germany's largest natural gas producer is based in Hanover and a company belonging to the Shell and ExxonMobil groups. Several hundred thousand tonnes of sulphur were until the end of 2019 transported annually in rail tank cars to the Brake terminal of Norddeutsche Erdgas Aufbereitungs-Gesellschaft (NEAG). The sulphur was exported either in liquid or solid form in bulk carriers, besides to Jorf Lasfar, for example, to Namibia, Israel or to the sulphuric acid factory of the Finnish Kemira group in Helsingborg. The main customers for liquid sulphur were Morocco and Tunisia. The liquid sulphur was stored in Brake in two tanks with a combined capacity of 36,000 tonnes. Since 1980, most of Germany's sulphur exports have been handled at the

220 m long Brake pier. The transhipment, however, declined as a result of decreasing sour gas production and a growing domestic sulphur market.

In 2020, Exxonmobil therefore sold the Brake terminal to US-based Saconix International. In cooperation with LogServ Logistic Services, a subsidiary of the port operator J. Müller, Brake, the new owner rebuilt the terminal for the storage and smelting of imported solid sulphur of annually 200,000 to 300,000 tonnes from 2022 on (Fig. 2.94).

2.10 OBO Carriers

The attempts to achieve a more economical operation of ships by their optional employment in either dry or liquid cargo transport requires a special design of ships, which are designated OBO carriers (derived from Oil Bulk Ore). Their construction has to take into consideration the widely varying specific weight of oil and other liquid cargoes in comparison with ore, which therefore has a comparatively small space requirement. The ships developed in the 1950s were in the beginning named combi carriers. They were laid out with separated ore holds of small dimensions, which were arranged in a high position to reduce the ship's *stiff* behaviour in rough seas and to prevent the cargo from shifting. Later, OBO carriers were designed in such a way, that they carried ore in only every second, shorter hold. This has the consequence of alternating stresses on the ship's structure due to the different cargo distribution, with up and down swelling forces by the buoyancy forces of the sea. The supposedly better economical operation of a bulk/oil carrier by avoiding ballast voyages

Fig. 2.94 The special sulphur tanker SULPHUR GENESIS of the Japanese shipping company Daiichi exported sulphur generated from the production of natural gas in Germany and was in 2022 still engaged in transports between western and north European ports (Ralf Witthohn)

was reduced by a number of technical disadvantages connected to the dual use of the cargo holds/tanks and the following need for tank cleaning, particularly required after oil transports. After the year 2000, there were only a few OBO carrier orders reported.

From 1996 to 1999, Korea's Hyundai Heavy Industries built ten OBO carriers for the Norwegian company SKS Shipowning. Named SKS TYNE, SKS TANA, SKS TWEED, SKS TRENT, SKS TAGUS, SKS TUGELA, SKS TORRENS, SKS TRINITY, SKS TIETE and SKS TANARO the 243.8 m length, 42 m breadth and 23.2 m depth ships carry 110,000 tonnes on a draught of 15.7 m. Their tanks are epoxy-coated. The shipping company designates the ships as LR2 carriers for product cargoes, Aframax carriers for crude oil transport and Baby Capers for bulk transport. The first four units were sold to Turkish owners in 2016/17, the SKS TUGELA broken up at Alang/India. SKS TIETE was from April 2019 on scrapped at the same place, along with the sister ship SKS TANARO.

At the beginning of 2018, SKS had listed two 121,000 dwt and four 110,000 dwt aframax class OBO carriers of the SKS TIETE type, ten product tankers of 120,000 dwt and six suezmax tankers of 159,000 dwt, which were, apart from the four meanwhile sold 110,000 dwt OBO carriers, still in the fleet 2 years later. The tanks of the suezmaxes are epoxy coated for the transport of refined products. The name of SKS, which belongs to the Kristian Gerhard Jebsen Group, is derived from a tanker pool founded in 1991 by the former Soviet Union's Sovcomflot, the Norwegian Kristian Gerhard Jebsen Skipsrederi (KGJS) and China's Sinochem to operate ten Hyundai newbuildings of 96,000 dwt. Instead of Sovcomflot and Sinochem, which left in 1996, the Chilean CSAV became a partner in KGJS from 1997 to 2004 (Fig. 2.95).

Among the few shipping companies still operating OBO carriers is Norway's Klaveness Combination Carriers by a fleet of 16 units, of which the owner

Fig. 2.95 Depending on the type of cargo transported—oil products, crude oil or bulk cargo—the meanwhile scrapped SKS TIETE was designated by the shipping company as a Long Range 2, Aframax tanker or Baby-Caper (Ralf Witthohn)

Fig. 2.96 The 82,500 dwt OBO carrier BARU can either carry liquids or dry bulk commodities (Arne Münster)

describes eight each as CABU combination carriers and CLEANBU combination carriers, resp. At deadweights from 72,500 to 89,500 tonnes the CABUs are described as being able to transport either caustic soda solution, liquid fertilizer and molasses as well as all types of dry bulk cargo. The CLEANBUs are of 82,500 dwt and designated LR1 product carriers when transporting clean petroleum products or heavy liquid cargoes and described as kamsarmax bulk carriers in the transport of dry bulk. The former ones are named BARCARENA, BANASTAR, BANGOR, BANTRY, BAKKEDAL, BALBOA, BAFFIN and BALLARD. They were built from 2001 to 2017 by Japan's Oshima and by China's Ouhua Zhejiang shipyard. The latter are named BARU, BARRACUDA, BARRAMUNDI, BALEEN, BANGUS, BAIACU, BASS and BALZANI. They were between 2019 and 2021 delivered by Yangzijiang shipyard in China (Fig. 2.96).

3

Industrial Products

The world's yearly production worth over USD 80 trillion is unthinkable without sea transport and the mass supply of raw materials. Imported ores are smelted into steel, copper, zinc and aluminium, which are then refined into high-quality industrial products, a large proportion of which are in turn exported (Fig. 3.1). After an increase of 4.6%, the world crude steel production of 2018 was figured at 1808 million tonnes, that of finished steel products at 1712 million tonnes. In 2020, the crude steel production still reached 1.864 million tonnes, which yet meant a reduction of 0.9% compared to 2019. A major, and still growing proportion of 57% (equivalent to 1053 million tonnes) was in 2020 contributed to China, up from 51% and 53% in the years before, while India faced a downturn by 10.6% to 100 million tonnes, as well as Japan, where 16% less was produced (83 million tonnes and South Korea with a loss of 6%, equivalent to 67 million tonnes. The EU put out 139 million tonnes, meaning a decrease of 12%.

Providing more than half of the world's crude steel demand, China, through subsidized prices, caused a strong market imbalance and declining shares of all other countries. At a high domestic consumption rate, China exports steel products all over the world. The preferential location of steelworks at ports facilitates the supply of ore, coal and coke as well as the removal of steel and by-products such as gypsum by shipping. Smaller lots of steel are part of the daily transport business of coastal cargo ships. The world's top steel producing companies in 2018 were Luxembourg-based ArcelorMittal by putting out 96 million tonnes, the China Baowu group by 67 million tonnes and Nippon

R. Witthohn, *International Shipping*, https://doi.org/10.1007/978-3-658-34273-9_3

Fig. 3.1 The 28,900 dwt heavy-load carrier AUDAX, built in 2016 at China's Guangzahou Shipyard and operated by Hong Kong-based ZPMC-Red Box Energy Services under the Curacao flag then, took part in the shipment of monopile foundations for an offshore wind farm from Germany to Taiwan in January 2020. The following year, the ship's ownership switched to Rotterdam-based Fortune MC Hercules Shipping and flag to Liberia (Ralf Witthohn)

Steel by 49 million tonnes in 2018. In 2020, China Baowu by 115 million tonnes overtook ArcelorMittal, which fell back to 78 million tonnes. The next rangs were hold by China's groups HBIS with 44 million tonnes and Shagang at an output of 42 million tonnes.

3.1 Steel Products

3.1.1 Steel Pipes for the New Kaiserhafen Quay

Only at the beginning of the week, a mobile shore crane begins to lift steel pipes from the cargo hold of the coaster PROSNA. The Maltese-flagged ship of a Polish shipping company has 2 days before arrived with steel pipes of 20 m length, 1.6 m diameter and 16 mm thickness. The pipes were loaded in Moerdijk, an industrial port located at the mouth of the Rhine at Hollands Diep. They are to form components of a combined steel wall consisting of sheet piling segments between the pipes. Their supplier is the ArcelorMittal steel group, which can load individual weights of up to 120 tonnes at its Moerdijk and Dintelmond terminals. The Moerdijk port, located south of Rotterdam, can be reached by coastal cargo ships and bulk carriers up to supramax size via the Maas and the Oude Maas. Three basins enable lively transhipment activities, even into smaller cargo vessels. The steel pipes from PROSNA are discharged at the western quay of Kaiserhafen III in Bremerhaven, where bananas had been imported for decades (Fig. 3.2).

For the renewal of the former banana quay, which was still founded on wooden piles as built in 1909, the pipes and intermediate segments form part of the new pile wall. As side effect of the new quay, rammed 9 m further inland, the manoeuvring of large car carriers is facilitated. The steel pipes were transported in several lots from Moerdijk to Bremerhaven. Three weeks after PROSNA, another coastal freighter with pipes arrives from Moerdijk, the 3800 dwt coaster MARTEN of German owner Hermann Lohmann from Haren/Ems, built according to the same design at a Slovakian shipyard.

After the fall of the iron curtain, Slovenske Lodenice in Komarno/Slovakia entered the western shipbuilding market by offering the series-produced PROSNA type, at first for West German, later also for East German and Russian customers with only slightly differing design characteristics. Many of the approximately 88 m length and 12.9 m breadth single-hatch ships were fitted with lowerable masts and wheelhouses, making them suitable for navigation on the Rhine, for example in steel exports of Rhineland manufacturers from Duisburg to major North European shipyards such as the Odense Staalskibsvaerft in Denmark. PROSNA was completed in 1998 as NORTHERN ISLAND and named WANI TOFTE, ANPERO, ANTORA and NORDICA, until the ship came under the management of Baltramp Shipping in Szczecin, as still recorded in 2020.

Fig. 3.2 The coastal cargo vessel PROSNA, managed by a Polish shipping company, took large steel pipes for a new harbour quay from Moerdijk to Bremerhaven (Ralf Witthohn)

The first ship of the Slovakian series started in 1991 was named RHEINFELS and registered for m.s. "Rheinfels" Domhardt River Liner KG in Haren/Ems. Only in 2016 the ship changed from the management of Germany's Wessels company in Haren/Ems to Turkish owner Nismar Shipping, which renamed the ship VON PERLE and entered it in the Cook Islands register. In November 2017, the vessel undertook a voyage from Antalya/Turkey to Cherson/Ukraine. This meant the return to the Black Sea, via which it had reached the northern European trade area after being built in Komarno and transited downstream on the Danube. The Wessels group initially received six "Rhein" type vessels from the Slovakian builder until 1997, Rohden Bereederung in Hamburg three units. In 1993/94, a series of six ships was commissioned by Deutsche Seereederei in Rostock, named RÜGEN, HIDDENSEE, POEL, USEDOM, VILM and FISCHLAND. Immediately afterwards, a batch of 16 newbuildings, starting with the NORTHERN LINANES, entered the bulk trading activities of Norway's Jebsen company in a joint venture with Russia's Northern Shipping in Murmansk. The majority of the ships were taken over by Wilson shipping company from Bergen/Norway in 2002 (Fig. 3.3).

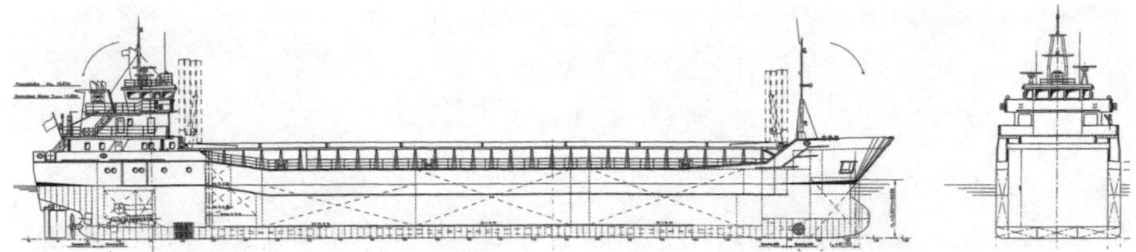

Fig. 3.3 Slovenska Lodenice built more than 100 units of this standard type, including the Wessels-managed NESTOR (Slovenska Lodenice)

3.1.2 Steel for Kota Kinabalu

The large revolving crane continuously picks up the steel coils produced at the nearby steel works from the storage area and lowers it into the holds of the container ship FESCO VLADIMIR. On its maiden voyage to the Far East, the vessel has been chartered by Rickmers Linie to deliver a part load of steel in Kota Kinabalu, Malaysia. The transport of steel products is symptomatic of the search for cargo by container ships, such as the Stocznia Szczecinska Nowa newbuilding delivered only a few weeks earlier in July 2009 amidst the shipping crisis caused by the 2008 financial crash. The employment opportunities in container shipping have become rare, after the market collapsed as result of the economic crisis. How precarious the situation for container carriers of this type will remain in the following years becomes as well obvious by the fact, that FESCO VLADIMIR, the 50th newbuilding of the series, remains the last one from Szczecin. Only the Vinashin Ha Long shipyard in Quang Ninh/Vietnam will still build two licensed units for Vietnam's Vinashin Ocean Shipping Co.

Usually, the steel products of ArcelorMittal's Bremen steelworks are, however, exported by bulk carriers of the handysize class, the majority of the cargo to Ravenna. In the Italian city, the largest plant of the steel processing Marcegaglia group is located over an area of 540,000 m^2. The steel rolls supplied from Bremen are treated by chemical commissioning, cold rolling, annealing, galvanizing and preservation. The branch in Ravenna also serves as a logistics centre for the entire group, which is headquartered in Mantua and which 6500 employees produced 5.6 million tonnes of steel products at 43 locations in 2016 (Fig. 3.4).

Fig. 3.4 The products of the ArcelorMittal steelworks in Bremen are mainly exported to Ravenna (Ralf Witthohn)

3.1.3 Geneva Handysize Bulkers in Steel Transport

Among 14 bulk carriers of Geneva-based Massoel Shipping in 2015, there were six of the handysize segment, in addition to six mini bulkers and two handymax ships. At the beginning of 2018 the massoel fleet consisted of eight ships. In addition to four LUGANO-class ships, these were the supramax carriers LUZERN and GLARUS and the handysize bulkers AARGAU and THURGAU. In contrast to the usually realized handysize designs with five holds and four revolving cranes placed in between, the 20,000 dwt bulkers AROSA, ANDERMATT, MARTIGNY and LUGANO purchased from abroad in 2006 have only four holds and three 30-tonne cranes. The ships were built at Sekwang Heavy Industries in Ulsan/South Korea for Bahamas-based Dockendale Ship Management (Dockship). Their main dimensions of 155.1 m length and 23.7 m breadth allow the ships to operate on the Great Lakes, where LUGANO, under its first name DS REGENT, was used to transport coal to the Montaup power station in Somerset/US. In December 2014, LUGANO carried a cargo of steel from Eregli in Turkey to Hamilton on Lake Ontario in Canada, from where the ship was then bound for Aratu in Brazil. In December 2017 the ship undertook a voyage from Argentina to Dakar/Senegal, and in February 2020 the LUGANO was underway from Castellon/Spain to Nador/Marocco. The ship is equipped with 30-tonne cranes from Tsuji for serving four holds. An MAN B&W diesel built by STX gives them a speed of 14 knots.

In 2020, the fleet still comprised eight units, this time six bulk carriers and two multipurpose cargo vessels, but the company contracted a law firm to sell them separately. AROSA and MARTIGNY were purchased by Egypt's Reknav

Navigation and became the REK NOBLE and REK ROYALE, resp. ANDERMATT changed hands to Piraeus-based New Kronos Star Maritime and was renamed AGIOS PORFYRIOS. LUGANO was registered for a Marshal Islands company as JAGUAR 1 (Fig. 3.5).

3.1.4 LNG Tank for WES AMELIE

The fitting of the first LNG burning propulsion system on a German container ship, named WES AMELIE, did not only require the conversion of the MAN 8L48/60B four-stroke main propulsion system to an MAN 51/60DF engine, but also the installation of an LNG tank. The 500 m^3 steel tank was supplied by TGE Marine Gas Engineering, a Bonn-based company specialising in the supply of gas systems. The isolated tank was, in this case, manufactured in South Korea, in April 2017 transported to Bremerhaven on the container ship MSC MAYA of Mediterranean Shipping Co. and directly lifted from the container terminal to the converting repair yard German Dry Docks by the floating crane ENAK of Bugsier-Reederei. After arrival of the Wessels owned WES AMELIE at the end of May, the tank was lifted into the forward cargo hold of the ship, which conversion was completed in August and thus took considerably longer than the first planned 30 days (Fig. 3.6).

The space required for the tank reduced the container intake of the ship from previous 1036 teu by 29 units, even though supports were fitted above the tank to use that space for the stowage of containers. Following the conversion, WES AMELIE was chartered for 1 year by the feeder operator Unifeeder in a service between Rotterdam and Baltic Sea ports, i.e. in an Emission Control Area (ECA) regulating reduced sulphur emissions. In November 2017, the shipping company announced that it planned to equip also the

Fig. 3.5 The 20,000 dwt LUGANO was for 12 years operated by a Swiss Massoel to carry all kinds of bulk cargo including steel products and then sold to a letterbox firm on the Marshall Islands (Ralf Witthohn)

Fig. 3.6 The LNG tank for the first German container vessel WES AMELIE using liquefied gas as fuel arrived from Korea in Germany on a container ship and was transported from the container terminal to the conversion yard by the floating crane ENAK (Ralf Witthohn)

sister ships WES JANINE, WES GESA and WES CARINA with an LNG system in the same way. In November 2019, MAN Energy Solutions and Wessels Marine announced, that the WES AMELIE main engine would in a first such project world-wide test the burning of synthetically produced methane gas (SNG). The gas is supplied by automobile manufacturer Audi's power-to-gas facility in Werlte/Germany and generated from wind energy. The project was to carried out in cooperation with the LNG transportation company Nauticor and the ship's charterer Unifeeder in spring 2020. Despite an improvement of market conditions, however, the Meppen district court initiated insolvency proceedings for the single-ship companies WES AMELIE, WES CARINA and WES JANINE in January 2021 due to claims by financing Nord/LB, which is backed by the federal states of Lower Saxony and Saxony-Anhalt. The ships were taken over by Elbdeich Bereederung and renamed ELBBLUE, ELBDEICH and ELBWATER (Fig. 3.7).

Fig. 3.7 The side view of WES AMELIE shows the position of the additional LNG tank installed behind the forecastle (Wessels)

3.2 Vehicles

Although the German-Jewish shipowner Arnold Bernstein used the first special ships, converted from warships, to transport cars as early as 1922, the design of purpose-designed newbuildings for the transport of vehicles only began with the introduction of the automobile as a means of mass transportation in the 1960s. Car transports were initially limited to one-way routes, such as Volkswagen exports to the United States. No suitable vehicle loads were available for the return voyage, so that combined car/bulk carriers were put into service for the alternate transport of vehicles and bulk cargoes. A common employment concept was to load VW cars on mobile hanging decks of bulk carriers, like on ships of the German shipping companies Schulte & Bruns, A. C. Toepfer and Schlüssel, which then loaded coal or grain for the return trip. When Japan became a car export country, the ships of the NATALIE BOLTEN type, in 1973 built at Lübeck for the Hamburg shipping company Aug. Bolten, were employed on a round-the-world route leading from Europe to North America and further on to the Far East (Fig. 3.8).

Car transporters, known in the industry as PCT (Pure Car Transporter) or PCTC (Pure Car Truck Transporter), operate according to schedules like container ships and multipurpose liners. The main routes are determined by the production sites, primarily located in Europe, Asia and the United States, but exports from countries such as Turkey or Central and South America also require corresponding special tonnage. Car shipping is also comparable with container shipping by the use of smaller feeder ships with smaller capacities, used to distribute vehicles, for example from North Sea to Baltic ports, which have been imported on deep-sea transporters from overseas. Employed are also short sea vessels, which distribute cars over secondary seas from production their sites, for example located in Spain, the UK or Finland. Time charter

Fig. 3.8 Besides cars, vehicle transporters carry a wide variety of rolling cargoes of any type, including railway wagons, trucks, tractors, busses, excavators and other special vehicles (Ralf Witthohn)

contracts are concluded between car manufacturers and shipping companies with a duration ranging from a a few months to several years. There are also long-term contracts signed between the car manufacturers and the port operators. Daimler-Benz and BMW primarily ship their cars through Bremerhaven, but also have them imported via the same harbour, as from Finland or the US. The BLG Logistics group and Borgward China agreed to cooperate in the export of new Borgward cars from their Beijing plant via the port of Tianjin to Bremerhaven, where a first shipment of 34 cars rolled from the car carrier MORNING LILY in April 2018. Volkswagen, on the other hand, has long-year ties with the port of Emden, but also exports to Scandinavia via Cuxhaven.

Believed to be the world's largest car handling port is Zeebrugge by offering berths for up to 16 carriers. In 2019, the Belgian port achieved a shipment volume of 2.96 million units, before a drop to 2.2 million vehicles was reported in 2020. In northern Europe, the German port of Bremerhaven ranked second by 2.1 million, but faced a decrease to 1.7 million vehicles in 2020 and 2021 each, while in Emden 1.34 million vehicles were handled in 2019, at an export rate of about 80% in the latter port. In 2021, the Emden figure fell below 1 million, not at least as a result of the temporary production stop at the VW works, which produced only 120,000 cars instead of 190,000 the year before. The most important recipient country of German automobiles in 2019 was the UK, followed by the United States. In 2020, 0.6 million cars from Germn production were shipped to Asian buyers, 0.4 million to the Americas and 0.04 million each to Australia and Africa.

About 1400 ro ro ships call every year at the Bremerhaven terminals, which offer parking spaces of 240 ha. The total parking capacity is 95,000 cars, 50,000 of which are parked in eight covered several decks high garages. In addition, approximately 1.15 million tonnes of high and heavy cargoes are handled annually. The quays reserved for vehicle handling comprise eleven berths with head ramps and 16 railway connections, the cars are worked on in three technical centres and in a painting hall. Dock workers carry out loading and discharging 7 days a week. They are taken on or off board in minibuses to drive the vehicles individually to their destinations. Likewise, the transhipment plannings have so far been done mainly manually. In order to make operational processes more efficient, a 3-year research project named Isabella and funded with EUR 2.6 million by Germany's Federal Ministry of Transport and Digital Infrastructure was started in 2017, and continued by a second phase in 2020. The project is intended to develop an interactive, simulation-supported planning and control of transhipment movements by using a touch table for recording the location of vehicles instead of an intuitive parking practice.

From March 2020 on, car handling was increasingly hit by the COVID-19 pandemic, which had caused a standstill of production in many car factories. World-wide disruptions of the services of the vehicle carrying operators continuously accelerated. Massively reduced volumes impacted schedules. Leading operators considered to offer ships of their deep-sea fleet as floating storage facilities. Especially the capacities of many US import ports were constrained. In Bremerhaven, the relation of the export to import shares, usually figured at about 2 to 1, was turned round (Fig. 3.9).

3.2.1 Scandinavian and Asian Dominance

The shipment of vehicles is dominated by Japanese, Korean and Scandinavian operators. The Asian shipping companies are usually branches of major universal shipowners, on their part branches of industrial conglomerates, such as NYK (Mitsubishi), Mitsui O.S.K. (Mitsui) and "K" Line (Kawasaki) in Japan or Hyundai in South Korea. In contrast, the Scandinavian shipping companies Wallenius and Wilhelmsen traditionally concentrate entirely on the car sector. Their fleet policy is similar to that of the large container shipping companies, as they primarily operate own vessels, but for long or short periods charter additional tonnage from owners, who do not maintain their own liner

Fig. 3.9 In January 2020, Finnish-produced Mercedes cars loaded in Uusikaupunki rolled 3 days later over the stern ramp of the Japanese „K"Line's car feeder THAMES HIGHWAY for transhipment by the same companie's deep-sea carrier POLARIS HIGHWAY, pictured right and destined for Singapore via western Europe ports, with arrival scheduled 1 month later (Ralf Witthohn)

network. A niche in car transport refers to the shipping of second-hand cars, mainly European end-of-life vehicles, to the Middle East and Africa (Fig. 3.10).

Almost all of the large transporters are designed in such a way that they can carry all types of vehicles, including particularly high or heavy units, technically referred to as high&heavies, in addition to cars. For this purpose, the ships are equipped with ramps laid out for high loads and with up to five light car decks that can be hoisted to create enough clear height for the high units.

The garage architecture of car transporters that is common today caused quite a stir when it was first realized on a transoceanic vessel, named DYVI ATLANTIC and delivered in February 1965 by Drammen Slip & Verksted in Norway to the Norwegian shipowner Jan-Erik Dyvi. Experts discussed whether the ship, which could thus be used exclusively for car transport, was not too dependent on one-sided export flows and whether it was not too much exposed to wind pressure due to its large depth. Employment of the 148.6 m length and 19 m breadth DYVI ATLANTIC was secured by a 5-year charter of German car producer Volkswagen, which provided 20,000 export cars yearly for shipment to North America. At the start of its maiden voyage

Fig. 3.10 Japanese vehicle transporters, such as NYK's LYRA LEADER, make up a substantial share of the world's car carrying fleet, besides Scandinavian and Korean operators (Ralf Witthohn)

Fig. 3.11 DYVI ATLANTIC was the first garage-like car transporter for use in ocean traffic (Ralf Witthohn)

the Baltimore-bound ship loaded 1344 VWs in Bremen within 7 h. Loading and unloading of the cars was carried out through four side doors arranged in pairs on top of each other at each of the ship's sides. The ramps were stored on the boat deck and could be hung in a lower or upper gate depending on the height of the quay (Fig. 3.11).

By a maximum capacity of 1380 cars on seven decks, the cargo weight of DYVY ATLANTIC amounted to about 1400 tonnes, so that the ship's deadweight capacity of 2620 tonnes included reserves for ballast and fuel. The 17 knots achieving ship was in service for 20 years, before being scrapped in Vigo in 1985. Even before this transoceanic prototype of the modern car transporter was commissioned, Dyvi had already received a smaller version of a garage ship for the England-Scandinavia route in April 1964. DYVI ANGLIA, built by Trosvik Verksted in Norway, featured a length of 87.8 m and a breadth of 14.7 m. The 15 knots fast, 499 gt vessel had a deadweight of 737 tonnes and could load about 450 cars on five decks. After delivery it was chartered by the Ford concern for 5 years to transport new cars from the Dagenham works in London to Copenhagen, Malmö and Stockholm.

Indeed, the high wind sensitivity of car carriers increased the more the larger their designs were developed. In strong storms they can often hardly be prevented from breaking loose of their lines and have to be kept at their moorings with the help of tugs permanently pushing the carriers to the quay. A storm-related accident with serious consequences occurred in May 2009, when the car carrier HÖEGH LONDON, after leaving the Bremerhaven lock, was pushed out of the fairway by a north-westerly storm despite tug assistance, so that it rammed three container ships moored at the container terminal and severely damaged them (Fig. 3.12).

There are further severe risks in the operation of car transporters. For example, there is—albeit in small quantities—explosive fuel in the tanks of the cars that roll on and off the ship under their own engine power. In the event of a fire on vehicle decks, its fighting is difficult. This was proven, when the car carrier COURAGE, the former Wallenius-owned AIDA built in 1991, and then operated by the US shipping company American Roll-on Roll-off Carriers (ARC) caught fire off the English coast in June 2015. The ship was able to call at Southampton under its own power. Discharge of the civilian

Fig. 3.12 Three of the to date largest car carriers, Höegh Ugland's HÖEGH TRAPPER, Wallenius-Wilhelmsen's THETIS and "K" Line's NIAGARA HIGHWAY met at the vehicle handling terminals of the Bremerhaven Nordhafen area (Ralf Witthohn)

and military vehicles, many of which had been destroyed by the fire, in Bremerhaven was carried out under visual protection from the public and in protective suits. Only 20 months later a fire broke out on the HONOR of the same shipowner, as well off the English coast.

Another danger car carriers are exposed to originates from their high architecture, which requires large amounts of permanent water ballast to maintain stability. Operating errors or other failures in the ballast system have caused several capsize accidents with spectacular consequences. In January 2015, the car transporter HÖEGH OSAKA was saved from complete capsizing only by quickly setting the ship on a sandbank between the Isle of Wight and Southampton. The loss of stability, which let heel the ship over to an inclination of 52 degrees, occurred due to incorrect weight specifications and errors of the crew in the loading and ballasting procedures. Only after a difficult salvage operation the ship, damaged by overturned cars and flooded by leaks, was successfully brought to Southampton. A similar stability accident occurred in July 2006, when the car carrier COUGAR ACE, owned by Japan's Mitsui O.S.K. Lines (MOL) and loaded with 4700 new Japanese cars, suffered a 90 degree list after a ballast failure on a voyage from Hiroshima to the US West Coast. In this case, too, the ship could be saved, but the new cars had to be scrapped.

In 2019, a difficult rescue operation was to be undertaken to free four crew members of a total of 23 from the Marshall Islands-registered vehicle carrier GOLDEN RAY of Isle of Man-registered Ray Car Carriers, which capsized off Brunswick shortly after having left berth. Built by Hyundai Mipo Dockyard the 200 m carrier had loaded new Kia and Hyundai cars produced in a Mexican factory as well as other vehicles for Middle East destinations. In January 2016, the car transporter MODERN EXPRESS of South Korea's Cido Shipping was caught in a severe storm on a voyage from Gabon to Le Havre in the Bay of Biscay. Near to capsizing, the ship had to be abandoned by the 22-strong crew. It drifted towards the French coast and, after initially unsuccessful towing attempts by the German tug CENTAURUS, could only be taken to Bilbao after several days. In December 2016, during a storm in the German Bight, part of the vehicle cargo loaded in Bremerhaven and Hamburg shifted on the car carrier GLOVIS CORONA. The ship anchored 4 days off the Weser estuary in a listed condition, because the draught having increased to 11.4 m did not allow to berth in Bremerhaven. Only after righting the ship by ballasting, GLOVIS CORONA reached port accompanied by tugboats to discharge the damaged vehicles (Fig. 3.13).

The generally high risk of a total loss concerning all types of ro ro ships due to their lack of sufficient leakage safety also affects car carriers. This became

Fig. 3.13 In September 2019, the 2 year-old vehicle carrier GOLDEN RAY capsized off the US port pf Brunswick and had to be broken up in place (Ralf Witthohn)

evident by the sinking of Wilhelmsen's TRICOLOR in December 2002 after collision in the English Channel, by the loss of HYUNDAI No. 105 after collision in May 2004 or the sinking of NORDIC ACE of Mitsui O.S.K. Lines, as well by collision, causing the death of 11 crew members in the North Sea in December 2012. The total losses of these relatively modern ships thus reveal the inadequate regulations regarding their leakage safety.

3.2.2 8000 New Cars from Japan and Korea

At the time of delivery of the 8000 car-capacity THEMIS, completed by Korea's Hyundai-Samho shipyard in June 2016, the ship's charterer and operator, the Swedish-Norwegian joint venture Wallenius Wilhelmsen, has already worked out its schedule for the coming months. After a short waiting period at sea, the newbuilding calls at Yokohama on 9 July, where the first loading takes 3 days time. New Japanese-built cars roll over the 12 m wide, 320 tonnes carrying angled stern ramp onto the cargo decks, of which 13 offer a total area of 66,370 m^2. In Singapore, where the newbuilding is registered on behalf of Copenhagen-based Maersk company, THEMIS is bunkered. The maximum fuel capacity of 3560 m^3

covers a range of 24,000 nm. To keep consumption low, the output of the MAN B&W type main engine has been reduced to 14,000 kW. In order to reduce sulphur oxide emissions, the vessel has been equipped with an exhaust gas scrubbing system. The next loading operations take place in the Japanese ports of Nagoya, Gamagori and Higashiharima, followed by the Korean destinations Kunsan, Pyongtaek and Masan, where also the first cars are being discharged.

THEMIS' next ports of calls are Piraeus, Bristol, Zeebrugge and Antwerp. In Bremerhaven, Zeebrugge and Southampton, cars of European production are simultaneously loaded, before the THEMIS crosses the Atlantic for Baltimore. Three months after the extension of the Panama canal, the waterway is able to accept the 36.5 m postpanamax-type THEMIS, which reaches Manzanillo/Mexico in mid-September. At the end of the month the ship moors in Canada's Port Hueneme, then in the US port of Tacoma, crosses the Pacific and arrives back in Yokohama on 10 October, after a 83 days lasting sailing around the world. The next voyage of the floating car park will again lead to Europe, then via South Africa and Réunion to Australia. In January 2017, the next circumnavigation of the earth to end in April is scheduled from Yokohama.

THEMIS was the fourth in a series of eight newbuildings lead by THERMOPYLAE, THALATTA and THEBEN. Initially, the registry named Toda Shipping of the Greek shipping company Oceanbulk as owner of the ship, later the entry changed to Copenhagen-based Maersk company. The next ships, TITUS, TRAVIATA and TANNHAUSER, were delivered in 2018 to 2020. The newbuildings, as well as the other Wallenius Wilhelmsen operated ships, are connected to a digital, cloud based onshore platform. This enables the use of data analytics to pinpoint improvements in energy efficiency (Figs. 3.14 and 3.15).

3.2.3 Record Holder HÖEGH DELHI

On the basis of a long-term charter contract signed with Norwegian operator Höegh, Norwegian shipowner P. D. Gram in 2007 took delivery of the car carrier HÖEGH DELHI from Uljanik shipyard in Pula/Croatia. At that time described as the world's highest capacity transporter, HÖEGH DELHI, later renamed VIKING QUEEN at the beginning of 2017, and the sister vessel HÖEGH BANGKOK can load up to 7034 cars on 13 decks of a total area of 58,819 m^2. However, the tonnage of 55,755 gt and a deadweight of 16,600 tonnes on 8.8 m draught of the ships remained below the figures of other vehicle carriers. The 199.9 m length, 32.2 m breadth ships are suited to accommodate vehicles of all kinds as well as goods that are loaded on trailers,

Fig. 3.14 Originally painted in orange-red, but later given a new paint scheme, THERMOPYLAE started a series of eight very large Pure Car Truck Carriers (PCTC) for Wallenius Wilhelmsen (Ralf Witthohn)

including breakbulk cargo. The decks 7 and 9 can be hoisted to two different positions. This increases the clear heights of the decks 6 and 8 below them. Depending on the position, the heights of deck 6 can thus be adjusted to 2.8, 3.15, or 4.9 m, while the height of deck 8 can be increased from 2.2 to 2.55 m or 4.3 m. Laid out for heavy units, deck 6 is designed for high individual loads of vehicles weighing up to 100 tonnes or 58.2 tonnes, resp.

The rolling cargo comes on board either via the angled stern ramp or over the side ramp. Both ramps are located on the starboard side of the vessel serving as access to deck 6. The angled stern ramp is laid out for an SWL of 100 tonnes. It has a passage width of 10.1 m and a height of 5.1 m. The side ramp is 22 m long and 6.5 m wide and designed for an SWL of 20 tonnes. In contrast to comparable ships with ramps installed continuously in the direction of travel, the ramps in the front and rear deck area are arranged on approximately the same ship length, so that the vehicles have to change direction twice to reach the next deck (Fig. 3.16).

Fig. 3.15 The 2019-delivered TRAVIATA, sixth in a series of eight car and truck carriers and similar to the THERMOPYLAE type, is fitted with a foldable, 320 tonnes carrying angled stern ramp lowered by a winch system (Ralf Witthohn)

3.2.4 More Capacity on Extended HÖEGH ASIA

In the beginning of 2018, Höegh Autoliners employed 53 vehicle transporters in its worldwide services. Among the largest were HÖEGH ASIA, HÖEGH TOKYO, HÖEGH SEOUL, HÖEGH BERLIN, HÖEGH NEW YORK, HÖEGH DETROIT, HÖEGH SHANGHAI, HÖEGH LONDON, HÖEGH ST. PETERSBURG, HÖEGH COPENHAGEN and the US-flagged ALLIANCE NORFOLK and ALLIANCE ST. LOUIS registered for the Wilmington Trust Co. One of the ships, named HUAL EUROPE, had run aground in a typhoon off the Japanese coast in October 2002 and become a total loss. At a length of 199.9 m, the ceu capacity of the first newbuildings amounted to 6300 units on delivery. In 2008, the ships were lengthened to 228.8 m, so that their capacity increased to 7850 ceu, corresponding to that of the more recent buildings of the series. The rolling cargo is distributed over 12 cargo decks. In order to ensure the stability of the towering ships with a weather deck distance of 32.6 m from the baseline, their ballast capacity figured at 9638 m^3. The ballast tanks are located in the double bottom and

Fig. 3.16 HÖEGH DELHI (picture) and HÖEGH BANGKOK had the highest capacity of all car carriers when being commissioned (Ralf Witthohn)

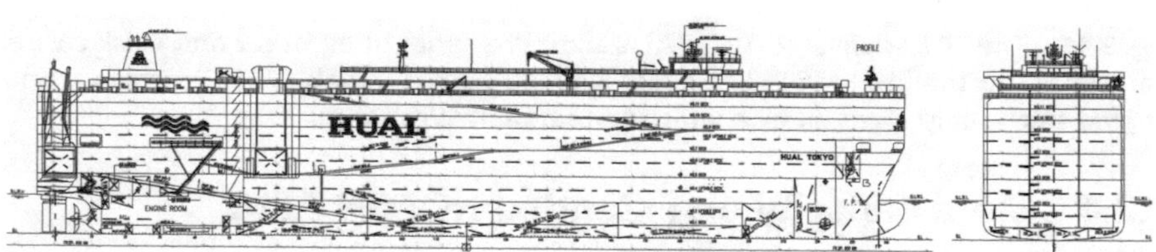

Fig. 3.17 The cross-section of the HÖEGH BERLIN shows the ballast tanks ending below the waterline, therefore offering a reduced leakage safety in the event of a collision (Höegh)

in the side tanks, which, however, only extend up to deck 3, although deck 5 at 14 m above the base was defined as freeboard deck. This means, that above deck 3, the 10 m draught ship is of a single-hull construction, resulting in a reduced level of leakage safety.

The HÖEGH fleet was in 2008 joined by the former MAERSK WELKIN as HÖEGH CHIBA and MAERSK WIZARD as HÖEGH KOBE, after the Danish A.P. Møller group had terminated ro-ro operation. Built 2 years earlier by Daewoo to a similar design but of only 179.9 m length, these vessels were also lengthened, to 199.9 m, thus achieving a capacity increase to 6000 ceu (Fig. 3.17).

3.2.5 New World Record Holder HÖEGH TARGET

HÖEGH DELHI and HÖEGH BANGKOK lost their first capacity rang in just a few months after delivery. From September 2007 on, Sweden's Wallenius Lines took delivery of the 231.6 m length FIDELIO, FAUST, FEDORA, ANIARA and OBERON of 8000 ceu capacity from Daewoo shipyard. Eight newbuildings of the THERMOPYLAE class, commissioned by Wallenius-Wilhelmsen between 2015 and 2020, have a further increased loading capacity of over 8000 units.

After only a short time again, THERMOPYLAE yet lost the title of the world's largest car carrier to HÖEGH TARGET, which entered service in July 2015. Höegh Autoliners specify a capacity of 8500 ceu for the carrier. After European, Japanese and Korean shipyards, now for the first time Chinese shipbuilders from Xiamen Shipbuilding Industry in the Hai Cang invest zone laid the keel of the next record holder. At a length kept below 200 m due to port restrictions in Japan, the breadth was increased from the panamax breadth to 36.5 m, the number of cargo decks to 14 offering a deck space of 71,475 m^2. In order to create large deck heights for vehicles up to 6.5 m high, five of the decks with a total area of 28,145 m^2 can be hoisted. At 375 tonnes SWL, the 45 m long angled rear ramp is laid out for large and heavy vehicle.

The sister ships of HÖEGH TARGET, named HÖEGH TRIGGER, HÖEGH TRACER, HÖEGH TRAPPER, HÖEGH TRAVELLER and HÖEGH TROTTER, joined the fleet in 2015 and 2016. Despite the introduction of the high capacity newbuildings Höeigh's fleet list had in April 2020 decreased to a number of 44 units by the decommissioning of elder ships (Figs. 3.18 and 3.19).

3.2.6 Fiats from Civitaveccia to Veracruz

In June 2015, Italy's Grimaldi group signed a contract about the delivery of five 7800-ceu ships costing a total of USD 300 million with China's Yangfan shipbuilding group, including an option to deliver seven more newbuildings. The ships were intended to be used for the transport of Fiat Chrysler cars from Civitaveccia to North and Central America. The pure car truck carrier newbuildings from China have the same main dimensions as the HÖEGH TARGET type built at Xiamen shipyard. However, the reduction to 12 instead of 14 cargo decks with a surface area of 62,000 m^2 resulted in a smaller capacity of 7742 cars. The rolling cargo reaches the sixth deck over the 7.6 m wide, 150 tonnes SWL stern ramp or via the 4.7 m wide, 15 tonnes starboard

Fig. 3.18 A large variety of rolling cargo, as seen in front of HÖEGH TRAPPER, can be transported by most of the vehicle carriers (Ralf Witthohn)

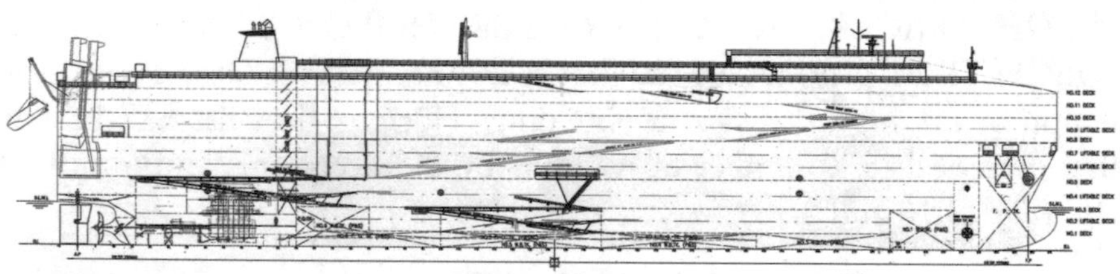

Fig. 3.19 From this 32.2 m breadth 12-deck design, Xiamen Shipbuilding developed the world's largest 36.5 m car carrier type HÖEGH TARGET featuring 14 decks for 8500 ceu (Xiamen)

side ramp adjustable to deck 6 or 7. All decks have a minimum clear height of 2 m. When decks 3, 5, 7 and 9 are raised by 1.7 or 2 m, the heights of the decks below increase accordingly, deck 2 from 2 to 4 m, deck 4 from 2.35 to 2.65 m or 4.35 m, deck 6 from 3 to 3.6 m or 5.3 m, deck 8 from 2.2 to 2.5 m or 4.2 m. Fixed ramps provide the connection between decks. The bulkhead deck is deck 6, which height of 15.3 m above the base creates a freeboard of 5.8 m at the maximum draught of 9.5 m. The ramps leading down from deck 6 are movable and, by hoisting, provide watertight leakage protection. The fixed ramp between decks 6 and 7 is fitted with a lockable door at both ends, so that watertightness is also achieved here (Fig. 3.20).

At 9.5 m draught, the 199.9 m length, 35.5 m breadth ships carry15,700 tonnes. The double bottom and forepeak can take 4500 m^3 of water ballast. An MAN B&W diesel engine of type 7S60ME-C8.5-TII putting out 12,312 kW at 101 rpm enables a service speed of 19.2 knots, the fuel capacity of 3600 m^3, stored in central tanks between the second and fifth deck, a voyage of 23,000 nm at a consumption of 48 tonnes per day. The 65,255 gt pure car truck carriers, which delivery started in 2018 by GRANDE TORINO and GRANDE MIRAFIORI, followed by GRANDE HOUSTON, GRANDE NEW JERSEY, GRANDE FLORIDA, GRANDE TEXAS and GRANDE CALIFORNIA in 2020/21, were intended to regularly call at 14 ports, namely Gioia Tauro, Civitavecchia, Livorno, Savona, Valencia, Antwerp, Halifax, Davisville, New York, Baltimore, Jacksonville, Houston, Tuxpan and Veracruz, but the shut-down of the Italian car industry took in the wake of the COVID-19 pandemic forced to partly revise the schedule and to deploy them to other routes.

Only a few weeks after the order, Grimaldi ordered three further, smaller ships of panamax breadth and a capacity of 6700 ceu from Yangfan. Jinling Shipyard in Nanjing, part of the shipyard group, delivered the first of the

Fig. 3.20 Italy's Grimaldi group contracted seven newbuildings of a 7800 ceu pure car truck carrier type for the export of Fiat cars and other cargo to America, but the shut-down of the Italian car industry took the just 2 months old GRANDE HOUSTON from Setubal, Tilbury, Antwerp and Rotterdam to Bremerhaven (Ralf Witthohn)

newbuildings, the 199.9 m length, 32.3 m breadth GRANDE BALTIMORA in July 2017, the sister ship GRANDE NEW YORK in October and the GRANDE HALIFAX in January 2018. On this transporter type, too, four of the decks can be raised, so that vehicles or other rolling cargo can be loaded up to a height of 5.2 m through the 150 tonnes carrying stern ramp. The 19 knots fast GRANDE BALTIMORA entered the weekly liner service of the shipping company, which had been established in February 2015 by the use of four ships with a total capacity of 25,000 cars from Gemlik/Turkey via the Italian ports of Gioia Taura, Civitaveccia and Savona to the North American East Coast and Antwerp. These Grimaldi carriers, too, were employed in other trades, like the GRANDE NEW YORK, which undertook sailings from the Far East to northern Europe in early 2022.

3.2.7 Mercedes Cars from Uusikaupunki

Grimaldi group's business relationship with China's Yangfan group stemmed from a ro ro ship order of Finnlines, the Finnish subsidiary of the Italian company. The order concerned eight FINNBREEZE freight ferries built between 2002 and 2012 by Jinling shipyard and which are also used for the transport of new cars. In 2013, ships of this type for the first time discharged Mercedes cars produced in Finland in Bremerhaven on their way to Bilbao and Santander. In October 2013, the cargo of the FINNSUN was damaged, when the ship was caught in hurricane in the German Bight. Part of the cargo stowed on the main cargo deck shifted and damaged 12 new cars. In 2017, a EUR 70 million conversion programme to be carried out until May 2018 on four of the ferries was launched at Remontowa shipyard in Gdansk. In November, the lengthening of the first ship, the 2012-built FINNTIDE, to 217.8 m by 30 m was completed increasing the ferrie's capacity by 1000 to 4213 lane metres. At the end of the month the vessel returned to the Uusikaupunki-Travemünde service, from where the next vessel sailed to Poland to be enlarged (Fig. 3.21).

In 2008, Finnlines' ro ro ferries operating in North Sea and Baltic Sea services moved from the old cargo terminals of their home port of Helsinki to the new port of Vuosaari, located east of the capital and constructed as the countrie's largest port project in history at costs of EUR 607 million. The container and ro ro handling facilities include 20 berths with a water depth of 11 m and are connected to the Finnish hinterland by a 13.5 km long railway and a shorter motorway tunnel (Fig. 3.22).

Fig. 3.21 The cargo of the FINNSUN was damaged in a strong storm in 2013 (Ralf Witthohn)

Fig. 3.22 The newly built port of Vuosaari offers freight ferries such as the FINNTIDE (left), passenger/car ferries and container feeders a total of 20 berths (Ralf Witthohn)

3.2.8 First Angled Stern Ramp on MADAME BUTTERFLY

Initially, car transporters had been loaded using side ramps, but in 1981/82 the first newbuildings of Sweden's Wallenius Lines were fitted with an angled stern ramp that could also carry heavy vehicles. These were MADAME BUTTERFLY, FIGARO, CARMEN and MEDEA delivered by Kockums shipyard in Malmö and scrapped between 2013 and 2016. At 198 m and 32.3 m the ships featured the car transporter length and breadth that remained common for a long time. They were laid out for 5230 cars. When loaded with 500 trucks of up to 6.5 m height, there was still space for 2380 cars. Of the same type with a slightly larger capacity were TRISTAN and ISOLDE ordered in 1983. The first newbuildings with a more aerodynamic round garage front, came from Japan's Hitachi shipyard as FAUST and FALSTAFF in 1985. In the same year the shipping companie's flag was hoisted on the Japanese car carrier KAIJIN MARU, after being renamed PARSIFAL, and on the purchased AEGEAN BREEZE, ARABIAN BREEZE, ASIAN BREEZE and BALTIC BREEZE (Fig. 3.23).

In 1991/1992 the 5870 cars accommodating newbuildings AIDA and OTELLO from Daewoo shipyard entered service. In 1994, a new series of 11 ships was launched as TITUS, TURANDOT, DON JUAN, DON CARLOS, DON PASQUALE, DON QUIJOTE, ELEKTRA, MANON, MIGNON, BOHEME and UNDINE offering capacities of 5870 cars. Five of the BOHEME class ships were lengthened by 28 m to 228 m in 2004, increasing their capacity to 7100 units. In the same year, Daewoo shipyard received four

Fig. 3.23 An angled stern ramp as first fitted to the MADAME BUTTERFLY in 1981 has since then remained an indispensable feature of all large vehicle carriers, including Wallenius' 2008-built OBERON (Ralf Witthohn)

of the then largest PCTC of 8000-car capacity for delivery in 2007/08, named FIDELIO, FAUST, FEDORA, ANIARA and OBERON. At panamax breadth their length exceeded the traditional 200 m mark for car carriers by figuring 231.6 m. In 2010, the 6300 units loading Japanese newbuildings PORGY and BESS of Toyohashi Shipbuilding were taken in long-term charters by Wallenius. From 2011 on, the 265 m length ro ro ships PARSIFAL and SALOME for 6000 cars came from Mitsubishi, the car and truck carriers CARMEN and FIGARO from Daewoo shipyard and the 6500-car capacity TOSCA from Mitsubishi (Fig. 3.24).

Fig. 3.24 PARSIFAL and SALOME were, unlike the traditional garage type car carrier types, commissioned as ro ro carriers with a larger deadweight capacity and an angled stern ramp of extraordinary 505 tonnes SWL (Ralf Witthohn)

3.2.9 Former Ro Ro Carriers under US Flag

Norwegian company Wilh. Wilhelmsen had some of its ro ro carriers, originally designed for the transport of containers on deck for a liner service to Australia, later converted into vehicle transporters. Their car intake was increased by the erection of, due to the visibility conditions, step-like garages above the weather deck and a small number of containers at the sides of the garages. The ships in question were named TAPIOLA, TOBA, TAMPERE, TALABOT and TOURCOING, built in Japan and Norway in 1978/79, and TEXAS, TAIKO and TAMPA, which had entered service in 1984. At the end of 2017, the only ship of this type still active was the 1996-built TARONGA. Its management under the name ENDURANCE and the US flag was in 2010 taken over by American Roll-on-Roll-off Carrier (ARC) in Wilmington, founded in the same year by Wallenius and Wilh. Wilhelmsen to cover US military transport needs that are mandatory to be carried out under the US flag. In January 2017, ARC deployed RESOLVE ex Wilhelmsen's TANABATA, ENDURANCE and FREEDOM ex TAKAMINE to carry out comprehensive military transports as part of the Atlantic Resolve manoeuvre in Eastern Europe, including the transport of 2500 military vehicles from Beaumont to Bremerhaven.

In early 2018 the US-Scandinavian joint venture, describing itself as the second largest US-flag carrier in international trades, operated ENDURANCE, RESOLVE and FREEDOM, HONOR ex TAKASAGO, INDEPENDENCE II ex TITUS, INTEGRITY ex OTELLO, PATRIOT ex AIDA and LIBERTY ex TOPEKA. In September 2019, ARC added ARC RESOLVE ex Wallenius' OTELLO to its fleet, in October 2019 ARC INTEGRITY ex FEDORA. On its first voyage the latter carried helicopters, vehicles, containers and support equipment of a US army brigade after participation in the NATO manoeuvre Atlantic Resolve from Rotterdam back to the US. In February 2020 the berthing of the ENDURANCE carrying about 1200 military vehicles of a US brigade was blocked at Bremerhaven by Greenpeace activists for about 1 h. The troop movements and transports within the NATO manoeuvre Defender Europe 20 in eastern Europe were carried out via a total of 14 sea- and airports in western Europe.

The shipping company of Wilh. Wilhelmsen, which can look back on a long history of transporting troops and military goods of the Allies starting during the Second World War, supported the UN mission to destroy chemical weapons from Syria in 2014 on the subsequently scrapped car transporter TAIKO as well as the US government in foreign military operations on several occasions. The military importance of vehicle transporters was underlined when Wilhelmsen concluded a 7-year contract about transport services for the Norwegian armed forces in March 2015. Within the contract, the shipping company and the NorSea Group, which is 40% owned by Wilhelmsen and 28% each by the Norwegian shipping companies Eidesvik and Simon Møkster, founded a joint venture that established nine supply bases along the Norwegian coast for national defence purposes (Fig. 3.25).

3.2.10 Former Russian Tank Transporters

Wallenius Wilhelmsen operated also smaller former ro ro carriers after conversion to vehicle transporters. The Russian twin-screw vessels NOVOROSSIYSK and SOCHI, built in 1994 and 1996 by Baltiyskiy Zavod in St. Petersburg, fitted with a 120-tonne stern ramp and thus suitable for the transport of military vehicles, were equipped with a stepped garage at COSCO Nantong shipyard in 2004/05. Carrying up to 1975 cars the ships were able to transport high & heavies on three decks. They were chartered out to Wilhelmsen by owner Eidsiva Rederi unter the names VINNI and VIBEKE to carry out short-haul services in the Far East, and later also in Europe. In 2014, Norwegian Car Carriers (NOCC), a merger of Eidsiva and Dyvi

Fig. 3.25 ENDURANCE is Wilh. Wilhelmsen's former TARONGA, built in 1996 as a roll-on-roll-off vessel and converted to a vehicle carrier by fitting a garage on the main deck (Ralf Witthohn)

Shipping from 2010, sold the ships, which original equipment had largely been replaced, for demolition (Fig. 3.26).

3.2.11 Vietnamese Premiere VICTORY LEADER

In April 2010, the first car carrier built in Vietnam was delivered by Ha Long shipyard of the state-owned Vinashin shipbuilding group to Ray Car Carriers under the name VICTORY LEADER. The Israeli shipping company is not a liner operator, but charters its ships out to established car carrier operators, in this case to Japan's NYK Line on a long-term basis. The 185.6 m length, 32.3 m breadth transporter can load 4900 cars on 13 decks. The second newbuilding from Vietnam, the sister ship VIOLET LEADER, also entered a charter with the Japanese shipping company after delivery in 2011. In 2019, VICTORY LEADER served as feeder vessel for United European Car Carriers (UECC), jointly owned by Nippon Yusen Kabushiki Kaisha (NYK) and Wallenius Lines. The fleet list of 1992-founded Ray Car Carriers, describing itself as the largest tonnage supplier of car carriers, comprised 52 car carriers in 2020 and in 2021. Built in Poland, Croatia, Vietnam and South Korea they offer loading capacities from 2200 ceu to 7700 ceu (Fig. 3.27).

Fig. 3.26 The Russian ro-ro vessel NOVOROSSIYSK, built in St. Petersburg in 1994 and suitable for military transport, became the car carrier VINNI after conversion in 2004 on Norwegian account (Ralf Witthohn)

Fig. 3.27 As the first Vietnamese-built vehicle transporter VICTORY LEADER of Israeli Ray Car Carriers was chartered to NYK and later engaged as feeder ship in Europe (Ralf Witthohn)

3.2.12 Jaguars from Immingham

After a long warning tone, the light red painted pilot boat NÜBBEL leaves the Cuxhaven port entrance between Alte Liebe and Steubenhöft at high speed to drop the pilot at the ladder of the roll-on roll-off vessel JUTLANDIA SEAWAYS arriving from Immingham. The freight ferry of Denmark's DFDS Tor Line, sailing under the British flag, has only a few hundred metres to proceed before mooring without tug assistance at the ro ro facility of the Cuxhaven Europakai. Shortly afterwards the stern ramp is lowered and the first trailers roll over the two-lane terminal ramp (Fig. 3.28).

Since 1997, the modern 200-tonne carrying Cuxhaven ro ro ramps have created a link to Europe's short-sea traffic scheme. The blue-painted freight ferries of DFDS Tor Line provided a connection on the so-called Elbe Bridge five times a week from the river quay of Cuxport to the UK. Actually, six sailings are made per week. In September 2017, the fifth million export car to the UK since 1982 was recorded. In August 2019, 15 years after the start of BMW car shipments, the two millionth car for the British market rolled on board the ro ro vessel JUTLANDIA SEAWAYS of Danish operator DFDS. Steel, paper, timber, building materials and project cargo at a great share received by rail are shipped to Immingham. New cars imported from England include Jaguar and Landrover models since 2009. In total, DFDS maintained 16 freight ferry lines across the North Sea, six in the Baltic, five across the British Channel and seven in the Mediterranean.

Fig. 3.28 The car transporter SCHELDE HIGHWAY of the "K" Line subsidiary KESS loads VW cars for Baltic Sea ports in the Amerika-Hafen of Cuxhaven (Ralf Witthohn)

The Cuxhaven car terminal is operated by Bremen-based BLG group and Cuxport with its shareholders Rhenus and HHLA. The terminal can accommodate up to 12,000 cars. Cars of BMW, Mercedes and VW brands are shipped as well. The car transporters of Japan's "K" Line subsidiary KESS load VW cars for Scandinavia, Russia and the Baltic States. United European Car Carriers (UECC) transports BMW vehicles to Southampton, the Spanish shipping company Suardiaz to Immingham. Once a week the freight ferry ML FREYJA makes a stopover at the Elbe estuary within its regular service from Bremerhaven and Harwich to Tallinn and Turku. The terminal has four berths with a total length of 1080 m. In contrast to the declining figures in Northern Europe, Cuxhaven increased its vehicle turnover by 15% in 2016 and handled 480,000 new cars, driven by a strong increase of imports. In 2017, the figure decreased to 476,000, and further to 423,000 in 2018. In 2019 vehicle handling declined for another time, in 2020 it reached 505,000. In November 2020, the Wallenius Wilhelmsen deep-sea car carriers TIGER and TORINO loaded 3500 commercial vehicles for Shanghai, Tianjin and Guangzhou.

Similar to container shipping, there are feeder and short-sea services in vehicle transport, in which smaller vessels are used. For distribution in northern Europe, Japan's "K" Line subsidiary, Bremen-based KESS chartered the car feeder ships ELBE HIGHWAY, SEINE HIGHWAY, DANUBE HIGHWAY and THAMES HIGHWAY from Ray Car Carriers on a long-term basis, and later acquired them. Built in 2005 by Stocznia Gdynia in a series of six ships of 148 m length and 25 m breadth, the high ice-class ships can on eight decks carry 1600 cars or vehicles of all kinds weighing up to 70 tonnes and loaded over two stern ramps. In 2015, exhaust gas purification systems were installed in the ships' enlarged funnel casings. In September 2017 Greenpeace activists boarded the ELBE HIGHWAY off the port of Sheerness to prevent the import of diesel-powered VWs (Fig. 3.29).

3.2.13 Kias from Koper to Tarragona

There are 780 kilometres of road from the Hyundai production site in the Nosovice special economic zone in the eastern part of the Czech Republic to the Slovenian port of Koper, and 730 kilometres from the Kia plant in Zilina, Slovakia, which is not far away. Nevertheless, the Korean industrial group Kia has decided to export its production destined for the Iberian Peninsula to Tarragona by ship from the beginning of 2016, instead of land transport. Its subsidiary Glovis Europe, founded in Frankfurt in 2006, organised the transport and commissioned Greece's Mediterranean Car-Carriers Line (MCCL) to ship about 50,000 Hyundai and Kia cars per year (Figs. 3.30 and 3.31).

At the beginning of 2018, MCCL operated five car carriers, named SEA HELLINIS, SEA PATRIS, SEA AMAZON, SEA AEOLIS and SEA ANEMOS, which were still active in 2022. Of the seven liner services previously operated, five were still maintained at that time. Two years later the

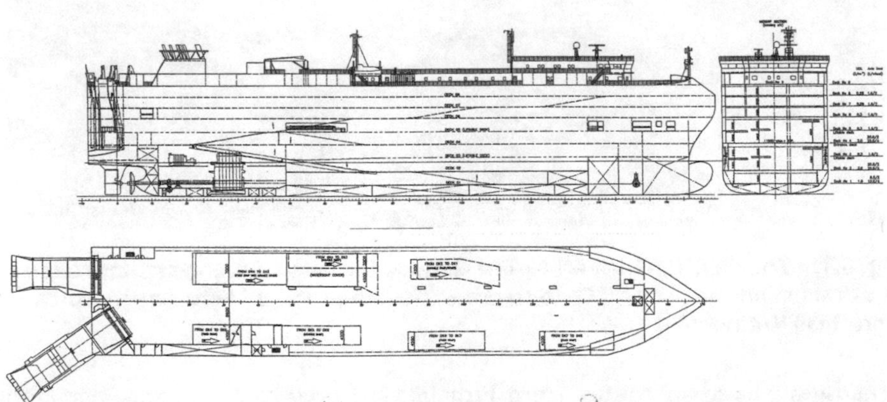

Fig. 3.29 The ELBE HIGHWAY type transporters can carry 1600 cars, their ramps are designed for heavy vehicles (KESS)

Fig. 3.30 The Greek-owned SEA HELLINIS transported Ford cars from Turkey to North Sea ports (Ralf Witthohn)

Fig. 3.31 The MEDITERRANEAN SEA of German owner Lauterjung was chartered by Greek shipping company MCCL to transport Turkish-produced Ford cars to North Sea ports (Ralf Witthohn)

schedules named six routes, from Piraeus via Koper to Tarragona, Barcelona and Piraeus, from Autoport (Istanbul) to Piräus, Tarragona, Zeebrugge, Antwerp, Amsterdam, Bremerhaven, Misurata and Benghazi, from Piraeus to

Alexandria, Port Said, Beirut, Tartous, Mersin and Limassol, from Piraeus to La Goulette, Djen Djen, Salerno and Marseille/Fos, Piraeus, from Piraeus to Derince, Novorossiysk and Constanta and from Piraeus to Misurata and Benghazi. The routes were still maintained in 2022.

In Istanbul, Ford Transit cars from the Gölcük factory are loaded, including those destined for North America, which are transhipped in North Sea ports. A liner service was discontinued from Incheon and Shanghai to Aqaba, Alexandria and Misurata. The most modern of the car carriers is SEA HELLINIS, a ship delivered in 2005 from Japanese Mitsubishi shipyard in Shimonoseki with a capacity for 4537 cars. MCCL also charters tonnage, like the MEDITERRANEAN SEA of Emden/Germany based, which participated in the transport of Ford cars to Bremerhaven in 2016. The ship represents a 4900-ceu type, built 18 times at Xiamen shipyard and 8 times at Yangfan shipyard. Elder is the 1984-built SEA PATRIS of 3742-car capacity, delivered in 1985 by Imabari Zosen for operation by Hamburg-based company Ahrenkiel as FRANCONIA in a "K"Line charter (Figs. 3.32 and 3.33).

3.2.14 Nissans from Sunderland

Following delivery 2010/2011, the car transporter CITY OF ST. PETERSBURG and CITY OF ROTTERDAM were employed to transport a maximum of 2000 Nissan cars from the English factory in Sunderland to European ports, and later also on Mediterranean routes. The Pure Car Carriers were constructed according to the high ice class IA of French class society Bureau Veritas, so that the ship's schedule could cover the Russian market. Loading and unloading is carried out via two stern ramps, a smaller one on the port side of the transom and a larger, angled one on the starboard side. In March 2022, Nissan suspended vehicle exports to Russia due to logistic challenges related to the Russian war against Ukraine.

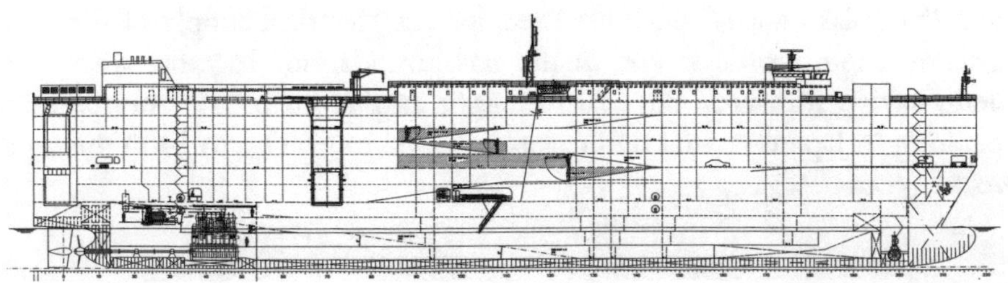

Fig. 3.32 Lauterjung's MEDITERRANEAN SEA represents a 4900-ceu type, 26 times realised by two Chinese builders (Lauterjung)

Fig. 3.33 35 years after delivery as FRANCONIA, SEA PATRIS was operated by its Greek owner in liner services from Piraeus (Ralf Witthohn)

The design of the car transporters delivered by the Kyokuyo Shipyard Corporation in Shimonoseki to Japan's Nissan Group and registered in Panama for Fair Wind Navigation paid particular attention to the phenomenon of air resistance, sometimes neglected in shipbuilding but of particular interest in this type of ship. Although efforts had already been intensified to give the initially angular architecture of car transporters a rounder and thus more streamlined shape, the architecture of the ships follows this idea more than all other designs. According to the builders, the spherical fore ship contour can reduce air resistance by up to a half and thus save fuel of several hundred tonnes annually. An engine developing almost 7000 kW is sufficient to accelerate the 140 m length and 22.4 m breadth car to a speed of 17 knots (Fig. 3.34).

Nissan's efforts to apply the fuel savings it aims to achieve in the construction of automobiles in a similar way to the car transporters produced by the group were also reflected in another newbuilding. Acquired from Shin Kurushima Shipyard in January 2012, the NICHIOH MARU was equipped with 281 solar panels, which are used for the electrical supply of the LED lights on the cargo decks and in the accommodation. Together with other energy-saving measures, the 170 m length and 26 m breadth newbuilding operating in Japan's coastal traffic, was reported to save up to 1400 tonnes of fuel per year.

Fig. 3.34 To reduce air resistance, the CITY OF ROTTERDAM was given a spherical front (Ralf Witthohn)

3.2.15 Buses for Havana

From 1998 to 2008, the Italian shipping company Grandi Navi Veloci employed the ro ro ferry VICTORY, at last in a liner service between Genoa, Barcelona and Tangier. Built in 1989 by the Mitsubishi shipyard in Shimonoseki the ferry measured at 17,113 gt. After being sold to the Mexican shipping company Baja Ferries, the ship, renamed CHIHUAHUA STAR, made an unusual detour via northern Europe, in order to compensate for the costs of the long transfer voyage. The vessel loaded buses destined for Havana in Paldiski/Estonia. Cargo transports to Cuba were not in demand from other shipping companies, because ships calling at Cuba at that time were locked out of US ports for half a year. After arrival in Mexico, the ferry, which had a capacity of 1200 passengers, started a regular service between La Paz, Mazatlan and Topolobampo. The 187.1 m length, 27 m breadth ship achieved a speed of 23.5 knots by two MAN diesel engines with a total output of 21,100 kW. The CHIHUAHUA STAR was renamed CARIBBEAN FANTASY in 2011 and employed in a service between Puerto Rico and the Dominican Republic. In August 2016 the ship caught fire due to the bursting of a fuel hose in the engine room off San Juan. 512 passengers and crew members had to be evacuated. In the following year the ship was scrapped in Aliaga/Turkey (Fig. 3.35).

3.2.16 Double-Deck Coaches for Ashdod

In April 2017, the cranes of the multipurpose cargo vessel NORMED ANTWERPEN hoist double-deck railway carriages from the Bombardier plant in Görlitz into the ship's cargo hold. The trains are destined for Ashdod, where in the following months a total of 60 of the railcars will be shipped in several lots from Neustädter Hafen in Bremen. The second party is taken over by the freighter NORJAN at the beginning of May. Both ships belong to the Brunsbüttel-based company Erwin Strahlmann and are chartered by Dutch Normed Line, which operates a liner service with a total of three ships from northern European ports to the Mediterranean Sea. Ports of call are Safi, Izmir, Diliskelesi, Gemlik or Derince, depending on available cargo (Fig. 3.36).

3.2.17 Loss-Making Investments in Multipurpose Cargo Vessels

Multipurpose cargo vessels are regularly used to transport special vehicles such as the railway wagons for Israel and similar project cargoes. Along with container ships, vessels of this type were among the preferred finance objects of German investment companies. Two of them, the 11,250 dwt ships NORMED ANTWERPEN and NORJAN, were delivered in 2007 within a ten-ship series from Korea's Daehan Shipbuilding, including the sister ships NORMED BREMEN, NORMED AMSTERDAM, BOSPHORUS, TEAM PARTNER and S. BOTHINA for German and Dutch shipping companies.

Fig. 3.35 The Japanese-built ro ro ferry CHIHUAHUA STAR took buses from Paldiski to Havana in 2008 (Ralf Witthohn)

Fig. 3.36 The German multipurpose cargo vessel NORMED ANTWERPEN took over a first batch of double-deck cars for Israel in Bremen (BLG)

The TEAM SPIRIT, chartered by Normed Lines as NORMED BREMEN after delivery in 2006, was no longer able to pay interest, when the worldwide shipping crisis continued. The ship was sold by forced auction to Foroohari Shipping from Stade/Germany in December 2014 at EUR 2.5 million at the Bremerhaven/Germany district court. This meant only a fraction of the new-building price, so that financing Bremer Landesbank Kreditanstalt Oldenburg had to give up the majority of its claims. After all, there were two parties interested in the ship, which had been arrested at the bank's request, first in Bremen and afterwards in Bremerhaven. The bank, as the principal creditor, asserted three mortgages for a total amount of approximately EUR 17.7 million. In total, the company owning the TEAM SPIRIT had more than EUR 18 million in debts. The representative of an Ahrensburg/Germany registered shipping company started the bids with EUR 1 million, and after 20 further bids in small increments of EUR 10,000 and EUR 20,000, the lot went to Bijan Foroohari, who put the freighter back into service under the name of BF TIMARU (Fig. 3.37).

The situation of unemployed ships was made clear in a letter from the Bremerhaven port captain read out by the bailiff. The port captain called

Fig. 3.37 The multipurpose general cargo ship TEAM SPIRIT was auctioned at Bremerhaven, when the shipping market did not allow to satisfy the creditors anymore (Ralf Witthohn)

upon the new owner of TEAM SPIRIT to have the ship removed from the port immediately or to apply for appropriate permits to stay. The long crisis had turned the Bremen ports into a collection place for bankrupt freighters at this time. Especially in the fishing port of Bremerhaven and Neustädter port in Bremen, the ships had been arrested due to financial problems of their owners.

Auction was also imposed on the 11,211 dwt multipurpose cargo vessel KATHARINA of Germany's Intersee Schiffahrt, which had been arrested in Neustädter port. Its management company had declared insolvency in September of that year. In November 2014, KATHARINA came under the hammer at the Bremen district court and was put back into service under the name MARMAUI by Hamburg-based shipping company MarConsult. On the same day, the 29,270 dwt bulk carrier LUEBBERT of Intersee, which was laid up with sister ship FRITZ in Neustädter Hafen, was as well up for forced sale. These were also purchased at auction by MarConsult to be operated as MARBIOKO and MARBACAN. As one of the largest shipping locations in Germany, Haren an der Ems was affected by a corresponding number of ship insolvencies. Already in February 2014, four ships of the local Intersee group, JULIETTA, NADJA, ANGELIKA and CELIA, were auctioned simultaneously at the Bremen court. Three of these ships were as well purchased by the shipping company MarConsult, which in the following months took advantage of the weak market to expand its fleet by 16 second-hand vessels.

Not all the ships for which an application for insolvency was filed at the Bremen district court were arrested in the port there. The 5575 dwt multipurpose cargo vessel FRANKLIN STRAIT owned by a fund of the same name with shares issued by Hamburg-based finance house König & Cie. had dropped anchor off Port of Spain/Trinidad under a charter of the US shipping company Bernuth. The arrest in the former pirate's nest became kind of symbol of the bankruptcy series, in which investors lost their investments as quickly as the victims of pirates once were robbed their belongings.

3.2.18 Auctioning VICTORIA

24 November 2014, room no. 100 of the Bremerhaven court: After the bailiff has unlocked the door, only a few people enter the largest room of the court, which has publicly announced the "Foreclosure of the steel cargo ship VICTORIA, registered in the shipping register of the Emden District Court sheet 5311, home port: Haren/Ems", as it is described. The bailiff declares the necessary regulations, sets a minimum bid of EUR 237,326 to cover the costs of the proceedings and the guarding of the ship, and figures the claim of the main creditor. Norddeutsche Landesbank Girozentrale Hannover, as the ship's financier demands EUR 12.8 million, another creditor from the Netherlands EUR 511,000. The market value of the ship has not been determined, but according to an expert opinion, which has been used as base to calculate the court fees due, the value is estimated at EUR 5.9 million.

Ten minutes later the actual bidding, which is limited to half an hour, begins. The managing director of the Hamburg shipping company Marconsult, accompanied by shipowner and sole shareholder Mad Dabelstein, goes to the bailiff, makes his offer and provides a security deposit of 10%. The bid is announced by the bailiff: EUR 8 million. A few minutes pass, before the bank representative goes to the bidder for a short interview. Thereupon the shipping company increases its bid to EUR 8.633 million. After the only bidder denied the question of a bid increase, the bailiff hits the hammer three times.

The date of the announcement is set 2 days later. The owning company of the 10,500 dwt cargo ship VICTORIA is now registered as Marmorotai Schifffahrtsgesellschaft GmbH & Co. KG in Hamburg. The ship is to be renamed THORCO CROWN indicating the Danish charterer Thorco Shipping. A week later, Marconsult also auctions the vessels KATHARINA and LUEBBERT at the Bremen district court, renames them MARMAUI and MARBIOKO and takes over the sister ship FRITZ for their Marbacan Schifffahrtsgesellschaft (Fig. 3.38).

Fig. 3.38 Shortly after the forced sale, VICTORIA entered a charter of Thorco Shipping under the name THORCO CROWN (Ralf Witthohn)

The auctioning of the 10,500 dwt VICTORIA was the second major event in the ship's history. In January 2004 delivered by China's Yichang shipyard, the ship was in May 2009 highjacked by Somali pirates and only released with its crew of 11 Romanians 2 months later. In 2004, the Antiguan-flagged vessel was given the charter name ONEGO MAAS and in 2019 for the first time carrying its company name MARMOROTAI. In December 2020, Bremen-based Candler Schiffahrt became manager of the ship being renamed ELBA.

3.2.19 Hundreds of KG Insolvencies

The forced sale of VICTORIA was the result of an insolvency, as it affected hundreds of so-called one-ship companies, once founded in the form of *Kommandit-Gesellschaften* (KGs) under German law, after the outbreak of the shipping crisis caused by the financial crisis of 2008. With high yield promises and tax benefits capital was collected from private investors according to the limited partnership model and invested in ship funds, in which about half of the money was supplemented by banks. When after the outbreak of the crisis the calculated rates failed to materialise and the ships imported losses or had to be decommissioned, the investors lost their money. At some of the co-financing banks, the bad loans grew into billions, even though they were able to save part of their investments as preferential creditors.

A special role in ship financing was played by Germany's Landesbanken, which entered the international lending business by abandoning their original function and caused horrendous losses for their owners, the federal states. HSH Nordbank mainly owned by the federal states of Hamburg and Schleswíg-Holstein, by a loan volume of EUR 20 billion stylizing itself as the world's largest provider of ship finance, passed on its non-performing loans to the public sector. After years of support, the state governments concerned in October 2015 reached an agreement with the EU commission on the assumption of loans in the amount of EUR 6.2 billion and the subsequent privatization of the bank by 2018. In July 2016 reports revealed, that 256 of the ships financed by bank credits figured at EUR 5 billion, but representing only half that market value, had been transferred to the states as collateral. In February 2018, the states of Hamburg and Schleswig-Holstein and Giroverband Schleswig-Holstein sold their 94.9% shareholding in HSH, to be renamed Hamburg Commercial bank, to the independent investment companies Cerberus European Investments, J.C. Flowers, GoldenTree Asset Management, Centaurus Capital and BAWAG or to funds initiated by them at around EUR 1 billion. The remaining burden for the states was estimated to be in the double-digit billion Euro range (Fig. 3.39).

3.2.20 Fresh Capital for Bankrupt Ship Investments

More of Germany's federal state banks became victims of their own business policy. In 2017 Norddeutsche Landesbank (NordLB), the bank of the states of Lower Saxony and Sachsen-Anhalt, took over Bremer Landesbank after loan defaults from ship financing had in 2016 accumulated to such an amount, that they could no longer be absorbed by the federal state of Bremen,

Fig. 3.39 The cargo vessel LUEBBERT on its way to Neustädter Hafen, where the ship was subsequently arrested for auction (Ralf Witthohn)

on its own suffering from an acute budget emergency. NordLB in turn was accused by 600 private investors of two multi-purpose cargo vessels of causing them considerable financial losses. The case concerned the multipurpose ships BELUGA FLIRTATION and BELUGA FASCINATION built in China in 2005 and 2007. The Hamburg-based issuing house Ownership Kapital had collected capital from private investors who contributed around EUR 10 million to each of the owning companies, NordLB around EUR 13.7 million each. The ships were chartered out to Bremen-based Beluga Shipping, but after the latter's bankruptcy could no longer cover their costs and were put up for forced auction in mid-2012. Ownership Kapital had worked out a sanitation concept that was yet not accepted by the bank. Instead, NordLB gave fresh capital to the buyer. In doing so, it avoided, that the vessels were bought at auction for a low price and that its own investment was lost. The ships were auctioned off at a price of around EUR 10 million each. Because of the priority of the bank loans, the private investors were left empty-handed. BELUGA FLIRTATION was from 2012 on registered by Pride of Paris Schifffahrt, from 2014 by Auerbach Schiff 2 as LEA AUERBACH, from 2021 by Russia's Sakhalin Shipping and since 2022 by Vostok Trade Invest as SASCO ALDAN (Fig. 3.40).

3.2.21 NATO Tanks from Norway

A mobile harbour crane lifts tank after tank from the hold of the heavy-lift carrier INDUSTRIAL CHARGER onto the quay of Bremerhaven Nordhafen. Even though the ship is equipped with two 200 tonnes lifting cranes and thus has a sufficient lifting capacity for the heavyweights, the faster working local crane is used. The tanks have participated in the NATO manoeuvre Cold Response 16 in northern Norway and are now to be returned to their home base.

Fig. 3.40 To secure its loan NordLB arranged the forced sale of BELUGA FLIRTATION, then managed by Auerbach Bereederung as MAPLE LEA (Ralf Witthohn)

The heavy-lift carrier INDUSTRIAL CHARGER was launched in 2000 under the name VIRGO J for Jüngerhans shipping company in Haren/Ems, Germany. Over a period of 8 years, the German owner received a series of 13 heavy-lift vessels from Portuguese shipyard Viana do Castelo under the type designation Century 8000, started by DIANA J. in 2000 and completed by JOHANN J. in 2008. The 119.8 m length, 20 m breadth ships of 7400 dwt dispose of a 71.6 m long and 15.8 m wide hatch, served by two 200 tonnes lifting cranes. Their container intake is stated as 506 TEU. VIRGO J started a long-term charter as INDUSTRIAL CHARGER by Industrial Maritime Carriers from New Orleans under the flag of the Marshall Islands and was later operated as OCEAN CHARGER, ZEA CHARGER and lately as CHARGER by Interscan Schifffahrtsges. in their fleet of 27 ships along with the sister ship CHALLENGER ex ZEA CHALLENGER (Fig. 3.41).

Further ships of the series entered into charters of the Intermarine group or were taken over by the operator, DIANA J as OCEAN ATLAS, HARUN J as INDUSTRIAL CHIEF, URSA J as INDUSTRIAL CHAMP, LIRA J as INDUSTRIAL CAPE, LUNA J as INDUSTRIAL CENTURY, POLLUX J

Fig. 3.41 The heavy-lift carrier INDUSTRIAL CHARGER ex VIRGO J transported US tanks from the Norwegian port of Hammarnesodden to Germany in March 2016 (Ralf Witthohn)

as OCEAN CRESCENT, AQUILA J as INDUSTRIAL DIAMOND, DRAGO J as INDUSTRIAL DESTINY, ORION J as INDUSTRIAL DAWN, VELA J as INDUSTRIAL DREAM and HENRICUS J as INDUSTRIAL DOLPHIN. In October 2017, the 7979 dwt JOHANN J, long-termed chartered as INDUSTRIAL DART, loaded military vehicles and containers within a few hours in Nordenham/Germany for Souda Bay, where Greek and NATO military bases are located. Industrial also employed the 12,500 dwt ships INDUSTRIAL STRENGTH, INDUSTRIAL SKIPPER and INDUSTRIAL SONG. Built in 2017 under the designation SDARI Ecotrader 12,500 the ships were equipped with two 250-tonne cranes at Shandong Huanghai shipyard in Rongcheng/China, three more were contracted by Taizhou Sanfu shipyard in Taizhou.

Jüngerhans received further, newly designed heavy-lift carriers from Sainty Marine shipyard in Nanjing/China. The 153.4 m length, 23.2 m breadth, 9.1 m draught ships of 14,288 dwt are equipped with two cranes of 400 tonnes SWL each from German manufacturer NMF positioned on port and a smaller one of 80 tonne lifting capacity on starboard side. The vessels have two cargo holds of 12.7 and 77.2 m length, resp., with a total volume of 19,680 m^3. The first unit named LIR J and delivered in 2010 was sold to charterer Industrial Services after 1 year for operation under the US flag as OCEAN FREEDOM. Sister ships are ERIS J, RAN J and SENDA J. Subsequently, Sainty realized a heavy-lift carrier design derived from the LIR J type, but featuring a nearly twice as large deadweight. The 25,240 dwt SE CERULEAN was in 2013 delivered to SE Shipping in Singapore, which operated a total of four heavy-lift carriers in 2022 (Figs. 3.42 and 3.43).

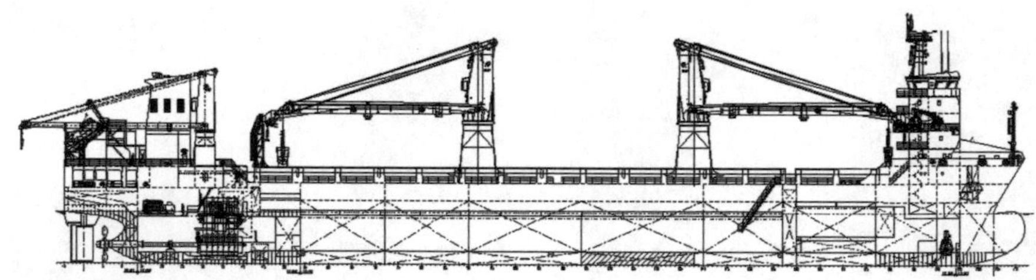

Fig. 3.42 The two 400-tonne cranes of LIR J can be coupled for lifts of up to 800 tonnes (Sainty)

Fig. 3.43 At 25,240 dwt, SE CERULEAN, in 2015 renamed CERULEAN, represents one of the larger heavy-lift carriers of this design type with a forward positioned superstructure and the cranes positioned at the ship's side (Ralf Witthohn)

3.2.22 Ecolift INDUSTRIAL FAME Renamed ZEA FAME

In January 2018, the construction of a series of nine 13,230 dwt, 1011 teu heavy-lift project carriers of the newly developed type Ecolift F900 for Intermarine started at Hudong Zhonghua Shanghai Shipbuilding and Huangpu Wenchong Shipbuilding in China, both shipyards of the state-owned holding CSSC. Six of the orders were placed with Hudong Zhonghua, three with Huangpu Wenchong Shipyard. Supervision of the construction and management of the 150 m length, 25.6 m breadth and 8.3 m depth vessels was carried out by Hammonia Reederei, a joint venture of the Hamburg shipping companies Peter Döhle and Ernst Russ with Paris-based private equity firm Ardian. The vessels are equipped with two portside-mounted heavy lift cranes of 450 tonnes SWL each, which can handle up to 900 tonnes in tandem without the use of stability pontoons. In their 26,000 m^3 cargo hold, the ships can transport a variety of cargo types worldwide, overheight cargo units with hatch covers removed. The container intake amounts to 1011 teu, of which 471 teu are stowed in the hold. The pontoon type tweendeck hatch covers can be positioned on two levels and also serve as grain bulkheads. The vessels have been built according to ice class E3 of Germanischer Lloyd. They are driven by a main engine of type MAN 7S40ME-B9.5 Tier II of 5750 kW output for a speed of 15.5 knots. Hammonia also developed a smaller, 99 m length design named CAR8500 and tailored to serve in the Caribbean. The six newbuildings from Wenchong were to be equipped with two 150-tonne cranes.

Planned as INDUSTRIAL FAME, INDUSTRIAL FRONTIER, INDUSTRIAL FUSION, INDUSTRIAL FALCON, INDUSTRIAL FAST, INDUSTRIAL FLASH, INDUSTRIAL FUTURE, INDUSTRIAL FOCUS and INDUSTRIAL FORTUNE the ships entered service from March 2018 under ZEA prefixes due to the forming of a joint venture by Intermarine and Bremen-based heavy-lift operator Zeamarine. Zeamarine employed the new vessel class in the transport of about 450,000 tonnes of heavy and oversized modules from Chinese ports to Port Hedland/Australia. The cargo was to be used within the BHP South Flank iron ore project in the Pilbara region. In the beginning of 2020, the ZEA FORTUNE and ZEA FUSION were engaged in the transports from Shanghai (Fig. 3.44).

In December 2019 Zeamarine clarified its business status by assuring, that the sole shareholder, Bremen-based Zeaborn group, would be committed to support the shipping arm, in which the activities of former Intermarine, Zeaborn Chartering and Rickmers Line had in 2018 been combined. In February 2020 Zeamarine, which had operated up to 75 heavy-lift carriers and intended to expand the fleet to 100 vessels by the end of 2018 with the help of new capital, gave up its US business, which was subsequently revived under the Intermarine name by new owners and took over several of the former Industrial ships. Furthermore, Zeamarine handed over nine multipurpose carriers to Hamburg-based United Heavy Lift (UHL). At that time, Zeaborn named chemical/oil tankers, container carriers up to the ULCC type, bulk carriers up to the newcastlemax class and multipurpose vessels as

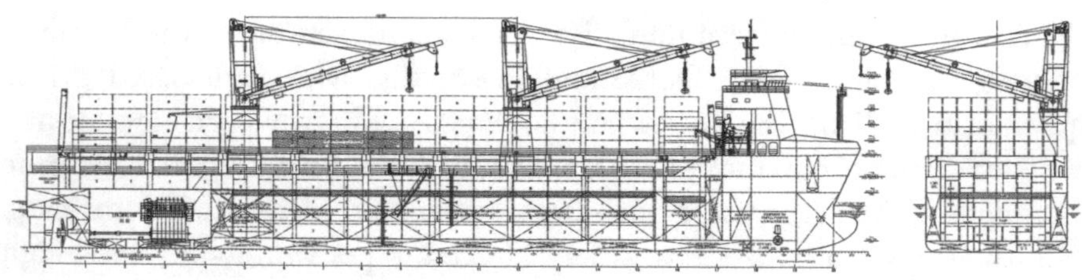

Fig. 3.44 The INDUSTRIAL FAME of type Ecolift F900 can couple two 450-ton cranes to create a lifting capacity of 900 tonnes (Hammonia)

belonging to the managed fleet of about 150 units In May 2020, Zeamarine filed for insolvency, in January 2021 Zeaborn Ship Management in Hamburg took delivery of the newbuilding UHL FAITH, another unit of the INDUSTRIAL FAME class and described as being the first of eight such ships. In early 2022, the UHL fleet list named 17 ships of the class, the UHL FELICITY, UHL FIGHTER, UHL FINESSE, UHL FREEDOM, UHL FIERCE, UHL FORCE, UHL FLAIR, UHL FAITH, UHL FORTUNE, UHL FALCON, UHL FUTURE, UHL FOCUS, UHL FLASH, UHL FAST, UHL FRONTIER, UHL FUSION and UHL FAME. Zeaborn still carried out the ISM management of four of the ships in a total fleet of 68 units (Fig. 3.45).

3.2.23 Van Carriers on RIA

A small episode of irony in July 2008 involving the German-owned coaster RIA revealed the deep restructuring of shipping. With its hatch covers being removed, the small ship carried two decommissioned van carriers from the

Fig. 3.45 After operation of nine units of the Ecolift F900 type, including ZEA FALCON ex INDUSTRIAL FALCON, by Zeamarine, some of the ships were employed in the carrying of modules for an Australian iron ore mining project (Ralf Witthohn)

Eurokai container terminal in Hamburg for scrapping to Bremerhaven. It was just the invention of the container, of all things, which largely deprived conventional coasters of their means of existence, but now the old container carrying vehicles provided the Elsfleth-homeported vessel with a special order. Built in 1960 by shipyard Gebr. Schürenstedt in Bardenfleth on the Lower Weser and owned from 1989 by Heinz-Georg Jacobs from Kirchhammelwarden, RIA made the last voyage under the German flag with feed from Husum/Germany to the inland port of Dörpen/Emsland in August 2012. A Dutch ship dealer acquired the ship and laid it up in Harlingen, where RIA met other old German coasters, such as HAMMELWARDERSAND, formerly registered in Brake as well, and STEENBORG from Rostock. From Harlingen RIA subsequently made its ever longest voyage to the new owner based in Douala/Cameroun under the name REHOBOTH. In February 2022, the ship, engaged in local food transports, was reported as being berthed in the port of Sao Tomé (Fig. 3.46).

3.2.24 Oslo Airport Express from Portugal

A special project transport was in July 2019 performed by the German coaster SCHILLPLATE managed by Leer/Germany-based Briese Schiffahrt. The

Fig. 3.46 Project cargoes such as decommissioned van carriers are rare cargo contracts for small coasters like RIA (Ralf Witthohn)

ship carried high-speed railway coaches built by Beasain/Spain-based CAF company from Pasajes/Portugal to the Bremerhaven ro ro terminal for trans-shipment to Norway. The 245 km/h achieving coaches of type Oaris were to be employed by the railway company Flytoget on a new rail connection between Oslo and the citie's airport. Discharge of the coaches at Bremerhaven was enabled by a temporarily built-up Liebherr caterpillar crane of 600 tonnes SWL. One of the first cargoes handled was a 333 tonnes weighing turbine lifted into an inland cargo vessel, the heaviest weight so far dealt with on the terminal.

The Gibraltar-flagged 3175 dwt coaster SCHILLPLATE was in August 2006 contracted with LISEMCO Haiphong shipyard and delivered 3 years later, in September 2009, by the Vietnamese builder within a series of eight such units named MITTELPLATE, SCHILLPLATE, BONACIEUX, HOHE BANK, ACCUM, CONSTANCE, ROCHEFORT and SCHILLIG handed over by 2012. Due to their single hold design, the ships dispose of a comparetively large cargo hold of 59.4 m length, 10.5 m breadth and 8.4 m height, which makes them suitable for the transport of large-volume project cargoes such as the Oslo airport express train (Fig. 3.47).

3.3 Industrial Plants

For a long time, the shipment of large and heavy industrial plant components was limited by the lifting capacity of the ship's own loading gear. Even though floating cranes were already able to perform heavy lifts in the ports of the manufacturing countries in the nineteenth century, shipments remained dependent on corresponding handling facilities in the recipient ports. In the past decades, a whole range of technological solutions has been devised to transport structures of almost unlimited weight and size over sea. This includes even large ships, drilling platforms or floating docks. Furthermore, newly developed deck transporters, dock and crane ships are developed and used for the erection and dismantling of large offshore structures (Fig. 3.48).

3.3.1 Wind Power Plants

The rapid expansion of offshore wind power has offered shipping a completely new business line. Large numbers of overseas and coastal cargo vessels are being provided to ship tower segments for the foundations, often from China, generator nacelles and rotor blades for both onshore and offshore plants. For

Fig. 3.47 Coaches of an Oslo express train were at the Bremerhaven ro ro terminal lifted from the cargo hold of the German coaster SCHILLPLATE, at the same time provided with bunker (Ralf Witthohn)

shorter distances, pontoon transports are occasionally chosen as the most economical form. Completely new technologies have been developed for the erection of offshore plants, for which installation ships working according to the jack-up principle and new types of cable layers have been designed. Large crane ships, fast crew boats, accommodation and guard ships also belong to the fleet of vessels for the installation and maintenance of offshore wind farms. In the ports, completely new industrial areas characterised by large production halls have been built up under high investment costs provided by companies mostly originating from the steel or energy sector. To enable the shipments, special heavy-lift quays at the water side have been provided, usually by the public sector. The rapid expansion was not successful in all cases. Especially in the offshore sector, the technological requirements of the wind power technology were underestimated. Wind farm operators and foundation builders gave up their businesses, the building of special offshore terminals was postponed or stopped by environmentalists. It was not until 2017 that a financial breakthrough was achieved, when for the first time offshore projects of the German energy company EnBW and Denmark's Dong Energy won the

Fig. 3.48 Once dependent on wind energy, shipping nowadays benefits from various transport, installation and maintenance tasks connected to offshore wind farms and carried out by a wide variety of auxiliary vessels including fast crew boats such as the pictured WORLD SCIROCCO (Ralf Witthohn)

tender without claiming public funding under Germany's renewable energy sources act.

In 2017, 13 new wind farms comprising 560 power plants of an average capacity of 5.9 MW and a total of 3.1 GW output were installed off Europe's coasts, increasing the capacity by 25% to 15.8 GW. 93% of the expansion was in the UK and Germany, which generated more than 75% of Europe's offshore energy. Eleven farms with 2.9 GW were under construction at the beginning of 2018, and investment decisions were made in 2017 for plants of 2.5 GW. However, offshore wind power since then suffered severe setbacks, especially off the German coasts, where only 32 turbines of 219 MW output were installed. In 2021, no new offshore plants are, due to the unfavourable political conditions, were erected. But the newly elected German government announced the will to expand the energy from offshore wind farms from 7.8 GW to 20 GW in 2030, and to 40 GW in 2040 (Fig. 3.49).

3.3.1.1 Towers from Shanghai, Blades for Mäntyluoto

Within the framework of a liner service between China and Europe maintained by the Chinese-Polish joint shipping company Chipolbrok, a cargo of tower segments for wind turbines is in Shanghai loaded on the hatches of the 30,435 dwt heavy-lift carrier CHIPOLBROK SUN. Seven weeks later the heavy-lift carrier

Fig. 3.49 A maximum of eight large rotor blades can be loaded on the hatches of the coaster TRANSVOLANTE, which was in 2018 sold to Russian owners and renamed ANYUY (Ralf Witthohn)

reaches the Niedersachsen quay of Brake/Germany, where the components are discharged within 3 days by harbour cranes. During its sailing upriver, the heavy-lift carrier passes the 4250 dwt coaster TRANSVOLANTE of Sweden's shipping company Transatlantic off Nordenham. The ship is loaded with rotor blades, of which no more than eight can be accommodated on the hatch covers. The transport is only possible with a raised wheelhouse, thus giving the ship's command a sufficient view forward. The blades are destined for Mäntyluoto, which is reached 4 days later via Kiel canal. In the outer harbour of the Finnish town, a harbour crane lifts the blades from the ship. On its voyage to Scandinavia, TRANSVOLANTE has passed the Blexen works of German steelmaker Dillinger Hütte, where four monopiles for the Borkum Riffgrund 2 wind farm have been loaded onto a pontoon shortly before. The piles will be towed into the North Sea by the Dutch tug EN AVANT 20 under assist of the local tug VB BREMEN (Fig. 3.50).

3.3.1.2 Rotor Blades from Spain, Turbine Houses for Nordsee 1

Only 2 days after the rotor blades have sailed downstream on TRANSVOLANTE, the 5555 dwt cargo vessel DRAGONERA of Brunsbüttel-based shipping company Erwin Strahlmann anchors on Blexen roads, bound for Brake/Germany. The ship transports rotor blades in the opposite direction from Aveiro/Portugal, where the rotor blade manufacturer Ria Blades, a company of the Senvion Group, has been

Fig. 3.50 Monopiles from a Nordenham-Blexen production site were transported on a pontoon to the offshore location Borkum Riffgrund 2 (Ralf Witthohn)

Fig. 3.51 On each voyage from Bremerhaven to Mukran a pontoon towed by tug BUGSIER 20 transported 15 components for the wind farm Baltic 2 (Ralf Witthohn)

established in 2007. While DRAGONERA is waiting for its berth in Brake, the 16,522 dwt heavy-lift carrier BBC SPRING of Leer-based Briese company takes over turbine houses with its own cranes in the Labrador port of Bremerhaven. In a shuttle service lasting over months, the vessel carries 54 turbines manufactured at the Bremerhaven Senvion plant to Eemshaven. The Dutch port serves as a base port for the construction of the Nordsee 1 wind farm. A few days earlier, the last components produced at the Adwen plant for the Baltic 2 wind farm were towed on a pontoon from the Bremerhaven fishing port to Mukran on the island of Rügen. Several sailings of the tugs BUGSIER 20 and ARION of Bugsier-Reederei are necessary to bring five foundation segments, turbine houses and rotor hubs on each trip to the Baltic destination (Fig. 3.51).

3.3.1.3 Wings for the Largest Power Plant

In December 2016, the deck carrier MERI of Finland's Meriaura company transports three rotor blades, described as the longest ones so far produced, from Esbjerg to Bremerhaven. The 88 m long blades are to be fitted to the world's first 8 MW wind power plant being erected in the next few months on the site of the former Luneort airfield. The turbine, manufactured by the Spanish-French company Adwen, represents a prototype that was to be installed in three French wind farms after successful testing. As part of the Spanish company Gamesa, Adwen had at this time yet already been taken over by market leader Siemens, and by this, like France's Areva group, got half the ownership of Adwen. Siemens, on its part, had developed an own 8-MW plant and ended production in Bremerhaven, after a large production plant was built up in near-by Cuxhaven.

The deck transporter for heavy and large cargoes MERI is the former AURA II built in 2012 by the Rauma site of STX Finland for Finnish owners. The main deck of the 105.4 m length, 18.8 m breadth 4359 dwt vessel is designed for loads of 7.6 tonnes/m². The diesel-electric propulsion system comprises three 1200 kW generators, which feed two 1400 kW electrical motors from Anhaltinisches Elektromotorenwerk Dessau, Germany, acting on two azimuth propellers and enabling a speed of 13 knots (Fig. 3.52).

Fig. 3.52 The deck carrier MERI, built in Rauma in 2012, transported the then world's largest rotor blades from Esbjerg to Bremerhaven in December 2016 (Ralf Witthohn)

3.3.1.4 Blades for Rostock

Participating in the transport of wind power plant components is another coastal cargo ship, the 3140 dwt RIX PACIFIC operated by Rix Shipmanagement in Riga. Shortly after purchase of the ship, the Latvian company won the contract about the transport of rotor blades between the German ports of Bremen and Rostock in December 2015. The 62.7 m long and 10.1 m wide single hatch of the 89.4 m length, 12.5 m breadth ship is large enough to stow nine blades on its covers. The RIX PACIFIC belongs to a design family named Combi Freighter and often realized by Dutch shipyard group Damen. Delivered to Armawa Shipping and Trading in January 1998 as EMMAPLEIN, the ship was taken over by the Briese company from Leer/Germany in 2004 under the name NEPTUN. Apart from wind turbines, the cargo range includes bulk cargo, steel, containers, forest products and general cargo. By operating a fleet of about 150 multipurpose general cargo vessels of small and medium size up to 18,000 dwt Briese is one of the main players in the shipment of wind farm components. Above all, ships of this operator were used when the German manufacturer Nordex in 2015 awarded the Danish logistics company DSV Air&Sea a contract to transport 150 rotor blades from Turkey to Bremen, Rostock and Kemi/Finland (Fig. 3.53).

Within the type family Combi Freighter the Dutch builder Damen Shipyards offered a wide variety of multipurpose cargo vessels over a period of several decades. In 2016, the types included freighters of 3850 dwt, 4850 dwt, 5500 dwt, 8200 dwt, 11,000 dwt and 14,000 dwt. The largest of the designs is of 146.3 m length, 20.1 m breadth and fitted with two cranes optionally

Fig. 3.53 On the hatch covers of the coaster RIX PACIFIC rotor blades were shipped from Bremen to Rostock (Ralf Witthohn)

ranging from 40 tonnes to 150 tonnes working load. Two units of this type were delivered by China's Damen Shipyard Yichang in 2012 as BBC PERU to German owner Eckhoff and as STELLAR MAESTRO to a Dutch Stellar Navigation. Of nearly the same length, but of only 18.3 m breadth is the Combi Freighter type 11,000, of which Heinz Corleis in Stade took over JORK and STADE in 2010/11, while Eckhoff commissioned THORCO COPENHAGEN and ROTES KLIFF, the latter chartered out as STX ALPHA and later as THORCO CLAIRVAUX (Fig. 3.54).

The 8200 dwt Combi Freighter type is represented by BEAUTRIUMPH, BEAUTROPHY, BEAUTRADER, BANDURA, TIDE NAVIGATOR, BEAUFORTE, ONEGO ARKHANGELSK, MARFAAM, ONEGO ROTTERDAM, ONEGO NAVIGATOR, ONEGO MARINER, TRITO NAVIGATOR, TRENT NAVIGATOR and TRADE NAVIGATOR. TRITO NAVIGATOR and its sister vessels operated by Forestwave Shipping at Heerenveen are equipped with two 60 tonnes lifting cranes. Another Chinese shipyard participating in the supply of multipurpose vessel to European owners is Dongfang. The builders completed a series of 130 m length, 16.5 m breadth general cargo vessels of 7900 dwt, of which the first one was delivered as MCL BREMEN in 2010. Of the same type are ANTJE, SIMONE, MCL TUNIS, BRIELLE, RHOON, BBC CHRSTINA, VERA, HARRY and HENRIK. Another Chinese builder, Shandong Baibuting at Rongcheng, contributed a 8000 dwt multipurpose cargo vessel type to the BBC fleet, realized by the 128.4 m length, 16.5 m breadth BBC BRISBANE and BBC BAHRAIN (Fig. 3.55).

Fig. 3.54 The Chinese-built 13,500 dwt BBC PERU is the largest of Damen's Combi Freighter types (Ralf Witthohn)

Fig. 3.55 The Combi Freighter 8200 TRITO NAVIGATOR is one of more than a dozen such ships built in China for European customers (Ralf Witthohn)

3.3.1.5 Jacket Foundations from Norway

Six months late, the first jacket foundations manufactured in Norway for the Nordsee Ost wind farm reach Bremerhaven in May 2012. The four substructures with the numbers 1 to 4 to support wind turbines of the 6-MW class have been lashed on a pontoon towed by the Danish BAMSE TUG. The foundations were created in Verdal, Norway, from where a total of 48 structures are to be transported to Bremerhaven (Fig. 3.56).

RWE Innogy, founded in 2008 as the renewables branch of Germany's energy concern, had the EUR 115 million costing jacket foundations manufactured by Aker Verdal, a subsidiary of the Norwegian construction and technology group Aker Solutions. Erndtebrücker Eisenwerk in Nordrhein-Westfalen/Germany, being paid a share of EUR 40 million, supplied the steel pipes for the foundations. RWE Innogy's own installation vessel VICTORIA MATHIAS carried the structures from Bremerhaven to the farm 33 km northeast of Helgoland and, by using its own crane, erected them in water depths of 22 m to 25 m. Each individual jacket foundation was anchored to the seabed at its four feet by separate piles. The foundations have a height of about 50 m and weigh 550 tonnes. The foundation covers an area of approximately 400 m^2. RWE Innogy awarded the wind turbines of the 6-MW class to turbine manufacturer Repower. In addition to the Nordsee Ost wind farm, RWE Innogy developed the Innogy Nordsee One farm. Initially planned as the largest offshore wind farm off the German coast with an installed capacity

Fig. 3.56 Jacket foundations built in Norway arrived in Bremerhaven on a pontoon towed by BAMSE TUG (Ralf Witthohn)

of around 960 MW, it was later reduced to 332 MW and commissioned end of 2017. The farm was built up 40 km north of the North Sea island Juist over an area of 150 km^2.

Off the north coast of Wales, RWE Innogy had since 2004 been operating the 60-MW North Hoyle wind farm and since 2009 the Rhyl Flats farm of 90 MW output. A third wind field off the Wales coast named Gwynt y Môr has an output of 576 MW. The company also held a 50% stake in the 504 MW Greater Gabbard wind farm off the southeast coast of England. Another major project planned with Statkraft in the UK was Triton Knoll, which projected capacity has been reduced from an initial 1200 to 900 MW. The farm Atlantic Array with a planned 1500 MW has, on the other hand, been abandoned, the Galloper field off Suffolk, which has been reduced from 500 MW to 336 MW, is to be inaugurated in 2018, while Dogger Bank was planned for 900 MW as part of a consortium. In Belgium, the company had a stake in the 325 MW Thornton Bank wind farm, in the Netherlands the Tromp Binnen project was abandoned. In September 2019, RWE transferred its Innogy shares to the German energy concern e.on and, in return, became control of the renewable energy sector of e.on, in addition to the renewable branch of RWE Innogy, now named purely Innogy. In April 2020, Innogy partnered with Asia Cement Corporation to continue the development of the

Chu Feng offshore wind project, which intended to participate in the next grid allocation round in Taiwan with a planned installed capacity of up to 448 MW off the northwest coast of the isle near Hsinchu City in the wind-rich Taiwan Strait.

3.3.1.6 Monopiles for Taiwan from Dillinger Hütte

On the rainy morning of 17 October 2019, the Dutch-flagged heavy-lift carrier JUMBO JUBILEE is being moored at the Asbestos quay in Nordenham/Germany. A few days later the ship proceeds downriver to the 200 m long heavy-lift quay of the Steelwind company, a subsidiary of the German steelworks Dillinger Hütte and built up in Nordenham-Blexen in 2014. In recent months Steelwind has prepared four of the world's heaviest monopiles for shipment to Mailiao/Taiwan. At 98 m length and 10 m diameter the foundations have a weight of 1870 tonnes. The steel plates rolled and welded to pipes have been transported to the work's quay from the Dillinger steelwork and its subsidiary Dillinger France in Dunkirk by coasters, including the 3577 dwt, UK-flagged EASTERN VIRAGE and the 3000 dwt INGEBORG PILOT of Norwegian owners, which carried the plates from Vlaardingen/Netherlands. For their transport to Taiwan two units each of the foundations are loaded into the cargo hold and on deck of the JUMBO JUBILEE. The lifts are carried out by two shore-based heavy-lift cranes able to lift up to 1600 tonnes at 45 m outreach when used in tandem. For carrying out the handling operations four German companies, Buss Port Logistics from Hamburg, Schmidbauer from Gräfeling, Thieling from Augustgroden and J. Müller from Brake, founded the Brake-based joint venture BMST Nordenham Heavy Logistics. The company is responsible for the discharge of the supplied steel plates from coastal vessels and the loading of the monopile components to the ocean-going ships, under contract of the Dillinger Hütte subsidiary Saar-Rhein Transportgesellschaft.

The monopiles are the first lot of a total of forty units of 60 m to 98 m length carried to Taiwan by ships of Dutch operator Jumbo Shipping and by additionally chartered heavy-lift carriers. The scheduled sailing time is 45 days, but dependent on various factors. The loading operations of JUMBO JUBILEE have to be interrupted after each pile, because the berth at the production plant is further deepened by dredgers. After completion of the loading operation and departure 24 October 2019, JUMBO JUBILEE left for Vlissingen anchorage for bunkering. Following the monopiles, 120 tower segments are to be expedited to Taiwan. They make up the foundations of 40 8-MW wind power plants produced at the Cuxhaven works of Siemens Gamesa for the Yunlin wind farm off Taiwan's west coast. The monopiles and tower segments from Blexen, being manufactured from

July 2019 until June 2020, are destined for Formosa Heavy Industries at Mailiao, which fabricates another 40 piles from the segments (Fig. 3.57).

A fortnight later, the second Jumbo Shipping carrier, named FAIRPLAYER, berths at Nordenham to load the next monopiles for Taiwan. At the same time, JUMBO JUBILEE has passed Gibraltar and is still only a few hours from Malta, sailing at speeds of 14 to 15 knots. The third heavy-lift carrier involved is FAIRMASTER. End of December 2019 a different type of heavy load carrier is participating in the monopile transports. The 15,630 dwt deck transporter CY INTEROCEAN I of Busan/South Korea-based Chung Yang Shipping loads the next lot. Within only 2 days four piles are loaded into prepositioned foundation supports and lashed on deck of the Marshall Island-flagged CY INTEROCEAN I (Fig. 3.58).

Commissioned in 2016, CY INTEROCEAN I and its sister vessel CY INTEROCEAN II are newbuidings of the Jingjiang Nanyang shipyard at Jingjiang/China. The ships of 152.6 m length and 40 m breadth are twin-propelled. In the Outer Weser, CY INTEROCEAN I passed the anchoring 28,899 dwt AUDAX. At 206.3 m length and 43 m breadth, the deck carrier is much larger. Its builder was China's Guangzhou Shipyard, which delivered

Fig. 3.57 In October 2019, the first four of 160 monopiles for a Taiwanese wind farm were loaded on the Dutch heavy-lift carrier JUMBO JUBILEE at Nordenham-Blexen (Ralf Witthohn)

Fig. 3.58 Two 2 × 800 tonnes lifting shore cranes were used to lift up to 1800 tonnes weighing monopile on deck of the FAIRPARTNER at Nordenham-Blexen (Ralf Witthohn)

the ship to Hong Kong-based ZPMC-Red Box Energy Services in 2016. The Curacao-registered transporter was the next heavy load carrier involved in the pile transports to Taiwan. In February 2020, FAIRPARTNER arrived at the Steelwind plant to load further foundation components. In March, FAIRMASTER was the first Jumbo Shipping unit that undertook a second transport to Taiwan.

At 145 m length and 26 m breadth heavy-lift carrier JUMBO JUBILEE has a hold capacity of 18,130 m^3 and a free deck space of 3100 m^2. The 2004-built ship is fitted with two 900 tonnes lifting mast cranes with a combined capacity of 1800 tonnes, and belongs with the sister ship FAIRPARTNER, to the seven units comprising fleet of Jumbo Shipping. The 144.1 × 27.7 m FAIRPLAYER and the sister vessel JUMBO JAVELIN, which was in February 2020 also engaged in the transports from Nordenham, are fitted with two Huisman mast cranes of the same SWL and disposing of deep-water lowering abilities. A helideck above the wheelhouse allows the landing of helicopters of up to Sikorsky type S-92 size. A Kongsberg dynamic positioning system controlling the use of two transversal thrusters forward, a retractable azimuth thruster aft and the main propulsion enables the 17 knots ship to undertake world-wide offshore operations on behalf of the energy industry. During subsea operations the FAIRPLAYER can set afloat two ROVs by a launch and recovery system (LARS) working to 3000 m depth. The forward positioned superstructure offers rooms for 75 persons (Fig. 3.59).

In 2018 Jumbo Shipping signed a letter of intent with China Merchants Industry Holdings about the construction of the 185.4 m length, 36 m breadth X-bow heavy lift vessel STELLA SYNERGY able to burn LNG as alternative fuel for delivery in early 2020. The newbuilding is to be equipped with a 2500 tonnes lifting mast crane and accomodation for 150 persons. In August 2019 Jumbo Shipping partnered with a biofuel supplier to test marine bio fuel on the FAIRPLAYER during an offshore decommissioning project in the North Sea.

Only several hours before the FAIRPLAYER departed for its second transport, another deck carrier, similar to the CY INTEROCEAN I, queued up in the long transport chain from Germany to Taiwan. The 20,157 dwt BIGROLL BERING entered the scene at Nordenham-Blexen. As a sister ship of the project carriers BIGROLL BISCAY, BIGROLL BEAUFORT, BIGLIFT BARENTSZ and BIGLIFT BARENTSZ, commissioned from 2015 to 2017, the ice-classed BIGROLL BERING belongs to the fleet of the Dutch Rolldock Shipping company. In April 2020, the Dutch-operated 18,680 dwt heavy-lift

Fig. 3.59 Besides for heavy-lift transport operations the FAIRPLAYER was especially designed for offshore work (Ralf Witthohn)

carrier HAPPY SKY joined the transports from Nordenham to Mailiao. In the same month the 15,016 dwt Korean deck transporter KOREX SPB NO.2 participated in the shipment of the offshore components by leaving Blexen on Easter Sunday for the Far East.

After a break Steelwind Nordenham resumed production in June 2021, when the supply of very large foundations for three windfarms was contracted. They referred to the location Arcadis Ost 1 off Rügen. to where 28 monopiles with weights of 1500 to 2000 tonnes, diameters of up to 10 m and maximum 107 m length are to be delivered to Belgian company Parkwind from 2022 on. 66 foundations without transition pieces and of up to 100 m length and 1200 tonnes weight for two more windfarms to be operated from 2024/25 on as Borkum Riffgrund 3 of 900 MW capacity and Gode Wind 2 (242 MW) north of Norderney were ordered by Denmark's Ørsted company, the former Dong enterprise (Figs. 3.60 and 3.61).

3.3.1.7 Open Hatch Carrier POSIDANA in Wind Power Components Transport

In January 2016, the general cargo ship POSIDANA discharged components for wind turbines carried from Taicang in China at the Bremerhaven

Fig. 3.60 In 2 days four monopiles weighing several thousand tonnes were loaded on the twin-propelled South Korean deck carrier CY INTEROCEAN I for transport to Taiwan (Ralf Witthohn)

Fig. 3.61 By its far forward positioned bridge, the project carrier BIGROLL BERING, seen upon arrival at Nordenham-Blexen in March 2020 to load offshore monopole foundations for Taiwan, offers an extensive deck area aft for the carriage of large volume cargo units (Ralf Witthohn)

container and later at the port's inward ro ro terminal. Built at Oshima Shipbuilding in 2008 for Norwegian, Bergen-based owner Westfal-Larsen, the POSIDANA and its sister ships PROVIDANA, PELICANA and PANAMANA are employed within the Saga Welco pool, which was established in October 2014 by partners Westfal-Larsen and Saga Forest Carriers from Tenvik/Norway, the latter being a subsidiary of Nippon Yusen Kaisha (NYK) in Japan. Westfal-Larsen had in 1962 been co-founder of Star Shipping and developed the open hatch concept mainly for the transport of forest products. In 1995, it established Masterbulk in Singapore. Saga contributed 32 red-painted vessels to the Saga Welco pool, Westfal-Larsen 20 similar, blue-painted ships. From 2017 to 2019, the combined fleet was joined by the newbuildings SAGA FREYA, SAGA FLORA and SAGA FAITH representing the Saga's earlier Future Class, which first vessels were delivered in 2012 and 2013 as SAGA FORTUNE, SAGA FUJI and SAGA FANTASY. In 2022, the Saga Welco pool disposed of 49 ships.

The companie's fleet capacity had already been enlarged by lengthenings of the OPTIMANA ships class at China's Chengxi shipyard in 2013, resulting in a deadweight increase from 50,470 tonnes to 54,200 tonnes. Shortly after conversion in May 2014, OKIANA loaded 51,700 tonnes of cellullose at Vancouver, the largest single cargo of this kind ever shipped in Canada. In the same year, OSHIMANA discharged 43,000 dwt cellulose at Qingdao within 27 h. The ship's 12 cargo holds have a total volume of 72,400 m^3 and are served by two gantry cranes of 68 tonnes lifting capacity each. In 2016 the operator started to fulfill a contract about the export of Brazilian cellulose for Fibria Celulos, one of the world's largest producers by delivering 5.3 million tonnes of cellulose yearly. Westfal-Larsen also contributed four units each of the types HARDANGER built in 1995/96, INDIANA of 1999/2000 and OPTIMANA. The share of Saga in the Saga Welco pool initially comprised seven vessels of the SAGA TIDE class commissioned from 1991 to 1996, the SAGA MORUS and SAGA MONAL of 1996/97, six units of the 1997 built SAGA BEIJA-FLOR type and nine newbuildings of the SAGA ADVENTURE type delivered after 2005 (Fig. 3.62).

Fig. 3.62 Otherwise mainly engaged in the forest products trade the general cargo vessel POSIDANA transported Chinese-built wind power plant parts to Bremerhaven in 2016 (Ralf Witthohn)

3.3.1.8 Turbine Towers on TIAN FU

1999-founded Nantong COSCO KHI Ship Engineering (NACKS) started the construction of a series of large multipurpose carriers in 2015. TIAN FU, TIAN LU, TIAN SHOU, TIAN LE, TIAN XI, TIAN ZHEN, TIAN QI and TIAN JIAN of operator COSCOL fly the Hong Kong flag. TIAN XI undertook its maiden voyage from China in the beginning of 2016 carrying towers and rotor blades to Rouen and Bremerhaven. The 190 m length, 28.5 m breadth and 15.8 m depth vessel achieves a deadweight of 38,100 tonnes and a container intake of 1015 teu. Four cargo holds served by portside fitted cranes with a maximum tandem load of 200 tonnes have a volume of 45,150 m^3. The ship's maximum speed is 15.4 knots. Within 2016 the sister vessels TIAN FU, TIAN LE and TIAN LU discharged wind generator tower segments at the Bremerhaven container terminal, provisionally used as offshore components handling quay (Figs. 3.63 and 3.64).

Europe's progressive development of the wind energy industry opened the highly subsidized Chinese steel industry another sales field. Danish wind generator producer Vestas reserved a separate storage area on the Bremerhaven container terminal, not fully occupied due to the container shipping crisis, for

Fig. 3.63 On one of its first voyages the 2016 built TIAN FU discharged turbine towers by help of its 100 tonnes lifting cranes at Bremerhaven (Ralf Witthohn)

Fig. 3.64 The shipment of tower segments for wind generators, like on the maiden voyage of the multipurpose carrier TIAN XI, was not economical, as the low draught of the ship revealed (Ralf Witthohn)

the import of tower sections from China. The export of the towers, on their turn, created shipment opportunities for Chinese carriers and were mainly undertaken by multipurpose cargo vessels of state-owned China Ocean Shipping Co. (COSCO), among ships of other nations. Their use, yet, was hardly economical, because the deadweight capacity of the ships remained far from being exhausted due to the fact, that the larger volume components had to be carried on the hatches.

The relatively large TIAN FU general cargo vessel type was in its deadweight capacity nearly doubled, when another class derived from a bulk carrier design was commissioned by COSCO Shipping from 2019 on. The first units of the 62,000 dwt class were named COSCO SHIPPING JIN XUE, COSCO SHIPPING ZHUO YUE, COSCO SHIPPING CHANG QING, COSCO SHIPPING GLORY, COSCO SHIPPING HONOR, COSCO SHIPPING PENG CHENG, COSCO SHIPPING XING WANG, COSCO SHIPPING HARMONY, COSCO SHIPPING VISION, COSCO SHIPPING GRACE and COSCO SHIPPING SINCERE, the latter due for delivery in 2022. Owned by the People's Republic of China or by Hong Kong-registered companies, the 201.8 m length, 32.3 m breadth ships were constructed by COSCO (Dalian) Shipyard. Different from the TIAN FU class their four cranes are positioned in the midship line and laid out to handle grabs (Fig. 3.65).

Fig. 3.65 The COSCO SHIPPING CHANG QING general cargo type, the keel of which was laid for a dozen units, resembles a bulk carrier design (Ralf Witthohn)

3.3.1.9 Aurich Wind Power Exports Under Wind Power

One of the most extraordinary innovative cargo ship designs refers to the rotor ship E-SHIP 1. The ship uses the rediscovered invention of the German engineer Georg Flettner, who had in the 1920s already developed the first rotor ship on the base of a physical phenomenon discovered by the physicist Heinrich G. Magnus in the middle of the nineteenth century. Aurich-based onshore wind turbine manufacturer Enercon signed a contract with Lindenau shipyard in Kiel about the construction of E-SHIP 1 to be primarily employed in the transport of Enercon wind power plants. Four 27 m high rotors of aluminium turned by electric motors should, together with other design measures, help to achieve substantial fuel savings. In the design phase calculations estimated, that at wind forces of 7 to 8 it would be able to travel at the same speed under rotor power only as with the ship's diesel-electric propulsion system, namely 17 knots. Initially installed seven Mitsubishi diesel engines were replaced by Caterpillar engines in 2013. Their exhaust gases are recovered by means of a steam turbine. The diesel generators supply the electrical propulsion motors self-manufactured by Enercon.

The considerable depth of the ship's hull results from the space-consuming cargo, which handling and stowage is optimised by using a special cassette system. The heavy wind generators roll on board through a stern ramp. Blades can also be handled by the ship's own gear. At 130 m length and 22.5 m breadth, E-SHIP 1 carries 10,000 tonnes on 9.3 m draught. Not only the underwater lines with bulbous bow and extended deadwood were optimized in the usual way. More than on all other ships built to date it was tried to design the ship's architecture above the waterline according to aerodynamic aspects. All corners and edges have been rounded off, including the roofing of the forecastle deck, bulwarks, and the elliptical layout of the only two decks high superstructure. Behind it, the cargo hold ventilators and gangways are integrated in such a way that they do not protrude, in order to decrease air resistance. The great effort involved in designing the individual components with the aim of reducing air resistance reduces the amount of energy needed to propel the ship. Initially, the newbuilding, which keel was laid in December 2007 at Lindenau shipyard and which was launched in August of the following year, was to be delivered in April 2009. But in November 2008, following financial problems at the Kiel shipyard, the ship was towed to Emden, where it was not completed until August 2010. Management of E-SHIP 1 was in 2014 taken over by Hamburg-based Auerbach Marine (Fig. 3.66).

E-SHIP 1 did not initially bring a further momentum to the use of wind power by shipping, as more units of the E-SHIP 1 type, which had been planned, were not implemented. In September 2015, Fehn shipping company in Leer revealed the intention to equip its 4200 dwt coaster FEHN POLLUX with a Flettner rotor of 18 m height and 3 m diameter, which was fitted in 2018. In February 2017, Finland's Viking Line announced, that it would have retrofitted its LNG-burning ferry vessel VIKING GRACE with a 24 m high rotor of 4 m diameter by 2018. Germany's Rörd Braren company contracted the 5035 dwt general cargo vessel ANNIKA BRAREN, which was delivered in April 2020 by Royal Bodewes shipyard at Hoogezand/Netherlands with a rotor of type Eco Flettner from a Leer/Germany-based engineering company fitted on its forecastle deck (Fig. 3.67).

While E-SHIP 1 is also employed in the transport of other cargoes than wind power components, Enercon is using chartered vessels to transport towers from its Landskrona works to wind farms. In January 2020, the 2380 dwt SIMON B of Stade/Germany-based GBS-Shipmanagement carried such segments from Sweden to Bremerhaven (Fig. 3.68).

Another, modest kind of renaissance of the wind power was tried by the Skysails technology equipped in 2008 to the 3700 dwt coasters MICHAEL A. and THESEUS, the latter named WILSON DUNDEE after 2014. The

Fig. 3.66 Four towering cylinders, a large depth and round shapes characterize the innovative architecture of the rotor ship E-SHIP 1 (Ralf Witthohn)

ships were equipped with a launching module including swell compensation on the forecastle for a kite sail of 160 m^2. Before, tests on the multipurpose cargo vessel BELUGA SKYSAILS of Beluga Shipping had revealed problems in heavy seas and regarding the long-term strength of the material. Under favourite wind conditions the sail of the MICHAEL A. generated a pull of eight tonnes, described as being equivalent to about 60% of the ship's maximum engine power.

3.3.1.10 Futuristic Project Carrier NORDANA SEA

The cargo vessel newbuilding, which is in March 2016 berthed at the quayside sector of the Bremerhaven container terminal reserved for handling wind turbine components, is characterized by a special architecture. On its foreship with a stem's negative rake and of a height twice as high as the freeboard has been set a five decks superstructure, located far forward. More common is the arrangement of the two cranes and the funnel on port side. Returning from its maiden voyage to Lavrion, NORDANA SEA leaves Bremerhaven after only 1 day for loading at Bremen, before sailing to the Greek port of Thisvi.

Fig. 3.67 The 2020-delivered coaster ANNIKA BRAREN has not only been fit with a rotor, but is occasionally employed in the transport of blades, like in May 2021 from Brake to Sweden (Ralf Witthohn)

Fig. 3.68 In January 2020, the small German-flagged coaster SIMON B transported Enercon tower segments from Landskrona to Bremerhaven, where they were discharged by a mobile crane (Ralf Witthohn)

Built by the Dutch shipyard Ferus Smit at its German site in Leer the project carrier NORDANA SEA is the third in a series of eight tweendeckers for Dutch shipowner Symphony Shipping. The first six units were built on the base of charter contracts with Denmark's Nordana company and thus named NORDANA SKY, NORDANA STAR, NORDANA SEA, NORDANA SUN, NORDANA SPIRIT and NORDANA SPACE. When Nordana retracted from the deal, new employments had to be found for the ships, which name prefixes were subsequently changed to SYMPHONY. The 122.5 m length, 17 m breadth and 10.7 m depth freighters have a deadweight of 10,600 tonnes on a draught of 7.8 m and achieve container intake of 698 teu. Their single cargo hold has a height of 10.7 m and is accessible by a 91 m long and 13.5 m wide hatch. The 13.5 knots running ships are equipped with two 85 tonnes lifting cranes. Number 7 and 8 of the series named SYMPHONY PERFORMER and SYMPHONY PROVIDER after delivery in 2017 are described as offshore supply vessels by the owner. Fitted with two Schottel rudder propellers, three bow thrusters and a dynamic positioning system they are purpose-built for carrying out offshore tasks (Fig. 3.69).

3.3.1.11 Diesel-Electrically Driven ABIS BILBAO

An innovative concept of a project carrier especially suited to transport wind power components was developed by the Dutch companies Shipkits and OAG on behalf of Harlingen-based ABIS Shipping. The smaller, 3900 dwt

Fig. 3.69 The carriage of large cargo units, in this case of wind power components, defined the design parameters of the project carrier SYMPHONY SEA ex NORDANA SEA, which transported wind turbine components from Bremen to Crotone (Ralf Witthohn)

version was realized between 2010 and 2012 at Partner shipyard in Szczecin and Hangzhou Dongfeng Shipbuilding under the names ABIS BELFAST, ABIS BERGEN, ABIS BILBAO, ABIS BORDEAUX, ABIS BREMEN and ABIS BRESKENS. The 90 m length, 14 m breadth, 12 knots vessels of the B class are conventionally driven by an MaK four-stroke engine, a widely used motor type in coastal shipping. This propulsion concept was replaced by a diesel-electric system in the larger D class delivered after 2012. It comprises six Scania diesel engines, which output is transferred through two electrical motors on two azimuth propellers for a speed of 12 knots. Another innovative equipment component is a dynamic positioning system that is able to keep the vessel at its position during offshore work by two transversal thrusters forward and the twin propeller arrangement. The cargo hold of the ships measures 80 × 12 m. The ABIS B class ships were in December 2016 sold by auction and taken over by Norway's Peak Project Carriers changing their first name to PEAK, the D class vessels by Dutch owner Amasus as EEMS DOVER, EEMS DUBLIN, EEMS DUISBURG and EEMS DUNDEE (Fig. 3.70).

3.3.1.12 Indian-Built HAPPY SKY Lifts 2 × 900 Tonnes

Though principally problematic because of the exposure to heavy seas, the designers of heavy lift carriers favour the forward position of the superstructure, because it offers more hold and deck space. In July 2013, Netherland's

Fig. 3.70 Developed from the ABIS B class, the diesel-electrically driven ABIS DUNDEE is 18 m longer (Ralf Witthohn)

BigLift Shipping commissioned the newbuilding HAPPY SKY, built by Indian shipyard Larsen & Toubro and equipped with two 900 tonnes lifting mast cranes from Dutch manufacturer Huisman. On its first four voyages the 17 knots vessel carried construction components for a port building project from Shanghai to Cape Lambert in Australia. The 154.8 m length and 26.5 m breadth HAPPY SKY has a lower cargo hold of 92 × 17.7 m and an upper hold of 98.4 m length. The hold is closed by flush deck hatch covers, creating a weather deck of 122.9 m length. To make different tween deck heights possible, the pontoon-like deck panels can be positioned in steps of half a metre difference (Fig. 3.71).

3.3.2 Bauxite Cracker for Kamsar

On 1 September 2017, the Liberian-flagged multipurpose cargo vessel KALVOE departs from Neustädter Hafen in Bremen with a cargo of special containers loaded in the ship's holds and on the hatches. The ship had already arrived 20 days earlier, but the cargo was delayed because permission for road transport of the

Fig. 3.71 The weather deck of the 155 m length HAPPY SKY offers space for cargo of up to 123 m length (Ralf Witthohn)

oversized containers was only granted after a delay of weeks. The containers are too large for rail transport and therefore have to be transported by road or waterway. The next port to be called by KALVOE is Las Palmas on Gran Canaria, but this is only a stopover. The port of discharge will be Kamsar in Guinea, in the world's largest bauxite loading port. The shipment is the third of nine transports in an EUR 100 million infrastructure project to upgrade a bauxite loading and processing facility in Kamsar.

Contractor is the Leipzig-based TAKRAF company, a member of the Tenova Group, which is active in the manufacture of open-cast mining equipment and transhipment facilities. TAKRAF is headquartered in Castellanza, Italy, and has subsidiaries on all continents. The transports with a total volume of 9000 tonnes are carried out from Germany, Shanghai, Dalian, Montreal, Houston and Durban. From there, various components are delivered. All transports are coordinated at the TAKRAF logistics centre in Lauchhammer, Brandenburg/Germany. An important part of the delivery is a wagon tippler which empties the railway wagons loaded with ore in the bauxite mines at the port of Kamsar. A plant for crushing the bauxite and a belt system, which transports the material to the loading quay is also being supplied. The ship loader comes from the Danish company FLSmidth. Operator and purchaser of the plant is the Compagnie des Bauxites de Guinée, which shareholders are the State of Guinea and the mining companies Alcoa, Rio Tinto and Dadco. A particular challenge of the project is its implementation during the continued operation of the existing plant, with keeping shutdowns as short as possible. Another TANKRAF transport in August 2014 dealt with two slat conveyors sent to Australia by barge and sea. The belts, each weighing 126 tonnes, were loaded into the inland vessel MARGIT in the Albert harbour of Dresden-Friedrichstadt/Germany and reached Bremerhaven via the Elbe, where they were taken over by the Norwegian ro ro vessel TONGALA.

The multipurpose cargo vessel KALVOE, chartered by TAKRAF for the transport of components of the bauxite loading facility, belongs to the fleet of Intership Navigation, founded in 1988 on Cyprus by Germany's Hartmann group. In 2017, the company employed a total of 45 multipurpose, bulk and cement carriers of various types and sizes. In 2020 the fleet still comprised 42 units. Of the KALVOE type, four more ships named BUGOE, KARKLOE, KROKSOE ex DELTADIEP and KORSOE were sailing under the companie's flag. The twin-hatch vessels were built by Rongcheng shipyard in China between 2010 and 2013. At 108.4 m length and 15.6 m breadth the deadweight of the geared version of the ships is figured at 5600 tonnes, while 5700 tonnes are reached, if no cranes have been equipped. The shipyard built a total of eight ships of this type for Hartmann (Fig. 3.72).

Fig. 3.72 The special containers lashed to the hatches of the multipurpose cargo vessel KALVOE carried components for a bauxite loading facility in Kamsar (Ralf Witthohn)

3.3.3 Gas Turbines for Hamitabat

From the US Gulf port of Houston via Donges on the Loire estuary, the heavy-lift carrier RICKMERS ANTWERP reaches Bremerhaven in May 2016 and moors at the east quay of the Kaiserhafen III dock. For decades, the Far East-bound cargo vessels of Hamburg Rickmers Line had loaded general cargo from the sheds named A, B and C at this dock, before the facilities had in the 1980s been abandoned in favour of car handling. Now the RICKMERS ANTWERP is to take over here components for a Turkish power station by its two 320-tonne port side cranes. From the Siemens plant in Berlin, the main component of a 600 MW gas turbine was transported on a barge to Bremerhaven. In tandem, the cranes of RICKMERS ANTWERP lift the 485-tonne turbine into the cargo hold, followed by the 465-tonne generator. The two heavy-weight colli are destined for the port of Tekirdag on the Marmara Sea. From there they will be transported to the 1200 MW Hamitabat combined cycle gas turbine power plant, which is to be modernized by equipping the new turbines. A total of four transports from Bremerhaven to Tekirdag take place within a couple of months, including one carried out by the sister ship RICKMERS HAMBURG, which has loaded two heavy pieces in February (Fig. 3.73).

RICKMERS ANTWERP was in 2002 built by China's Xiamen shipyard under supervision and according to the regulations of the classification society Germanischer Lloyd. The ship was awarded the class notation "100 A5 E with freeboard 4,318 m, G IW NAV-O SOLAS-II-2,Reg.19 C2P56, Multi-Purpose Dry Cargo Ship, Strengthened for heavy Cargo, Equipped for Carriage of Containers MC E AUT". According to this notation hull and

Fig. 3.73 Siemens gas turbines were transported to Tekirdag on ships of the RICKMERS HAMBURG type (Ralf Witthohn)

engine of the ship meet the design rules of the lowest ice class E of Germanischer Lloyd. The freeboard of 4.3 m is the result of the freeboard calculation, issuing a maximum draught of 11.2 m based on a ship's depth of 15.5 m depth. The letter G of the notation means that the holds have been reinforced for the use of grabs, the designation IW (in water survey) allows divers to inspect the hull without the need of docking of the ship. The SOLAS designation indicates the permitted transport of dangerous cargo, the formula C2P56 gives information about the leakage stability. AUT stands for the automated operation of the engine room without watch. The 192.9 m length, 27.8 m breadth RICKMERS ANTWERP has a light ship weight of 12,323 tonnes and a deadweight capacity of 29,912 tonnes. The container intake amounts to 1864 teu. A 16,785 kW developing two-stroke engine of MAN B&W type 7S60MC-C made under licence at Hudong accelerates the ship to a speed of 19.4 knots. In 2008 the management of the vessel switched from Cyprus-based Columbia Shipmanagement to Rickmers Reederei at Hamburg.

Nine newbuildings of the RICKMERS HAMBURG type, named Superflex by their owners, were supplied by the Chinese shipyards Xiamen, Jinling and Shanghai between 2002 and 2004 for an eastward-bound round-the-world service. The Marshall Islands-flagged ships with Polish and Croatian captains and officers and Philippine crews circumnavigated the globe in about 126 days and called at 20 ports. Named Pearl String Service it ran from Northern Europe through the Mediterranean Sea, the Suez Canal to Asia and further across the Pacific Ocean to the North American West Coast, through the Panama canal, the Gulf of Mexico and the US East Coast back to Europe. Rickmers had operated a conventional liner service to the Far East jointly with Hapag-Lloyd from 1969 to 1974. Subsequently it was controlled by Hapag-Lloyd until 2000 and only then bought back by Rickmers. To replace older and chartered ships, Rickmers further developed a new multipurpose cargo vessel type, initially ordered by the Cyprus-based Schoeller Holding, into an even more dedicated heavy lift carrier. The ships are equipped with four heavy lift cranes of 45-, 100- and two 320-tonne lifting capacitiy, the latter of which able to be coupled to 640 tonnes SWL. At 183.6 m length and 27.8 m breadth the 19.5 knots achieving five-hatch ships feature a deadweight of 30,151 tonnes on 11.2 m draught. In addition to the RICKMERS HAMBURG type, Rickmers chartered one of the Schoeller vessels, named CAPE DARBY, as RICKMERS HOUSTON in 2008.

At the beginning of 2017, Bremen-based Zeaborn group took over the majority of the activities of Rickmers Line, the liner shipping branch of the long-established Hamburg-based company, which had run into financial difficulties and declared insolvency in May 2017. As consequence of the take-over of Rickmers by Bremen-based Zeaborn, the round-the world ships were renamed ZEA DALIAN, ZEA HAMBURG, ZEA JAKARTA, ZEA NEW ORLEANS, ZEA SEOUL, ZEA SHANGHAI, ZEA SINGAPURE and ZEA TOKYO, their management switched to Zeaborn. When this company went insolvent in turn, the first ships of the type were sold for scrap in 2020.

Of the Superflex class a total of 32 newbuildings with various crane variants were contracted on both European and Chinese account between 2001 and 2011. The series started by CAPE DARBY, followed by CAP DON and CAPE DYER, from Xiamen shipyard for the Schoeller group. Dalian shipyard participated in the Schoeller series by supplying CAPE DENISON, CAPE DORCHESTER, CAPE DONNINGTON, CAPE DELGADO, CAPE DELFARO and CAPE DARNLEY, equipped with two 100-tonne and two 50-tonne cranes in 2002/03. Some units were employed in the liner services of Hamburger based operator MACS. In July 2015, for example, MACS briefly chartered VICENTE, which had been delivered to Schoeller in 2002 under the name CAPE DORCHESTER, from Hamburg-based NSC Shipping for its Europe-South Africa service. On this route, the ship met its sister ships CAPE DARBY and CAPE DENISON, which were sailing for MACS as RED CEDAR and BRIGHT HORIZON.

The Chinese-Polish Joint Stock Shipping Co. (Chipolbrok) ordered WLADISLAW ORKAN, CHIPOLBROK SUN, CHIPOLBROK MOON, LEOPOLD STAFF, ADAM ASNYK, CHIPOLBROK STAR,

PARANDOWSKI, CHIPOLBROK GALAXY, KRASZEWSKI and CHIPOLBROK COSMOS, in 2017 renamed QIAN KUN. The ships were equipped with the same crane capacity as the Rickmers vessels (3.74 and 3.75).

3.3.4 700-Tonne Lifting Capacity for Heavy Cargoes

In addition to its 30,000 dwt heavy-lift carriers, Rickmers Line chartered smaller tonnage on the Europe-Asia leg of its round-the-world-service, in 2016 the 12,800 dwt HHL ELBE, equipped with two 120-tonne cranes, and in 2015 NORDANA EMMA, NORDANA EMILIE and NORDANA THELMA of the same type. Already in 2013 hired were RICKMERS CHENNAI ex PACIFIC WINTER ex BELUGA PRECISION and RICKMERS MUMBAI ex BALTIC WINTER ex BELUGA PROJECTION of German owner Heino Winter These ships represent an 18,500 dwt heavy-lift project carrier type designated Volharding 18,000 and originally contraced by Bremen-based shipping company Beluga Shipping at Qingshan Shipyard in Wuhan. Built ten times, the ships were equipped with two 400-tonne cranes that can be combined to 800 tonnes SWL to serve a 86.1 m long hold, while a third, 120-tonne crane covers the first, 25.2 m long hatch (Fig. 3.76).

Fig. 3.74 The comprehensive crane equipment makes the WLADYSLAW ORKAN and its sister vessels best suitable for the shipment of heavy project cargoes by the Chinese-Polish joint shipping company Chipolbrok (Ralf Witthohn)

Fig. 3.75 For a short period Hamburg liner operator MACS chartered CAPE DORCHESTER as VICENTE for its South Africa service. The ship had already sailed as CSAV GENOA, CCNI ANTARTICO and was in 2020 registered as YU RONG for Sea Ray Shipping Co. at Qingdao (Ralf Witthohn)

Fig. 3.76 Built as Beluga's P1 type BELUGA PROMOTION by Qingshan shipyard at Wuhan, HHL VENICE joined the Hansa Heavy Lift fleet in 2011, in 2019 the Zeamarine company as ZEA GULF and became the US-flagged OCEAN GLADIATOR the following year (Ralf Witthohn)

After Beluga's insolvency, the ships were put into service by other shipping companies, four by Winter, one as PRIMA DORA by Marshall Islands-registered Braveheart Shipping, two as OCEAN GIANT and OCEAN GLOBE by US-based Intermarine and three by Beluga's Hamburg-based successor company Hansa Heavy Lift under the names HHL LISBON, HHL VENICE and BBC SPRING. A dozen newbuildings of an enlarged version of about 20,000 dwt with the type designation MP P-2 were also planned for Beluga, then built between 2009 and 2015 by Hudong Zhonghua Shipyard in Shanghai and delivered as HHL RIO DE JANEIRO, HHL VALPARAISO, HHL MACAO, HHL HONG KONG, HHL RICHARDS BAY, HHL TOKYO, HHL LAGOS, HHL FREMANTLE, HHL NEW YORK and HHL KOBE on behalf of Hansa Heavy Lift. Two similar vessels, equipped with two 450-ton cranes, were used in project shipping after their commissioning in 2015 by Intermarine in Houston as INDUSTRIAL GRAND and INDUSTRIAL GLORY (Figs. 3.77 and 3.78).

3.3.5 Power Plant Components for Siberia

In August 2016, 3 years after the bulk carrier NORDIC ORION, a European multipurpose cargo vessel sailed through the Northwest Passage from Lianyungang/China to Baie Comeau/Canada for the first time. By chosing this 3750 nm long route, the multi-purpose cargo vessel AFRICABORG shortened the usual voyage by 40%. The vessel of the Dutch shipping company Wagenborg was built according to the Finnish/Swedish ice class 1A, equivalent to the IACS ice class Polar 7. Out of environmental reasons, gas oil was the only fuel burnt during the voyage. From 2006 to 2012, Wagenborg received 21 ships of this 17,300 dwt type from Hudong-Zhonghua shipyard. Bremen-based Beluga Shipping also took over four ships of the same type under the names BELUGA GENERATION, BELUGA GRAVITATION, BELUGA GRATIFICATION and BELUGA GRADUATION.

The Beluga company was as well engaged in transports through the Northwest Passage by deploying their 12,800 dwt ships BELUGA FRATERNITY and BELUGA FORESIGHT. Under assistance of the Russian

Fig. 3.77 During loading operations of Senvion wind turbine components at Bremerhaven in November 2017 a floating counterweight was installed on the starboard side of the HHL TOKYO (Ralf Witthohn)

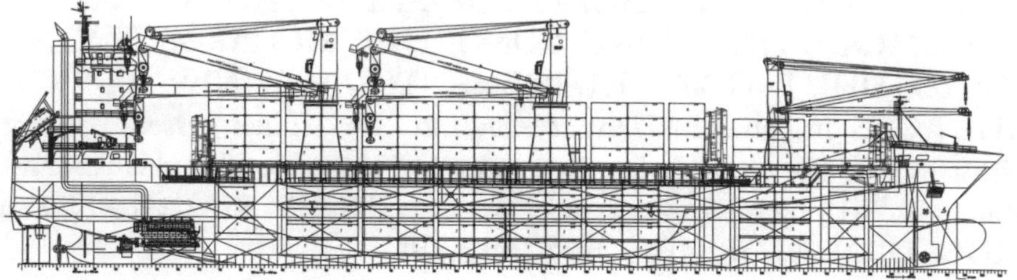

Fig. 3.78 A crane capacity of 800 tonnes distinguishes the P-2 design of the BELUGA PRESENTATION type (Beluga)

nuclear-powered icebreakers 50LET POBEDY and ROSSIYA the ships transported power plant components weighing 200 tonnes to 300 tonnes from Ulsan/South Korea to Nowy Port/Jamburg at the mouth of the river Ob in August and September 2009. The two ships subsequently continued their sailing round Novaya Zemlya to load 6000 tonnes of steel pipes for Nigeria in Arkhangelsk. The route through the Bering Sea, Laptev Sea, Vilkitzki Strait and Kara Sea, which is open during a short navigation period due to global warming, is 3000 nm shorter than the route via the Suez canal. According to the shipping company, this saved a total of 200 tonnes of fuel worth USD 100,000 and USD 20,000 for each day of the reduced voyage time, totalling to 300,000 dollars per ship. The first such voyages of non-Russian cargo vessels required the permission of the Russian authorities, which yet had not approved a voyage of the BELUGA FORESIGHT when first planned in 2008. Another voyage from Dalian to Rotterdam via the northeast route was carried out in September 2013 by the 19,150 dwt vessel YONG SHENG of China's COSCO.

In August 2016, the heavy-lift carrier SENDA J of Germany's Jüngerhans company, chartered as INDUSTRIAL FAITH by US company Intermarine, sailed the Northeast Passage with the support of the Russian nuclear-powered icebreaker YAMAL. The vessel, built in accordance with Germanischer Lloyd's ice class E3, took 27 days from Shanghai to St. Petersburg, 9 days less than on the Suez route (Fig. 3.79).

3.3.6 Shipyard Crane on the Hook

Due to the lack of a sufficiently powerful German specimen, the Dutch floating crane MATADOR 3 was in November 2012 sent from the Netherlands to Bremen to remove the jib of the old shipyard crane of Bremer Vulkan. The

Fig. 3.79 The first Chinese ship on the Northeast Passage was the general cargo vessel YONG SHENG (Erich Müller)

crane part was set down a few hundred metres further downstream. Immediately afterwards, the floating crane picked up the substructure of the hammer crane, which had not been moved since the shipyard went bankrupt in 1996. The total crane weight of almost 1000 tonnes did not allow its handling in one hoist. Therefore it was divided into two parts of 750 tonnes and 250 tonnes weight, resp.

The crane of MATADOR 3 is installed on a pontoon of 70 m length and 32 m breadth. Its lifting capacity depends on the outreach. The maximum load of 1800 tonnes is reached at the smallest possible outreach of 5 m, with the main hook at a position of 40 m height. The largest outreach of the main hook is 35 m, at which 550 tonnes are still liftable. The maximum working height when using the upper hook is 75 m, at which 900 tonnes can be hoisted. The farest outreach is 65 m at a load of 160 tonnes. When the two hooks of the crane are combined, the maximum lifting capacity at 50 m is 200 tonnes, at 10 m 1675 tonnes are achieved. The floating crane is manned by a crew of eight. Though fitted with two propellers, the crane is pulled by a tug during transite over sea (Fig. 3.80).

3.3.7 Piggyback Ships and Docks

Ships, docks and oil platforms are among the heaviest single colli shipped. To be able to take them on board, the dock principle is the usually applied handling technology. This means, that the transport ship takes water ballast, until it is submerged so far, that the floating object can be manoeuvred above the deck of the transporter's deck, which is then raised by pumping out the

Fig. 3.80 The Dutch floating crane MATADOR 3 disassembled the old hammer crane of Bremer Vulkan (Ralf Witthohn)

Fig. 3.81 RWE group's wind farm installation vessel VICTORIA MATHIAS was transported from Korean builder Daewoo for completion to Bremerhaven on deck of the Norwegian deck carrier EAGLE (Ralf Witthohn)

ballast. On such so-called semi-submersible heavy-lift transporters, cargo with large dimensions can even protrude beyond the side or rear of the carrier. Smaller dock ships transport the cargo in their hold, which cargo hold may remain open for high objects. As a rule, these types of dock ships and deck carriers have forward positioned superstructures to provide a clear view from the bridge (Fig. 3.81).

3.3.7.1 SWATH Type Pilot Tenders for Houston

Arriving from Conakry via Gibraltar end of August 2017, the heavy-lift carrier OCEAN GLORY enters Bremen Industriehafen through the Oslebshauser lock and moors at the rarely used quay of the Kap Horn street. One day later, the pilot tender newbuilding HOUSTON arrives at the berth, coming from its builder Abeking & Rasmussen in Lemwerder at a distance of just 5 nm. With the two stronger units of its three heavy-lift cranes, OCEAN GLORY hoists the 26.5 m long SWATH vessel onto the rear part of hatch 2, where it is subsequently lashed. The following day the sister ship BAYOU CITY finds place in front of the HOUSTON. In the next morning, OCEAN GLORY leaves Bremen for Houston, which is reached after a fortnight amidst the catastrophic consequences of a flood caused by hurricane Harvey (Fig. 3.82).

By the SWATH pilot tender newbuildings HOUSTON and BAYOU CITY the Houston pilots organization replaced older ships of the same SWATH design principle and named identically. The US pilots consider double-hulled vessels to be able to keep the port of Houston during adverse weather conditions 6 days longer open per year than single-hulled vessels would do, thus avoiding millions reaching economic losses by a halt of shipping. The first Houston SWATH ships were also the model for an own development line of shipyard Abeking & Rasmussen in Lemwerder/Germany.

Fig. 3.82 After 2 days of loading and lashing work, the US heavy-lift carrier OCEAN GLORY transported the pilot tenders HOUSTON and BAYOU CITY from Bremen to the US Gulf (Ralf Witthohn)

Starting in 1999, the German builders completed the 25 m length tenders DÖSE, DUHNEN and GRODEN as well as the 50 m length pilot station ship ELBE and the reserve ship HANSE for Germany's Elbe pilots. On the Weser river, the station ships WESER and the tender WANGEROOG came into service, for the Ems, the tender BORKUM. Subsequently, the Belgian and Dutch pilots also contracted SWATH vessels at the shipyard, the tenders CETUS and PERSEUS for Rotterdam, as well as the station ship WANDELAAR and the tenders WIELINGEN, WESTDIEP and WESTERSCHELDE for operation off the Scheldt estuary. The hydrographic research vessel JACOB PREI was exported to Estonia, the crewboat NATALIA BEKKER delivered to the German offshore company Bard. Furthermore, a SWATH luxury yacht was built, as well as minesweepers for the Baltic States. The circle in the history of SWATH pilot ships was closed when the tenders HOUSTON and BAYOU CITY were delivered in 2017. They were the only ones from the yard's production that needed a piggyback transport to the customer.

3.3.7.2 Four-Masted Barque PEKING from New York to Germany

On a July evening in 2017, the dock ship COMBI DOCK III, coming from La Guaira, sails under the Verrazano Bridge to anchor just behind on the Hudson River. Another seven miles away the four-masted barque PEKING had been berthed in the South Street Seaport Museum of Manhattan for 40 years. But the sailing ship has already been moved from its place of rest to the Caddell Drydock in Staten Island in September 2016 to be prepared for its last great voyage. Two days after arrival of the COMBI DOCK III at New York, PEKING, its yards being dismantled, is towed over the stern into the flooded hold of the transport vessel. It takes 4 days to lash and support the slender hull of the barque. Then, the piggyback ride across the Atlantic begins. The four-master has been decaying more and more over the recent years, until Hamburg people developed the plan, to take the legendary windjammer back to Hamburg, where it had been built by Blohm & Voss in 1911.

As the PEKING's poor condition did not allow to tow the ship, the return could only be undertaken successfully in a dock ship. For this purpose was the dock ship COMBI DOCK III of Bremen shipping company Harren & Partner chartered. The 11 days lasting voyage ended in the port of Brunsbüttel/ Germany, where PEKING was floated out of the dock ship and towed on its own keel to Wewelsfleth. There, Peters-Werft carried out the repairs and restauration. Of the costs for the transport and repair estimated at EUR 120 million, 26 millions were spent from tax funds. The full-rigger PEKING was once employed under the flag of the Hamburg shipping company F. Laeisz in the saltpetre trade from Chile and in the transport of wheat from Australia. During its Atlantic crossing, one of the last existing barques became a—historic—transport object itself.

COMBI DOCK III belongs to a series of four dock ships which Lloyd Werft in Bremerhaven delivered from 2008 to 2010 on behalf of the shipping company Combi Lift, a joint venture of the companies J. Poulsen in Denmark and Harren & Partner in Bremen. The special carriers are able to load single cargo pieces of up to 700 tonnes weight over the 18 m wide stern ramp or to float in cargo of up to 4.5 m draught through the stern door. Cargo handling is also possible by three cranes with a maximum lifting capacity of 700 tonnes. The 169.4 m length, 25.4 m breadth twin-propelled ships achieve a deadweight of 10,480 tonnes. COMBI DOCK II, extended to 179.6 m, was equipped for service as offshore construction ship BLUE GIANT before completion. In 2011 an accommodation block to be used as living quarters for offshore workers was set on the ship, now named OIG GIANT I. Following a conversion on behalf of the Norway's Offshore Installation Group (OIG) founded by Harren with support of the US bank Goldman Sachs at capital costs of USD 500 million, COMBI DOCK IV performed similar tasks as OIG GIANT II from 2011 on (Fig. 3.83).

Fig. 3.83 In July 2017, COMBI DOCK III transported the 106-year-old four-masted barque PEKING from New York to Brunsbüttel (Ralf Witthohn)

3.3.7.3 Survey Vessel TAGU SUPPLIER to Jamaica

At the beginning of May 2017, the Dutch heavy-lift carrier STATENGRACHT, moored at Labradorkai of the Bremerhaven fishing port, hoists the survey vessel TAGU SUPPLIER onto the ship's hatch. TAGU SUPPLIER is equipped for investigations of the seabed. Its Bremerhaven-based owner Tiefbau Gesellschaft Unterweser (TAGU), a subsidiary of the building company Ludwig Freytag, has commissioned the Bremen company S.S.C. Seacargo Service to transport TAGU SUPPLIER, an inflatable boat and other equipment to Jamaica where the survey vessel was still reported anchoring in July 2021 (Fig. 3.84).

In addition to project cargoes such as the survey vessel TAGU SUPPLIER, the main cargo types transported by STATENGRACHT refer to forest products and breakbulk. From Bremerhaven, the ship of the Dutch shipping company Spliethoff set course for Finland to load paper before continuing its voyage to the US Gulf and Caribbean ports. Spliethoff's S-class general cargo vessels can load their cargo with help of cranes or by their side-loading equipment. For this purpose, the doors of the side gates fold up so that forklift trucks can place paper rolls, pulp, coils and various types of breakbulk on the ship's elevators. On the decks of the ship, in turn, forklift trucks are used to

Fig. 3.84 The survey vessel TAGU SUPPLIER made a voyage to Jamaica on deck of the Dutch general cargo vessel STATENGRACHT (Ralf Witthohn)

carry on the cargo. Thanks to their box-like cargo space, the ships are also suitable for the transport of bulk cargoes (Fig. 3.85).

3.3.7.4 Tanker Newbuilding on Deck

In June 2007 the non-propelled transport barge FAIRMOUNT FJELL reaches Valetta on Malta on the hook of a highsea tug. So far, the barge is only equipped with two azimuth propellers at the stern and a wheelhouse. During the next months Malta Shipyards will carry out an extensive conversion of the barge. On behalf of the Dutch company Fairmount Heavy Transport generators and a ballast system are installed to rebuild the barge into a semi-submersible heavy-lift transporter (Fig. 3.86).

The first order of FAIRMOUNT FJELL in 2009 was spectacular. With the help of the semi-submersible carrier, the Italian tanker newbuilding MARETTIMO M was launched at the CNT shipyard in Trapani. FAIRMOUNT FJELL and its sister ship FAIRMOUNT FJORD were afterwards employed to transport offshore components from Lobito to Daewoo shipyard in South Korea. The carrier had in 2000 been built by China's Jinling shipyard as the 25,000 tonnes carrying barge FJELL for Norway's Boa Offshore company. After Fairmount merged with the Dutch shipping company Royal Boskalis Westminster, the carrier became part of the Boskalis fleet.

3.3.7.5 Repair Dock on TRANSSHELF

The cargo handling principle of the Barge or LASH carrier, which partly submerges to take on board its cargo in a floating state, was applied to transport even complete ships and docks. Those are yet not carried in cargo holds, but on the open deck free of restricting structures. One of these semi-submersible heavy-lift carriers was ordered by the Soviet Ministry of Gas from the Wärtsilä shipyard in Turku in 1987 under the name TRANSSHELF. The 173 m

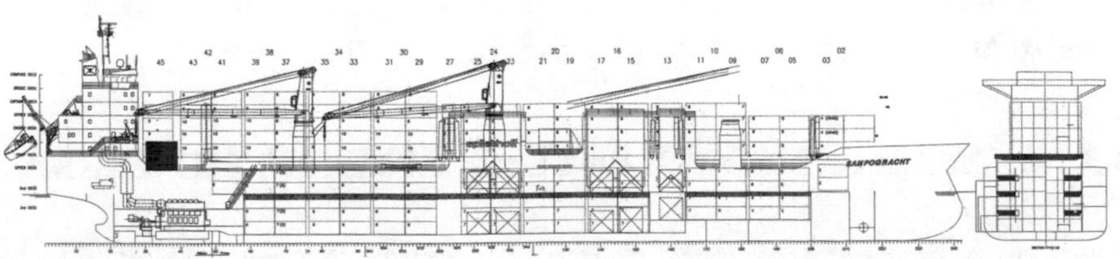

Fig. 3.85 Side-loading facilities make Spliethoff's STATENGRACHT class usable for a large variety of cargo types, including paper rolls, breakbulk etc (Stocznia Szczecinska)

Fig. 3.86 Malta Shipyards converted the barge FJELL into the semi-submersible transporter FAIRMOUNT FJELL (Ralf Witthohn)

Fig. 3.87 The Russian, later Dutch semi-submersible transport ship TRANSSHELF carried the floating Dock 2 of Bremer Vulkan on its deck (Ralf Witthohn)

length, 40.1 m breadth vessel was employed on the international market, and for example transported the repair dock 2 of Bremer Vulkan after its bankruptcy in 1995 to a foreign buyer. In 2004 the Dutch shipping company Dockwise acquired the transport ship and employed it without name change. After a voyage reported from Vancouver to Brunei Bay, the vessel was registered as having entered a lay-up state in February 2021, a year later it was reported as staying in Singapore (Fig. 3.87).

3.3.7.6 Yachts on DEO VOLENTE and EEMSLIFT HENDRIKA

Based on a European network of booking agents, Monaco-based Starclass Yacht Transport maintains a regular seasonal transport service for large motor and sailing yachts. During the summer months the chartered Dutch multi-purpose cargo vessels DEO VOLENTE and EEMSLIFT HENDRIKA call at ports between Norway and Turkey. Served on the base of a three-weekly sailing schedule are Bergen, Bremerhaven, Rotterdam, Poole, Gibraltar, Alcudia, Nice, Imperia, Valetta, Istanbul, Lavrion, Split and other ports on demand. The idea to carry yachts to the Baltic and Mediterranean Sea had first been realised in 1987, when a shipment took place from Harlingen to Malaga, later between Harlingen and Kiel, but also transatlantic, to the Middle and Far East. The business was in 2000 purchased by the Dutch shipping company Spliethoff, but re-founded in 2002 at Monaco. After using ordinary cargo vessels in the beginning, DEO VOLENTE was in 2013 converted to a special yacht carrier, like later EEMSLIFT HENDRIKA. The ships' crews are specially trained for handling the high-value yachts.

With the help of their two cranes the ships are able to handle the yachts independently from shore-based cranes. DEO VOLENTE is a 2007-built small heavy-lift carrier, which hull was supplied by Partner shipyard in Poland and fitted out by Hartman Marine. Hartman Seatrade at Urk/Netherlands is also the owner and operator of the 3750 dwt vessel (Fig. 3.88).

The hull of the 4200 dwt heavy-lifter EEMSLIFT HENDRIKA was built at the same Polish site within a series of 11 such vessels and completed by

Fig. 3.88 In August 2019, DEO VOLENTE arrived from Norway to discharge sailing and motor yachts with help of its own cranes in Bremerhaven (Ralf Witthohn)

Shipkits at Groningen in 2015 as ABIS ESBJERG for Dutch Abis Shipping. In April 2021, all 12 crew members had to be rescued by helicopter off the Norwegian coast after part of the cargo had shifted and caused a heavy list of the ship in storm. The first ship of the class was built in 2013 as ATLANTIC DAWN for Hartman Shipping. Subsequently, ABIS DUSAVIK, ARCTIC DAWN, INDIAN DAWN, ABIS DUNKERQUE and ABIS ESBJERG entered service until June 2015. The 111.6 m length and 16.8 m breadth single-hatch vessels with variable transverse bulkheads and tween decks are equipped for worldwide operation with two NMF cranes, each capable of handling 150 tonnes. They achieve double the lifting capacity in tandem operation. The "smooth" weather deck consisting of 11 pontoon decks is designed for almost complete use across the full width of the ship. Underneath the deck are passage ways with pilot ports on both sides of the ship, so that the loading area is only limited by the crane columns, the funnel and the free-fall boat placed behind it. As usual on Dutch ships, the hatch covers are moved by a trolley. The ships are driven by an MaK diesel engine, which is reduced via gears to the shaft with a controllable-pitch propeller allowing a speed of over 17 knots on a draught of 5 m. In August 2014, INDIAN DAWN loaded a cargo of large pipes and then, using only one of the cranes, took over a hoovercraft vessel destined for Mumbai in Southampton.

At the beginning of 2017, following the insolvency of Abis Shipping, ABIS DUSAVIK, ARCTIC DAWN, INDIAN DAWN, ABIS DUNKERQUE and ABIS ESBJERG joined the management of Dutch Amasus shipping company and were renamed EEMSLIFT NADINE, EEMSLIFT NELLI, EEMSLIFT ELLEN, EEMSLIFT DAFNE and EEMSLIFT HENDRIKA (Fig. 3.89).

3.3.7.7 Fish Farm Serving Boat CLAYOQUOT SERVER for Canada

In 7 October 2019, the general cargo vessel STAR LUSTER departs from Bremen bound for Wilmington. On the hatches are lashed side by side a catamaran type sailing boat and another special catamaran named CLAYOQUOT SERVER. The boat is a fish farm serving vessel built in Croatia for Njord Marine Services of Campbell River/Canada. After transite through the Panama canal the schedule of the STAR LUSTER includes the west coast ports of Caldera, San Diego, Stockton, Longview, Squamish, New Westminster and Kitimat. After discharge in the port of Squamish, CLAYOQUOT SERVER takes up its work in the bays of Vancouver Island (Fig. 3.90).

Fig. 3.89 EEMSLIFT HENDRIKA, like DEO VOLENTE, is able to handle the transported yachts by their own cranes, also by using a special anti-heeling pontoon before set into the water by crane (Ralf Witthohn)

3.3.7.8 Airbus Transporter VILLE DE BORDEAUX

As a prerequisite for the Airbus A380 aircraft construction programme started in 2006 and divided between different European sites, Airbus has on a long-term basis chartered several ro ro carriers for the transport of large aircraft parts. The first and largest newbuilding was in 2004 the ro ro vessel VILLE DE BORDEAUX featuring a deadweight of 5290 tonnes. The ship was delivered by Jinling shipyard in Nanjing and is operated by the shipping company FRET/Cetam, a joint venture between France's Louis Dreyfus Armateurs (LDA) and the Norwegian shipping company Leif Höegh. To accommodate the aircraft parts, the 154.2 m length, 24 m breadth, 21.9 m depth vessel was designed with a 22 m wide and 14 m high stern door to the 6720 m^2 large cargo deck. The ship is accelerated to 21 knots by two main engines of 16,800 kW total output acting on two controllable-pitch propellers. The 21,528 gt ship carries Airbus parts from factories in the UK, Germany, Spain and France to Bordeaux, from where they are transported on barges up the

Fig. 3.90 The fish farm serving vessel CLAYOQUOT SERVER was transported on the hatch of the general cargo vessel STAR LUSTER to Vancouver Island (Ralf Witthohn)

Fig. 3.91 The 5290 dwt ro ro vessel VILLE DE BORDEAUX was built in China to transport Airbus parts to Bordeaux

river Garonne to the Langon river port and then on special road transporters to Toulouse (Fig. 3.91).

In December 2008 and April 2009, the joint Norwegian-French shipping company took over further newbuildings for employment by Airbus, the ro ro carriers CITY OF HAMBURG and CIUDAD DE CADIZ, from Singapore Technologies Marine shipyard. At 126.5 m length and 20.6 m breadth, these newbuildings have a tonnage of 15,643 gt and a deadweight of 3500 tonnes, making them smaller than VILLE DE BORDEAUX. Two eight-cylinder four-stroke engines with an output of 4000 kW each power the twin-propelled ships to a speed of 15 knots. They are as well equipped with a stern ramp for rolling cargo handling. For the transport of plane components in large containers between Einswarden/Germany and Hamburg, a small special newbuilding with a stern ramp has been used since 2009, the 2670 dwt deck carrier KUGELBAKE of Cuxhaven-based shipping company Otto Wulf. The ship is less dependent on favourable weather conditions than pontoon transports carried out before. In February 2019 the Louis Dreyfus subsidiary LD Seaplane chartered the Italian-built and -flagged ro ro vessel WEDELLSBORG of Denmark's Nordana Line for transports of complete A320 aircraft and Mirage engines dedicated to the US Army from Montoir/France to Mobile/US (Fig. 3.92).

3.3.7.9 Ariane Transporters MN TOUCAN and MN COLIBRI

Similar special transports as undertaken by the Airbus carriers are made by two ro ro vessels of the French shipping company Compagnie Maritime Nantaise (MN) between Germany, France, Russia and French Guyana. The shipping company started transporting Ariane rockets from Bremen and Le Havre to its launch site in Kourou in the 1980s. Since 2009, the transports included components for the Soyuz launchers from St. Petersburg to Kourou. Initially, the German 7190 dwt special ship ARIANA, built in 1988 by Sietas

Fig. 3.92 The 3500 dwt ro ro ships CITY OF HAMBURG and CIUDAD DE CADIZ, built in Singapore in 2008/09, were modelled on the larger VILLE DE BORDEAUX (Ralf Witthohn)

shipyard for German owner Heinrich, had been used for this purpose. The 115.3 m length, 20 m breadth and 12.3 m depth MN TOUCAN, delivered in 1995 by the Dutch IHC shipyard, carries 7840 tonnes on 5.5 m draught. The similar MN COLIBRI was built at Dutch Merwede shipyard in 2000. In addition to these two special ships, MN also operates the con ro ships MN CALAO and MN TANGARA for the French Ministry of Defence. They were delivered by Hyundai Mipo shipyard in 2013. A third ship, named MN PELICAN and built in 1999 by Norway's Fosen shipyard, was in 2016 chartered out to Roscoff/France-based Brittany Ferries and was in early 2022 serving as freight ferry on the Poole-Bilbao route (Fig. 3.93).

3.3.7.10 Zhen Hua's Crane Carriers

Among special heavy load carriers for the transport of large cargo units of high weights are the ships of China's Zhen Hua company, which is exporting its container cranes to ports in the whole world. In 2020, the leading supplier of such port equipment disposed of 30 such carriers, which have been rebuilt from tankers. For the transport of several container bridges simultaneously the main deck level to be lowered and high ballast capacities to be created by using the former oil cargo tanks. The handling of the cranes is carried out on rails leading from the ship's deck to the quay. This method demands permanent trimming and some experience especially in tidal ports.

In 2006, the former 81,000 dwt tanker MARINE RENAISSANCE, built by ASTANO shiphyard in El Ferrol/Spain in 1983, began to transport

Fig. 3.93 MN TOUCAN has been transporting Ariane rockets from Bremen and Nantes to Kourou since its construction in 1995 (Ralf Witthohn)

Shanghai-built container cranes to world-wide customers as ZHEN HUA 13. The ship was broken up in 2018. The 234 m length, 40 m breadth 48,412 dwt ZHEN HUA 23, which in March 2012 carried the first four Postpanamax container cranes from Shanghai to the new Wilhelmshaven container terminal via the Cape of Good Hope, is the former 80,000 dwt tanker RICH DUCHESS, 1986 built at Japan's Kasado Dockyards and converted to a heavy load transporter in 2007. In October 2021, ZHEN HUA 12 carried four container cranes from the wharf of ZMPC's Changxing brach to New Orleans. Even elder is the 45,323 dwt ZHEN HUA 10, built in 1981 as 86,843 dwt tanker EVA at ASTANO. The largest of the ZHEN HUA carriers is the 1998 Samsung former 104,000 dwt tanker P. ALLIANCE, which since its conversion to ZHEN HUA 7 achieves a deadweight of still 48,127 tonnes. The ship made the headlines in November 2020, when 14 crew members were kidnapped in the Gulf of Guinea. In early 2022, ZPMC operated 26 heavy load carriers and the 147,490 gt crane ship ZHEN HUA 30 (Fig. 3.94).

3.4 Dangerous Goods

The share of cargo classified as dangerous is high and includes supposedly relatively safe goods such as dry ice, coal, fish meal, paints or fertilizers. Special safety regulations apply to their carrying, indicated by a red flag or red light in accordance with the dangerous goods classes of the IMDG Code put in force

Fig. 3.94 Built in Spain in 1983, the former 81,000 dwt tanker MARINE RENAISSANCE transported Shanghai-built container cranes to world-wide customers as ZHEN HUA 13 from 2006 until its demolition in 2018 (Ralf Witthohn)

by the IMO. The transport of explosives and ammunition requires increased attention. In Germany, such transports are mainly carried out through the privately owned port of Nordenham by US ships.

3.4.1 Chemicals for Japan

In January 2015, the first block train with 42 containers roll from the BASF plant in Ludwigshafen to the Jade Weser Port in Wilhelmshaven. The containers carry products from the world's largest chemistry producer and are destined for Japan. In Wilhelmshaven, they are lifted from the wagons at the rail terminal and transported by van carriers directly to the container bridges for loading them onto the MAERSK EMDEN. The container ship, chartered by Maersk Line from the Hamburg-based Rickmers group, will call at the Japanese ports of Kobe, Nagoya and Yokohama, which are reached after 6 weeks within Maersk's AE1 service. However, the expectation of at least slightly improving the capacity utilization of the under-employed Jade Weser terminal in the long term by an estimated yearly volume of 5000 BASF containers is dampened after only 3 months, when BASF switches to another port (Fig. 3.95).

MAERSK EMDEN is one of eight neopanamax vessels built by Korea's Hyundai shipyard for the Rickmers group. RUBY RICKMERS, PEARL RICKMERS, AQUA RICKMERS, COCONEE RICKMERS, LEO RICKMERS, SCORPIO RICKMERS, LIBRA RICKMERS and TAURO RICKMERS were renamed MAERSK EMDEN MAERSK EDINBURGH MAERSK EINDHOVEN, MAERSK ESSEN, MAERSK EDMONTON, MAERSK ELBA, MAERSK ESSEX and MAERSK EVORA under a

Fig. 3.95 The Hamburg container ship MAERSK EMDEN took delivery of BASF containers for Japan, which were yet for only a short time exported via Wilhelmshaven (Ralf Witthohn)

long-term charter arrangement with Maersk Line. The 366.4 m length, 48.2 m breadth ships carry 142,100 tonnes on 15.5 m draught. Their nominal cargo intake amounts to 13,092 teu. Powered by a 12-cylinder Wärtsilä engine of 68.640 kW output they achieve a speed of 24.3 kn. In 2022, the ships were still sailing under their Rickmers maiden names in the initial Maersk charter, but owned by Chinese interests, named Minsheng Financial Leasing, BoCom Leasing or Bank of Communications. The ships belong to a series of 53 newbuildings delivered by the Korean shipyard (Fig. 3.96).

3.4.2 Ammunition for Nordenham

Instead of the destination port's name five crosses are marked in the AIS signal of the transport vessel FLICKERTAIL STATE. The ship has left the US port of Sunny Point on 27 April 2017. Thus, it is not made public that the ship is bound for Nordenham/Germany. Shortly before arrival on 11 May, the vessel's AIS signal is switched off. Only the tugboat BRAKE, which is berthed as a safety measure next to the FLICKERTAIL STATE, indicates that a transport of dangerous goods of military origin is taking place. While the cargo handling work on the FLICKERTAIL STATE at Nordenham is proceeding under high safety precautions including fire hoses permanently spilling water, its sister ship GOPHER STATE sets as well course from Sunny Point for the Weser on 29 April, with arrival scheduled 16 May. The US army still has three vessels of this vessel type under its command. They were in 1968/69 built at Bath Iron Works in the US as container ships C.V. LIGHTNING, EXPORT LEADER and C.V. STAG HOUND (from 1987 CORNHUSKER STATE) for American Export Isbrandtsen Lines and in 1987 converted for the transport of ammunition and military equipment in short-term operational readiness. For this purpose, the ships were on starboard side equipped with two double heavy-lift cranes, which can be operated in tandem mode (Fig. 3.97).

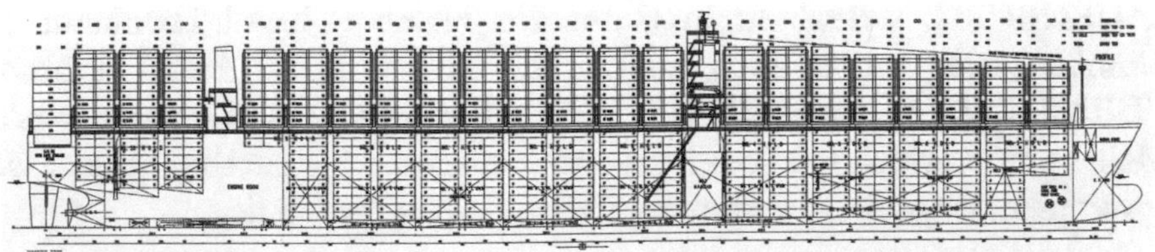

Fig. 3.96 Of the MAERSK EMDEN type, Korean shipyard Hyundai built 53 units, eight of them for Germany's Rickmers group (Rickmers)

Fig. 3.97 The US military transporter FLICKERTAIL STATE was involved in the shipment of explosives from and to Nordenham in 1990 and 2017 (Ralf Witthohn)

The former container ships belong to the fleet of the United States Maritime Administration (MARAD) and are part of the Ready Reserve Force. Twenty-seven years earlier they had in Germany caused nationwide interest, when they transported 395 tonnes of poison gas grenades of the US Army to the Johnston Atoll in the Pacific Ocean for destruction. Only on this occasion did it become known that the ammunition had also reached Germany via Nordenham in 1958 and 1967.

In December 2019, another ship, the geared container vessel MAERSK VALENCIA, transported military goods from Sunny Point/US to Nordenham. The 23,482 dwt ship is the former German-owned container vessel HANSA LAUENBURG, series-built in China for Hamburg-based Leonhardt & Blumberg company in 2006, sold to US owners in 2013 and renamed VIKING EAGLE. The ship was in 2017 taken over by the Danish Maersk/Møller group, which has been long-term contracted for the shipment of US government cargo on Maersk-owned vessels under the US flag.

3.4.3 Former Soviet Ro Ro Carriers Under US Flag

Military transports via Nordenham were also carried out by a ro ro ship originating from a Soviet building series of Chernomorskiy shipyard in Nikolayev. It comprised four ships of a decidedly high-tech type, the 20,075 dwt vessels KAPITAN SMIRNOV, KAPITAN MEZENTSYEV, INZHENER YERMOSHKIN and VLADIMIR VASLYAYEV, delivered to Odessa-based Black Sea Shipping Co. in 1979/80 and 1987. For the prospective handling of the heaviest and most modern tanks the ships were equipped with a stern ramp. To ensure fast transport in times of political crisis, the 227.3 m length, 30.1 m breadth twin-propelled carriers were fitted with a gas turbine propulsion system for a speed of 24.5 knots.

Two of the vessels, KAPITAN SMIRNOV and INZHENER YERMOSHKIN, were in 1997 sold and entered in the register of St. Vincent and The Grenadines under the names G.T.S. KRISTA and G.T.S. KATIE. G.T.S. KATIE was to return Canadian military equipment from the Kosovo to Canada in 2000 under the command of a Russian captain. However, because of allegedly unpaid bills, the ship stayed away from port, whereupon it was picked up and towed in by the Canadian navy. A few months later, the ship was scrapped in Alang/India, like its sister ship and 1 year later, KAPITAN MEZENTSYEV, meanwhile renamed BALAKLEYA.

The 1987-built VLADIMIR VASLYAYEV underwent a comprehensive conversion in 1998 as the second unit of the “Enhanced Maritime Prepositioning Force” of the US Marine Corps at the US shipyards Bender in Mobile and Tampa Bay in Tampa. The rebuilding was based on a design of Finland’s Kvaerner Masa. While retaining the gas turbine system, the conversion included a lengthening of the hull to 263.4 m, the fitting of a garage on the main deck, reinforcement of the stern ramp and the equipment of three cranes. Renamed LCPL ROY M. WHEAT the US ship handled military goods in Nordenham at the end of a voyage from Jacksonville in October 2016 (Fig. 3.98).

3.5 Other Industrial Products

3.5.1 Sewage Sludge from the East Frisian Islands

Of the hundreds of conventional coasters from the pre-container era, only a handful is still active off the North Sea coasts in 2020. The German-flagged

Fig. 3.98 The former Soviet tank transporter VLADIMIR VASLYAYEV handled ammunition in Nordenham in 2016 as US ship LPCL ROY M. WHEAT (Ralf Witthohn)

HELGOLAND of company Karl Meyer at Wischhafen, delivered in 1964 by Hamburg shipyard J. J. Sietas under the name of HELIOS II, serves as an island supplier. The 1400 dwt MERIDIAN owned by captain Timo Janßen in Wittmund and built in 1969 by shipyard Martin Jansen in Leer/Germany as CASTOR, sometimes interrupts its sailing due to a lack of orders. As one of only a few coasters under command of its owner, the Wilhelmshaven-registered ship transports the most diverse types of cargo between North Sea and Baltic Sea ports in decidedly niche markets: bulk goods, such as animal feed from rapeseed meal or grain from southern Denmark, Schleswig-Holstein and Poland to silo facilities in Oldenburg and Bremerhaven, sewage sludge from the East Frisian Islands to Hamburg, general and project cargoes such as Caterpillar MaK engines from Rostock to the shipyards. The ship's range of operation is limited, as is the case with comparable ships, because the ISPS code introduced and the associated investment costs made the operation of small ports uneconomical, so that they failed as transhipment places. In 2021, MERIDIAN was sold to Koole Meridian at Vijfhuizerdijk/Netherlands and renamed KOOLE 74. In January 2022, the ship undertook a voyage from Willemstad/Curacao to Paramaribo and Paranam/Suriname under the flag of Saint Kitts and Nevis (Fig. 3.99).

The decline of European shipbuilding is well detectable from the order scheme of coastal ships, which had over centuries been delivered by domestic shipyards but are increasingly contracted with non-European builders. One of those unusual origins refers to five newbuildings taken over by German owners from Israel Shipyards at Haifa, first time mentioned in 1965 as builder of sea-going vessels. At 90 m length and 15.3 m breadth the 5193 dwt IMINA, renamed MANDRAGA in 2015, BLUE CARMEL ex THEDA, renamed CALOBRA in 2016 and MARIAN R. in 2021, BLUE NOTE, BLUE TUNE and TRIPLE S are single hatch vessels. Their 54.6 m long and 12.6 m wide

Fig. 3.99 Captain Timo Janßen from Wittmund/Germany secured his existence with rape, project and sewage sludge transports of the coaster MERIDIAN, also involved in engine and crane shipments, before the ship's sale to a Dutch owner and its employment in the Caribbean (Ralf Witthohn)

Fig. 3.100 Haifa was the place of build of the BLUE TUNE and four sister vessels (Ralf Witthohn)

cargo hold of 5777 m^3 volume is able to take all kinds of cargo, including timber and grain or 120 teu containers. Classified by Germanischer Lloyd the ship series was built according the society's lowest ice class E and fitted with an MAN B&W four-stroke engine of 2040 kW output enabling a speed of 11.5 knots (Fig. 3.100).

3.5.2 Back from Russia

The termination of Germany's coaster era was interrupted, when in 2016 Elsfleth-based JEB Bereederung of Christopher-Niels Severin, in addition to two further acquisitions, took over the Russian coasters KENTO, KERET, TULOS, SUOYARVI and KOVERA from the dissolved fleet of Petrozavodsk-baseed Onegoship. The ships hoisted the German flag under the names LILY-B, PAULIN-B, MIA SOPHIE-B, KIRSTEN-B and SARAH-B. Four of the 2300 dwt vessels had in 1994/95 been delivered by Arminius Werke in Bodenwerder/Germany, located on the Upper Weser, while the builder of SUOYARVI was cited as Onega Arminius Shipbuilding in Petrozavodsk. In Bodenwerder, a larger series of ships with the names ONEGO, VYG, SEG, MEG, SYAM, KENTO, KERET, TULOS, KOVERA and KELARVI had been built for White Sea-Onega Shipping from 1991 onwards and, after a difficult transfer on a tidal wave artificially created by the reservoir of the Edertal dam, were towed to the German coast for completion. In 1996, the hulls of two larger 2900 dwt coastal motor ships named IRTYSH 1 and BALTIC SKIPPER were built at Bodenwerder and completed at the shipyard's sister company Cassens in Emden, as the last of almost 20 such seagoing vessels. KIRSTEN-B was in 2019 sold to Dutch owners and hoisted the Chilean flag as LOGIMAR II. JEB Bereederung instead acquired the 4620 dwt Dutch-built coaster PAUL and in 2021 the 5535 dwt MOON STEP, a 1997 newbuilding of Don Cassens shipyard at Aksay/Kazakhstan. They were renamed INGE-B and BARBARA-B, resp. KIRSTEN-B was in 2019 sold to Dutch owners and hoisted the Chilean flag as LOGIMAR II. JEB Bereederung instead acquired the 4620 dwt Dutch-built coaster PAUL and in 2021 the 5535 dwt MOON STEP, a 1997 newbuilding of Don Cassens shipyard at Aksay/Kazakhstan. They were renamed INGE-B and BARBARA-B, resp (Fig. 3.101).

Fig. 3.101 The coaster KENTO, built in 1994 in Bodenwerder/Germany for Russian customers, returned to the Weser 22 years later and was renamed LILY-B by JEB Bereederung in Elsfleth (Ralf Witthohn)

4

Container Transports

The success of the container as transport mean is based on its suitability for accommodating a wide variety of goods while at the same time it is standardising the transport method and thus able to combine sea, road, rail and inland waterways transportation. The shipped goods comprise a large range of agricultural, forestry and industrial products, including chemical, pharmaceutical and electronic goods as well as car parts, complete vehicles, machinery, food, clothing, dangerous goods, refrigerated cargo in reefer or liquids in tank containers. In this way, sawn timber, chipboard and laminate are transported from Europe to China, Australia, South America and the Middle East. Peat loaded in bags is exported in containers from the Baltic States, granite from India and China. Millions of tonnes of waste paper and plastic waste, which for a long time had been shipped from Europe, the United States and Japan are not any longer accepted by China, since the country has banned such imports since July 2017.

Container shipments as counted in 20-ft equivalent units (teu) increased by 3.8% in 2018, down from 6% growth the year before. This meant a strong decrease of growth figures after double-digit expansions in the 2000s and less than half the 5.8% average annual growth rate recorded over the two decades before. In 2019, the containerized trade expanded at a rate of only 1.1% totalling to 152 million teu. Following the outbreak of the COVID-19 pandemic, China, as motor of container trades accounting for one third of global volumes, expected a negative full year decline by up to 5% in its ports for 2020, only the second such decrease counted over the five decades before. In fact, the global container trade fell by only 1.1% to 149 million teu year on year. Container ships made up, by 282 million deadweight tonnes, 13% of the world's fleet in 2021.

R. Witthohn, *International Shipping*, https://doi.org/10.1007/978-3-658-34273-9_4

In February 2020, the reduction of volumes, which co-incided with a scrubber retrofitting programme and extended yard stays of about 120 ships at the time, had amounted to about 16% compared to the pre-year. As precautionary measure the member lines of the THE Alliance consortia raised contributions to a contingency fund of a total of USD75 million. The participants of the fund, arranged as cash reserve in case of an insolvency of a partner, are Hyundai Merchant Marine from South Korea joining the alliance 1 April 2020, ONE formed by three Japanese carriers, Germany's Hapag-Lloyd and Taiwan's Yang Ming Line (Figs. 4.1 and 4.2).

Fig. 4.1 Delivered in June 2018 as the tenth of eleven 20,569 teu newbuildings from South Korea's Daewoo shipyard for Denmark's Maersk Line, MUMBAI MAERSK suffered a grounding off the German isle of Wangerooge in February 2022 und could only be salvaged 2 days later (Ralf Witthohn)

Fig. 4.2 In the wake of the COVID-19 pandemic, the member lines of the consortia THE Alliance raised contributions to a contingency fund, arranged as cash reserve in case of an insolvency of one of its partners, of which the three major Japanese companies NYK, Mitsui-OSK and "K" Line are operating container vessels like the 8110 teu, in 2008 built by Japan's Imabari shipyard (Ralf Witthohn)

The strategic goal of the major liner shipping companies is to establish the most comprehensive transport network possible in order to offer customers all routes for the distribution of their goods. The shipping companies often operate their own ships on the main lines and charter vessels on the secondary lines. Almost always, joint services are set up, which were called consortia at the beginning of container shipping and have more recently been formed as alliances of various types. The liner shipping companies involved contribute their own or chartered tonnage. However, they also have the option of renting slots on other shipping lines' vessels without providing own ships. In this way, they are able to create transport offers that cover the demand at lower investment costs. A substantial share of shipping companies, counting about 670 firms managing or owning container ships worldwide, does not maintain their own liner services but charter their ships out to about 450 liner companies. The world container ship fleet comprised 5100 units with a total loading capacity of 20.3 million teu in 2018, 56% of which was chartered by the lines operating companies. Of these, many only act on regional markets. The number of large liner shipping companies operating internationally in consortia or cooperations was subject to an ongoing concentration processes and had shrunk to just under a dozen by 2018, and even lower than ten by 2020. In their aim to constantly optimise the liner service networks, schedules, ship sizes and port rotations are permanently changing, so that the following extracts from main routes container traffic can only reflect momentary stocktakings.

Despite the container crisis, the order activities increased in the course of 2017 by newbuildings contracts with a total capacity of 672,000 teu, after 280,000 teu were ordered in 2016. In 2019, a total of 101 container vessel newbuildings for delivery between 2020 and 2022 were contracted in only three shipbuilding countries, namely South China (60 units), Korea (31) and Japan (10). The largest capacity ships contracted were 23,000 teu newbuildings for Swiss/Italian Mediterranean Shipping Co. (MSC) to be built by Daewoo and for Evergreen to be built at Samsung in Korea and Hudong in China, though the vast majority of box carrier orders referred to feeder vessels of only 1000 to 2500 teu capacity. In March 2020, 378 container ships of 2.4 million teu capacity, equivalent to 14% of the world fleet, were inactive, many of them due to on-going scrubber refits or an reduced schedule in connection with the COVID-19 pandemic, which was expected to impact particularly container shipping. Due to the tonnage oversupply, the market rates nearly remained on a historic low level. The average rate for the transport of a 20-ft container from Shanghai to northern Europe in 2018 was figured at USD 820, which meant a 6.2% year-on-year decrease. In 2010, the rate had stood at 1800 Dollars, more than twice that much. The rate from Shanghai to the US West Coast was reported as 1700 Dollars, anyhow an increase of 17% related to the pre-year, but still considerably less than the 2300 Dollars paid in 2010. A container shipping boom was triggered in the second half of 2020, when consumer goods were increasingly demanded instead of services as a result of the Corona crisis. In an overheated market, there was a shortage of tonnage and containers, which led to shipping delays of several weeks and strong profit increases of the operators. During the second half of 2020, the rate levels rose considerably and in January 2021 reached USD 4452 per teu and USD 8900 per feu from Shanghai to northern Europe. As a consequence of the COVID-19 pandemic with shut-down container ports, interrupted transport chains and the lack of containers, the global container rates continuously grew further to USD 10,300 in September 2021 and fell slightly to USD 9400 until the end of the year. The sharp increase of rates resulted in unprecedented earnings for the operators, of which Maersk reported a profit before taxes of USD 19.7 billions and Hapag-Lloyd expected up to USD 9.5 billions for 2021. Another COVID-19 lockdown at Shanghai in February 2022 again caused a far-reaching global supply chain chaos, in the receiving countries worsened by too small port workforces and containers blocking the terminals, this letting outsiders like the German discounter Lidl invest in own container ships.

In contrast to Maersk Line, all other big operators had increased or hold their fleet capacities according to 2021 figures. MSC's box intake grew to 3.9 million teu (up from 3.8 million a year before), China's COSCO to 3.0 (up from 2.9), France's CMA CGM to 3.0 (up from 2.7), German Hapag-Lloyd stayed at 1.7 million, as Japan's newly formed ONE did by 1.6 million teu. Taiwan's Evergreen increased its tonnage to 1.3 million teu (up from 1.1), Yang Ming's fleet remained unchanged at 0.6 million teu, compared to 2018. In the beginning of 2022, MSC for the first time surpassed Maersk by fleet size, then disposing a capacity of 4.31 million teu, equivalent to a 17.1% share, compared to Maersk with 4.29 million slots. At this time, the CMA CGM fleet offered 3.22 million teu, COSCO 2.93 million, Hapag-Lloyd 1.75 million teu, ONE 1.53 million teu and Evergreen 1.47 million teu.

As market leader for decades, Denmark's Maersk Line had operated container ships with a cargo capacity of 4.1 million teu at the beginning of 2021, slightly down form 4.2 million teu at the beginning of 2020. In February 2020, Maersk secured a new sustainability-linked revolving credit facility of USD 5 billion through a syndicate of 26 selected banks, the first refinancing arranged after transformation from a diversified conglomerate to a global container logistics company. Adjustment of the credit margin under the facility is to be based on Maersk's progress to meet its target of reducing CO2 emissions per cargo moved by 60% until 2030. This is more ambitious than the IMO target of 40% by 2030, both aims based on 2008 figures. In 2019 Maersk had announced its commitment to becoming carbon neutral by 2050. In February 2021, A.P. Møller-Maersk declared to operate the first carbon neutral feeder vessel of 2000 teu using e-methanol or bio-methanol by 2023. All of the companie's newbuildings would in future be fitted with dual fuel technology and operate either on low sulphur fuel oil or carbon neutral without determining the fuel chosen for the latter. In August 2021, Maersk contracted eight carbon-neutral container ship newbuildings with Korean builder Hyundai. Four options were realized in the beginning of 2022, though the company had to admit, that the supply of green bunker needing several hundred thousand tonnes of e-methanol yearly would be hardly to achieve (Fig. 4.3).

Since the container shipping companies, which—apart from the Chinese ones—are primarily privately owned, had steered themselves into the crisis by overestimating market opportunities and over-expanding capacities, the profitability of the tonnage deployed became all the more important for financial survival. As a result, further growth in ship sizes but lower speeds to reduce fuel costs became the trend.

Fig. 4.3 A major part of globalization is maintained by so called mega carriers, which largest ones were in 2019 ships of the 23.656 teu type MSC MINA. In early 2020 the 400 m length, 61 m breadth ship served in the Asia-Europe loop AE-10/Silk and in 2021/21 in the AE-11/Jade service from Asia to the Med under the 2M agreement of Maersk Line and Mediterranean Shipping Co. (Ralf Witthohn)

Dependent on their special design parameters, container ships have shown their operational vulnerability under quite different aspects. Most spectacular were the repeated accidents caused by structural failures, as they occured on MSC CARLA, which broke into two parts on a voyage from Le Havre to Boston in 1997, on MSC NAPOLI experiencing the same fate in 2001 and on the Japanese MOL COMFORT in 2008. Among the many dangers to which container ships are exposed, explosions and fire belong to the most critical ones. In 2019 a 19 days lasting fire broke out in a deck container on Hapag-Lloyd's YANTIAN EXPRESS and spread to other containers some 650 nm off the Canadian coast, causing a highly difficult salvage operation. Investigations of Germany's Federal Bureau of Maritime Casualty Investigation (Bundesstelle für Seeinfalluntersuchung, BSU) assumed, that coconut charcoal may have self-ignited, as it had already happened on MSC KATRINA in 2015 and LUDWIGSHAFEN EXPRESS in 2016. Explosions and fire also caused major accidents on HANJIN PHILADELPHIA in 2002 and MSC FLAMINIA in 2012. All vessels, except MSC KATRINA, were German-owned (Fig. 4.4).

Fig. 4.4 A structural failure caused the breaking of MSC CARLA in two parts, of which only the aft one could be towed to Las Palmas (Erich Müller)

Often, liner shipping companies have historical preferences for certain trades with business relations dating long back to the pre-container era. This is accompanied by a demarcation of interests, which in Germany could be seen in the shipping companies' founding names, like Deutsche Afrika-Linien, Hamburg-Südamerikanische Dampfschifffahrtsgesellschaft (Hamburg Süd) or Hamburg-Amerikanische Packetfahrt Actien-Gesellschaft (HAPAG, today Hapag-Lloyd). While Deutsche Afrika-Linien has remained mostly true to its original trade commitment until today, Hapag-Lloyd and Hamburg Süd expanded their areas of interest early on, primarily by buying up shipping lines, so that they ranked among the ten leading global carriers. Hamburg Süd had realised its strengths above all in the South America trade and its red colour is still carried to that continent, even after the company was sold by Germany's Oetker conglomerate to A.P. Møller/Maersk Line in 2017.

From the 1960s onwards, large terminals were built for the handling of containers. In addition to man-operated, rail-mounted gantry bridges for ship-to-shore operation on the quay side, van carriers, also named straddle carriers, or instead of them rail cranes, and above all a large amount of space were required. The first semi-automated terminals, where unmanned vehicles, so-called AGVs (Automated Guided Vehicles), were used to transport containers were built in Rotterdam-Maasvlakte in 1993 onwards and in 2002 at Hamburg-Altenwerder. The gantry cranes at the terminals of Maasvlakte II opened in 2015 are as well unmanned. The first automated terminals in Asia

Fig. 4.5 As the ninth in a series of 16 mega carriers, then representing the world's largest container ships, the 23,656 teu, Daewoo-built MSC SIXIN took up the 2M Far East—Europe AE-10/Silk service in December 2019 (Ralf Witthohn)

were inaugurated in Shanghai and Qingdao in 2017, one in Melbourne in the same year. The terminal at Jebel Ali/Dubai is also partially automated.

In 2020, Asian ports accounted for a throughput of 537 million teu, equivalent to 65% of the world total, followed by Europe by 117 million teu and North America, where 61 million teu were handled. Shanghai, Singapore, Ningbo-Zhoushan, Shenzhen, Busan, Hong Kong, Guangzhou and Qingdao are the terminals with the highest throughput rates in the Chinese economic area. In 2017, Shanghai handled more than 40 million teu for the first time, 42 million in 2018 and 43.5 million in 2020, when Singapore reached 36.6 million, Ningbo-Zhoushan 28.7 million teu. Rotterdam is the most important European container port in tenth place by a throughput of 14.4 million teu in 2018, ahead of Antwerp and Hamburg (Fig. 4.5).

4.1 Europe: Far East Route

On Christmas Eve, the Panama-flagged MSC NELA leaves the berth of MSC Gate at Bremerhaven and continues its maiden voyage from the Far East to Gdansk. Not to impair the last handling operations of the MONACO MÆRSK moored at the most northern berth of the terminal, the MSC vessel passes the whole whole 5 km long container quay at the slowest possible speed. The broad funnel, which architecture has been designed to accommodate the new SOx gases cleaning scrubbers, is, 1 week before the new IMO rules are put in force, yet emitting

Fig. 4.6 After its Christmas maiden call at Bremerhaven 2019, the then world's largest container vessel type MSC NELA set course for Gdansk. The 23,656 teu ship was employed in the AE-10/Silk service of the 2M coopeation together with seven more MSC ships of the same class and five 18,340 TEU vessels of Maersk's MARY MAERSK type maintaining the port rotation Bremerhaven, Gdansk, Bremerhaven, Rotterdam, Tanjung Pelepas, Shanghai, Xingang, Qingdao, Kwangyang, Ningbo, Shanghai, Yantian, Tanjung Pelepas, Algeciras, Bremerhaven (Ralf Witthohn)

considerably large grey-white clouds—instead of those deep dark clouds that left elder MSC vessel over decades. At a capacity of 23,656 teu MSC NELA is the seventh of 16 units of the actually world's largest container ship class, named megamax-24 by its operator and built by South Korean shipyards Samsung and Daewoo (Fig. 4.6).

After delivery in November 2019 to Beijing-registered Minsheng Financial Leasing, with China Minsheng Bank as major shareholder, MSC NELA started its long-term employment by Mediterranean Shipping Co. (MSC). The newbuildng joined the AE-10/Silk service of the Maersk/MSC consortia 2M, which is at this time was maintained by the same class vessels MSC GÜLSÜN, MSC SAMAR, MSC LENI, MSC MIA (all of 23,756 teu capacity), MSC MINA, MSC ARINA and MSC SIXIN (of 23,656 teu) with the MARY MÆRSK, MARIE MÆRSK, MAYVIEW MÆRSK, MOGENS MÆRSK and MARIT MÆRSK of 18,340 teu capacity. The 399.8 m length, 61 m breadth MSC NELA reaches a speed of 22 knots, achieved by an 11-cylinder MAN B&W engine of model G95ME-C9.5 and 59,250 kW output. 2024 plugs for 40 ft. boxes provide a reefer volume of about 137,000 m^3.

The North Europe-Far East route, on which three major shipping groups are still active after the waves of mergers in 2020, is regarded as the high street of container shipping. The groups are the 2 M cooperation, Ocean Alliance and THE Alliance. The 2 M agreement on East-West routes between Maersk Line of Denmark and Mediterranean Shipping Co. of Svitzerland/Italy, initially concluded for 10 years, started in January 2015. The Ocean Alliance, which was set up for 5 years with an option of extension in April 2017, involved the Hong Kong shipping company OOCL, China's COSCO, Taiwan's Evergreen and the French CMA CGM, including the acquired Singapore company APL. Also in April 2017, the Japanese shipping lines "K" Line, MOL and NYK together with Taiwan's Yang Ming began their cooperation in THE Alliance with their German partner Hapag-Lloyd, joined by the Saudi UASC following its takeover by Hapag-Lloyd. This alliance as well covers only East-West routes.

4.1.1 20,000 Teu Ships in Large Numbers

Container shipping on the Elbe reaches a peak on 21 December 2014, when 14 large container ships pass Cuxhaven on their way to and from Hamburg within only 5 h. Whenever possible, they are to be dispatched before the Christmas holidays so that no idle, cost-generating port berthing times are incurred. The container ship armada includes four ships of Germany's largest liner shipping company Hapag-Lloyd, the incoming MONTREAL EXPRESS from Canada and the outgoing OSAKA EXPRESS, HAMBURG EXPRESS and LEVERKUSEN EXPRESS bound for the Far East. Another ship of a Hamburg-based shipping line, Hamburg Süd's CAP SAN ARTEMISSIO, is coming in from the East Coast of South America. CSCL SOUTH CHINA SEA of China Shipping is on its maiden voyage from the Far East. Hamburg's leading position in the East Asia trade in Germany is underscored by the passage of further Asian mega containerships, such as the Taiwanese YM UTMOST, the Chinese COSCO TAICANG, the Arabic AL ULA, as well as the Japanese MOL BRIGHTNESS and NYK HERCULES. The French CMA CGM CALLISTO is also operating in the Far East service. For CMA CGM, the medium-sized German container ships E.R. CALAIS sails to the Caribbean and the SANTA REBECCA to the West Coast of South America (Fig. 4.7).

By a distance of 7700 nm between Rotterdam and Singapore, the length of the Europe-Far East route requires the deployment of the largest container ships currently in service. They offer maximum cargo capacities of more than 20,000 teu. This means that the ships have a capacity almost 30 times greater

Fig. 4.7 Giants' meeting on the Elbe on 31 October 2013: Sailing in the wake of the large container vessel COSCO DEVELOPMENT (13,092 teu) of China Ocean Shipping Co. is another mega carrier from the Far East, the 14,074 teu carrying CSCL SATURN of China Shipping Container Line, followed by the 16,020 teu capacity CMA CGM ALEXANDER VON HUMBOLDT of French CMA CGM, one of the largest container carriers at this time, on maiden voyage, and by the 13,808 teu THALASSA HELLAS of Athens-based shipping company ENESEL, which has chartered out the ship to Taiwan's Evergreen. Two more of a total of eight ships pass within only 2 h time, named MSC SHENZHEN of 7850 teu and E.R. PUSAN of 6000 teu. Despite their Asian names reference, they actually belong to the Hamburg-based shipping E. R. Schiffahrt and are chartered out to MSC. Japan's largest shipping company NYK contributes the 13,208 teu newbuilding NYK HELIOS to the fleet parade. Seven years later, the capacities of the largest carriers employed on the Far East—Europe run are increased to 24,000 TEU. COSCO and CSCL have merged to China Shipping, while E.R. Schiffahrt has meanwhile given up business in course of the container shipping crisis (Ralf Witthohn)

than that of the first full container newbuildings that came into service for overseas shipping 50 years ago. The development in size induced a renaissance of the two-island vessel, which had already been realized much earlier in the architecture of tankers, bulk carriers and general cargo vessels. Whereas on these ship types the relocation of the engine room aft primarily served to make better use of the cargo capacity in the middle, widest part of the ship, the shifting helped to improve the visibility conditions on the mega container ships. By the geometric advantages, this concept allows a higher stowage of deck containers in front of the wheelhouse, while the stowage behind the command bridge is not at all restricted for visibility reasons. Further benefits from the aft arrangement of the engine room result from a shorter propeller shaft, which is costly to manufacture, as well as from a reduced space requirement of the shaft tunnel in the hold area, advantages which weigh twice as much in a double propeller system as fitted to the largest Maersk carriers (Fig. 4.8).

Fig. 4.8 Among mega class container ships, the twin propulsion concept has only been realised on ships of the MILAN MÆRSK type, which by 20,568 teu was the highest-capacity newbuilding when commissioned in November 2017 in the 2M service AE-1/ Shogun. In January 2020, the ship had switched to the AE-5/Albatross loop of the 2M cooperation, on which it still sailed 2 years later (Ralf Witthohn)

4.1.2 Nine Europe-Far East Services of the 2M Agreement

In the beginning of 2022, the 2M cooperation maintained nine Europe-Far East services, six of which sailed from north European ports, three more from the Mediterranean. By Maersk, they were designated AE-10, AE-5, AE-7, AE-55, AE-1 and AE-6, the ones from the Mediterranean AE-11, AE-12 and AE-15. Their corresponding MSC names were Silk, Albatross, Condor, Griffin, Shogun, Lion, Jade, Phoenix and Tiger. The AE-5/Albatross employed 13 ships of between 18,340 and 20,568 teu capacity calling at Rotterdam, Bremerhaven, Göteborg, Aarhus, Bremerhaven, Wilhelmshaven, Tanger Med, Singapore, Shanghai, Dalian, Tianjin, Busan, Ningbo, Shanghai, Tanjung Pelepas and Rotterdam. In February 2022, these were MADRID MÆRSK, MUNICH MÆRSK, MURCIA MÆRSK, MILAN MÆRSK, MAASTRICHT MÆRSK, MSC RIFAYA, MAREN MÆRSK, MERETE MÆRSK, MORTEN MÆRSK, METTE MÆRSK, MARGRETHE MÆRSK and MARIBO MÆRSK. At the same time, the AE-7/Condor used 12 ships with capacities from 16,652 to 20,568 teu, namely the MONACO

MÆRSK, MSC LEANNE, MATZ MÆAERSK, MARCHEN MÆAERSK, EBBA MÆAERSK, EVELYN MÆAERSK, ELEONORA MÆAERSK, EDITH MÆAERSK, ESTELLE MÆAERSK, MSC HAMBURG and MSC NEW YORK. The port rotation of this loop included Ningbo, Shanghai, Nansha, Yantian, Tanjung Pelepas, Colombo, Tanger Med, Wilhelmshaven, Hamburg, Antwerp, Londonn Gateway, Le Havre, Tanger Med, Salalah, Khalif, Jebel Ali, and Ningbo.

From March 2017 Hyundai Merchant Marine joined 2M for a period of 3 years, after which the Korean operator switched to THE Alliance on the east-west lanes. The 2M cooperation had in 2017 already operated six northern Europe/Far East routes. Two years later, the loops were still maintained, but the largest tonnage capacity was not anymore used on the AE-5/Albatross loop, but on the AE-10/Silk service employing the largest ships, the 23,756/23656 teu MSC GÜLSUN, MSC SAMAR, MSC LENI, MSC MIA, MSC MINA, MSC ARINA, MSC NELA and MSC SIXIN, plus the 18,340 teu ships MARY MÆRSK, MARIE MÆRSK, MAYVIEW MÆRSK, MOGENS MÆRSK and MARIT MÆRSK. The total weekly capacity amounted to 19,950 teu. The port rotation on this route named Bremerhaven, Gdansk with bunkering in Kaliningrad, Bremerhaven, Rotterdam, Tanjung Pelepas, Shanghai, Xingang, Qingdao, Kwangyan, Ningbo, Shanghai, Yantian, Tanjung Pelepas, Algeciras, Bremerhaven. In February 2021 the service was served by 13 Maersk vessels of 18,000/21,000 teu, and still was a year later.

The AE-5/Albatross, which was in early 2020 maintained by twelve 20,568 teu ships of the MUNICH MÆRSK class supplemented by the 19,437 teu MSC MIRA, was 2 years on up-graded to 13 ships with weekly capacity of 18,800 teu. Even more capacity, 20,568 teu per week, had in 2021 been employed in the AE-7/Condor service carried out by ten Maersk and three MSC ships. In early 2022, the AE-5/Albatross still used 12 Ships offering 18,000 teu week. The ports called were Ningbo, Shanghai, Nansha, Yantian, Tanjung Pelepas, Colmbo, Tanger Med, Wilhelmshaven, Hamburg, Antwerp, London Gateway, Le Havre, Tanger Med, Salalah, Khalifa Seaport, Jebel Ali, Ningbo. Set up in March 2019, the AE-1/Shogun string had employed 16 ships from 13,000 teu to 19,400 teu intake. The pendulum service was combined with a Far East to US West Coast leg, thus affording a duration of 112 days from Felixstowe, Rotterdam, Bremerhaven, Tangier, Salalah, Cai Mep, Hong Kong, Yantian, Xiamen, Los Angeles, Yokohama, Ningbo, Shanghai, Xiamen, Tanjung Pelepas, Colombo to Felixstowe. In 2022, the Pacific string was abandoned, so that only still 11 ships with a weekly capacity of 16,600 teu was still needed. The suspended AE-2/Swan loop regularly used Maersk M-class vessels of 18,340 teu carriers and such of MSC offering

Fig. 4.9 Under the 2M partnership between Maersk and MSC, the 19,400 teu vessel MSC ANNA, chartered from Ship Finance International since its commissioning in December 2016, served the AE-1/Shogun, in early 2020 the AE-2/Swan line and 2 years later the AE-15/Tiger loop from the Mediterranean to the Far East (Ralf Witthohn)

19,300/23,700 teu and calling at Rotterdam, Felixstowe, Antwerp, Rotterdam, Tangier, Algeciras, Singapore, Hong Kong, Shanghai, Qingdao, Ningbo, Tanjung Pelepas, Rotterdam.

Even 17 ships of between 9700 teu and 14,300 teu had in 2020 deployed to the AE-6/Lion service, which was, like the AE-1, a pendulum line combining a Europe-Far East leg with a trans-Pacific string and lead from Antwerp to Oakland. In 2022, the service maintained by 11 ships completed its round voyage in Tanjung Pelepas (Fig. 4.9).

4.1.3 Quantum Leap by EMMA MÆRSK

The major share of the 17,000 teu weekly capacity of 12 ships employed in the AE-7/Condor service was from 2017 to 2020 contributed by seven of eight ships of the 18,340 teu EMMA MÆRSK type, the EMMA MÆRSK, ESTELLE MÆRSK, ELEONORA MÆRSK, EBBA MÆRSK, ELLY MÆRSK, EDITH MÆRSK and EUGEN MÆRSK together with the larger MORTEN MÆRSK. In 2020, their schedule named the ports of Rotterdam, Hamburg, Antwerp, London-Gateway, Le Havre, Tangier, Salalah, Abu Dhabi, Jebel Ali, Ningbo, Shanghai, Nansha, Yantian, Tanjung Pelepas, Tangier, Rotterdam. When launched in 2006, the introduction of EMMA MÆRSK was a quantum leap in Far East shipping. Its container capacity described by the shipping company at 11,000 teu meant an increase of 43% compared to the previously largest container carriers of the GUDRUN MÆRSK class with a declared cargo capacity of 7700 teu. The increase in efficiency associated with this capacity growth alone provided a significant competitive advantage.

At the same time, the shipyard and the shipping company implemented a number of innovations which further increased the economic efficiency of the ship and thus enabled a more environmentally friendly operation. The technological lead, which benefitted from the cooperation between the shipbuilder at Odense and the Copenhagen ship operator within the A.P. Møller/Maersk group was put forward by the management as an argument for maintaining the shipyard in Odense, which yet operated at a comparatively high cost level. Nevertheless, this cooperation came to an end with the phasing out of the newbuilding series, and Maersk Line received the following even larger M-class vessels from South Korea. The generational change brought about by the commissioning of EMMA MÆRSK was also reflected in the growth in deadweight from 115,700 dwt of the previous series to 158,200 tonnes meaning a 37% increase. Apart from the beginnings of container shipping, there had never been such a major development step in the design. This is all the more remarkable as shipbuilders, shipping companies and classification societies were already in a race for the title of the world's largest container ship in previous years (Fig. 4.10).

Fig. 4.10 The construction of EMMA MÆRSK meant a leap in capacity that was only achieved in the early days of container shipping (Ralf Witthohn)

It was not until the christening of EMMA MÆRSK in September 2006 that the shipyard, which was part of the A. P. Møller group then, announced the main dimensions of the series featuring a length of 397.7 m and a breadth of 56.4 m. While this represents an 8% increase in length compared to the 367 m long GUDRUN MÆRSK, the old breadth of 42.8 m was extended by 32%—with the wanted effect of an extraordinary improvement in stability. The freeboard grew from 4.6 to 6.7 m, the depth from 24.1 to 30.2 m. The latter is an expansion of 25%, which makes it possible to stow ten layers of high-cube containers in the holds. Despite the disproportionate changes in the ship's dimensions, the characteristic Odense builders' lines optimised by CFD calculation were largely retained in EMMA MÆRSK. This included a long bulbous bow extending 15.4 m forward of the forward perpendicular, a forecastle raised to the beginning of the third hold and a vertical transom stern with an almost elliptical contour. The extraordinary increase in breadth required a new lines design without a parallel midship part. Another design change, also resulting from the large width of the ship, relates to the extended, closed bridge of EMMA MÆRSK, that have to be carried by stable tubular supports.

The design decision for the substantial widening of the ship with the aim of increasing the cargo capacity became necessary not least because both the length and the draught of the ship of 16 m reached or even exceeded the limits of the fairway accesses to some of the ports served in the Europe-Far East service. For example, the turning point in the Weser had to be extended from 400 to 600 m. This was done only shortly before the first call of EMMA MÆRSK at Bremerhaven in September 2006, after the planning approval procedure had been completed a few days earlier by the authorities. Since there was not enough time left for completing the necessary dredging measures, the arrival, which had initially been scheduled for early in the morning at low tide, was shifted by 1 day to the tide expected during daylight. However, the average draught of the ship was only 10.6 m due to the small amount of cargo loaded. The forecast already made when EMMA MÆRSK was commissioned, that for the time being—in contrast to the previous classes—a further increase in length would hardly be possible for new designs, was confirmed over the following one and a half decades. In this respect, innumerable press publications, in which the term "ever larger container ships" became a familiar term especially in connection with fairway problems, were based on incorrect facts.

The ships of the EMMA MÆRSK type were given the notation ✠A1, Container Carrier, E, ✠AMS, ✠ACCU, SH-DLA, SHCM by the classification society American Bureau of Shipping. It included the execution of a Safe Hull Dynamic Load Analysis (SH-DLA) and a Safe Hull Condition Management. The ship was designed under the ABS Enhanced Hull Construction Monitoring Program. This meets the requirements for an inspection to be carried out without docking (UWILD). The gross measurement on delivery was 170,794 tons, the net measurement 55,794 tons. The ligt ship weight was quantified at 60,600 tonnes, the maximum displacement at 218,800 tonnes. After the raise of the bridge, the tonnage increased to 171,542 gt, but the deadweight capacity decreased due to the additional superstructures and the raising of the lashing bridges. EMMA MÆRSK flies the Danish flag and is registered in Taarbaek (Fig. 4.11).

While the shipyard and the shipping company initially only made a statement about the nominal slot capacity, which is of little significance for the actual performance of the ship, an estimated figure of 11,300 loaded teu containers weighing 14 tonnes each could be derived from the deadweight of 158,200 dwt. This is close to the first official figure of 11,000 teu, taking into account fuel reserves. The French experts from Alphaliner corrected their first position analysis on the basis of seven deck layers of 6588 teu below and 6924 above deck, i.e. a total of 13,512 teu, to 6399 teu below and 7904 teu on deck, resulting in a total of 14,300 teu on the basis of nine layers on the hatches after commissioning of the ship. This figure, which strongly depends on the line of sight conditions, was increased to 16,810 teu in 2016, when the navigation bridge and the lashing bridges between the hatches were raised. The latter allowed the stowage of up to two additional layers of containers on the hatches. This shows the problems of an exact capacity indication for container ships without a precise definition of, inter alia, the various container

Fig. 4.11 ELEONORA MÆRSK showing lashing bridges after being raised (Ralf Witthohn)

lengths and heights, and that accurate size comparisons of container ships without such a basis therefore remain inadequate.

At a ship's breadth of 56 m and with side tanks of container width, the stowage geometry allows the ship to be loaded with 20 rows of containers in the holds, a number which was achieved for the first time. A maximum of ten tiers can be transported in the holds. The number of container rows on the hatches and on deck is 22, the number of tiers on deck was initially specified as seven. Up to the sixth 40-foot bay, the lashing bridges were the height of a container, behind it they were two container tiers high. After extension by one container height in each case, up to nine tiers could be stowed on deck. A patented lashing system developed by the shipyard enables two container tiers stowed above the lashing bridges to be secured with the aid of lashing rods.

The number of 40-ft bays in the holds totals to 22, 13 of which are in front of the engine room. 13 bays can also be loaded on the hatches in front of the superstructure. Behind the superstructure there are ten 40-ft bays, one more than in the holds, the extent of which below the main deck is restricted along the length of a bay by the aft winch deck. A further reduction in the cargo space is due to the main engine, which length of 27.3 m is longer than that of the superstructure and therefore means the loss of space over half the length of a 40-ft hatch. In addition, the bunker tanks were no longer located in the outer skin area in accordance with the guidelines which were to become effective 1 year after commissioning. The total volume of the cargo space amounts to 355,191 m^3.

The visual appearance of EMMA MÆRSK differs significantly from the predecessor class because of a changed longitudinal arrangement of the engine room and the superstructure above. On GUDRUN MÆRSK the number of deck container bays in front of the superstructure is 16 and six behind the superstructure, i.e. only one 40-ft bay less than on EMMA MÆRSK, which by 30 m increased length would have allowed two additional bays. Measured by the length of the hold, the superstructure of EMMA MÆRSK is at 59% of the ship's length, while on GUDRUN MÆRSK it is 76% from forward. Of the reasons given for moving the superstructure three bays forward compared with the previous ships is the better trim and the better visibility conditions over the containers stowed in front of the superstructure. The line of sight was also favored by raising the bridge superstructure compared to the forerunner. Initially, the EMMA MÆRSK superstructure was one deck higher, so that it comprised—including the wheelhouse—12 decks. From 2016 onwards it was extended by an additional "hayloft" to 13 decks. The cargo holds are each closed by four pontoon covers which can be moved by container gantry cranes and which were manufactured to the yard's own design according to a proven pattern by the Loksa branch in Estonia.

The main engine of the EMMA MÆRSK class is a 14-cylinder two-stroke diesel motor of type Wärtsilä RT-flex96C, which achieves a maximum of 80,080 kW at 102 rpm. This was the first of its kind and the most powerful ship engine built to date. It was manufactured by Doosan in Korea and drives a six-bladed fixed propeller weighing 131.5 tonnes and measuring 9.6 m in diameter, supplied by MMG in Waren/Germany. Power is transmitted via a shaft of about 120 m length manufactured by the engine builder. The ship reached a speed of over 26 knots, which could yet only be achieved by additional power from auxiliary systems. The exhaust gases of the main engine served as a supplementary source of energy by feeding a recovery system that achieved 8500 kW. The gases pass through an economizer generating steam to operate a turbine generator. The system also includes an ABB turbo-generator driven by part of the exhaust gases supplied to the turbocharger. The electrical power generated by the two generators was used to drive two Siemens Siship Boost type engines of 2 × 9000 kW output directly acting on the propeller shaft. Due to reduced speed requirements as a result of the slow steaming development during the shipping crisis, the shaft motors were put out of operation (Fig. 4.12).

Fig. 4.12 The exhaust system of EMMA MÆRSK includes a recovery system of 8500 kW output (Ralf Witthohn)

Maersk Line described the EMMA MÆRSK as one of the most environmentally friendly container ships, mainly because of the exhaust gas recovery system, which fuel savings also meant a reduction in emissions. Fuel consumption is as well reduced by the stabiliser system, which helps to reduce rolling motions and thus also protects the cargo. A new type of digital lift control system ensures, that the stabilisers are effectively adjusted and prevents them from entering the range of cavitation. A further fuel saving of 1200 tonnes per year is reportedly made possible by the underwater silicone coating reducing resistence. The basic electrical supply is provided by five auxiliary diesel engines of 5 × 4320 kW coupled with Siemens generators. The air supply of the engine room is guaranteed by energy-optimized fans with power savings of up to 20%. Positive effects of the high efficiency of the fan type are the reduction of exhaust emissions as well as a reduction of the noise level. The ship's heavy fuel oil tanks were laid out for a total volume of 16,920 m^3, the diesel oil capacity was initially specified as 405 m^3, the lubricating oil capacity as 652 m^3. The fresh water capacity is figured at 753 m^3.

An important factor in minimising costs refers to manning. The minimum crew number for the ships of the EMMA MÆRSK class was specified by the shipping company to be 13 persons accommodated in air-conditioned rooms. The low crew number, necessary for safe operation, was made possible by extensive automation with machine and cargo monitoring by more than 8000 constantly controlled signals. In fact, during the maiden sailing, 30 crew members were on board. Among them were so-called Maritime Officers, who, according to the Danish training concept, experienced both nautical and technical education. The officers were for 6 months specially instructed in the shipping company's training centre in Odense about their work on EMMA MÆRSK and got to know their new ship during the building phase. When steering the ship they can make use of two radar units. A third radar antenna is located on the forward lamppost. In addition to state-of-the-art radio, telex and telefax equipment, a powerful satellite device is used for communication.

In view of the ship's length of almost 400 m, four Rolls-Royce transversal thrusters were installed to improve the manoeuvring characteristics. They are each driven by an electric motor with an output of 1750 kW turning a controllable pitch propeller with a diameter of 2400 mm. Two each are installed forward and aft and help to reduce the number of assisting tugs during docking manoeuvres. Under good weather conditions, two tugs with bollard pulls of 60 and 75 tonnes were therefore sufficient to turn EMMA MÆRSK on the Weser within a quarter of an hour during its maiden visit.

The background of the gearless auxiliary motors was the size limit of the main engine. In order to achieve the output necessary for the designed service speed, two shaft motors were required. At the same time, they were flexible consumers of the energy obtained from exhaust gas recovery and also supported the start-up of the main engine. The power output of the booster engines to the propeller shaft was continuously variable in the speed range of the main engine. In addition, the boosters, which were designed as 16-pole three-phase synchronous motors with 39 rpm to 104 rpm, could serve as emergency function in the event of failure of the main engine. The highly sophisticated propulsion and energy system of EMMA MÆRSK could yet not prevent the ship from suffering a serious accident and almost a total loss due to a relatively simple manufacturing defect combined with a fundamental design deficit (Fig. 4.13).

Fig. 4.13 As result of the flooding of the shaft tunnel following a leak in one of the two aft transversal thruster tunnels, the EMMA MÆRSK ran into acute danger (Ralf Witthohn)

4.1.4 Near Loss of EMMA MÆRSK

On 1 February 2013, EMMA MÆRSK is bound from Tangier to the Far East with a cargo corresponding to 13,537 teu, of which 7112 units are empty. The ship sails on a southerly course off the Suez canal entrance. The two forward transversal thrusters are supporting manoeuvring, the two at the stern are running with the propeller blades in zero position. At 8 pm the anchor has been hoisted on the deep water anchorage, at 9.34 pm the pilot is taken. 7 min later a leakage in the shaft tunnel is detected. To prevent flooding of the engine room, the—supposedly—watertight bulkhead to the engine room is closed another 9 min later. The captain hopes to reach the Suez container terminal under own power. But within 25 min the shaft tunnel is completely flooded by 14,000 m^3 of seawater. Due to the high water pressure reaching a column of 8.9 m, the cable and fan ducts between the shaft tunnel and the main engine room leak. At 22.05 h the main engine is threatened with an imminent breakdown. In addition, water enters hold 21 due to a defective cable duct in the rear emergency exit of the shaft tunnel.

The convoy ships travelling behind EMMA MÆRSK, the first of which is an LNG tanker, stop after being informed by the traffic control. When the requested tugs are initially unable to bring EMMA MÆRSK under control because of broken lines, the main engine is stopped and cannot be restarted at 22.55 h due to the water ingress. After dropping the anchors at 0 o'clock, the ship is successfully brought close to the container quay by the help of the shifting wind. Due to the failure of the electrical power supply, which has occurred in the meantime due to the flooding of the auxiliary diesel engines, the mooring lines have to be deployed manually. The mooring manoeuvre is not completed before 4.46 am.

Although the leaky bulkhead between the shaft tunnel and the engine room was designed to be watertight, it was primarily intended to limit the capacity requirements of the CO2 fire-fighting system of the engine room and not to improve the leak stability of the ship. Thus not being prescribed by SOLAS regulations, the leakage and the ingress of water into the cargo spaces behind the engine room have a decisive effect on maintaining the intact stability. The inspection after the accident revealed, that the welded joint between the vertical flange guiding the motor of the transversal thruster and the horizontal tunnel tube of the first of the two stern thrusters had been completely torn off. Also, three of the four blades of the Rolls Royce TT2400RRM controllable pitch propellers had broken off at the transition from blade to base and were lost. The base plate of the propeller was broken. The same damage was found on the aft stern thruster propeller. The propellers had been replaced in 2006/2007.

After extensive investigations it could be determined that the radius in the throat between blade and base of the propellers was no longer 25 mm but only 10 mm due to changes in the manufacturing technology. It was therefore too small to guarantee the strength of the propeller. This led to fatigue fracture of the blades and, due to imbalance, to destructive centrifugal forces on the support plate of the propeller, finally to the fracture of the flange on the tunnel tube with the serious consequences of leakage. Until the accident was resolved, Maersk Line banned the use of the aft transverse thrusters on all EMMA MÆRSK class vessels, though no further damages were found during diver inspections. After unloading all containers in Port Said and temporarily sealing the leak, EMMA MÆRSK was towed to Palermo, where the vessel was repaired within 5 months (Fig. 4.14).

Fig. 4.14 The flooding of the shaft tunnel caused by a leak in the stern thruster pipe endangered the EMMA MÆRSK considerably (Ralf Witthohn)

4.1.5 400-Meter Ships Only Conditionally Maneuvrable in a Storm

The vulnerability of the largest container ships to adverse natural conditions or accidents is demonstrated again only 2 years after the serious accident of EMMA MÆRSK in the Suez canal. At the beginning of January 2015, an approaching storm front prompts Maersk Line to change the sequence of ports on the maiden voyage of the 18,270 teu triple-E newbuilding MORTEN MÆRSK. The ship has to call at the new deep-water terminal of Wilhelmshaven directly on the North Sea coast, instead of Hamburg, which is located far up the Elbe. On the narrower part of the Elbe river between Glückstadt and Hamburg, ships over 360 m length are not allowed to operate in winds of more than six Beaufort. One week later, a western storm again affects the timetable of one of the shipping company's Triple-E ships. MERETE MÆRSK is for 2 days unable to depart from the quay of the North Sea Terminal in Bremerhaven. Only after the storm has abated, the four strongest tugs available are able to support the leave of the ship, which is manoeuvring with two propellers and its transversal thrusters for more than 1 h, before being able to sail for Rotterdam (Fig. 4.15).

There, the weather conditions have already by more than 2 days delayed the schedule of the CSCL GLOBE of China Shipping Container Line on its maiden voyage. A few days later, Maersk Line's long-term chartered MAERSK EMDEN of the German Rickmers group is forced to an emergency anchoring off Wremen in the Outer Weser due to an engine failure shortly after having left the Bremerhaven container terminal. The Bremerhaven harbour tugs RT PIONEER, RT CLAIRE, RT ROB and WESER come to the aid of the vessel. Due to the unforeseen anchorage the fairway is closed for larger vessels. The incoming container ship MAERSK BROOKLYN has to turn round, as well as the bulk carrier GOYA bound for Nordenham. Only after a few hours can the MAERSK EMDEN machinery is

Fig. 4.15 The large wind-exposed area of the almost 400 m long MERETE MÆRSK prevented the vessel from leaving the Bremerhaven container quay for 2 days time (Ralf Witthohn)

working again, so that the initially planned return to Bremerhaven and a difficult turning manoeuvre in the narrow fairway is no longer needed to be undertaken. Instead, the two tugs, which could not have offered effective assistance under the conditions of such a storm anyway, accompany the vessel on the further voyage to Wilhelmshaven.

The navigational challenges that can only be overcome to a limited extent in an emergency are similarly evident when the French CMA CGM JULES VERNE, commissioned in 2013 as the world's largest container ship, loses one of its many tonnes weighing bow anchors with the chain during the same storm period off the Elbe estuary. Only weeks later, anchor and chain are salvaged by the Belgian salvage ship SMIT ORCA after 2 days of search. For 5 days in January 2016, CSCL INDIAN OCEAN, a sister ship of the CSCL GLOBE, is stuck in the silt of the Lower Elbe, after an additionally installed safety system named Safematic has blocked the steering gear due to a technical failure. Before entering the Elbe, the CSCL INDIAN OCEAN has anchored in deep water during a storm, but lost its starboard anchor after a break of the shackle. This gives evidence, that the emergency use of the anchor, for example intended in case of an engine failure, can hardly offer safety in stormy weather. Under adverse conditions, the quest for the greatest possible economic efficiency by maximizing the size of container ships is obviously reversed, if the forces of nature bring ship traffic or cargo handling to a standstill (Fig. 4.16).

Another ultra-large container vessel casualty, in this case with consequences on the world's trade, occurred in March 2021, when the Japanese-owned

Fig. 4.16 During storms, conventional anchor systems of large container ships are overtaxed, as was shown by the breaking of the chain of the CMA CGM JULES VERNE, which was lost off the Elbe estuary (Ralf Witthohn)

EVER GIVEN of Taiwan's Evergreen company ran aground in the Suez canal and blocked it for 6 days leaving hundreds of ships in waiting positions. In early 2022, investigations into the accident were still lasting on, but the reason for the grounding was assumed to be a combination of failing manoeuvres and wind impact. Remarkably, the 20,124 teu capacity ship had already caused another severe accident in strong winds, when it hit and nearly sunk the local Hamburg ferry FINKENWERDER at Blankenes in February 2019.

4.1.6 Maersk's Triple-E Vessels

The specially conducted simulator training for ship's officers and pilots of the first newbuilding of Maersk's triple-E class laid out for 18,270 teu was already an indication of the borderline character of this container ship type. Their design was completely newly developed with a forward-shifted superstructure and a twin-engine system. The designation triple-E awarded by the shipping company stands for the attributes economy of scale, efficiency, environmentally improved. According to their information, the ships emit 50% less CO2 per container transported, without any comparison being made. This would be achieved, inter alia, by a further developed exhaust heat recovery system. This feeds, via generators, electric auxiliary drives of 2 × 2000 kW at the shaft, with which the EMMA MÆRSK class had already been equipped. The assumed speed for the ships of the new class was around 22 knots. It is achieved by a twin-propulsion system of MAN B&W diesel engines type 7G80ME-C9.5 with a power output of 32,970 kW each, acting without gearbox on two propellers.

Still relying on the application of the diesel engine, A.P. Møller/Maersk Line together with the UK-based classification society Lloyd's Register in October 2019 published the findings of a joint study on ships' fuel alternatives, citing alcohol, biomethane and ammonia as the best-positioned fuels to reach zero net emissions. The study considers the shipping industry's transition to alcohol-based solutions as still to be defined, while the introduction of biomethane could rely on existing technology, but would remain a challenge due to the emission of unburned methane. The use of ammonia, on the other hand, is looked upon as risky because of its toxicity, whereas the Korean Register considers its use to have a high probability of being commercialized.

Twenty newbuildings of the 399.2 m length, 59.0 m breadth, 30.3 m depth and 16 m draught MÆRSK MC-KINNEY MØLLER class were delivered in two series of ten units each between 2013 and 2015 to Maersk Line by Daewoo shipyard. The 194,915 tonnes carrying ships were followed by a further developed class of smaller 58.6 m breadth, but of a larger depth of 33.2 m. Their container intake was increased to 20,568 teu, not least due to the larger depth of the hold allowing a 12th layer below deck, an advancement of the superstructure by two bays and higher lashing bridges. The increased draught to 16.5 m resulted in a deadweight of 210,019 tonnes. Initally, eleven units of the MADRID MÆRSK type were delivered at a price of 160 million US dollars per ship, another five units were additionally contracted. Sister ships in the series are the MUNICH MÆRSK, MOSCOW MÆRSK, MILAN MÆRSK, MONACO MÆRSK, MARSEILLE MÆRSK, MANCHESTER MÆRSK, MURCIA MÆRSK, MANILA MÆRSK, MUMBAI MÆRSK and MAASTRICHT MÆRSK, the latter delivered as last ship in January 2019 (Fig. 4.17).

Fig. 4.17 By its 20,568 teu capacity the MUNICH MÆRSK type set a new record for only a few weeks (Ralf Witthohn)

4.1.7 Eleven Europe-Far East Services of Ocean Alliance

In early 2021, the six plus four weekly Europe—Asia connections offered by 2M corresponded to nearly the same weekly frequency maintained by competing Ocean Alliance, as the latter consortia maintained seven Far East services to Northern Europe and four to the Mediterranean. In early 2022, the five plus three weekly Europe—Asia connections offered by 2M were surpassed by the weekly frequency of competing Ocean Alliance, which operated seven Far East services to Northern Europe and four to the Mediterranean. The alliance's prime service named NEU1 was launched in April 2017 and employed the then largest container carriers OOCL HONG KONG, OOCL GERMANY, OOCL JAPAN and OOCL UNITED KINGDOM of 21,413 TEU together with the 18,000 teu ships CSCL GLOBE, CSCL PACIFIC OCEAN, CSCL INDIAN OCEAN, CSCL ATLANTIC OCEAN, CSCL ARCTIC OCEAN of China Shipping Container Line. As some of the OOCL ships were only commissioned in the course of the year, smaller 10,000/11,000 teu capacity units were progressively replaced until January 2018. In early 2022, eleven ships of 19,000 to 21,400 teu were serving the route, which included the ports of Felixstowe, Gdansk, Wilhelmshaven, Zeebrugge, Singapore, Yantian, Shanghai, Ningbo and Xiamen by using six vessels of the OOCL HONG KONG and two each of the COSCO SHIPPING GALAXY and CSCL ARCTIC OCEAN types. This port rotation circumvented the need to navigate the Elbe, where the CSCL INDIAN OCEAN had ran aground in February 2016 after rudder failure and remained grounded for several days (Figs. 4.18 and 4.19).

Not relying only on own container ships the Ocean Alliance partner CMA CGM supplemented its initial contributions to the Far East services by chartered newbuildings. These included 15,052 teu neo-panamax ships, of which Singapore-based Eastern Pacific Shipping had contracted a total of 22 newbuildings including options for delivery from 2019 to 2022 by South Korea's Hyundai Samho yard. The first four units named ATLETICO BAY, CLEARWATER BAY, COMPTON BAY and FAMALICAO BAY took up service as CMA CGM ARGENTINA, CMA CGM MEXICO, CMA CGM PANAMA and CMA CGM CHILE in 2019. The 366 m length, 51 m breadth ships are fitted with scrubber systems, the following units of the series, also to be chartered by CMA CGM, were to be equipped with LNG burning main engines. CMA CGM ARGENTINA and CMA CGM PANAMA were in the beginning of 2020 employed in the Ocean Alliance's NEU 3 loop, together with ten larger, CMA CGM-owned ships of 16,000

Fig. 4.18 The 19,000 teu CSCL PACIFIC OCEAN temporarily moved from Hamburg to Germany's only deepwater port of Wilhelmshaven as part of Ocean Alliance's Europe-Far East service (Ralf Witthohn)

Fig. 4.19 Of the total 11 newbuildings of the 21,237 teu type COSCO SHIPPING UNIVERSE, the Ocean Alliance partner COSCO initially contributed four, later two vessels, to the Far East-North Europe service NEU1 (Ralf Witthohn)

teu to 21,000 teu capacity. The port rotation was Southampton, Dunkirk, Hamburg, Rotterdam, Southampton, Algeciras, Port Kelang, Xingang, Busan, Ningbo, Shanghai, Yantian, Singapore, Southampton The CMA CGM ARGENTINA switched to the Ocean Alliance Far East—US East Coast route, which it still served in 2022, while the CMA CGM PANAMA took part, along with the CMA CGM MEXICO and CMA CGM CHILE, in the

Fig. 4.20 Having started on the Far East—North Europe lane, the Singapore-owned, Malta-flagged 15,000 teu class COMPTON BAY, chartered after delivery as CMA CGM PANAMA and fitted with an SOx scrubber system from the beginning, was in 2022 engaged on an unusual Far East—US West Coast—Far East—US East Coast route (Ralf Witthohn)

Ocean Alliance Far East—US West Coast—Far East—US East Coast service with the port rotation Port Kelang, Singapore, Laem Chabang, Cai Mep, Yantian, Los Angeles, Yantian, Cai Mep, Singapore, Port Kelang, Colombo, Halifax, New York, Norfolk, Savannah, Charleston, Port Kelang (Fig. 4.20).

At the end of 2019, the CMA CGM order list comprised a number of 23 newbuildings with an aggregated capacity of 365,000 teu and contracted at a price of USD 2.82 billion. This, in face of the then onlasting container shipping crisis impressing number, included nine 23,000 teu, five 15,000 teu and five 15,000 teu carriers, all to be delivered by China's CSSC shipyard group. Alliance partner OOCL, an affiliate of COSCO Shipping Holdings, in March 2020 ordered five 23,000 teu ships to be built by shipyards Nantong COSCO and Dalian COSCO at price of USD 156 million per ship.

4.1.8 Cascade Effect Displaces 8000 Teu Ships

The two-island ships of CMA CGM, COSCO, CSCL and OOCL with cargo capacities of around 20,000 teu or even more replaced single-island newbuildings with a capacity of the 8000 teu class in the Europe-Far East trade. These had entered service only a few years before. The largest series of ships in this size category was built at Hyundai shipyard. From 2004 to 2011, the builder delivered 37 units, plus 11 newbuildings of a longer version of up to 9500 teu in 2006. This size class, also developed by other Korean shipyards, had in turn replaced 5000 teu panamax ships as the former workhorses in container shipping. The first Hyundai newbuilding of the 8200 teu class was CONTI

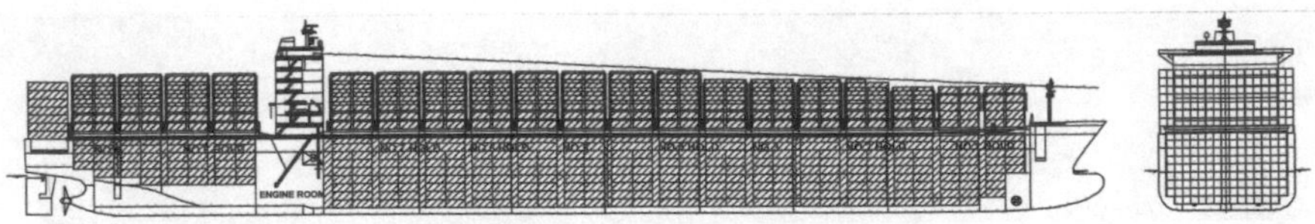

Fig. 4.21 The 8200/8500 teu Hyundai series type E. R. TIANAN of E. R. Schiffahrt was among the world's largest container ships when delivered (E. R. Schiffahrt)

EVEREST contracted by Munich-based Conti shipping company and managed by Buxtehude/Germany-located NSB company in August 2004, followed in the next months by CONTI MAKALU and CONTI ANNAPURNA. At that time, the vessels were among the world's largest container ships (Fig. 4.21).

Other shipping companies added newbuildings of this type to their fleets. Hammonia Reederei (Peter Döhle) from Hamburg commissioned HAMMONIA JORK, in 2018 chartered out as MSC VALENCIA for temporarily employment on the Europe-Australia route and in 2020 serving the north Europe—Middle East link of MSC. HAMMONIA HAMBURG was in 2018 employed as MSC PARIS in a Europe-South Africa service and in 2020 in MSC's Africa Express from the Far East, while HAMMONIA BREMEN for several years sailed as MSC BILBAO in an India-US East Coast service. Hamburg-based E. R. Schifffahrt received E. R. TIANAN, which was as well chartered by MSC as SM CHARLESTON in 2018, and E. R. TEXAS, which was employed as PARSIFAL by Israel's ZIM. Further units were E. R. TIANSHAN, E. R. TIANPING, E. R. TOULOUSE, E. R. TORONTO, E. R. TOULON and E. R. TRIESTE. Other customers of this type were Taiwan's Yang Ming Line and Israel's ZIM, each with four units, CMA CGM, MSC and China's COSCO with eight units (Figs. 4.22 and 4.24).

Conti group's CONTI EVEREST entered a CMA CGM charter under the name CMA CGM HUGO until 2015. The carrier was afterwards temporarily without employment until it was taken up by MSC from February to July 2016. From August 2016 Hapag-Lloyd chartered the vessel, in 2017 CMA CGM took it up again. At the end of 2017 the vessel was employed under its christening name in a service of the Ocean Alliance from the Red Sea to the Far East, as the largest of eight vessels with cargo capacities from 5700 teu upwards. In 2018 MSC chartered the ship for its Europe-Australia-Far East-India line named Nemo, and, after an intermezzo by the Shipping Corp. of India, again from 2020 on, serving a Europe-Red Sea-India loop in the beginning of 2022. The sister ship CONTI MAKALU ended its long-term charter in 2016 under the name MSC TEXAS for MSC, which took the ship after a period without operations again in 2017 for a 2M service from Asia to the US

Fig. 4.22 Delivered in 2006 as the 16th of 37 8200 teu postpanamax carriers from Hyundai shipyard, HAMMONIA BREMEN was 14 years later still employed by Mediterranean Shipping Co. as MSC BILBAO, in 2020 within the Indus Express from the Middle East to the USEC (Ralf Witthohn)

Fig. 4.23 CMA CGM took over a newbuilding initially contracted by the Munich-based Conti group and put more of the 8200 teu ships into service, including the CMA CGM DON GIOVANNI (Ralf Witthohn)

East Coast. CONTI MAKALU was the smallest ship on this line, which is maintained by ten ships of up to 9400 teu capacity. In 2020 the ship had joined its sister CONTI EVEREST in the Nemo service. In 2022, CONTI MAKALU was chartered by MSC for a 2M/ZIM service from the Far East to the US Gulf. CONTI ANNAPURNA was employed as PACIFIC LINK by CMA CGM until 2016, then by Hapag-Lloyd, Maersk Line and the Shipping Corp. of India, which employed it in 2017 under the ship's original name in the Himalaya Express between India and Europe set up jointly with MSC. Of the total of eight ships in this service, MSC provided seven, which cargo capacities of 11,700 to 13,400 teu significantly exceeded those of CONTI ANNAPURNA. In 2020, the ship served a Far East—North America east Coast line of the Japanese alliance ONE (Ocean Network Express) with partners Hapag-Lloyd and Yang Ming, and in 2022, ONE employed the CONTI ANNAPURNA between the Far East and the Americas west coast in a joint service with Israel's ZIM (Fig. 4.23).

At an overall length of 334 m, the CONTI EVEREST type has nine holds, which are separated by transverse bulkheads and of which seven are located in front of the engine room. At 28.8 m length each of the holds can accommodate two 40-ft bays, with the exception of 44.9 m long hold 1, which was laid out to accommodate three 40-ft bays. The holds accommodate up to 15 rows next and nine tiers of containers. The ship's breadth of 42.8 m allows to load 17 rows on the hatches, to be stowed up to seven tiers high. The maximum intake of ISO standard containers handled by shore-based container gantry cranes is 8189 teu, calculated on a maximum of seven tiers on the hatches. 3820 teu of the total are transported in the holds, 4369 teu on the hatches and on deck. If only six tiers on deck are counted, the deck capacity is 3927 teu, totalling to a deck capacity of 7747 teu. Stability is ensured up to a figure of 6310 teu, homogeneously loaded with 14 tonnes (at 45% VCG of the container height) and with full bunkers. At an aft trim of 1 m, the IMO visibility regulations still allow the loading of 7702 teu at maximum draught. At a draught of 14.5 m, the deadweight of the CONTI EVEREST amounts to 101,662 tonnes. Initially classed by Germanischer Lloyd the ship switched to Bureau Veritas in 2016.

In the nine cargo holds, seven tiers of containers of 8.5 ft. height and above that two layers of 9.5 ft. can be loaded. Only the front bays of holds 5, 6 and 7 are designed for the optional sequence of stowage of 8.5-ft and 9.5-ft containers. Extra-long 45-ft containers are transported on all hatches above the second layer. Holds 1, 2, 3, 4, 8 and 9 are designed for the transport of dangerous cargo according to SOLAS definition. They are ventilated twice an hour using explosion-proof fans. The ventilation of the other holds is designed for 70 m^3/min per 40-ft reefer container. Electrical connections are provided for loading 500 40-ft refrigerated containers in the two lowest deck tiers and 200 40-ft units in the holds. The hatch openings of 12.6 m length and 38.3 m width are each closed by three pontoon covers made of welded steel structures. Only the same length covers for the forward holds have differing widths of 17.9, 27.9, and 33.5 m. The hatch covers, which individual weight does not exceed 41 tonnes, are designed for a stack load of 90 tonnes for 20-ft units and of 120 tonnes for 40-ft containers. Between the hatches and at the stern, a total of 20 lashing bridges of about 2.8 m height with 550 mm wide walks are provided at the heights of the hatch covers and the base of the second container tier. Foundations for twist locks and lashing eyes are provided on the hatch covers.

An automatically operating anti-heeling system is used to compensate listing during loading and discharging operations. It consists of four compensation tanks, the ballast side tanks 4 and 5 on the starboard and port side. They are connected by a steel pipe. A reversible, electrically operated screw pump of 1000 m^3/h capacity is fitted in the cargo hold, controlled from the central ship's office. There is a control unit with a heeling indicator, protocol diagram and alarm triggering in case of too large inclinations as well as maximum and minimum tank levels. The total capacity of the ballast water carried in the double bottom and side tanks is 26,000 m^3. The ballast system for the double bottom and side tanks consists of two main pipes laid through the double bottom ballast tanks. The bilge and ballast pump has a capacity of 1000 m^3/h, the ballast pump of 1000 m^3/h, and a suction pump driven by the bilge pump 80 m^3/h. In an emergency, the heeling compensation system can be operated by a ballast pump and remote-controlled valves. The bilge pumping system for the cargo holds, consisting of two main lines, passes through the double bottom tanks and is operated by two pumps, the bilge and fire pump and the bilge and ballast pump. In each hold there are two bilge wells at the rear corners and an additional one in hold 1. In addition, an air-driven diaphragm pump of 25 m^3/h capacity is installed in the holds 1 and 8, which are equipped for dangerous cargo. The forward pump delivers to a tank complying with the IMDG Code (International Maritime Dangerous Goods) of about 30 m^3 volume in the lower part of the breakwater, the aft pump delivers to a 10 m^3 tank in the steering gear compartment.

CONTI EVEREST is powered by a low-speed two-stroke Hyundai/B&W crosshead diesel engine of type 12K98MC. Putting out 68,670 kW at 94 rpm (MCR, Maximum Continuous Rating) and 61,800 kW at 90.8 rpm (NCR, Nominal Continuous Rating), the motor acts directly on a fixed-pitch propeller. The prospected service speed at the design draft was 25.4 knots at NCR, while at the maximum draught it was to be 24.5 knots under the same conditions. At NCR, the main engine consumption was figured at approx. 248.8 tonnes of heavy fuel oil per day (hfo up to 700 cST at 50 °C). Calculating the total capacity of the fuel tanks including settling and day tanks at 10,400 m^3, the ship's sea endurance was figured at 38 days, the range at 23,000 nm. The main engine, for which reversal the starting air is used, was equipped with four sets of turbochargers and air coolers. The design conditions for the engine room are based on the assumption of a maximum seawater temperature of 32 °C and an engine room temperature of 45 °C. Hyundai's shaft system consists of three intermediate shafts and the tail shaft, supported by a total of four supporting bearings. The shaft system is arranged in such a way that the tail shaft can be pulled into the engine room after dismantling the propeller. The propeller, made of a nickel-aluminium alloy, rotates clockwise when viewed from the rear.

The propulsion plant is designed to be operated from an unmanned engine room under normal sea conditions. The engine room contains a control room with the console for the main engine and a low voltage switchboard. Start, stop, reversing and speed of the main engine can be remotely controlled from this room electro-hydraulically or by an engine telegraph from the wheelhouse. An automatic stop of the main engine is triggered by, among other causes, overspeed, drop in lubricating oil pressure at the main bearing or camshaft and overheating of the thrust bearing. The pumpability of the fuel stored in the lower side tanks 2, 3, 4, 5, 6 and 7 is achieved within 48 h by heating coils with a pipe diameter of 50 mm laid on the bottom. The diesel oil tanks of an initial 500 m^3 capacity are bunkered by two shore connections located next to the HFO connections. The fuel was processed by separators with a capacity of 9200 l/h, one of which was in standby mode and one for separating MDO. The main engine is supplied by two feed pumps from 3 days tanks, one originally laid out for low sulphur HFO and one for MDO. Two circulation pumps feed the fuel to the injection pumps of the main engine via two steam-heated preheaters and the visculator. The two main lubricating oil pumps each provide 1480 m^3/h. The starting air for the main engine is generated by four air-cooled main compressors of 360 m^3/h capacity each.

The electrical energy for the on-board supply is provided by five diesel generators of two power classes. Two 9L27/38 four-stroke engines of 2940 kW output each at 720 rpm are coupled with two main generators of 2800 kW output. Three 7L27/38 auxiliary diesels provide 2200 kW each. One main generator is sufficient for normal sea operation, if no reefer containers are to be supplied. When reefer containers are to be cooled, four generators have to be operated. All five generators are in use during bow thruster operation. The gas oil consuming emergency diesel generator from STX, which is also used as harbour generator, has an output of 550 kW at 1800 rpm. It is automatically started, when the normal AC voltage of 440 V is lost, and switched off again 5 min after the normal power supply is restored. The on-board network is supplied by a medium-voltage electrical system. The heat requirement is covered by a 6000 kg/h capacity auxiliary boiler from Aalborg, initially laid out to burn heavy oil up to 600 cSt and generating a steam pressure of 7 kg/cm^2 at a feed water temperature of 60 °C. The exhaust gas boiler, which uses the exhaust gas heat of the main engine, has an output of 4500 kg/h. The central cooling water system for the main engine is designed for a seawater temperature of 32 °C and a fresh water temperature of 36 °C. The three main sea water cooling pumps of 1520 m^3/h capacity each provide 55% of the required flow rate. They take the coolant from two sea chests, one of which is arranged as a high-suction and one as a low-suction unit, and feed it to two central heat exchangers of 25 million kcal/h each, thus covering 55% of the required capacity. In addition to the main engine, the fresh cooling water system supplies a total of 25 consumers, including generators, lubricating oil coolers and the air conditioning.

4.1.9 Private and Public Capital

The Munich-based Conti group financed the newbuildings from Hyundai as part of an USD 1.35 billion investment programme comprising a total of 18 container vessels. Two more of the vessels ordered by Conti were acquired by CMA CGM as CMA CGM VIVALDI and Mediterranean Shipping Co. (MSC) as MSC RACHELE. By the building programme, the long-time established initiator of ship investments increased the slot capacity of its fleet to over 300,000 teu. The construction of the Conti ships was based on the experience of one of the most productive shipyards, which also delivered the highest number of such large container ships.

To finance the, according to the prospectus, EUR 73.3 million costing CONTI EVEREST, Conti offered private investors to contribute to the venture by promising them, that 60% of the borrowed capital would be repaid at the end of a 10-year charter. For the investors, there would also be a "high level of tax security", in particular by taking into account the non-inclusion limit of § 2b EStG of the German tax law and other favouring tax regulations. The investors could choose between two investment alternatives. One proposed quarterly distributions forecast at 9 or 7% with a low tax burden due to the tonnage tax and should begin by charter entry including an advance distribution of 4% on early payments. The alternative intended an immediate priority distribution of 7% over the entire term. The latter arrangement provided for a special right of termination 15 years after the start of the charter with a repayment of 100% of the investment. Shareholder funds made up EUR 31.8 million of the project, outside capital USD 24 million plus 2.8 million yen. The distributions in 2011 amounted to 4% or 7%, the amount of the capital repayment in that year was figured at EUR 53.9 million. In 2015, CONTI EVEREST was in a charter income pool of 15 ships, the distributions amounted to 0% and 7%, resp., the capital repayment was EUR 57.4 million of prospected EUR 68.1 million. In March 2017 the Hamburg-based shipping company Claus-Peter Offen took over the Conti financing group, which had been founded in 1970 as Cosima-Reederei and which at the time of the takeover still had 30 container ships with a cargo capacity of 208,000 teu, eight product tankers and 29 bulk carriers in their fleet (Fig. 4.24).

Fig. 4.24 Of the lengthened CONTI EVEREST type version, the 9500 teu COSCO NINGBO was one of eight units delivered to COSCO (Ralf Witthohn)

For a long time, the financing of shipbuilding projects, which can reach investment levels of USD 1.4 billion for large cruise liners, had been regarded as the prime challenge of their realisation. It lost this controlling function when new sources of finance were found and eventually contributed to a lasting disruption of the market, thus triggering—not unlike the devastating subprime effects of the US housing market—a deep shipping crisis. Already after the almost complete loss of their ships due to the war, German shipowners were from 1950 onwards able to use non-shipping capital for newbuildings with tax relieves. After the 1960s, the corporate form of a partnership shipping company, named Partenreederei, was used. Within a few years, the Brake merchant Helmut Meyer was one of the first to build up a fleet of smaller dry-cargo ships that could be employed worldwide. More than 500 people acquired shares in these ships, with Meyer acting as manager (*Korrespondentreeder*). Even when corporate profits failed to materialize, expansion continued, until the fleet comprised 28 ships at its peak in 1971. The continued existence of the shipping firm was subsequently secured by ship sales, the last of which marked the end of the company in 1983.

At that time, the acquisition of capital from private investors in public ship funds had become common practice. Like the Conti group, another pioneer, Norddeutsche Vermögen in Hamburg, invested primarily in container ship projects to be chartered out on the international market. Finally, there were hundreds of German capital collectors. The KG company system gave German shipping companies a dominant position in container shipping and contributed significantly to the oversupply of tonnage causing a decline of rates. HSH Nordbank at Kiel, the bank of the German states Hamburg and Schleswig-Holstein, waived debts of EUR 570 million from Norddeutsche Vermögen in 2016. While individual investors often lost their capital, the long-established companies active in container shipping, both liner operators or family-run businesses, remained largely supported by the interplay between politicians and banks.

4.1.10 Similar 8000 Teu Types from Samsung, Daewoo and Hanjin

In parallel with the CONTI EVEREST type, Hyundai built a series of 26 ships with almost identical main dimensions and an only slightly higher deadweight of about 103,000 tonnes to 101,000 tonnes. But thanks to some design steps the container intake was increased from 8750 teu to 9650 teu. The first ship in 2005 was the 335.5 m length, 42.8 m breadth, 24.5 m depth and 14.6 m draught COLOMBO EXPRESS for Germany's Hapag-Lloyd. The Hamburg-based shipping line received 14 newbuildings of this type until 2010 and chartered two more from Hamburg owner C. P. Offen. Other recipients were the liner operators Maersk and Hanjin (Fig. 4.25).

From the design point of view, there was a high degree of permeability between the project departments of the competing Korean builders Hyundai, Samsung, Daewoo and Hanjin. Between 2003 and 2010, Samsung completed a series of 29 units, lead by the OOCL SHENZHEN and completed by OOCL LUXEMBOURG for the Hong Kong shipping company OOCL, which put a total of 16 units of that vessel class into service. From 2004, the Conti group was involved in this series type and took over eight ships, the 14th unit of the total series being CONTI CORTESIA, followed by CONTI CHAMPION, CONTI CHARMING, CONTI COURAGE, CONTI CHIVALRY, CONTI CONTESSA, CONTI CONQUEST and CONTI CRYSTAL. The ships were initially chartered out on a long-term basis to Taiwan's shipping company Evergreen. After 10 years of service, LT CORTESIA was without employment, before the ship in 2017 entered a charter of Japan's NYK between the Far East and the US East Coast via the Suez Canal. In 2020, CONTI CORTESIA supported MSC's Indus Express. EVER CHAMPION, EVER CHIVALRY, ITAL CONTESSA, EVER CONQUEST and HATSU CRYSTAL continued employments in Evergreen services. The former CONTI CHARMING, which switched to the US company MC-Seamax, was taken up by OOCL in 2017 for a Far East US West

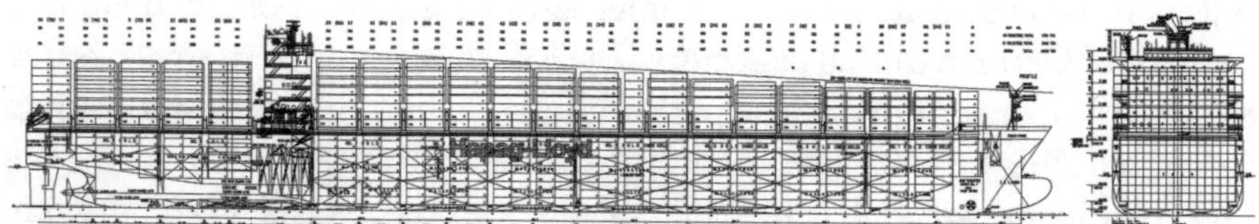

Fig. 4.25 The Hyundai type COLOMBO EXPRESS differs only slightly in main dimensions from the shipyard's 8000 teu class, but achieves an intake of several hundred teu more (Hapag-Lloyd)

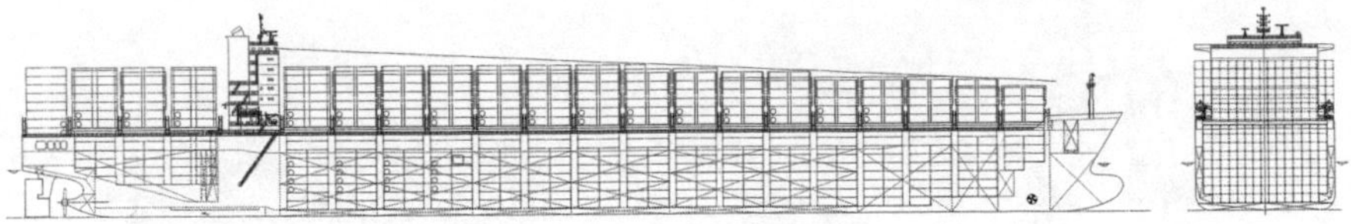

Fig. 4.26 The Samsung designed 8000 CONTI CORTESIA was much similar to the equivalent Hyundai type (Conti)

Coast service and in 2020 participated in a Far East—US East Coast service maintained by 2M and Israel's ZIM company (Fig. 4.26).

Another design similar to the Hyundai and Samsung types was built 23 times at Daewoo shipyard between 2006 and 2010, featuring a capacity of 8400 teu at 332 m length, 43.2 m breadth, 20.2 m depth and 14.5 m draught. The shares issuing capital house Norddeutsche Vermögen in Hamburg received ten of the ships as NORTHERN JULIE, NORTHERN JADE, NORTHERN JAVELIN, NORTHERN JUBILEE, NORTHERN JAGUAR, NORTHERN JUVENILE, NORTHERN JUPITER, NORTHERN JUSTICE and NORTHERN JAMBOREE, others came into service for the shipping companies MSC and Maersk. The series included MAERSK STRALSUND delivered in 2006. Initially, the ship had been commissioned by Norddeutsche Vermögen in Hamburg to be operated by subsidiary Norddeutsche Reederei H. Schuldt and for employment under the name P&O NEDLLOYD MARYLIN for the competing Grand Alliance.

Private investors were offered stakes from USD 15,000 onwards in the MARYLIN STAR fund established under Germany's tonnage tax regime by a Commerzbank subsidiary. A charter rate of USD 39,150 US dollars was quoted for the first 10 years, with an option for five more years of charter and the purchase of the ship. The funding company offered investors the prospect of an annual distributions of 6%, rising to 20% in 2022, and a total distribution of 216.5% including proceeds from the sale. The ship was operated under the Liberian flag by the Hamburg-based shipping company Blue Star, a company of the shipping company P&O Nedlllloyd. Following the purchase of P&O Nedlloyd by Maersk, the ship was integrated into the Maersk Line schedule. As a tribute to Maersk's East German shipyard location, it was given the charter name MAERSK STRALSUND by the Danish group. Under this name, the ship sailed in early 2018 with ten ships of the same or similar capacity in a 2 M service between the Far East and the US East Coast. In 2020, MAERSK STRALSUND was, in ownership of the China Construction Bank, still performing a Maersk Line charter on the Far East—Australia route, in 2022 Maersk employed the ship in a East Asia—South Africa service named Safari (Fig. 4.27).

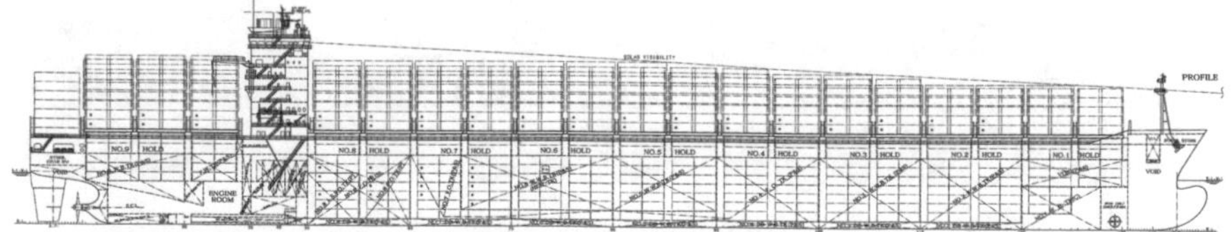

Fig. 4.27 Initially chartered by Maersk through the purchase of P&O Nedlloyd, the Daewoo-built MARILYN STAR was still serving the network of the Danish operator in 2020 as MAERSK STRALSUND (Blue Star)

A fourth Korean builder of 8000 teu class ships was Hanjin, which in 2005 and 2006 delivered nine units, five of them, named MSC LUCY, MSC RITA, MSC MAEVA, MSC JUDITH and MSC VITTORIA, to MSC and four of them to C.P. Offen from Hamburg for a long-term charter as MSC BUSAN, MSC BEIJING, MSC TORONTO and MSC CHARLESTON.

4.1.11 Eight Europe-Far East Services of THE Alliance

Five liner services from the Far East to northern Europe and three to the Mediterranean were kept up by the container consortia THE Alliance at the end of 2017. In April 2020, one of the services, designated FE5, was combined with a Far East—US West Coast service to a pendulum service. A similar pendulum, named FP1, had already been established in April 2019. It was served by 13 container ships of the 9000 teu class contributed by the Japanese venture ONE (Ocean Network ONE) of partners NYK, "K "Line and Mitsui-OSK. The ships were named NYK ALTAIR, NYK OCEANUS, NYK ORION, NYK OCEANUS, NYK ORPHEUS, NYK VEGA, NYK VENUS, NYK VESTA, NYK VIRGO, HENRY HUDSON BRIDGE, ONE HANNOVER, ONE HAMBURG and ONE HONG KONG. This line to Rotterdam, Hamburg, Le Havre, Southampton, on which Hapag-Lloyd and Yang Ming had chartered slots, concentrated on the Japanese ports of Kobe, Nagoya, Shimizu, Tokyo and as well calls at Singapore. On the Far East to North America leg to Los Angeles and Oakland, the Ocean Alliance members COSCO, Evergreen and APL reserved slots.

In 2022, the lines of THE Alliance from North Europe to the Far East numbered five connections, a new one was to be launched in April to South East Asia with calls at Laem Chabang, Cai Mep, Singapore, Colombo and Jeddah on the outward-bound voyage as replacement for the FP2 loop. In the FE2 service a dozen 20,000 teu vessels of ONE and Hapag-Lloyd had in the beginning of 2020 been running to mostly Chinese destinations by rotating between the ports of Southampton, Hamburg, Rotterdam, Jebel Ali, Singapore, Busan, Qingdao, Shanghai Hong Kong, Yantian, Singapore and Southampton, with calls at Le Havre to be resumed and Port Kelang added from April 2020 on. Japan's MOL provided six 20,182 teu vessels of the MOL TRIUMPH type, for a short time the world's highest capacity type, while Hapag-Lloyd put in six former ships of the purchased United Arab Shipping Co (UASC) of similar capacity (Figs. 4.28 and 4.29).

The FE2 service used 14 ships of up to 20,200 teu in 2022, six ship from Hapag-Lloyd, four from ONE and one from HMM. They berthed at Qingdao, Busan, Shanghai, Ningbo, Yantian and Singapore. The FE3 between Rotterdam, Hamburg, Antwerp, London Gateway, Piraeus, Singapore, Hong Kong, Xiamen, Kaohsiung, Yantian and Rotterdam had in January 2020 been operated by ten 14,000 teu class carriers, four of Yangming's YM WELLHEAD type, four ONE ships plus Hapag-Lloyd's MADRID EXPRESS ex MSC FABIOLA, chartered from Peter Döhle, and NYK FALCON. In 2022, HMM deployed seven 16,000 teu newbuildings of 2021, Hapag-Lloyd three 13,000/15,000 teu units to the FE3 string. Up to 24,000 teu carrying ULCCs delivered from April 2020 on as HMM HAMBURG, HMM ALGECIRAS, HMM HELSINKI, HMM COPENHAGEN, HMM LE HAVRE, HMM GDANSK, HMM ROTTERDAM, HMM SOUTHAMPTON, HMM OSLO and HMM ST PETERSBURG plus one ship each of MOL and Hapag-Lloyd made up the fleet of the FE4 service including calls at

Fig. 4.28 When comissioned in 2017, the 20,170 teu MOL TRIUMPH was considered as the world's largest capacity container vessel (Ralf Witthohn)

Fig. 4.29 In February 2022, Hapag-Lloyd contributed the AL MURAYK to the FE2 loop of THE Alliance (Ralf Witthohn)

London-Gateway, Algeciras, Qingdao and Busan. Before, the FE4 had used ten former UASC ships of 14,000 teu to connect northern Europe with Chinese ports (Fig. 4.30).

From the beginning, THE Alliance had maintained three Mediterranean—Far East links. The MD1 from western Mediterranean ports employed ten 13,000 TEU class ships, of which four were provided by ONE member Mitsui and six by Hapag-Lloyd. In the beginning of 2020, MILLAU BRIDGE, MANCHESTER BRIDGE, MANHATTAN BRIDGE, MILANO BRIDGE, AIN SNAN, SOUTHAMPTON EXPRESS, PARIS EXPRESS, NEW YORK EXPRESS, HONG KONG EXPRESS and ESSEN EXPRESS sailed from Barcelona, Valencia, Tangier and Genoa to Damietta, Jeddah, Singapore, Hong Kong, Qingdao, Busan, Shanghai, Ningbo, Shekou and back to Singapore, Jeddah and Barcelona. The MD2 employed ten similarly sized units, seven from ONE, two from Hapag-Lloyd and one from Yang Ming. And the MD3 connected the east Mediterranean terminals of Ashdod, Istanbul, Yarimca-Izmit, Izmir and Mersin with ports in China Korea and Singapore by ten 9000 TEU ships, of which six were supplied by Yang Ming and four by Hapag-Lloyd.

In 2022, THE Alliance still maintained three Med services. The MD1 and MD 2 each employed seven Hapag-Lloyd and four ONE ships with total weekly capacities of 13,600 teu and 15,000 teu, resp. The MD3 with a total weekly 12,500 teu volume used seven Yang Ming ships of the YM W-class and two Hapag-Lloyd ships, LUDWIGSHAFEN and LEVERKUSEN

Fig. 4.30 Ten 24,000 teu newbuildings of the HMM SOUTHAMPTON class were from 2020 on employed on the FE4 loop of THE Alliance (Ralf Witthohn)

EXPRESS. Independently from its alliance membership Hapag-Lloyd is to start a new South China-North Europe express line in April 2022 as replacement for a slot agreement with 2M by using eight panamax ships and offering transit times of 27 days from Dachan Bay (Shenzen) to Hamburg as only ports.

4.1.12 Direct Australia Services from Europe

Although Australia and New Zealand were among the first containerised trade lanes in 1970, most of cargo imports to the Fifth Continent are actually undertaken by transshipments. From Europe only a few direct container connections are maintained, including the joint Nemo service of MSC and CMA CGM, set up in September 2019 by using 15 ships of the 8000/9000 teu size. Within scheduled 98 days, APL DANUBE, CMA CGM RHONE, APL PHOENIX, APL DETROIT, MSC PAMELA, MSC SINDY, MSC CHANNE, E.R. TOKYO, SEAMAX GREENWHICH, EUROPE, MSC RACHELE, CONTI EVEREST, CONTI MAKALU, CONTI CONTESSA and APL AUSTRIA called at London, Rotterdam, Hamburg, Antwerp, Le

Havre, Fos, La Spezia, Genoa, Gioia Tauro, Damietta, Port Louis, Réunion, Sydney, Melbourne, Adelaide, Fremantle, Singapore, Colombo, Gioia Tauro, Valencia, London. Involved as CMA CGM partner was Hapag-Lloyd marketing the service as EAX Europe Australia Express. The three European operators were still engaged in 2022 in the service, which used 14 ships with a total weekly capacity of 9200 teu.

Another liner connection between Europe and Australia leading through the Panama canal is maintained by the French companies CMA CGM and Marfret. Taking the same time of 14 weeks as the Nemo service, special reefer container vessels of the 2260 teu capacity SEATRADE ORANGE type are operated together with mostly chartered ships. Pacific Ocean ports called are Papeete, Noumea, Brisbane, Sydney, Melbourne and Tauranga.

4.2 North Atlantic Route

In the first transoceanic container trade area, where container ships—with a loading capacity of less than 1000 teu—were used from the 1960s on, all major international shipping companies have services organized. The ships' capacities range from 3500 teu to the postpanamax ship class. In 2020, the 2 M cooperation operated four main lines from North Europe to the US East Coast/US Gulf and three from the Mediterranean. The number was reduced to three and two lines, resp., in 2022. The TA1 then employed the 4800 teu ships SAFMARINE MAFADI, MAERSK KANSAS, MAERSK IOWA, MAERSK OHIO and MAERSK MONTANA, which fly the US flag, and the chartered Panama-flagged ORCA 1. Their port rotation was Antwerp, Rotterdam, Bremerhaven, Norfolk, Charleston, Houston, Norfolk, Antwerp. The TA2, which had in 2020 used the MAERSK GAIRLOCH and MAERSK UTAH, together with the 5900/5700 teu capacity SM TACOMA and MSC KRYSTAL of MSC, for serving the ports of Bremerhaven, Felixstowe, Antwerp, Le Havre, New York, Baltimore, Norfolk, Savannah and New York, employed the MSC ELODIE, NORTHERN JUBILEE, SEAMAX GREENWICH, MSC UBERTY, CONTI CORTESIA and SANTA ROSA offering a total weekly capacity of 8300 teu in 2022. At this time, the port rotation had changed to Bremerhaven, Antwerp, Le Havre, Newark, Norfolk, Baltimore, Charleston, Savannah, Newark, Bremerhaven (Fig. 4.31).

Within the TA3, five 5500 teu to 6700 teu carrying MSC ships and the 4800 teu MAERSK GIRONDE had sailed to Charleston, Vera Cruz, Altamira, New Orleans, Mobile and Freeport. In 2020, the service was upgraded to six ships of between 5500 and 8800 teu and the inclusion of

Fig. 4.31 Panamax container ships of the MAERSK MONTANA type were long-time employed in the 2M North Atlantic services under the US flag (Ralf Witthohn)

New York. In 2022 seven ships with a total weekly capacity of 7200 teu maintained the line from Antwerp, Rotterdam and Bremerhaven to New York, Veracruz, Altamira, New Orleans, Mobile, Freeport and Charleston, the MSC ANCHORAGE, MAERSK SEMBAWANG, NORTHERN MAJESTIC, PORTO CHELI, MSC MARINA and SEALAND ILLINOIS. The given up TA4 had deployed three MSC and Maersk ships each of 4200 teu to 7000 teu from North Europe to New York, Savannah, Port Everglades and Charleston (Fig. 4.32).

Concerning the Med-North America schedule scheme, the TA5 had in 2020 expedited via Sines seven ships of between 6600 teu and 9200 teu from the terminals at Algeciras, Valencia, Gioia Tauro, Naples, Livorno, Genoa, Valencia and Algeciras to New York, Baltimore, Norfolk, Savannah and Charleston, the MSC AMALFI, SAN VICENTE, C HAMBURG, MSC AGADIR, MSC JUDITH, MSC VITTORIA and MAERSK KARLSKRONA. In 2022, the rotation was described as including Tanger Med, Valencia, Gioia Tauro, Naples, Leghorn, Genoa, Valencia, Algeciras, Sines, New York, Baltimore, Norfolk, Savannah, Charleston, Tanger Med. At the same time the TA6 employed seven similar capacity units to Freeport, Miami, Vera Cruz, Altamaria, Houston, New Orleans, Miami and Freeport. The service connected western Med ports via Sines with Freeport, Veracruz, Altamira, Houston, New Orleans, Miami and Freeport by deploying the MSC CANDICE, MSC ALTAMIRA, EUROPE, SEAMAX DARIEN,

Fig. 4.32 Originally delivered in 2002 as CMA CGM RAVEL to French owners, the 6734 teu PORTO CHELI was from 2021 on chartered by Maersk from Greece's Costamare Shipping for employment in the TA3 Northatlantic service (Ralf Witthohn)

MAERSK SENANG, METHONI, SEALAND WASHINGTON and MAERSK KINGSTON with an average weekly volume of 7800 teu (Fig. 4.33).

4.2.1 Colombian Bananas for Europe

Maersk Line, without its 2M partner MSC, in December 2017 reorganised its Colombia Express to be operated by container vessels with high refrigeration capacities. The 2550 teu MAERSK NOTTINGHAM, RIO TESLIN, HAMMONIA EMDEN, HAMMONIA HUSUM and HSL SHEFFIELD had 600 reefer plugs and could thus transport almost half of their cargo refrigerated. The schedule of the banana carriers named the ports of Rotterdam, Felixstowe, Newark, Charleston, Turbo, Santa Marta, Portsmouth, Antwerp, Hamburg, Rotterdam. In June 2019, the Colombia Express was incorporated in a new North Europe to north and west coast South America line of Maersk Line and Hamburg Süd employing eight 4500/3830 teu ships, the LEXA MAERSK, LICA MAERSK, SAFMARINE NOMAZWE, SAFMARINE NOKWANDA, POLAR MEXICO, POLAR ECUADOR and POLAR COSTA RICA. These ships called at Antwerp, Rotterdam, London-Gateway, Hamburg, Cartagena, Manzanillo, Balboa, Callao, San Antonio, Balboa, Manzanillo and Antwerp. In 2020, the service was maintained by eight ships of the 3100 teu MAERSK BULAN class. And in 2022 were employed

Fig. 4.33 In February 2022, MSC deployed the 9400 teu MSC AMALFI to the TA5 service from the Med to the US East Coast/Gulf (Ralf Witthohn)

POLAR ARGENTINA, GARDINER, HAMMONIA HUSUM, MAERSK NESTON, CAP BEATRICE and MAERSK NEWBURY (Fig. 4.34).

Shortly before Hamburg Süd had been taken over by Denmark's A.P. Møller/ Maersk Line, the company had ordered eight 3830 teu reefer container ships from China's Jiangsu Yangzijiang shipyard. The ships have an extraordinary high proportion of 1000 electrical connections for reefer containers to transport fruit, mainly bananas, from Mexico and Central America to Northern Europe. Of the 40-ft containers generally used for this purpose, the ships can accommodate 1915 units, so that the reefer share of the cargo amounts to 50%. The first two newbuildings, named POLAR MEXICO and POLAR COLOMBIA, were delivered in 2017. Being renamed DIEGO GARCIA and JOAO DE SOLIS ex POLAR COLOMBIA, they initially started in a cabotage service of Hamburg Süd's Brazilian sister company Alianca along the Brazilian coast under the countrie's flag. Eleven months after delivery in September 2017, POLAR MEXICO switched from the intra-Brazilian trade service to the Central America-Europe route under Singapore flag, the port rotation including Altamira, Vera Cruz, Progreso, Big Creek, Manzanillo, Moin, Cork, Tilbury, Antwerp and Bremerhaven before integration into the above mentioned Central America north and west coast line. Sister ships are POLAR ECUADOR, POLAR COSTA RICA, POLAR CHILE, POLAR PERU, POLAR ARGENTINA and POLAR BRASIL (Fig. 4.35).

A.P. Møller in Singapore became the owner of the first four newbuildings contracted by Hamburg Süd. POLAR COSTA RICA was taken over by Hamburg Süd in Shanghai only 2 days before the companie's sale took effect on 1 December 2017. The next two ships were chartered under the names

Fig. 4.34 For its Colombia Express Maersk Line has chartered HAMMONIA HUSUM from a Hamburg owner, to transport mostly fruit from Central America in reefer containers painted white for thermal reasons (Ralf Witthohn)

POLAR CHILE and POLAR PERU from the Hamburg shipping company Peter Döhle, two others under the names POLAR ARGENTINA and POLAR BRASIL from Greece's Costamare shipping company. The 230 m length, 37.3 m breadth, 22 knots achieving vessels were equipped with two revolving cranes covering only part of the cargo holds.

For their part, Mediterranean Shipping Co. operated a service to the US East Coast and Panama with a connection to Ecuador at the end of 2017 by six ships accepting Maersk containers, MSC VIDHI of 5800 teu capacity, MSC CARMEN of 4800 teu and MSC ANISHA R., MSC ZLATA R., MSC VIDISHA R. and MSC JULIA R. of 4100 teu. They called at the ports of Antwerp, Bremerhaven, Rotterdam, Boston, New York, Philadelphia, Cristobal and Philadelphia. In December 2018, MSC merged the Panama service with the Panama-Ecuador feeder service. The tonnage deployed was extended by one vessel and harmonised to 4100 teu vessels by adding the MSC KATYA R., MSC VAISHNAVI and MSC ARUSHI R. The new port rotation was Antwerp, Bremerhaven, Rotterdam, Boston, New York, Philadelphia, Freeport, Guayaquil, Puerto Bolivar, Paita, Cristobal, Antwerp. The postpanamax vessel MSC VIDHI took over feeder duties in the Baltic

Fig. 4.35 POLAR PERU and seven sister vessels are especially laid out to transport a large share of fruit including bananas (Ralf Witthohn)

Sea, MSC CARMEN switched to a North Europe—Levante service, and in 2020 served the Canada-Gulf Bridge link, in 2021 on a Mediterranean to West Africa route (Fig. 4.36).

MSC CARMEN, delivered to MSC in 2008 from Romania's Daewoo Mangalia shipyard, is an example of the evolutionary development in shipbuilding, but also an illustration of an extraordinary conversion measure to adapt a type to changing conditions. The 275 m length and 32.2 m breadth 4872 teu ship was the second extended version of a prototype, which had been delivered in 1995 under the name CARIBBEAN SEA to the Conti group for operation by NSB Niederelbe Schiffahrt at Buxtehude/Germany. This type was four times exercised for Conti. At panamax breadth it has a length of 245 m resulting in a cargo capacity of 3681 teu. Six further newbuildings of the series were extended to 259 m by an additional 40-ft container length, thus achieving a corresponding capacity of 4113 teu. The second increase in capacity was achieved from 2006 onwards by increasing the ship's length by a further 40-ft bay. MSC LORENA was the first of such 14 vessels to reach 4860 teu at 275 m length. For the ships's breadth the Panama canal dimensions remained the limit. The length to breadth ratio increased from 7.6 to 8.0, and then to 8.5, resulting in a capacity increase of 14 tonne weighing containers, but with the undiserable effect of a steady reduction in stability. Consequently, the intake of 14-tonne containers from 2560 teu over 2807 teu to 3160 teu did not grow equivalent to the nominal capacity.

Fig. 4.36 In early 2022, eight 4100/4900 teu vessels including MSC VAISHNAVI R. took part in MSC's stand-alone service from North Europe to Boston, Philadelphia, Charleston, Freeport/Bahamas, Puerto Bolivar (Ecuador, Guayaquil, Paita, Balboa, Cristobal and Moin

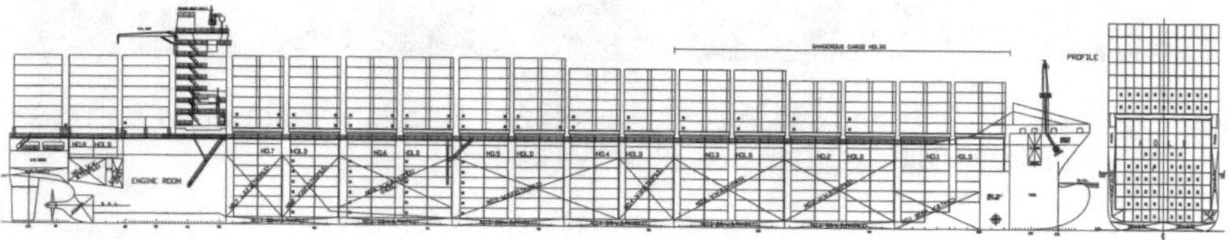

Fig. 4.37 MSC GENEVA's state before widening of the hull (Conti)

After the lifting of the old panamax limit, NSB as the operator of three of the ships, namely MSC GENEVA, MSC LAUSANNE and MSC CAROUGE, decided to have them widened to 39.8 m and lengthened to 283 m in 2015 as part of one of the most unusual ship conversions carried out. The L/B ratio was by this measure reduced to 7.1, while the capacity increased for another time to nominal 6336 teu and to 4945 teu of 14 tonnes weight, this meaning an increase of 72% compared with the nominal loading capacity of the very first design and of 93% in comparison with the intake of 14 tonnes containers (Figs. 4.37 and 4.38).

4.2.2 Hapag-Lloyd with Asian Partners

After its implementation in 2016, THE Alliance offered a total of five North Atlantic services. The 4600 teu capacity sister ships YM EVOLUTION, YM ESSENCE, YM ENLIGHTENMENT and YM EXPRESS of Taiwan's Yang Ming Line were employed in the AL1 loop at the end of 2017. The YM EVOLUTION was later replaced by ONE's BUDAPEST BRIDGE. In January 2019, the German port of call of all the alliance's North Atlantic

Fig. 4.38 The widening of three units of the MSC CAROUGE type produced special lines necessary to find a transition from the broader midship to the old forward and aft ship parts (Ralf Witthohn)

services was shifted from Hamburg to Bremerhaven, the rotation then including Rotterdam, Hamburg, Antwerp, London, Norfolk, Philadelphia, New York and Halifax. In April 2021, the AL1 was suspended, the remaining loops adopted. The four-ship fleet for the second service was in the beginning of 2020 provided by Hapag-Lloyd with one own and two chartered vessels plus the MOL GRATITUDE from ONE. The ships ran from Le Havre, London, Rotterdam and Hamburg to New York, Charleston and Savannah. The latter two were cancelled to take over Norfolk and Philadelphia calls from the AL1. In the third service of the German-Asian cooperation, five Hapag-Lloyd-owned 3200 teu CHARLESTON EXPRESS type ships together with the 4250 teu RIO GRANDE EXPRESS sailed to US ports. In the fourth transatlantic service to the US Gulf, five container ships with cargo capacities of 5500/6000 teu were deployed by Hapag-Lloyd, one as MOL GLIDE by ONE. In March 2021, Hapag-Lloyd still deployed three units, Yang Ming two and one was contributed by ONE.

In 2022, THE Alliance still operated three US East Coast services, plus one to the US West Coast. The AL4 was run by six ships, Hapag-Lloyd's HAMBURG BAY, DIMITRA C, SANTA VIOLA and CHACABUCO plus MOL GRATITUDE and NYK DAEDALUS of ONE. They connected Le Havre, London Gateway, Antwerp and Hamburg with Veracruz, Altamira and Houston. The AL2 to New York, Norfolk and Philadelphia was supported by five, mostly chartered Hapag-Lloyd ships, named CROATIA, TORRENTE, AL SAFAT, CAPE PIONEER and PALENA. The AL3 to New York, Norfolk and Philadelphia was kept going by Hapag-Lloyd's US-flagged 7300 teu ships DELAWARE EXPRESS, COLORADO EXPRESS, HUDSON EXPRESS, POTOMAC EXPRESS and MISSOURI EXPRESS, calling at Charleston, Savannah and Norfolk.

Another service, which runs to the US West Coast with CMA CGM hiring slots, was in 2020 maintained by ten panamax vessels contributed by ONE, eight units originating from NYK and two from MOL. Their port rotation was Southampton, Le Havre, Rotterdam, Hamburg, Antwerp, Savannah, Cartagena, Balboa, Los Angeles, Oakland, Seattle, Vancouver, Oakland, Long Beach, Balboa, Cartagena, Caucedo, Savannah, Southampton. In 2021, the fleet employed comprised eleven units. In 2022, a fleet of 11 ships from ONE plus Hapag-Lloyd's DALLAS EXPRESS served the route (Fig. 4.39).

For decades, Hapag-Lloyd has organised two Canada services with the Hong Kong shipping line OOCL, with MSC involved in one of them. The 4000 teu ships MSC ALYSSA and MSC SANDRA called at Antwerp, Bremerhaven, Le Havre, Liverpool and Montreal at the end of 2017, together with the 2800 teu OTTAWA EXPRESS of Hapag-Lloyd and the OOCL BELGIUM. In the beginning of 2020, MSC contributed the 5000 teu class MSC ELENI and MSC LEIGH, 2 years later the similar-sized MSC MARIA CLARA, MSC JERSEY and MSC SAO PAULO supporting the OOCL ST LAWRENCE and OTTAWA EXPRESS. The 4400/4000 teu vessels TORONTO EXPRESS, MONTREAL EXPRESS, OOCL MONTREAL and QUEBEC EXPRESS linked Southampton, Antwerp, Hamburg and Montreal still in 2022 (Fig. 4.40).

4.2.3 Two Ocean Alliance Services

The Ocean Alliance consortium operated two Atlantic services from northern Europe in 2020. Between Southampton, Antwerp, Rotterdam, Bremerhaven, Le Havre, New York, Norfolk, Savannah and Charleston operated the 8000/8500 teu vessels COSCO PHILIPPINES, COSCO VIETNAM, CMA CGM OTELLO, EVER LIFTING and OOCL ASIA. To the US Gulf CMA CGM and OOCL were deploying APL NORWAY, CMA CGM CHATEAU D'IF, CMA CGM WHITE SHARK, CMA CGM TARPON, BUXCLIFF and BRUSSELS with cargo capacities of 5100 teu to 6900 teu. In 2022, the US East Coast run was maintained by EVER LASTING, COSCO

Fig. 4.39 The Europe-US West Coast run is dominated by panamax tonnage supplied from Japan's consortia ONE, including the 4922 teu carrying NYK RUMINA (Ralf Witthohn)

Fig. 4.40 OOCL MONTREAL is OOCL's contribution to the Canada service jointly operated with Hapag-Lloyd (Ralf Witthohn)

PHILIPPINES, CMA CGM TOSCA, OOCL ATLANTA and OOCL SEOUL. The US Gulf connection was covered by the 6000/7000 teu ships BUXCLIFF, CMA CGM MUSSET, CMA CGM LAMARTINE, APL NEW JERSEY, APL MINNESOTA and BRUSSELS (Fig. 4.41).

Autonomously from consortias, Independent Container Line operated a long-established service from Antwerp, Southampton and Cork to Philadelphia and Wilmington, in 2022 employing the 3100 teu carriers INDEPENDENT HORIZON, INDEPENDENT QUEST, INDEPENDENT PRIMERO and INDEPENDENT VISION.

Fig. 4.41 In 2000 commissioned by Germany's Conti Reederei as BRÜSSEL, the 6078 teu carrying BRUSSELS has been chartered by Hong Kong operator OOCL for employment on the North Europe-US Gulf route (Ralf Witthohn)

4.3 North-South Traffic

In 2008, the 2510 teu type SSW Super25 developed by SSW Schichau-Seebeck Shipyard in Bremerhaven/Germany still met the capacity requirements of many north-south routes. The shipbuilder tried to reduce construction costs by having the hulls of three units of a six-ship order from Hamburg-based E. R. Schiffahrt in 2002 delivered by Daewoo Mangalia shipyard in Romania, but the whole contract could not be made profitable despite the hulls being purchased cheaply. The hull of one of two more Super25 units for Projex shipping company in Hamburg under the names HERMES and ULYSSES was in 2006 also awarded abroad, to Crist shipyard in Gdansk. The fact that this type of ship, equipped with three container cranes and a powerful two-stroke engine for a speed of 22 knots, was in great demand was demonstrated by the fact that the design was purchased by HDW shipyard in Kiel, where another six ships were built for Münchmeyer & Petersen in Hamburg between 2004 and 2006. China's Xiamen shipyard as well realised copies of the SSW design and delivered five newbuildings to Martin Lauterjung's Sunship Schiffahrtskontor in Emden, starting in 2008 by CITY OF XIAMEN. Nevertheless, the order sequence for the SSW Super25 type from Germany via Romania and Poland to China gives evidence about the vagabond's character of order flows and their shift to Asia, first of all caused by the development of personnel and subsidized steel costs. Renamed MSC HERMES, the HERMES in 2022 fulfilled a charter of MSC in a US East

Coast/Gulf/East Coast of South America service jointly operated with ONE. The sister vessel MSC ULYSSES was chartered by Maersk and employed under the name AS PATRICIA in a Middle East-India-Indian Ocean islands-South Africa service jointly operated with CMA CGM (Figs. 4.42 and 4.43).

Like in the Europe-Far East services, in which the capacities of the largest container ships grew three to fourfold within 15 years after the turn of the millennium, the cargo capacity in the North-South trades as well increased from long-time usual 2500 teu to 8000 teu and more. The alliance agreements on the east-west routes did not cover the north-south lanes, for instance from Europe to South Africa and South America or between the east and west coasts of North and South America.

In 2017, the 2 M partner MSC established a joint service with Hapag-Lloyd from North Europe to South America's east coast. The operators employed nine vessels of 8700 teu to 9400 teu intake. Their schedule named the ports of Rotterdam, London, Bremerhaven, Hamburg, Antwerp, Le Havre, Sines, Rio de Janeiro, Santos, Navegantes, Buenos Aries, Montevideo, Rio Grande do Sul, Navegantes, Paranagua, Santos, Rio de Janeiro, Salvador, Pecem, Rotterdam. Hapag-Lloyd used three carriers of the acquired UASC named UASC AL KHOR, UASC ZAMZAM and UASC UMM QASR sailing along with CCNI ANDES, which in 2018 was replaced by the Turkish-owned HUNGARY ex HANJIN HUNGARY. In 2022, Hapag-Lloyd deployed one 8000 teu ship, the TENO.

Two of the five ships initially supplied by MSC to the Europe—SAEC service, named MSC BRUNELLA and MSC LILY, were built by Dalian Shipbuilding in Dalian, MSC ELODIE by New Century in Jinjiang. They were part of a newbuilding series for MSC comprising a total of 14 units owned by China International Marine Containers. The Chinese company chartered them out to MSC on a long-term basis. Delivery of the 300 m length, 48.2 m breadth type of 8800 teu capacity was started by MSC LILY

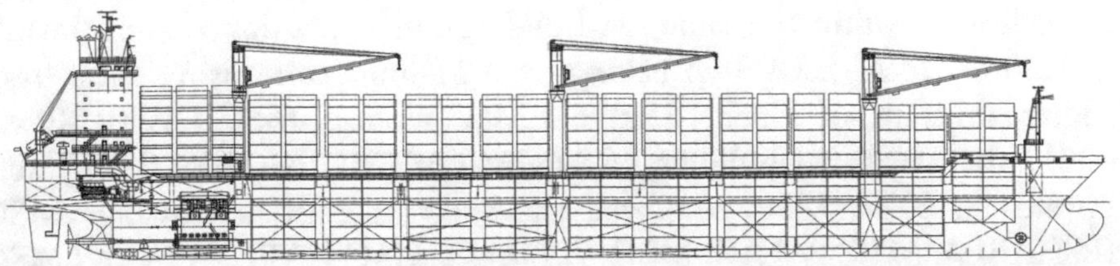

Fig. 4.42 The 2006-built 2510 teu container vessel HERMES from SSW Schichau-Seebeck Shipyard for Hamburg-based Projex company was laid out for the capacity requirements on many north-south routes (SSW)

Fig. 4.43 After delivery, the charter agreements of HERMES included contracts with CMA CGM, Hyundai, MOL, KMTC, Hamburg Süd, Evergreen, OOCL, Maersk, Tehama, Simatech and COSCO (Ralf Witthohn)

in 2014 and completed by MSC CHANNE the following year. After delivery in May 2015, MSC ELODIE commenced a liner service operated by MSC and Hamburg Süd with slot offers to Maersk Line and Hapag-Lloyd between the Far East and the east coast of South America before switching to the Europe—east coast service. MSC BRUNELLA, which was delivered in October 2015, was after the opening of the enlarged Panama canal employed between North Europe and the west coast of South America. For this route the ship type is predestined thanks to a high number of 1462 reefer connections. Their breadth of 42.8 m makes the vessel a so-called wide beam ship allowing the stowage of 19 rows of containers on deck and of 17 rows in the hold (Figs. 4.44 and 4.45).

In the beginning of 2020, the tonnage contributed by MSC had switched to MSC SOFIA CELESTE, MSC CATERINA, MSC BRANKA and MSC DESIREE. MSC SOFIA CELESTE and MSC CATERINA had in 2015 been delivered within the same newbuilding series mentioned above, while the 9400 teu MSC BRANKA belongs to a 21 units comprising series from Jiangnan shipyard, of which 16 entered MSC services. MSC DESIREE was in 2017 the fourth newbuilding of a quintet of 9400 teu ships from Jinhai shipyard. In 2022, MSC employed the 8800 to 9400 teu carriers MSC MICHAELA, MSC PALAK, MSC ATHENS, MSC CATERINA and MSC SOFIA CELESTE, Hapag-Lloyd the 9000/9100 teu ships UASC AL KHOR, UASC ZAMZAM, HUNGARY and the 6600 teu MEHUIN. In a joint

Fig. 4.44 MSC ELODIE started in 2015 in a Far East—South America east coast service of MSC and Hamburg Süd offering Maersk and Hapag-Lloyd slots, switched to the Europe—SAEC link in 2017 and had in the beginning of 2020 anchored off Dalian for several months, before joining the TA3 of 2 M and in 2002 the TA2 (Ralf Witthohn)

Fig. 4.45 MSC LILY, which after delivery in 2015 began on a pendulum route from the US west coast to Asia and the US east coast, was five years later employed on the Asia—South America east coast line (Ralf Witthohn)

service from the Western Mediterranean to South America's east coast, MSC in 2022 deployed the 4000 teu to 9400 teu ships MSC DIEGO, MSC ALBANY, MSC ADELAIDE, MSC BRUNELLA, MSC ATHOS, MSC AJACCIO and MSC SILVIA, Hapag-Lloyd the 8000 teu TENO (Fig. 4.46).

Within the MSC fleet, the 14 MSC LILY type ships owned by China International Marine Containers are part of a series of in total 34 newbuildings built at shipyards in China and Korea. They represent the bosphorusmax size and are as such similarly classified as neopanamax vessels. The remaining newbuildings of the 8800 teu series were built by Jinhai Heavy Industries, Jiangnan Changxing Shipyard and Korea's Sungdong Shipyard for the shipping companies Costamare, Zodiac, Eastern Pacific and MSC. Among them was the 299.9 m length, 48.2 m breadth MSC AGRIGENTO from Sungdong Shipyard in 2013. By 8772 teu, the ship has, due to its aft-arranged superstructure, a lower maximum capacity than two-island designs of the same main dimensions. But by an intake of 7100 units the ship has a nearly equal intake of 14 tonnes containers. At 110,800 tonnes on a draught of 14.5 m, the deadweight capacity corresponds to that of the two-island types. The first vessel of the Sungdong series, named MSC ATHENS, together with its sister ships MSC AGRIGENTO and MSC ADELAIDE. participated in an MSC service from northwest Europe to Central America and to the West Coast of South America, which was launched in 2016 after the opening of the new Panama locks. The second vessel, MSC ATHOS, operated between the Mediterranean and the west coast, while the MSC ABIDJAN was chartered to MOL and sailed as MOL ABIDJAN between Asia and the east coast of South America. Initially, the 2013-built MSC ABIDJAN had been employed in a service from the Far East to South Africa and Europe. In 2020, the ship was one seven 9000 teu class ships, named MSC JEONGMIN, MSC NAOMI, MSC AGRIGENTO, MSC ADELAIDE, MSC AGADIR and MSC ANTIGUA, sailing in the Mediterranean—South America east coast

Fig. 4.46 In 2022, Hapag-Lloyd's tonnage share in the Europe-South America east coast service jointly operated with MSC included two former UASC ships, the UASC AL KHOR and UASC ZAMZAM (Ralf Witthohn)

service maintained together with CSAV TRAIGUEN of Hapag-Lloyd and offering Israel's ZIM slots (Figs. 4.45 and 4.47).

4.3.1 Along the American East Coasts

Though the long-time speculations about a possible merger between the German liner companies Hapag-Lloyd and Hamburg Süd never materialized and were in the end terminated by the sale of Hamburg Süd to Maersk, there was in the beginning of 2020 still a formal cooperation maintained on some routes, incuding in the so called New Tango service between the east coasts of North and South America. The line had been established in 2014 by the operators Hamburg Süd/Alianca, Hapag-Lloyd/Norasia and NYK. After the leave of NYK the following year, the two German names, with slots taking Maersk, were still marketed in the beginning of 2020. Hamburg Süd contributed the 6600 teu CAP ANDREAS, Hapag-Lloyd the chartered SM HONG KONG and CARDIFF and Maersk the MAERSK MEMPHIS, NORTHERN MAGNUM and NORTHERN MAGNITUDE. The two latter 7000 teu ships had after delivery to Hamburg-based financier Norddeutsche Vermögen in 2003 co-incidentally been chartered by Hapag-Lloyd as LOS ANGELES EXPRESS and BANGKOK EXPRESS and employed in a trans-Pacific service of the Grand Alliance. The six-ship fleet connected the ports of New York, Philadelphia, Norfolk, Charleston, Jacksonville and Port Everglades with Santos, Buenos Aires, Rio Grande, Itapoa, Santos, Rio de Janeiro, Salvador and Pecem. In 2022, the tonnage supplied by Maersk were MAERSK KARACHI, MONTE AZUL, MONTE OLIVIA, MONTE ALEGRE and MONTE VERDE still sailing under the Hamburg Süd brand along with Hapag-Lloyd's MAIPO and DUBLIN EXPRESS (Fig. 4.48).

Fig. 4.47 The extension of the Panama canal locks made it possible to use neopanamax carriers such as the MSC ABIDJAN in north-south trades to both coasts of South America (Ralf Witthohn)

A sister vessel of the NORTHERN MAGNUM and NORTERN MAGNITUDE, the 2001 Daewoo built BUXCOAST, was the largest vessel that MSC was in February 2020 using on US east coast—Caribbean—South America east coast line maintained together with Maersk between New York, Norfolk, Baltimore, Charleston, Savannah, Port Everglades, Caucado, Santos, Buenos Aries, Montevideo, Rio Grande, Navegantes, Santos, Rio de Janeiro, Salvador, Suape, Caucado, Freeport and New York. The chartered 6900 teu BUXCOAST, a finance product of Germany's meanwhile dissolved Conti group, was employed together with the 5500 teu to 5900 teu ships LEO C, SM VANCOUVER, MSC CAROLINA, MSC VIDHI, MSC FIAMETTA, MSC CORUNA and MSC VIGO. Two years before, MSC had deployed the 5500/6700 teu ships MSC BARCELONA, MSC CADIZ, MSC KRYSTAL, MSC MARGARITA, MSC MARTA, MSC MARINA, MSC MAUREEN and MSC VANESSA. Then, MSC had still offered slots to Hamburg Süd, Hapag-Lloyd and Israeli ZIM. At 40 m breadth and 270.3 m length, MSC BARCELONA and MSC CADIZ represent a design in 2011/12 realised six times at Romania's Daewoo Mangalia shipyard for Hamburg shipowner Claus-Peter Offen. The type achieves a capacity of 4390 14 tonnes boxes. MSC ALICANTE, MSC BARCELONA, MSC CADIZ, MSC MADRID, MSC CORUNA and MSC VIGO were delivered to an MSC charter from June 2011 on. At a draught of 13.5 m, they carry 74,456 tonnes and achieve 22.4 knots. MSC's still maintained its stand-alone service in 2022 by nine carriers of between 5550 to 6700 teu (Fig. 4.49).

Fig. 4.48 Under its maiden name NORTHERN MAGNUM the former Hapag-Lloyd-chartered LOS ANGELES EXPRESS was in 2020 a Maersk Line contribution to the east coast service maintained with Hamburg Süd and Hapag-Lloyd (Ralf Witthohn)

4.3.2 HANJIN COPENHAGEN in Tsunami

On 11 March 2011, the tsunami triggered by the Tohoku seaquake hits the German container ship HANJIN COPENHAGEN at its position 150 nm from Yokohama, the vessel's port of destination. While more than 18,000 people die in the up to 38 m high wave and the Fukushima nuclear power plant is so severely damaged that a catastrophic disaster is caused, the wave passes under the ship without any damage. The ship yet has to anchor off Yokohama due to the port's closure, but is later handled with some delay. At this time, the 5600 teu capacity HANJIN COPENHAGEN, owned by Munich-based ship financing company Conti and managed by NSB Niederelbe Schiffahrt in Buxtehude, is employed with other ships of the company in a joint transpacific service of Hanjin and COSCO between US west coast ports, Yokohama, Hong Kong and Yantian. Following Hanjin's bankruptcy in 2016, the ship was chartered by Maersk Line, before in 2017 Hapag-Lloyd leased it for a liner service from the US Gulf to the East Coast of South America. In 2020 the ship was, under its original name CONTI DARWIN, by Maersk's subsidiary Sealand Asia used in an inner-Asian service between Chinese and Indonesian ports (Fig. 4.50).

The postpanamax carrier HANJIN COPENHAGEN was part of an 18-ship series built by Hanjin Heavy Industries, the first seven of which were delivered to group-owned Hanjin shipping company. The other newbuildings were commissioned by the German finance investors Conti in Munich and GEBAB in Meerbusch. HANJIN COPENHAGEN was given its maiden name CONTI DARWIN only after the Hanjin bankruptcy. At the same time, the intake of the 279 m length and 40 m breadth vessel was increased

Fig. 4.49 MSC BARCELONA was operated between the east coasts of North and South America (Ralf Witthohn)

from 5600 teu to 6100 teu by constructional measures. The CONTI DARWIN took up charters by Maersk, Hapag-Lloyd and X-Press Feeders and was in 2022 employed by Sealand Asia/Maersk in a China-South East Asia service including Indonesian destinations.

4.3.3 5500 Teu Carriers from Samsung

One of the first large series of postpanamax carriers finally comprising 33 newbuildings was between 1999 and 2008 built by Korean Samsung shipyard. The ships were initially employed mainly in the Europe-Far East trade, but later switched to shorter routes including even to such as the north European feeder network. Introduced by the 5500 teu carrying duo OOCL NEW YORK ex E.R. HONG KONG and OOCL SHANGHAI ex E.R. SHANGHAI, the type's capacity was enlarged to a 6000 TEU version, of which E. R. SEOUL, E. R. PUSAN, E. R. LONDON, E. R. AMSTERDAM, E. R. FELIXSTOWE, E. R. BERLIN, E. R. FRANCE, E. R. KOBE, E. R. LOS ANGELES, E. R. CANADA, E. R. INDIA, E. R. DENMARK and E. R. SWEDEN were put into service by Germany's E. R. Schiffahrt. Eight more ships of the 5500 teu class entered service under the names CMA CGM BELLINI, CMA CGM CHOPIN, CMA CGM MOZART, CMA CGM PUCCINI, CMA CGM ROSSINI, CMA CGM STRAUSS, CMA CGM VERDI and CMA CGM WAGNER. Neptune Orient Lines (NOL) in Singapore received APL ENGLAND, APL SCOTLAND, APL HOLLAND and APL BELGIUM, while Mediterranean Shipping Co. employed MSC

Fig. 4.50 One of the worst natural disasters could not affect HANJIN COPENHAGEN in March 2011 (Ralf Witthohn)

VIDHI, MSC MARGARITA, MSC FIAMMETA, MSC KRYSTAL, MSC ORIANE and MSC SORAYA as owned or chartered units.

In their architecture, the 277 m length, 40 m breadth, 67,500 dwt carrying ships are characterized by high container cell guides on the aft deck, which make laborious manual lashing work redundant. For constructional reasons the guides cannot be arranged above the holds closed with hatch covers. Even before the 5500 teu type programme of Samsung was finished, ships of a new and larger 323 m length, 42.8 m breadth 8000 teu type were built. In 2006, OOCL ASIA was the ninth of 29 such newbuildings to start its first sailing in the EU3 loop of the Grand Alliance consortium. Three Chinese shipping companies used the container ship type in their liner services, OOCL from Hong Kong, the China Shipping Container Line and Taiwan's Evergreen Marine (Fig. 4.51).

4.3.4 Wide-Beam Types for Vinnen

In anticipation of the Panama canal extension, shipping companies and shipyards took the opportunity to design container ships that exceeded the old panamax breadth of 32.3 m, thus to combine the advantages of a smaller draught with a greater stability. Only a few years after the introduction of container traffic, the main dimensions of the largest newbuildings, deployed in the Europe-Far East service from 1972 onwards, had already reached the panamax limit. The 287.7 m length, 32.3 m breadth ships of Hapag-Lloyd's 48,750 dwt HAMBURG EXPRESS class that were put into service at that time had a loading capacity of 3010 teu or 2272 teu of 14 tonnes. The almost maximum utilisation of the Panama restriction resulted in a length-to-breadth ratio of 8.9—with all its associated consequences on important design parameters and properties such as draught, stability, ballast capacity, speed, engine power, course stability and manoeuvrability. The good market chances for wide-beam ships kept on even during the COVID-19 crisis, when such tonnage was hardly available and could secure high daily charter incomes of around USD 18,000 for a 5000 teu carrier.

From 2011 to 2013, Denmark's A. P. Møller Group received a total of 22 so-called wide-beam vessels of the MAERSK CONAKRY type from Hyundai shipyards. At a length of 249 m far below the old panamax dimension, the ship's breadth was fixed at 15 times a container breadth, at 37.4 m and thus two container widths above the former panamax dimension. The length-to-breadth ratio shrunk to 6.6, resulting in a reduced draught of 13.5 m and a corresponding deadweight of 61,614 tonnes. Their slot capacity of 4496 teu

Fig. 4.51 The 6000 teu carrier E. R. LONDON was one of 13 Samsung built vessels for E. R. Schiffahrt and, after charters as P&O NEDLLOYD VESPUCCI and MSC GEMMA, found employments for Hapag-Lloyd in a North Europe-South America East Coast service in 2016, in 2020 in a Med-US Gulf loop and in 2021 in an Asia-America service of Wan Hai under its christening name. Meanwhile owned by US interests and renamed GSL VIOLETTA, the ship sailed in a Maersk service between the Far East and Seattle with stops at Dutch Harbour/Alaska in 2022 (Ralf Witthohn)

is exceeding that of the HAMBURG EXPRESS class by almost 50%. Due to the relatively low draught, the crane-equipped newbuildings are particularly suitable for secondary routes with smaller ports and shallow fairways. 13 vessels of the MAERSK CONAKRY type were in service at the end of 2017, and still in 2021, in a Far East-West Africa service established in 2014 by Maersk, its subsidiary Safmarine and CMA CGM with calls at the ports of Shanghai, Ningbo, Nansha, Singapore, Tanjung Pelepas, Cape Town, Lagos-Apapa, Lagos-Tincan, Onne and Xiamen.

Two gearless ships of the in total 27 newbuildings comprising series were delivered to tradition-rich Bremen shipping company F.A. Vinnen under the names MERKUR HARBOUR and MERKUR PLANET. The renunciation of cranes increased the intake to 4622 teu, or 3645 teu of 14 tonne containers, thus achieving 1.6 times the intake of the first large container ships. After sale to Maersk in 2018 and being renamed RHINE MAERSK, the former MERKUR HARBOUR was on behalf of the Maersk subsidiary Sealand Americas deployed in a 2019-established feeder service from Balboa to the South American west coast ports Iquique, Arica, Valparaiso, San Vicente and

Callao, together with three more, similarly sized ships. MERKUR PLANET became RHONE MAERSK and was in 2022 still employed on a Middle East to South and West Africa route with calls at Jebel Ali, Mundra, Mumbai-Nhava Sheva, Colombo, Pointe Noire, Tema, Cotonou and Durban (Fig. 4.52).

In 2013 Vinnen took over two newbuildings of another wide-beam type designated SDARI 3800 from Shanghai Shipyard. At 227.9 m length to 37.3 m breadth, the MERKUR FJORD and MERKUR OCEAN featured an even smaller length to breadth ratio of 6.1. Following delivery the two Vinnen ships took up a Hamburg Süd charter in liner services between the Far East and the US west coast to Australia/New Zealand together with another vessel of the same type. At the end of 2017 they served a link established by Hamburg Süd together with the Asian partners COSCO, MOL and NYK between Japan and China to New Zealand by a total of six vessels of equal size. Of these 3820 teu wide-beam vessels, a total of ten units was built, including four for the Brazilian Hamburg Süd subsidiary Alianca and one each for the Hamburg shipowners Rehder and Vilmaris. In the beginning of 2020, both ships sailed in a CMA CGM charter, MERKUR FJORD from the Far East, and in 2022 from West Europe, to West Africa and MERKUR OCEAN from the Mediterranean to the Far East, and in 2022 from the Med to West Africa.

Previously, the old Panama lock restrictions, combined with the container geometry and hydrodynamic conditions, had resulted in only a small number of container ships of the 4000-teu class at all. An interesting variant of the wide-beam vessel was represented by four newbuildings of the RHL CONSCIENTIA type of Shanghai Shipyard for Hamburger Lloyd. They are

Fig. 4.52 The relatively large breadth favours the intake of the wide-beam vessel MERKUR HARBOUR, in 2018 sold by Bremen based shipping company Vinnen to Maersk (Ralf Witthohn)

another example of the old practice of increasing loading capacities by lengthening, in this case to 259.8 m and a corresponding intake of 4620 teu. The ships were deployed in charter services from the Far East to West Africa, the Middle East and Australia (Fig. 4.53).

The trend towards wider carriers was as well reflected in orders for an eleven-ship series placed by the German shipping companies Hansa Shipping, Hammonia Reederei, Peter Döhle, Dietrich Tamke and Gerd Ritscher at Jiangsu New Yangzi and Jinling Yizheng Shipyard. The type was subsequently changed from an order for panamax carriers of 294 m length and 32.3 m breadth to a type of 255 × 37.3 m. The lead ship HAMMONIA ISTRIA, delivered in September 2013, has an intake of 4957 teu and achieves 58,000 dwt on 12.5 m draught resulting in 3676 teu homogeneously loaded with a 14 tonnes weight. The vessels, which are mainly operated in the Asian region, thus did not reach the nominal performance of today's optimised panamax vessels, such as the two-island type CHOAPA TRADER of 5294 teu/65,550 dwt of German owner Hermann Buss. But thanks to their greater stability they surpassed that one in the key figure of loaded containers, which is only 3330 teu of 14 tonnes weight on CHOAPA TRADER. For a further comparison: another panamax ship, the 294 m length, 32.3 m breadth and 13.5 m draught HS HUMBOLDT built in 2004 in China for Hansa Treuhand in Hamburg, features 4990 teu/67,679 dwt, but had an even slightly lower loading capacity of 3280 teu of 14 tonnes weight. This gives evidence how the extension of the Panama width restriction enabled the choice of a smaller length/width ratio leading to actually higher actual container intakes (Fig. 4.54).

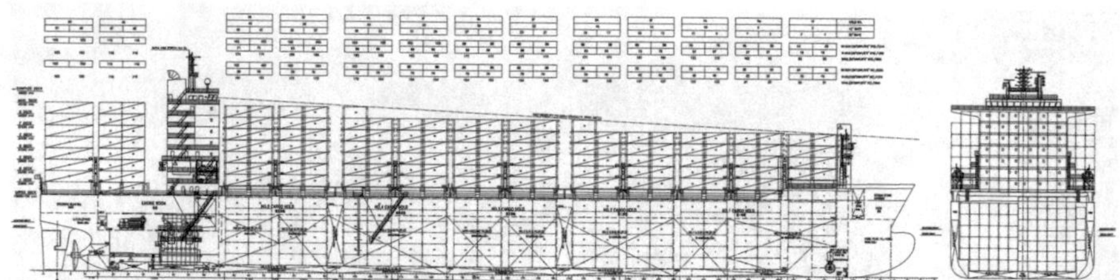

Fig. 4.53 MERKUR OCEAN represents the type of a wide-beam ship that can carry 13 rows of containers in the hold and 15 on the hatches (Vinnen)

4.4 Feeder and Short-Sea Traffic

On all seas, a special branch of container shipping has been developed for the further distribution of containers from deepsea carriers or for their supply. It is carried out by smaller container ships thus providing a connection between main ports, so-called hubs, and secondary ports. In 2020, more than 140 feeder services existed only in northern Europe. A particularly close-meshed feeder network exists between the North and the Baltic Sea, which has, however, lost some of its intensity since the 2M cooperation included Gdansk in the schedule of its largest container ships employed on the Far East route. Ports in the Caribbean, on the African and Asian coasts are as well served by numerous feeder services. The transhipment from the large container ships takes place because the smaller ports cannot accommodate them for navigational reasons or because the cargo volumes are too small for economic transport in large vessels. Participants in feeder traffic are often specialised shipping companies which collect the containers from various liner companies. There are yet also deep-sea operators that maintain their own feeder services. A distinction is to be made between feeder and short-sea shipping, which is as well carried out by smaller container ships, but originally transporting containers on shorter routes under inclusion of feeder cargo.

Among the major shipping companies maintaining their own Baltic feeder services is China's COSCO, which in 2022 employed the 2017 built, 1924 teu capacity DELPHIS GDANSK and DELPHIS BOTHNIA between Hamburg, Zeebrugge and St. Petersburg. The ships belong to a quartet of ice-classed ships commissioned in 2016 by Antwerp-based Delphis company, founded in 2004 and acquired by Belgian shipping group Compagnie Maritime Belge (CMB) in 2015. Delphis designates the newbuildings as kielmax ships and describes them as having a 40% higher capacity than the largest feeders, which before transited the Kiel canal (Fig. 4.55).

Fig. 4.54 The nominal slot intake of the two-island vessel CHOAPA TRADER benefits from the better visibility conditions from the forward located wheelhouse, while the ship's panamax breadth reduces its stability and thus its intake of loaded containers (Ralf Witthohn)

Delphis, on its turn, was involved in north and west European feeder shipping by its Hamburg-based subsidiary Teamlines. By the end of 2017, 39 container ships were sailing for Teamlines to Norway, Sweden, Denmark, Finland, Poland, Russia, the Baltic States and Spain. In February 2019, the shareholders of Team Lines Deutschland decided to discontinue all the companie's business activities citing deteriorating market conditions in north Europe as reason.

Another major liner operator, France's CMA CGM, maintained two feeder services to the Baltic, one loop from Hamburg and Antwerp to St. Petersburg by weekly deployment of the 1440 teu vessel A LA MARINE chartered from Delphis and the own 2550 teu newbuilding CMA CGM LOUGA. Russia's FESCO ESF, COSCO, Hapag-Lloyd, Evergreen, Unifeeder and X-Press Feeders are taking slots. CMA CGM's second loop connects Wilhelmshaven and Hamburg with St. Petersburg using the 2487 teu feeders CMA CGM PREGOLIA and CMA CGM NEVA, only able to reach the Baltic via Skagen when in laden condition (Fig. 4.56).

4.4.1 Collective Transporter Unifeeder's

In northern Europe, the Danish shipping company Unifeeder operates feeder lines between Germany and Scandinavia, from Benelux ports to Scandinavia,

Fig. 4.55 Deployed by China's liner major COSCO to St. Petersburg, the Hong Kong-flagged DELPHIS GDANSK and DELPHIS BOTHNIA represented the largest feeder newbuildings using the Kiel canal in the beginning of 2020 (Ralf Witthohn)

to the Iberian peninsula and to the UK. Other lines run from North Sea ports to St. Petersburg, Finland and Baltic ports. At the end of 2017, the shipping company listed 50 German and Dutch owned container ships with capacities of up to 1700 teu in its schedule, some of them sailing on behalf of other charterers. In February 2020, the Unifeeder fleet had reached a number of 67 units sailing on 20 routes. The links to Scandinavia were maintained by 19 feeders, Portugal was called by five, the UK by seven vessels, Russia and the Baltic states by the remaining ships. Unifeeder also names the A LA MARINE in its schedules as one of the largest ships, sailing from Antwerp and Hamburg to St. Petersburg.

In 2020 jointly operated by Unifeeder and CMA CGM was the 1404 teu open top feeder vessel HEINRICH EHLER owned by Heinz Ehler from Otterndorf/Germany. On the loop from Bremerhaven and Hamburg to Ust-Luga, Kotka, Tallinn and Rauma the major lines OOCL, COSCO and Evergreen were offered slots, X-Press Feeders from Hamburg to Ust-Luga. HEINRICH EHLER sailed along with the sister vessel VERA RAMBOW and the 1600 teu ships BALTIC TERN, BALTIC SHEARWATER and CALISTO. Among the largest ships were the 1638 teu ship BALTIC PETREL, the around 1430 teu carrying AMERDIJK the HEINRICH EHLER and the similarly sized OOCL RAUMA and THETIS D, which maximum draught of 9 m still allows the regular passage through the Kiel canal. The first three vessels operated a weekly service from Germany to Poland, Latvia and Finland. The ports of call were Wilhelmshaven, Bremerhaven, Hamburg, Gdynia, Bremerhaven, Hamburg, Riga, Kotka, Wilhelmshaven and Bremerhaven. On the Unifeeder vessels the X-Press Feeders, belonging to Sea Consortium in Singapore, and the CMA CGM rented slots (Fig. 4.57).

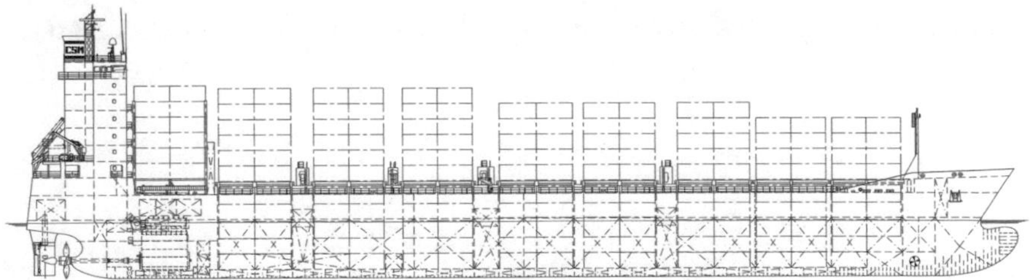

Fig. 4.56 The 1440 teu type A LA MARINE, built 16 times at Wolgast/Germany-based Peene Werft between 2006 and 2009, is one of the largest feeders operated in the Baltic Sea (Peene Werft)

4.4.2 Open-Top Type from Meyer Werft

The Unifeeder- and CMA-chartered BALTIC TERN and BALTIC SHEARWATER were, along with the sister ships REINBEK and FLOTTBEK, built in 2005 by Germany's Meyer Werft under the names EILBEK and BARMBEK. The Papenburg builders, in a slump of orders after the cruise shipping crisis following the 9/11 terror attack, devoted themselves to container shipbuilding, which was still promising at the time. They signed the contract with the shipping company Hansa Hamburg Shipping International to build a completely new designed type with partially open holds and under high ice class requirements. The ships were initially intended for the route from Montreal to North Europe of Canada's CP Ships, but later switched from the transatlantic to European services, where they were over the years hired by Delphis, Hapag-Lloyd, Hamburg Süd, MSC, Team Lines and Unifeeder. REINBEK was in early 2022 serving the Singapore-headquartered X-Press Feeders group as BALTIC FULMAR linking Zeebrugge, Rotterdam, Dunkirk and Gijon, FLOTTBEK as BALTIC PETREL Antwerp and Rotterdam with St. Petersburg.

A key design feature of the first container ship type from Meyer Werft is the partial omission of hatch covers, which had already been implemented in a number of other container ship designs since the early 1990s. The Meyer design, however, was the first time based on a farer reaching idea. The cargo holds were divided lengthwise into three sections by two bulkheads. The middle and largest part of the hold was left open, while the outer holds were closed with hatch covers. This innovative solution prevented water from penetrating into the open holds even at large heeling angles. At 15.1 m depth and a maximum draught of 9 m, the ship's freeboard has a considerable height of 6.1 m. Out of this reason, the design manages with a relatively low longitudinal hatch coaming in the area of the open holds. Only the forward cargo area

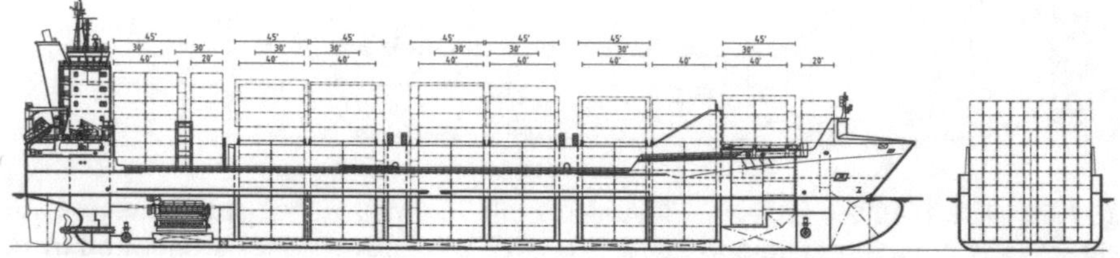

Fig. 4.57 In 2008, the 1421 teu open top ship HEINRICH EHLER marked the start of the construction of the last series of feeder types designed by Hamburg shipyard J. J. Sietas (Sietas)

is protected by a high transversal coaming designed in the shape of a breakwater (Fig. 4.58).

The arrangement of the outer holds creates the possibility of block stowage of the containers. Furthermore, they are suitable for containers with sensitive cargoes, for which below deck transport is required and which can be loaded in the same bay as the customer's other containers. It is then less often necessary to move the land-based container gantry cranes during cargo handling. In addition to the rules of the International Convention on Load Lines, the lack of hatch covers demands the application of the IMO regulation MSC 608. This required model tests with defined wave heights and sea state spectra. The regulation also specifies the conditions in case of cargo hold flooding and the required pump capacity reserves. On the basis of the Load Lines convention, the ships were calculated as "with hatch covers". Their maximum draught resulted from the respective results of the model tests, the stability considerations and the freeboard calculation.

At a ship's breadth of 27.2 m, the main section of the EILBEK shows a hold divided by longitudinal bulkheads with space for two rows of containers in each of the two outer enclosed spaces and for five rows in the central open space. The hull was constructed as a double skin construction. The side holds are constructed with their smooth walls facing the side tanks, while the stiffeners of the inner longitudinal bulkheads are set outwards. Due to the relatively large bilge radius, a tank was arranged in the bilge area to take up the stowage space for the lowest outer container. There is thus still space for four tiers in the outermost container row of the enclosed hold, while the inner row is stowed in five layers (Fig. 4.59).

In the open hold, the number of container tiers is also five. A maximum of seven layers of containers are stowed on top of this. They are supported by stoppers arranged at deck level due to the limited number of containers which can be stowed on top of one another. In total, the number of tiers in both the

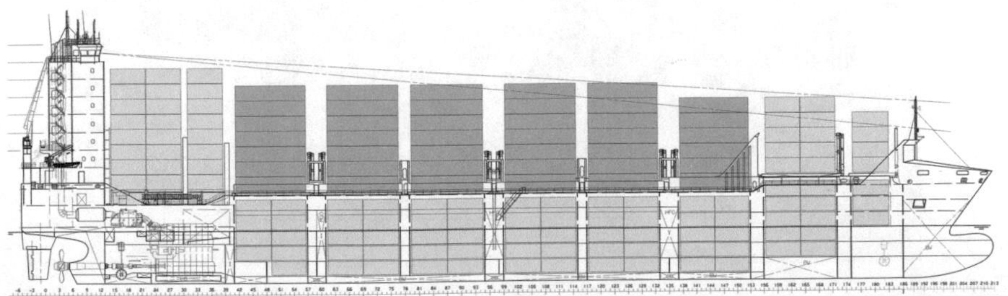

Fig. 4.58 The key design feature of the first container ship type built by Meyer Werft is the omission of hatch covers in the main cargo hold section (Meyer Werft)

open and closed hold area figures a maximum of 12. According to the visibility rules, this results in a container capacity of 1604 teu. It consists of 1186 20-ft and 207 40-ft units. The stability of the vessel, homogeneously loaded by 14 tonnes units, is guaranteed up to 1090 teu. The high degree of loading flexibility includes the possibility of accommodating 2.5 m wide containers as used for Europe-normed pallets, 45-ft containers and up to 311 reefer containers. At a draught of 9 m, the ships' deadweight amounts to 15,950 tonnes.

As with most designs of open-top container vessel types, the forward holds are closed by hatch covers. The foremost hold is designed for a 20-ft bay and the one behind for a 40-foot bay. The two forward hatches are surrounded by

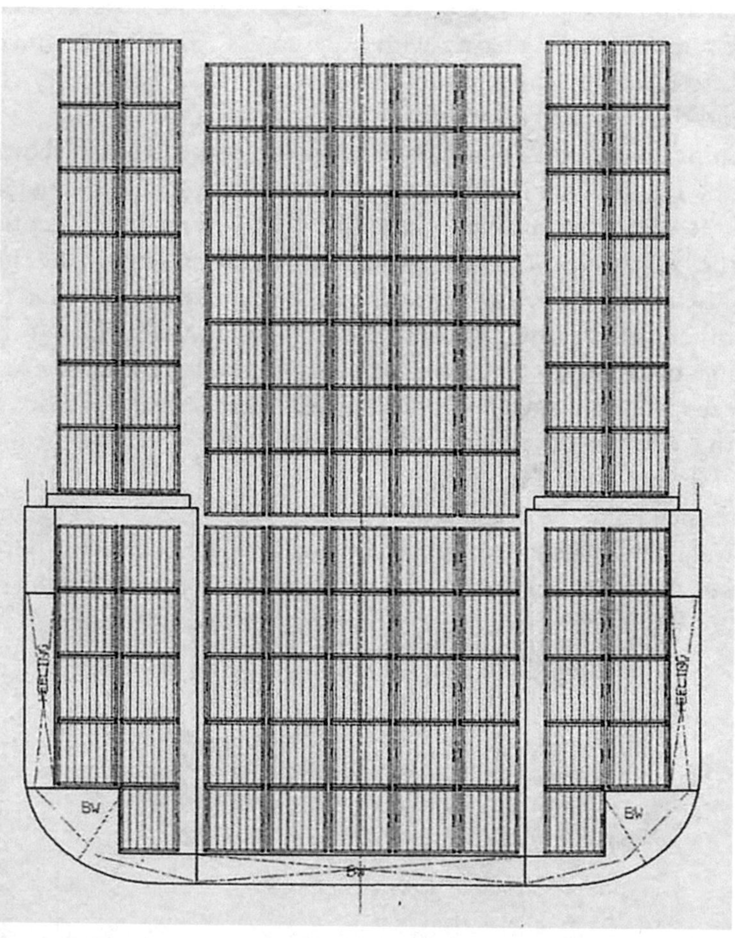

Fig. 4.59 The cross-sectional drawing of the Meyer Werft feeders reveals an open hold section for five rows of containers and two outside ones closed by hatch covers (Meyer Werft)

conventional coamings in the area of the raised forecastle deck. Behind them, the open, cell frames-fitted cargo hold section provides six 40-ft bays distributed over four holds. The first and last bay are arranged in one hold each, two bays each in the two remaining holds. Above the engine room, the lowered main deck accommodates a 20-ft bay and a 40-ft bay at a maximum of nine tiers. In total, the ship's length of 169 m allows the arrangement of eight 40-ft bays and two 20-ft bays. To protect the anchor and mooring winches, the forecastle deck is covered. The breakwater at the aft edge extends to the height of the second container tier.

The ships were classified by Germanischer Lloyd according to the regulations of ice class E4 and meet the requirements of the Swedish-Finnish ice class 1A Super. The 12,640 kW developing main engine of the type MAN B&W 8S50MC-C acts directly on a controllable pitch propeller, enabling a speed of 20 knots. In order to improve manoeuvring characteristics, the ships are equipped with a patented rudder and two transversal thrusters, one each below the first cargo hold and at the stern. The forward thruster has an output of 950 kW, the stern thruster of 650 kW. The far aft positioned superstrucutre with its front bulkhead arranged on the aft engine room bulkhead comprises seven superstructure decks and the wheelhouse. It offers 17 single cabins for the crew and six double cabins for 12 passengers (Fig. 4.60).

4.4.3 Chinese-Built Container Feeder Types

Triggered by the loss of the customer Contship Container Lines to Bremer Vulkan, Rickmers Werft in Bremerhaven went bankrupt in 1986—1 year after its 150th anniversary. Compared to other shipyards, the shipyard had in 1980 delivered its first cellular container ship, the 580 teu carrying TAURIA of type RW39, to Jork/Germany based owner W. Harms later than the competitors. Despite its low capacity, the ship went into a service from Europe to Australia in 1982 under the name ARABIAN EAGLE for Eagle Container Lines, a service of the Contship group. It was followed by the same type of ship, named GOTHIA for Harms again, ESTETURM (ELMA CINCO) for Julius Hausschildt, WEJADIA for Harms, CHAMPION and PREMIER for Projex in Hamburg, MARIVIA for Herm. Dauelsberg in Bremen, LILIENTHAL, LEERORT, ESTEBRÜGGE, HEIDE, HUSUM, DÖRTE, VANELLUS and as the last newbuilding of the shipyard in July 1986 BRITTA THIEN, all for German shipowners. In response to market demands Rickmers then developed the larger container ship type RW49 of 1002 teu intake, of

Fig. 4.60 Only the two outer container rows of the open-top container vessel type FLOTTBEK type are stowed on hatch covers (Ralf Witthohn)

Fig. 4.61 The container ship PATRICIA RICKMERS, initially chartered out as VILLE DE LUMIERE to French shipping company CMA, was the last newbuilding of Rickmers Werft for the own group in 1985. Demolition took place in Alang/India in 2012 (Ralf Witthohn)

which HEIKE and PATRICIA RICKMERS were completed in 1985, OLANDIA and BIRGIT NABER in the following year (Fig. 4.61).

The Rickmers shipping branch continued to exist and in search of reasonably priced container ships found what they were looking for in neighbouring Poland. Based on the PATRICIA RICKMERS drawings, Stocznia Szczecinska built further units of the type, by the Polish builders designated as B-183. The type developed into one of the longest series of smaller container ships starting in 1991 by NOBLE for Hamburg-based Projex company and ending after 27 newbuildings by SELANDIA for Anglo Eastern in Hong Kong. Two ships of this type were also built at Vinashin shipyard in Quang Ninh in 2004 and 2005 under a Vietnamese order. The Szczecin shipyard took over other another container ship design of Schlichting shipyard in Travemünde, named B-186 and B-186L. They could accommodate 1354 teu in its basic and 1684 teu in the lengthened version. Of 15 newbuildings for mainly German customers, the first ship in 1994 was the NORDPOL for Reederei "Nord", the final ship in 1998 delivered as HANS SCHULTE to Bernhard Schulte in Hamburg. The type B-170 with an intake of 1730 teu was created by further extension from 179 to 184 m. Of this type, 50 newbuildings were delivered, mainly to German clients. The series was launched in 1995 by the ELISABETH RICKMERS. Two ships of this design, too, were built by Vinashin in Haiphong in 2012/2013.

Another example of the transfer of technology to Asian shipyards was the further construction of the BV 1000 container ship type, which Bremer Vulkan delivered to Contship after taking over Rickmers' customer. In the course of its bankruptcy in 1996, Bremer Vulkan tried to make a last capital out of its once leading know-how in container shipbuilding by selling construction documents. In 1998, Shanghai Edward shipyard built SIGGA SIF, which in 2003 became the MERITO to the German shipping company tom Wörden in Oldendorf near Stade. In 2017 the ship was chartered by MSC, which still employed it in an inner-Greek feeder service in 2020. Remarkably, Rickmers received the only sister ship MARINE RICKMERS built in Shanghai in the same year and registered it for the Marick Shipping Co. in Liberia, later for the Rickmers Reederei in Luxembourg, which finally brought the ship under the German flag in 2009 and gave it to the Asian Spirit Steamship Co. Ltd (Spirit of Kolkata GmbH & Co. KG). At the beginning of 2020, the ship operated under the name BARRIER ex SPIRIT OF DURBAN in charter of Durban-based Ocean Africa Container Lines under Portuguese flag in a South Africa-Namibia service and was managed by Jebsen Shipping Partners in Jork/Germany (Fig. 4.62).

The Asian Spirit Steamship Co., which belongs to Bertram Rickmers' Brick Holding, launched a newbuilding programme on its own in 2016 for similarly sized feeder ships with a loading capacity of 1162 teu. SPIRIT OF KOLKATA and SPIRIT OF CHENNAI were taken over from China's Mawei Shipyard in 2019. The ships were managed by Jebsen in the beginning of 2020. The Mawei order list included six further newbuildings of the type for Rickmers, including SPIRIT OF CHITTAGONG, as well as two more for Marlink Schiffahrtskontor. In addition, the Asian Spirit Steamship managed

Fig. 4.62 For more than 30 years Rickmers managed the MARINE RICKMERS, later named NORASIA HAMBURG and built in China according to a design of Bremer Vulkan shipyard, which once had delivered the first units to a former customer of Rickmers Werft (Ralf Witthon)

the 714 TEU vessel SPIRIT OF DUBAI, built in Turkey according to Danish plans and chartered out for an employment in a Mediterranean feeder service.

4.4.4 Maersk's Ice-Class Feeders

Following the most unusual voyage undertaken by a container vessel so far, the newbuilding VENTA MÆRSK arrives in September 2018 at the Bremerhaven container terminal, fully loaden with a cargo of containers stowed up to three tiers high on the hatches and partly filled with frozen fish from the Aleutian Islands. The 3600 teu capacity ship is the first of seven units built at China's COSCO Zhoushan shipyard for all-year feeder operation in the Baltic Sea and therefore built according to high ice-class rules. This enabled the VENTA MÆRSK to sail through the North East passage from its Chinese builders to Europe on a route 2800 nm shorter than the usual way via the Suez canal.

Five of the 200 m length, 35.2 m breadth ships, named VOLGA MÆRSK, VAYENGA MÆRSK, VENTA MÆRSK, VUOKSI MÆRSK and VAGA MÆRSK, were in 2020 employed by the Maersk subsidiary Sealand Europe & Med in a weekly Germany-UK-Finland-Latvia-Lithuania-Russia-Poland loop rotating between the ports of Bremerhaven, Felixstowe, Kotka, Riga, Wilhelmshaven, Bremerhaven, Gdansk, Klaipeda, St. Petersburg, Rauma, Gdansk and Bremerhaven. In 2022, the ports of call were Bremerhaven, Felixstowe, Kotka, Gdansk, Klaipeda, Bremerhaven, St. Petersburg, Gdansk, Bremerhaven. Two ships of the 19 knots achieving type fast ran on a direct line from Rotterdam to St. Petersburg. After Maersk Line, like other major

Fig. 4.63 Seven ships of the VAGA MÆRSK type have been designed and built for operating in the Baltic. The first one, VENTA MÆRSK, was due to its high ice class able to undertake the positioning voyage from China through the North East passage (Ralf Witthohn)

shipping companies, suspended container shipments to Russian ports in the Baltic Sea and the Far East in the wake of the Russian war against Ukraine, VISTULA MAERSK and VAYENGA MAERSK switched to a Maersk/CMA CGM Europe-Canada service (Fig. 4.63).

5

Ferry Traffic

In continuation of road or rail connections, ferries operate mainly on the marginal seas and to islands. They are operated according to fixed timetables and often in a shuttle service between two ports. Established ferry routes are being maintained on the Baltic and North Sea, in the Mediterranean and Black Sea, but also off the coasts of other continents. Most of these ferries transport both commercial vehicles and passengers with their private cars and are therefore described as passenger and car ferries, in short ro pax ferries (Figs. 5.1 and 5.2).

5.1 Freight Ferries

In addition to ferries combining the transport of passenger and freight, there are on many routes employed ferries carrying cargo only. These ships often have accommodation for truck drivers, whose number is yet limited to 12 berths for classification reasons. Otherwise, the vessels would have to be considered as passenger vessels with adequate design conditions and safety equipment. A number of classic ferry routes exist between the UK and the European mainland. These include the short routes across the English Channel. However, also longer routes to the UK make sense, as they give truck drivers the opportunity for extended periods of rest. Germany's Flensburger Schiffbau-Gesellschaft (FSG) has specialised in the construction of ferries since 2000 and completed more than 50 such ships, thus achieving a global market leadership on this sector (Fig. 5.3).

R. Witthohn, *International Shipping*, https://doi.org/10.1007/978-3-658-34273-9_5

Fig. 5.1 The world's largest passenger ro ro vessels COLOR MAGIC and COLOR FANTASY maintained a daily connection between Oslo and Kiel, until in March 2020 the COVID-19 pandemic stopped, as most other passenger ferries, the ships, which resumed service in February 2022 after another suspension in December 2021 (Ralf Witthohn)

Fig. 5.2 The Baltic Sea is crossed by countless ferry routes. Off Helsinki, Tallink's STAR, seen departing for Tallinn, meets Viking Line's GABRIELLA arriving from Stockholm (Ralf Witthohn)

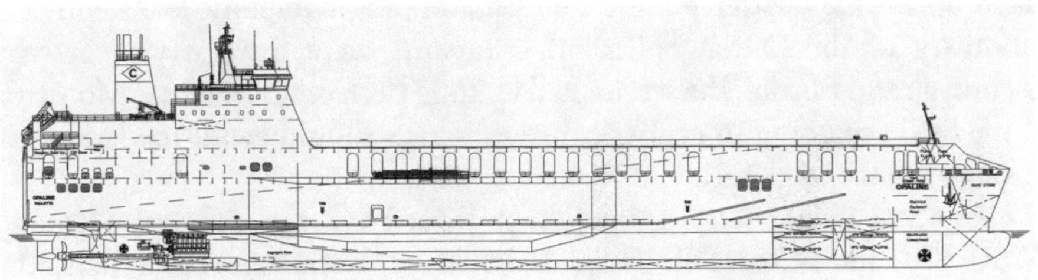

Fig. 5.3 By designing and building freight ferries such as the 2900-lane-metre OPALINE in 2010, FSG achieved a market leadership in the construction of this vessel type (FSG)

5.1.1 BORE SONG on Route to England

For the line from Teesport/UK to Zeebrugge/Belgium, UK's P&O Ferries in 2017 chartered the freight ferry BORE SONG. In 2020/2021, the ship provided a connection between the two ports six times a week with departure times around 8 pm in Teesport and 9 pm in Zeebrugge and arrival times at 1 pm in the port of destination. In early 2022, BORE SONG was running between Tilbury and Zeebrugge. As UK's main ferry operator P&O Ferries operated five routes in 2017, between Calais and Dover by 23 daily crossings, from Hull to Rotterdam and Zeebrugge by 7, resp. 6 sailings, from Cairnryan to Larne by six departures and from Liverpool to Dublin, which were connected 12 times per week. In 2022, the operator maintained three more lines from Teesport and Tilbury to Zeebrugge and from Teesport to Rotterdam.

The sister ship of BORE SONG, named BORE SEA, was leased in 2016 by Compagnie Luxembourgeoise de Navigation Roll-on-Roll-off (CLdN Ro-Ro) and took up service on the Rotterdam-Leixoes route. In total three ships operating there, of which PALATINE and CATHERINE called at

Santander as well as Leixoes, allowed three weekly departures in each direction with a voyage time of 64 h. The CLdN shipping company, known as Cobelfret, also offers connections from Rotterdam to Dublin and Killingholme. Furthermore, the shipping company's timetables included the ports of Zeebrugge, Esbjerg, Hirtshals, Gothenburg and London. In 2020/2021, BORE SEA undertook regular sailings between Hanko and Lübeck, in 2022 the Dutch-flagged ferry connected Hanko with St. Petersburg and Paldiski in a charter of Finnish company Transfennica, since 2002 part of the Amsterdam-based Spliethoff group.

BORE SEA and BORE SONG were delivered in 2011 by Flensburger Schiffbau-Gesellschaft (FSG) to Bore, the shipping branch of the Finnish Rettig group. After completion, the Finnish-flagged BORE SEA had initially taken up a time charter for the Finnish shipping company Transfennica, a subsidiary of the Dutch Spliethoff company, on a liner service between Zeebrugge and Bilbao. The service had in 2007 been established as "Motorway of the Sea" according to an EU concept, which should avoid the land route through France (Fig. 5.4).

Like other freight ferry newbuildings from FSG, the high ice class ships BORE SEA and BORE SONG are designed for the transport of trailers, high & heavy units, project cargo, cars, other vehicles and containers of all kinds. The ship type designated by the operator as Bore RoFlex is based on the

Fig. 5.4 The cargo ferry BORE SEA was in 2016 chartered for the Rotterdam-Leixoes link, which was still served in 2020 (Ralf Witthohn)

versatility of its cargo decks. Three of the six decks, the tank deck, the main deck and the weather deck, are fixed decks. Three additional car decks, one in the hold and two between main deck and weather deck, can be raised. This creates flexible deck heights that can be adapted to the cargo. Access for the rolling cargo is through a very wide stern ramp at the level of the main deck, from which fixed or movable ramps lead to the other levels. Even though the design was based on the requirements of the shipping company, the shipyard took the MAZARINE ro ro vessel type previously built four times for the Belgian shipping company Cobelfret as base for the design. Its main dimensions were adopted unchanged, so that the series effect of shipbuilding production was largely retained.

Similar to container ships, which designs are dependent on the dimensions of the standardized cargo units in terms of ship breadth, cargo hold depth or hatch sizes, the dimensions of the vehicle units to be transported form the basis for determining the loading deck dimensions in roll on roll off carriers. Subtracting the side tanks, the breadth of BORE SEA and BORE SONG allows the arrangement of seven spurs of trailers, each 2970 mm wide, on the tank deck and eight rows on the main and weather deck. On the weather deck a gangboard of 1.2 m remains on each side of the ship. The tracks on the tank deck are reduced by the descent from the main to the tank deck on the starboard side and by the foreship contour. On the tank deck, extending from the forward engine room bulkhead to the forward arranged fuel tank, the total lane metres amount to 549, corresponding to a capacity of 42 Mafi trailers. The clear height of the hold is 6.55 m. When the car deck is lowered, this height is divided into 4.6 m between the tank deck and the car deck and 2 m above. This car deck offers an area of 1665 m^2. The tank deck is designed for a load of 5 t/m^2, the car deck carries 0.18 t/m^2.

The main deck, which acts as freeboard deck and therefore can be made watertight by a cover over the downward ramp, offers 1078 lane metres for 82 Mafi trailers. Its area is only restricted by the fixed access ramp to the weather deck located on port side next to the engine casing. The main deck has a clear height of 7.4 m in its aft section up to frame 101.5, where there are no car decks. The second car deck, which extends from frame 101.5 to frame 209 and has an area of 1710 m^2, can accommodate 2.4 m high vehicles when lowered. For the main deck below, 4.15 m remain, but only up to frame 138.5. The forward area of the hold from frame 138.5 to the bulkhead is additionally divisible horizontally by the first car deck, in such a way that heights of 2.m on the two lower decks and 2.4 m on the uppermost deck result. This first car deck has an area of 1000 m^2. The weather deck offering 1236 lane metres for 93 Mafi trailers has the largest area. It includes the driveway from the stern ramp as parking space. The uppermost loading deck is only limited by the gangways on both sides and the engine and fan casings on port aft. The main deck carries 2.7 t/m^2, the weather deck 1.2–1.8 t/m^2, the two car decks each 250 tonnes in total. The total area of the movable car decks adds up to 4375 m^2. Of lane meters are a total of 2900 m available. The total Mafi capacity is 217 units. The ship's deadweight capacity on summer freeboard amounts to 13,625 tonnes.

Rolling cargo reaches the cargo decks via the two stern ramps designed for units weighing 2 × 93.5 tonnes. They have a total width of 21.25 m. The much larger starboard ramp to the main deck is 15.6 m wide, 15 m long and fitted with 3 m long flaps. Within the main ramp, the 4.4 m wide side ramp to the weather deck is located on port side. In continuation of this access, the fixed ramp has a width of 4.35 m and an angle of inclination of 6°. The fixed 4.38 m wide inner ramp to the tank deck has the same angle. The ramp systems were supplied by TTS Marine. The capacity of the heeling tanks working during handling operations is 1460 m^3. The ventilation system of the cargo holds is designed for 20 air exchanges in port and 10 air exchanges per at sea.

The cross-section shows a double hull vessel with side tanks of 2.7 m breadth up to the main deck and of 1.2 m width above. They serve as ballast tanks, together with the double bottom, which is limited to a height of 1.65 m for reasons of centre of gravity, for a total volume of 4660 m^3. The fuel tanks, which are separated from the outer skin, are concentrated in the fore ship area. Only the three upper decks of the superstructure are reserved for accommodation and wheelhouse, below there is a garage at weather deck level. On the upper deck are foundations for accommodating 60 40-ft and 45-ft load/load-off containers, 20 30-ft and 46 20-ft containers with stack weights of 61 and 48 tonnes respectively. 50 electrical connections are available to secure the power supply of reefer units. Containers can be loaded in two layers on all three fixed decks of the ship.

In order to provide sufficient height for the aft positioned main engine, a V-engine was chosen. The four-stroke diesel engine of type Wärtsilä 12V46F-CR (common rail) produces a CSR output of 12,000 kW at 600 rpm. It acts through a reduction gear on the four-bladed stainless steel variable-pitch propeller of 5.8 m diameter. On the design draught of 7.05 m and a CSR power of 81% (9720 kW) minus a shaft generator power of 300 kW, a service speed of 19 knots is reached. At this speed the ship will achieve a range of 9500 nm based on a daily consumption of about 43 tonnes and a fuel capacity of 1180 m^3. At a trailer draught of 6.3 m and 100% CSR, the vessel consumes about 36 tonnes per day at a speed of 18 knots (Fig. 5.5).

Fig. 5.5 Through the stern ramps the rolling cargo of the BORE SONG reaches the main, weather and tank deck (Ralf Witthohn)

5.1.2 Lengthening of PRIMULA SEAWAYS

The freight ferry PRIMULA SEAWAYS arrives at Bremerhaven Nordschleuse on 7 July 2016 with a formidable, only scantily repaired bow damage. The ship has suffered meanwhile corroding deformations of the bulwark in December 2015 in a collision with the car carrier CITY OF ROTTERDAM off the Humber estuary. But the final repair of the damage is not the reason for calling at Bremerhaven. The 199.8 m length PRIMULA SEAWAYS is to be extended in a complex shipbuilding operation. In the months before, a steel construction company has built the additional 30 m long hull section for PRIMULA SEAWAYS in the hall of an insolvent foundation builder for offshore wind turbines and carried it onto a barge by help of a rail transport system. To discharge the 1500 tonnes weighing section from the barge, a complicated action is necessary. The section is pulled over two beams into a floating dock of the repair yard German Dry Docks, which after lowering allows section to float up. At the same time, the cutting work through the cargo ferry in the floating dock of Lloyd Werft has begun. The hull is cut in way of the overhanging superstructure, first vertically and then horizontally at a distance 10 m below the superstructure. The zigzag cut avoids the need to cut through ramps inside the ship.

Only 9 days after arrival of the ferry, the actual lengthening operation begins. The floating dock of Lloyd Werft is lowered, the detached forecastle is pulled out of the dock and the newly built section is floated in. Then the dock is raised, all three parts are hydraulically pulled together on a 140 m long sliding way and welded together. On August 19, the operation, which took a total of 43 days, is completed, having lasted 12 days longer than originally planned and therefore costing the shipyard a loss. PRIMULA SEAWAYS undertakes trial runs in the North Sea and, after a short stopover in Cuxhaven, the next day sets course for Gothenburg to resume liner service to Brevik/Norway and Ghent/Belgium together with the sister ships BEGONIA SEAWAYS and MAGNOLIA SEAWAYS. After enlargement, the 229.8 m length PRIMULA SEAWAYS has a quarter more cargo capacity of 4650 lane metres for 307 trailers (Figs. 5.6 and 5.7).

Fig. 5.6 The new hull section, built in a hall, was towed to the docked PRIMULA SEAWAYS after having been transported on a pontoon and from this floated up in a floating dock (Ralf Witthohn)

Fig. 5.7 After floating in the additional section and pulling together of the three ship parts they were welded together (Ralf Witthohn)

In the case of increasing cargo volumes, capacity adjustments such as those of PRIMULA SEAWAYS are much quicker and easier to implement than the realisation of a newbuilding project. Consequently, ship enlargements represent a technical measure that has been undertaken over decades. Similar lengthenings had already been carried out by MWB shipyard in Bremerhaven in 2009 on the sister ships TOR FREESIA, TOR BEGONIA and TOR FICARIA. Within the conversion series the separated forebody of TOR FREESIA suffered a stability accident. A total of six of the ships were delivered between 2004 and 2006 by Flensburger Schiffbau-Gesellschaft. Another two units of the cargo ferry type featuring a loading capacity of 4076 lane metres for 262 trailers were ordered in May 2016. Customer was the owner of the Flensburg shipyard, the Norwegian Siem Group. The ships were chartered out with a purchase option to DFDS under the names GARDENIA SEAWAYS and TULIPA SEAWAYS. After delivery in July and September 2017 the vessels operated between Immingham and Rotterdam. In 2014 DFDS had received the 33,313 gt cargo ferries ARK GERMANIA and ARK DANIA with a lane length of 3000 m on three decks from Volkswerft Stralsund with a delay (Figs. 5.8 and 5.9).

One and a half years after the extension of the PRIMULA SEAWAYS, the passenger and car ferry PETER PAN was enlarged in the floating dock of shiprepairer German Dry Docks in Bremerhaven. The ferry had been built in 2001 by SSW shipyard in Bremerhaven and operated by Hamburg-based TT

Fig. 5.8 During the lengthening action of the TOR FREESIA the cut-off foreship lost stability (Ralf Witthohn)

Fig. 5.9 DFDS' German-built and Danish-flagged ro ro carriers ARK DANIA and ARK GERMANIA have been used for military transports of the Danish and German government (Ralf Witthohn)

Line under the Swedish flag between Travemünde and Trelleborg. In a complicated shipbuilding operation the ferry was extended by a 30 m long section to 220 m. This increased the capacity, distributed over three decks, from 2600 lane metres for 169 trailers to 3000 m lane metres for 195 trailers. The hull and superstructure were separated in a zig-zag cut, and the stern was moved aft on sledges, after the overhanging decks of the forecastle had been supported. The dock was then flooded and the aft section floated out, the new middle part floated in. The new section had been built by Pella Sietas shipyard in Hamburg and towed to Bremerhaven on its own keel (Figs. 5.10 and 5.11).

5.1.3 Odense's FSG Variant PAQIZE from Istanbul to Trieste

The FSG design of the DFDS ferries, measured at 32,289 gt, was used elsewhere in a somewhat smaller version of about 29,000 gt, when Odense Staalskibsvaerft was looking for new customers after having lost the container ship orders from Maersk Line, part of the joint parent company A. P. Møller. From 2009 to 2012, the Danish shipyard built MAAS VIKING, HUMBER VIKING, WESSEX, MERCIA, STRAIT OF MESSINA, CRAGSIZE, STRAIT OF MAGELLAN and BERING STRAIT according to the

Fig. 5.10 Before being floated out from dock, the cut-off stern of the combi ferry PETER PAN was pulled aft on sledges (Ralf Witthohn)

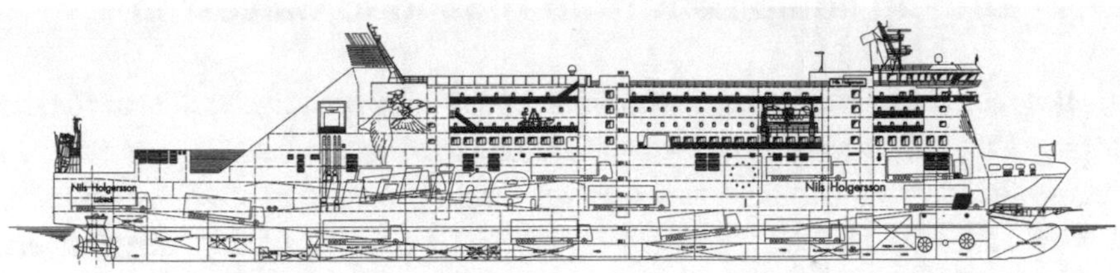

Fig. 5.11 NILS HOLGERSSON, the sister ship of PETER PAN, transports up to 169 trailers and 744 passengers on the Travemünde-Trelleborg route (SSW)

Flensburg plans as the last vessel type in the yard's long history. The 193 m length and 26 m breadth 29,429 gt buildings WESSEX and MERCIA delivered to British customers sailed under the names QEZBAN and PAQIZE for the Turkish logistics company EKOL between Istanbul, Trieste and Lavrion in Greece from 2013 onwards. In February 2021, DFDS employed the PAQIZE for their Tuzla-Patras-Trieste run, after a fire had damaged their GALLIPOLI SEAWAYS, and in December used the ship on their Cuxhaven-Iimmingham line, before it returned to the Mediterranean for operation between Mersin and Trieste in the beginning of 2022. The QEZBAN changed hands to Polish Ocean Lines and was operated by Stena Line as POL MARIS between Killingholme/UK and Rotterdam in May 2021, which route it still served in February 2022 (Fig. 5.12).

Fig. 5.12 The ro ro ferry PAQIZE, in Odense built as MERCIA according to a German design, created a freight connection between Turkey, Italy and Greece in 2013 and was end of 2021 for a short time employed on a North Sea route, before it returned to the eastern Mediterranean (Ralf Witthohn)

5.1.4 Drilling Equipment from Finland to Galveston

Releasing shrill warning tones, one special vehicle after the other rolls over the stern ramp of the Italian ro ro vessel ML FREYJA. During the handling operation in January 2018, the vehicles present a whole range of models of the Swedish manufacturer Sandvik, including the rubber-tyred 21-tonne LH 621 large-capacity excavator to the track-driven, partly remote-controlled horizontal drilling rigs Ranger DX700, Ranger DX800, Leopard SI560 and Tiger DG700. Some of the fire-red painted vehicles are destined for the US company Carmine-Romco in Carmine. They are parked in Bremerhaven to be transferred to a car transporter bound for Galveston. Right next to it, tractors from the Fendt company and, not far away, ten shunting locomotives of the Russian type TGM8KM loaded on flats for Cuba are waiting to be transported onward. While the special vehicles are driven from the main deck of ML FREYJA to their storage sites, dock workers are chauffeuring new Ford Transits backwards over the sidewards ramp of the ro ro ship onto the weather deck. Just 3 days earlier, the Turkish-built vans have been discharged from the car transporter PASSAMA of the German shipping company F. Laeisz and are now being distributed to England and the Baltic States. After eleven and a half hours, loading and unloading of ML FREYJA is completed, and the ship sets course for Harwich, which is reached after 18 h.

The ro ro vessel ML FREYJA, registered in Venice, transported the vehicles on its first voyage at the end of December 2017 from Finland via Cuxhaven to Bremerhaven in charter of Harwich/UK-based Mann Lines. The shipping company from eastern England transports yearly 500,000 tonnes of cargo comprising containers, semi-trailers, machinery, new cars, wood products, paper and steel between the UK, the northern continent, the Baltic States and Scandinava, partly for further transport to Russia and Central Asia. In Cuxhaven ML FREYJA had already taken over new cars from Austrian and south-German production for Estonia and Finland. The newbuilding replaced the Swedish ro ro vessel STENA FORETELLER in the 25 years existing liner service. In 2022, ML FREYA rotated between Paldiski/Estonia, Turku, Bremerhaven, Rotterdam, Harwich, Cuxhaven and Paldiski.

After delivery in June 2017 to Venetia's Visemar group, ML FREYJA had initially been employed in a 6-month sub-charter of Svenska Orient Lines between Gothenburg and Zeebrugge. The ship was designed by Trieste-based engineering firm NAOS and built at the Visentini shipyard in Porto Viro, Italy's largest private shipyard south of Venice. The ship was built according to the regulations of the Italian classification society RINA and awarded the ice class IA, which is indispensable for winter sailing to Finland. The 191.4 m length and 26.2 m breadth ship offers 2800 lane metres and can stow 414 container teu in two tiers on the weather deck. Two MAN B&W four-stroke engines with a total output of 12,000 kW accelerate the ro-ro vessel via two Rolls-Royce controllable pitch propellers to a maximum speed of 24 knots. In 2016, Visentini delivered the similar, but at 179.4 m shorter ro ro ships WEDELLSBORG and FRIJSENBORG to the Danish shipping company Dannebrog. Their lanes have a length of 2478 m. NAOS also designed two 210 m length, 28 m breadth ro-ro vessels with a capacity of 3084 lane metres, delivered in 2018 from China's Jinling shipyard to the Australian logistics group Toll as TASMANIAN ACHIEVER II and VICTORIAN RELIANCE II. Accelerated to 20.5 knots by two MAN B&W 9S40ME B9.5 engines of 2 × 10,215 kW, the ships, equipped with a scrubber system and ready for gas equipment, serve the Bass Strait between Victoria/Australia and Tasmania by connecting Melbourne with Burnie (Fig. 5.13).

Fig. 5.13 With rolling cargo loaded over the stern ramp and containers on deck the Italian-flagged, fast ice class ro ro vessel ML FREYJA maintained a weekly five-country service of UK-based Mann Lines in 2022 (Ralf Witthohn)

5.2 Ro Pax Ferries

The combination of passenger and freight transport, which is questionable from a security point of view, has been developed for economic reasons. It requires the classification of the vessels, also known as ro pax ferries, as passenger ships that have to be equipped with appropriate rescue facilities. Although their design principle, according to which vehicles roll onto the ship through gates via ship-owned or land-based ramps, corresponds to that of car carriers, the actual cargo does yet not consist of vehicles but of freight loaded on trailers, trucks or railway wagons. This results in a higher weight, which limits the number of loading decks with the effect of a much lower ship architecture. Ferries—especially freight ferries—are, however, also frequently used for the transport of new cars because of their rolling transhipment capabilities. Similar to the cruise industry nearly all passenger transport on ro pax ferries was temporarily suspended from March or April 2020 on due to the outbreak of the COVID-19 pandemic.

5.2.1 World's Largest Combined Passenger and Car Ferries on the Finland Route

In April 2021 Helsinki-based Finnlines announced the launch of the first of three 5800 lane metres hybrid ro pax vessels to be named FINNECO I, FINNECO II and FINNECO III at China's Nanjing Jinling shipyard as part of an EUR 500 million newbuilding programme to be commissioned from the end of 2021 on. In February 2022, commissioning of the series had yet not started. The programme includes the ro-pax vessels FINNSIRIUS and FINNCANOPUS starting employment on the Finland- Sweden route in 2023.

At preliminary 64,575 gt, these ships will supersede the companie's actual FINNSTAR class measuring 45,923 gt and employed on one of the longest ferry routes connecting Germany on the western Baltic coast with Finland in the east of the marginal sea. One of the longest ferry routes connects Germany on the western Baltic coast with Finland in the east of the marginal sea. The large cargo volumes resulting from Germany's function as Finland's most important trading partner and the length of the route were the base for the construction of the then world's largest ferries for the combined transport of passengers and cars. Helsinki-based Finnlines, owned by Italy's Grimaldi group, operate three of the ships on the 29-h crossing between Vuosaari near Helsinki and Travemünde allowing six scheduled departures weekly, while two ships are employed on other routes. FINNSTAR, FINNMAID, FINNLADY, EUROPALINK and NORDLINK were ordered in 2006 for a total of EUR 500 million from Italy's state-owned Fincantieri shipyard group. They were designed for a trailer capacity of 4200 m and 500 passengers. Due to the late delivery of the ferries by the Castellamare di Stabia site, Fincantieri granted a compensation of EUR 15 million. In March 2020 Finnlines temporarily suspended all passenger transport into Finland due to the COVID-19 pandemic, including the traffic from Travemünde and from Kapellskär. At the same time, cargo transportation was continued as usual, because the restrictions did not affect truck drivers. In early 2022, the transport of passengers had to be carried out according to constantly updated rules (Fig. 5.14).

At the beginning of 2018, Finnlines announced that the ownership of EUROPALINK would be transferred from the Grimaldi group to its subsidiary Finnlines and that the vessel would switch from the Livorno-Palermo to the Malmö-Travemünde route after being equipped with an exhaust gas cleaning system. On this route, the ferry had already been employed together with its sister ship NORDLINK after being commissioned. In early 2022, EUROPALINK was crossing the southern Bothnia Sea from Naantali/Finland to Kapellskär/Sweden joining NORDLINK, which had changed from the Malmö-Travemünde to the Naantali-Langnäs-Kapellskär service between Finland and Sweden in April 2018 following change of ownership and renaming to FINNSWAN.

The design of the FINNSTAR class takes into account the shipping company's demand for a higher passenger capacity than cargo share compared to

Fig. 5.14 As one of five Italian-built Finnlines ferries FINNMAID represents the world's largest ro pax type (Ralf Witthohn)

the previously deployed Hansa class ships. The increase of the trailer capacity from 3200 to 4200 m means an increase by over 30%, the passenger capacity from 114 to 500 persons by 440%. The greater focus on passenger transport, however, is at the expense of the deadweight capacity, which fell from 11,600 tonnes of the Hansa class to 9650 tonnes of the new ships. On the basis of a displacement of 26,800 tonnes, the weight of the "light ship" is cited as 17,000 tonnes. Unlike on the previous ships, the greatly enlarged passenger area has been located in the foreship area, and instead on the lowest of the four passenger decks there is a restaurant and other public rooms on the upper deck. Cargo handling has been extended by an option, which is yet not used on the Germany-Finland route. While on the older ships the roll-on/roll-off cargo was exclusively handled through the stern door on to the three cargo decks, the new class has access to a total of four cargo decks at the stern and at the bow (Fig. 5.15).

The newbuildings have 203 cabins for a total of 575 passengers. The maximum permitted passenger capacity is yet limited to 500. 57 cabins are located on deck 7 and offer space for 147 passengers. 44 cabins on deck 8 offer 148 beds, and on deck 9 there are 100 cabins for 280 passengers. Nine cabin classes, all of which are air-conditioned, offer floor spaces from 10 to 38 m^2. Two large luxury suites have windows to the front and side. The basic equipment of all cabins includes shower/WC, telephone and satellite TV. Cabins from 16 m^2 in size are equipped with refrigerator and safe. The most reasonably priced interior cabin is designed for six persons and comprises two rooms

Fig. 5.15 Replacing the FINNHANSA class by the Italian-built FINNSTAR type ships meant the introduction of a new ro pax concept with a stronger focus on passenger transport (Ralf Witthohn)

with a total size of 20 m², a connecting door and two baths. The most expensive cabin is a 38 m² suite with four windows. Specially designed cabins of 18 m² are available for three handicapped passengers. Some cabins have laminate flooring instead of carpet, especially for use by allergy sufferers. Leisure facilities are concentrated on deck 11. These include saunas, whirlpools and the fitness room with panoramic windows. The conference room, buffet restaurant, two bars, Hansa Lounge, Sailor's Shop and the play area for children as well as video games and slot machines are also located on the same deck (Fig. 5.16).

The sun deck behind the bridge is also designed as a landing pad and can accommodate helicopters up to the Super Puma size with a weight of 8.6 tonnes. All public areas are monitored by recording video cameras. The ships comply with the SOLAS regulations of the "Consolidated Edition 2001". According to this, sufficient rescue equipment is required for 540 persons (500 passengers and 40 crew members). This is met by 300 available places in each of the partially closed lifeboats—two each on the starboard and port side—with a capacity of 150 persons each. Six life rafts can each take 25 persons and are launched with the help of davits. The rescue concept provides for three assembly stations, namely in the stairwell on decks 9, 8 and 6.

The equipment of the vessels includes special ro ro systems with hydraulically operated ramps and gates, including a bow gate at main deck level (deck 3) with a three-part bow ramp of 19.4 m length (+2 m flaps). It creates a passage width of 5.6 m. Above, the bow visor provides access to cargo deck 5. The one-piece main stern ramp to deck 3 has a length of 12.5 m (+3 m) and

Fig. 5.16 The junior suite on the FINNSTAR class ships can accommodate up to four people on 32 m² (Ralf Witthohn)

provides a passage width of 23 m. The stern ramp above, consisting of two parts, allows an 18 m wide access to the upper deck areas. The connection from the main deck to the lower deck is created by two fixed ramps, which are arranged in the front and rear area of the midship line. They are closed by two 52 m long, two-part ramp covers suspended at the sides. The entrance of the rear ramp is 4.2 m wide, the front ramp has a width of 3.7 m. In their closed position these covers ensure the watertightness of the main deck. Another fixed ramp provides the connection between loading decks 2 and 3, which is also closed by a movable cover 26 m long and 3.3 m wide and which in the closed position provides additional cargo space. Between decks 5 and 7, a liftable tipping ramp of 50 m length and for vehicles up to 4.4 m wide operates on the port side. When folded up, it ensures that the deck is weathertight (Fig. 5.17).

The total lane length is distributed over 371 m on the second deck, 1457 m on the third deck, 1534 m on the fifth deck, 1116 m on the seventh deck and 302 m on the eighth deck. For trailers, of which a maximum of 276 units can be loaded, are 4256 lane metres available. They are secured by means of specially designed lashing trestles. There are 100 plugs for reefer cargoes. The drivers have the opportunity to check the cooling of their trucks once a day. The cargo decks are designed for distributed loads of 2 tonnes/m^2, with the exception of main deck 3, which carries 4 tonnes/m^2, and the hoistable deck 8 laid out for 0.3 tonne/m^2. The rear, lower part of the superstructure is

Fig. 5.17 Seen from forward is the uppermost cargo deck of the FINNSTAR including hoisted decks for car transport and an open deck area for trailers behind (Ralf Witthohn)

designed as a garage, where mainly cars and motorcycles of passengers can be parked on deck 7. The aft lock is provided by two doors suspended at the top. A triple-divided, liftable car deck with a movable access ramp on the port side is used for the horizontal variation of this deck area, for example for the optional placement of cars or buses. On starboard side a fixed ramp leads to the fixed part of deck 8. Directly adjacent to this garage on port side of deck 7 is the access to the reception, where passengers arriving with their own vehicle enter the hotel area. Passengers without a vehicle are driven to the entrance in front of the reception by shuttle buses of the shipping company. Although equipped with bow and stern ramps, the ships in Helsinki and Travemünde are loaded and unloaded via the stern ramps.

An automatic system has been installed to compensate for heeling during loading operations. Furthermore, the ferries are equipped with stabilizers to reduce ship movements in rough seas. The propulsion system of the ferries is of conventional design. It comprises four Wärtsilä 9L46 diesel engines. Their power of 10,400 kW each is transmitted to the two propeller shafts through two double reduction gears in a ratio of 3.327. The maximum speed is 26.7 knots, the service speed 25 knots. Consumption was initially specified as 199 g/kWh heavy fuel oil with a maximum of 1.5% sulphur. In 2015, the three vessels sailing between Finland and Germany were retrofitted with scrubbers during a shipyard overhaul in Poland, but the ships burnt diesel oil in German waters, where emissions of the installed Ecospray system into seawater was not allowed.

During the crash stop, the ship came to a standstill within 2 min and 49 s over a distance of 1173 m. As was to be expected with such a high engine power, vibrations were noticeable, which occurred on FINNSTAR in the area of the navigation bridge, especially in shallow water, but became significantly less after the propellers were replaced. The two gearboxes are equipped with power take-offs (PTO) for two shaft generators. In addition to the shaft generators, three diesel engines of 1200 kW output each serve as power supply. The high manoeuvrability of the ferries, which despite their size usually manage without tug assistance in ports, is ensured by active rudders. Two electrically driven transversal thrusters of 2 × 2000 kW forward act as manoeuvring aids (Fig. 5.18).

Fig. 5.18 Four main engines act through two double reduction gears on two propellers on the conventionally driven ferry FINNSTAR (Ralf Witthohn)

5.2.2 Superfast Ferries in Europe and Australia

A seven and half hour night trip takes the car and passenger ferry SKANIA across the southern Baltic Sea and its main shipping route between Bornholm and the Swedish south coast, before it reaches the breakwaters of the Swinoujscie harbour entrance. Poland's Unity Line employs SKANIA together with POLONIA on the 93 nm long route from Ystad to Swinemünde. Passengers and truck drivers are offered a total of four departures daily from each port. Built in 1995, the 20 knots running POLONIA can accommodate 918 passengers in 74 quadruple, 46 triple and 92 double cabins. The main deck can accommodate vehicles over a length of 604 m, the upper deck on 1112 m, thus offering space for 150 cars and 120 trucks or 600 m of rail wagons. The 27 knots reaching SKANIA carries 1397 passengers, of which 600 find place in 196 cabins. Three cargo decks of 260, 837 and 838 m length offer space for 260 cars and 50 trucks on three loading decks. At the beginning of 2018, Unity Line also operated the rail ferries JAN SNIADECKI and KOPERNIK on the same route, while the combined ferries GRYF and GALILEUSZ together with the rail ferry WOLIN served the Swinoujscie—Trelleborg route. KOPERNIK ex ROSTOCK was in 2019 sold to Mediterranean

Fig. 5.19 The 1995-built SUPERFAST I entered a service of Poland's Unity Line on the Swinouscie—Ystad route under the name SKANIA in 2008 (Ralf Witthohn)

Ships Breaking, sold on to Levante Ferries and in February 2022 still laid in Piraeus as SMYRNA due for conversion for a ferry service between Thessaloniki and Izmir (Smyrna in ancient Greece) (Fig. 5.19).

When the overland routes between central and south-eastern Europe were cut by the Yugoslavian wars from 1991 onwards, Greek's Attica group of the Panagopoulos family of shipowners commissioned Schichau Seebeckwerft in Bremerhaven to design a fast ro pax type, which should be able to make the Adriatic route between Greece and Italy within 1 day. In order to meet the shipping company's requirement for high continuous speed, the shipbuilder, experienced in ferry shipbuilding over decades, undertook innovative development and design work, in which the choice and arrangement of the propulsion system was of decisive importance.

The propulsion concept did not only have to solve the problem of a restricted engine room height due to the vehicle guidance through the stern ramp, which is a general problem to tackle in ro ro ship design. In the case of the 23,663 gt Superfast newbuildings this posed a particular challenge when the towing tests revealed, that a very high engine power of over 30,000 kW with correspondingly large engines would be necessary. V-engines were therefore selected, which are usually preferred on ro ro ships due to their principally lower design. The four main Sulzer 12AV40S engines achieve a total output of 31,680 kW. Two of them each act on a controllable pitch propeller through couplings and a double reduction gear on the two propeller shafts. The engines were supplied by the "Zgoda" works in Swietochlovice. They gave the 173.7 m length, 24 m breadth ships a speed of 27 knots, which had rarely reached by conventional ferries until then and made them the forerunners of a fast ferry generation.

In order to keep handling times as short as possible, the ferries can be loaded and unloaded through the bow or stern. The bow ramp has a length of 14 m and a width of 5 m. The two stern ramps are 9 m long and 8 m, resp., 5 m wide. Shortly before the completion of SUPERFAST I and SUPERFAST II in 1995 the sudden sinking of the ferry ESTONIA in the Baltic Sea causing the death of 852 passengers and crew members raised deep concern about the principal reliability of bow doors and let the builder retroactively reinforce the closing devices of the door. After delivery, the first duo was deployed between Patras, Igoumenitsa and Ancona. The ship' classification, initially carried out by the companies Hellenic Class of Shipping (HRS) and American Bureau of Shipping (ABS), changed over the years and was in 2017 taken over by the Norwegian-German society DNV-GL (Fig. 5.20).

The bankruptcy of Schichau-Seebeckwerft in the wake of the Bremer Vulkan collapse let Attica order follow-up constructions at Kvaerner Masa Yards in Turku/Finland. There, the SUPERFAST III and SUPERFAST IV, enlarged to 194.3 m in length, 25 m breadth and 29,067 gt, were put into service in 1998. The number of cylinders of the Sulzer engines was increased from 12 to 16, so that the total engine power was increased to 42,240 kW, their speed to 28.5 kn. Afterwards, Attica ordered again in Germany. From 1999 onwards, the shipping company had the 32,728 gt SUPERFAST V, SUPERFAST VI, SUPERFAST VII, SUPERFAST VIII, SUPERFAST IX and SUPERFAST X built at Howaldtswerke-Deutsche Werft (HDW), according to a once more lengthened design of 204 m. HDW, however, was only able to complete the construction programme late due to problems with the propulsion system. Two further newbuildings, SUPERFAST XI and

Fig. 5.20 After delivery in 2001 and start of the Patras-Ancona ferry service, the propulsion system of SUPERFAST V underwent a refit at Lloyd Werft in Bremerhaven in January 2002 (Ralf Witthohn)

SUPERFAST XII, were delivered by Lübecker Flenderwerft in 2002. On these newbuildings four 12V46 Wärtsilä engines, each producing 12,000 kW at 500 rpm, drive two shafts with Rolls Royce propellers.

The background to Attica's expansive fleet policy was the plan to extend the fast ferry idea, born out of necessity for the Adriatic route, to other ferry links in the North Sea and Baltic Sea, such as the Rostock-Helsinki connection. In 2008/2009, the shipping company awarded Nuovi Cantieri Apuania in Marina di Carrara the order to build two further combined ferries, which were again christened SUPERFAST I and SUPERFAST II. With similar main dimensions of 199.1 m and 26.6 m, these ferries require only about half the engine power—24,000 kW—to reach a speed of 24.2 knots. The ships have 102 cabins, seats for 938 passengers and a vehicle capacity of 2505 lane meters. In 2017 Attica employed, in addition to these latest ships, the Lübeck buildings SUPERFAST XI and SUPERFAST XII. The ferry operator served the main route Patras-Igoumenitsa-Corfu-Ancona and the lines Patras-Igoumenitsa-Corfu-Bari and Patras-Igoumenitsa-Venice, the latter together with Greek's Anek Lines. The SUPERFAST XII was in 2018 sold to Italy's Grimaldi company and was running between Palermo and Livorno in early 2022 under the name CRUISE AUSONIA (Fig. 5.21).

While SUPERFAST I, SUPERFAST II and SUPERFAST XI were still active on their traditional Greece-Italy route in 2022, the other Superfast ferries had left their original routes. After operating as EUROSTAR ROMA from 2004, the first Superfast ferry 4 years later became the SKANIA of Unity Line. The first SUPERFAST II, after an employment as SPIRIT OF TASMANIA III in Australia, operated from 2006 as MEGA EXPRESS FOUR on behalf of Corsica Ferries on the Toulon-Ajaccio line, the

Fig. 5.21 After the takeover of the Rostock-Helsinki route by Tallink, the 2001-built SUPERFAST VIII sailed there under the Estonian flag and from 2011 as STENA SUPERFAST VIII between Belfast/Ireland and Cairnryan/Scotland (Ralf Witthohn)

SUPERFAST III from 2002 as SPIRIT OF TASMANIA II between Melbourne and Devonport in Australia. In May 2016, the ferry rolled in heavy seas to such an extent that vehicles on the loading deck broke loose causing severe cargo damage. SUPERFAST IV also moved to Australia in 2002 as SPIRIT OF TASMANIA I. SUPERFAST V was renamed CAP FINISTERE in 2010 serving the Portsmouth-Bilbao route in 2016, in 2022 under the name GNV SPIRIT. SUPERFAST VI as CRUISE OLBIA sailed on the Civitaveccia-Olbia line in 2016, but in 2022 was again engaged on the Ancona-Patras link as Grimaldi's EUROPA PALACE. SUPERFAST VII and VIII operated as STENA SUPERFAST VII and VIII on the short distance Belfast-Cairnryan link. SUPERFAST X operated as STENA SUPERFAST X between Dublin and Holyhead, but in 2020 switched to the French flag as A NEPITA for employment by Corsica Linea from Marseille to Ajaccio. SUPERFAST IX in 2008 started a service as ATLANTIC VISION for Marine Atlantic between Argentia and North Sydney on Nova Scotia, which it still maintained in 2022.

5.2.3 SUPERSPEED I and SUPERSPEED II Across the Skagerrak

At noon time, SUPERSPEED 1 departs from the quay in Hirtshals, turns round by 180°, passes the long breakwaters of the harbour entrance a quarter of an hour later and heads straight for Kristiansand at an almost constant course of 295°. The ferry connection between Skagen, the northernmost headland of Denmark, and the southernmost Norwegian port is, by 70 nm, the shortest connection between the two countries. At an average speed of 25 knots, SUPERSPEED 1 can complete the voyage in 4 h. The high speed and a discharge and loading time of only 1 h allow two trips per day in each direction. While the fast ferry crosses the Skagerrak, vehicles roll onto the sister ship SUPERSPEED 2 in the neighbouring dock of Hirtshals, from where the SUPERSPEED 2 leaves for Larvik, 88 nm away. This ferry, too, makes four crossings a day. Both ships start their service in Kristiansand and Larvik at 8 a.m., arriving at 24 h and at 2 h from the last crossing. After the outbreak of the COVID-19 pandemic, Color Line Cargo continued to operate the route, including the carriage of drivers, while passenger transports were initially restricted (Fig. 5.22).

Although attempts have been made to increase the attractiveness of Baltic Sea ferry traffic by increasing ship speeds, at the latest since the service entry in 1977 of the gas turbine-driven, over 30 knots achieving FINNJET, operators still maintain their services on the basis of comparatively moderate

Fig. 5.22 The 27 knots achieving SUPERSPEED 1 and SUPERSPEED 2 provide fast connections between Norway and Denmark for passengers, cars and trucks (Aker Yards)

turnaround times. For example, the cruise ferries COLOR FANTASY and COLOR MAGIC, also built by Aker Yards in 2004 and 2007 for the Kiel-Oslo route, are able to maintain their timetable at a service speed of 22 knots. The conventional ferries PETER WESSEL and CHRISTIAN IV built in 1981/82 sailed at speeds of 19/20 knots, until they were replaced by the SUPERSPEED ferries.

SUPERSPEED 1 and SUPERSPEED 2 were in February and June 2008 delivered by Aker Yards in Finland to Norway's Color Line. The 34,231 gt ferries were designed for a speed of 27 knots on the routes Larvik-Hirtshals and Kristiansand-Hirtshals, while carrying up to 1928 passengers, 117 trailers or 764 cars. The ships were built as part of a EUR 900 million investment programme, which included a rebuilding of the port facilities. Due to the relative shortness of the Skagerrak routes, the absolute gain in time remains small despite the increase in speed by about one third compared to the replaced ships. Only the reduction in the number of night trips has significantly shortened the overall travel times. On the old ships, they had been extended to a duration of up to 13 h in order to offer acceptable departure and arrival times outside night time. The high speed requirement resulted in an extremely slender lines design of the 213 m length, 25.8 m breadth, 20.4 m depth ferries with a block coefficient of 0.5961. The "light ship" weighs 14,682 tonnes. Propulsion is performed by four Wärtsilä 9 L46 engines installed side by side and each putting output 9000 kW at 500 rpm. They are separated from two Flender GVL 1850 double reduction gears by a transverse bulkhead. Outside the hull the shafts are supported by two bearings. The propeller shafts drive two Rolls Royce Kamewa controllable pitch propellers. The design of the two SUPERSPEED ferries did not only include an increase in ship speed but also an acceleration of handling procedures due to the fundamental changeover to ro ro handling taking place simultaneously on two levels instead of one.

Finnish equipment supplier MacGregor, later part of Cargotec, delivered the ramp, gate and deck systems of the ferries as well as the additional port-related upper access to the ro ro facilities and the passenger gangways. The ro ro components of the vessels include the bow gate, the bow ramp and gate for the lower cargo deck, a forward gate to the upper loading deck, the folding bow ramp, the stern ramp and gate to the lower cargo deck, a tiltable inner ramp between the main cargo decks 3 and 5 and a total of 12 movable parts of the liftable car deck 6. Six of these can be used as inclined access ramps. The watertight bow door, weighing about 40 tonnes, is divided in the centre line. Each of the two sections is moved parallel to the hull by double-acting hydraulic cylinders when opening or closing. The self-locking closures of the bow door ensure that it remains closed in case of hydraulic failure. The 18.7 m long and 4.5 m wide bow ramp consists of two main sections and an additional folding end which is permanently connected to section 2. It provides the connection from the main deck to the shore. During the closing process section 2 detaches from section 1 and is secured below deck by hydraulic devices. When closed, section 1 of the ramp forms a watertight door in the collision bulkhead. Section 1, which is attached to the main deck, is opened by directly acting hydraulic cylinders which rotate the section until the two ramp sections are connected and extended. The maximum roll trailer load of the bow ramp—as well as the stern and inner ramp—is 40 tonnes. Three-axle road trailers can weigh up to 44 tonnes, tractors up to 26 tonnes.

The design with two fixed main cargo decks, each with its own entrance and exit on different levels, as well as an additional retractable car deck accessible from the upper fixed deck, allows flexible loading variations. On a maximum of 2036 lane metres, of which 1007 m extend on the lower deck and 1029 m on the upper deck, a maximum of 117 trailers, trucks or buses can be placed in seven rows of 3 m width and of 4.6 m height. In this loading case, the hydraulically and rope-operated car deck of 330 mm constructional height and a total area of 2750 m^2 is in the upper position. In the lower position there exists a free deck height of 2 m and a track length of 1035 m for 204 cars with an individual maximum weight of 1.7 tonnes. The deck consists of a total of 12 panels—six on each side of the central casing—with lengths ranging from 20.1 to 23.8 m. The first, fourth and sixth panel can be tilted to serve as an ascent and descent to the car deck. The fourth panel can be tilted forward or backward to provide access or exit if only part of the car deck is to be used. With a surface load of 0.2 t/m^2, each panel can lift cars with a maximum weight of 27 tonnes into the transport position.

The ship can load a total of 764 cars in 2.4 m wide rows over a length of 3609 m, of which 1418 m for 284 units are offered on the first cargo deck and 231 on the second deck at a length of 1156 m. Examples of further loading variants are 89 trucks and buses together with 284 cars, or 60 trucks and buses together with 525 cars. The rope-operated inner ramp of 44.6 m length and a 3 m travel width between decks 3 and 5 carries a maximum of 100 tonnes. When closed, the ramp ensures the weathertightness of the upper deck. It can be lowered either at the front or rear, thus allowing access from the bow or stern.

Above the cargo decks are three passenger decks 7, 8 and 9. A major share of the public areas are the lounges and bars of the so-called Bluefins with an area of 1018 m^2 and 365 seats extending over the levels of decks 7 and 8 at the aft end of the superstructure. On the seventh deck the first Street leads from there to the so-called Centrum, in the rear area of which the passenger access from land is located. Adjacent to the Centrum are the fashion (100 m^2) and delicatessen shops as well as the duty-free shop (768 m^2). On deck 8 the way leads from the Bluefin to the business class with 93 seats. Internet, television, radio and films are available here. In front of the Bluefin is the cafeteria area with 204 seats and the buffet restaurant with 312 seats. The Voyager class in the aft section of deck 9 offers 261 seats in four rows of three seats each. In the central area of this deck there are 71 crew cabins, and in front of them 54 cabins for truck drivers. These two accommodation areas are separated by a passage extending across the entire width of the ship, at the ends of which there are ports in the shell plating. The total passenger capacity is stated as 1928 persons. The crew comprises 14 officers and 57 crew members. Four closed boats and an evacuation system serve as rescue means (Fig. 5.23).

Fig. 5.23 The fast ferry character of SUPERSPEED 1 and SUPERSPEED 2 isreflected in their architecture (Aker Yards)

5.2.4 Toys from Hal Far

Since 1974, billions of toy figures have seen the light of day from hundreds of injection moulding machines at the Maltese Playmobil factory. Because of the low wage level, the German company headquartered in Zirndorf/Bavaria chose the location in the south of the Mediterranean island, first in Bulebel and later in Hal Far near the Marsaxlokk container port. Even at the Malta airport, guests are presented Playmobil figures in a play corner. In the Hal Far industrial area, which is only a few kilometres from the airport and freeport, there is furthermore an adventure park run by the company, which by its 900 employees is one of the largest employers in Malta. The island location limits the export of the toys to be done by air or by sea. The latter includes exports via the free port of Marsaxlokk as well as transport on the ferries calling at the island (Fig. 5.24).

One of the ferry companies that have included Malta in their schedule is Italy's Grimaldi group. Between 2007 and 2010, the group had a series of eight ferries built by Nuovi Cantieri Apuania in Marina di Carrara/Italy. The first ship that entered service in a bare-boat charter of Grandi Navi Veloci (GNV) under the name CORAGGIO was followed by AUDACIA, TENACIA, SUPERFAST I, LISCO MAXIMA, SUPERFAST II, FORZA and ENERGIA. CORAGGIO started a regular service between Genoa, Tunis and Malta and later switched to the Genoa-Palermo service. In 2020, the vessel, renamed ATHENA SEAWAYS in 2013, was operated between Klaipeda and Karlshamn in charter of Danish ferry line DFDS. In 2022, the ATHENA SEAWAYS still sailed the Baltic Sea on a ferry link from Klaipeda to Kiel under the Lithuanian flag. Of the other sister ships, AUDACIA, delivered in September 2007 to Grimaldi, initially served the route Livorno—Palermo. After sale to China's Rizhao Haitong Ferry the vessel switched in 2014 as RIZHAO ORIENT to a line from Rizhao in China to Pyeongtaek on the Korean west coast, which it still served in 2022. TENACIA sailed between

Fig. 5.24 The container port of Marsaxlokk eases the rapid export of products from the near-by Maltese industrial park Hal Far (Ralf Witthohn)

Barcelona and Palma de Mallorca in early 2016, and did so 6 years later, after a temporary job on the Palermo-Genova-Naples run in 2021. SUPERFAST I and SUPERFAST II are long-term employed on Greek's Superfast Ferries homeroute between Patras, Igoumenitsa and Bari. FORZA was in 2020 engaged on a Spanish line from Mahon to Palma de Mallorca, and in 2022 connected Civitaveccia, Palermo, Naples and Genova on behalf of Naples-based ferry operator SNAV of the Grimaldi holding. At the same time, the Grimaldi-owned ENERGIA served as REGINA SEAWAYS under the Lithuanian flag in a DFDS service between Dunkerque/France and Rosslare/Ireland.

As typical ro pax ferries, the newbuildings combine the simultaneous transport of passengers and vehicles. The maximum number of passengers on CORAGGIO is 473, the number of vehicles 190 trailers plus 113 cars. Passengers are accommodated on decks 5 and 6 below the bridge in the forward area of the ship. The vehicles roll on the same level via two stern ramps of different widths onto the main loading deck 3 and via an inner ramp directly onto the upper deck 4, which forward area extends to below the superstructures. The lower decks 1 and 2 are reached from the main cargo deck by internal ramps. The passenger access is also located at the stern, on the very outside on the starboard side of the larger stern ramp. A staircase is connected to it aft, from which passengers can reach deck 4, until a corridor along the boardside leads them to the foreship. The main cargo deck offers 13 trailers to be placed behind each other in eight rows. The same number can also be parked next to each other on the upper deck 4, while there is still room for 12 trailers in the length. The high-speed lines design resulted in a service speed of 23 knots at an engine output of 24,000 kW. During the test runs CORAGGIO achieved 26 knots (Fig. 5.25).

Fig. 5.25 After delivery, the ro pax ferry CORAGGIO started service on the route Genoa—Tunis—Malta (Ralf Witthohn)

5.2.5 Battery-Powered Double-Ended Ferry GLOPPEFJORD

On 30 November 2017, Turkey's Tersan shipyard in Altinova/Yalova delivers one of the first battery-powered ships to the Norwegian shipping company Fjord1. Before the start of a regular ferry service between Anda and Lote, GLOPPEFJORD has to undertake a voyage of almost 4000 nm from the Black Sea to the Norwegian fjords. The special propulsion system of the innovative ferry is also its handicap for this ferry trip, as the regular battery operation allows only short distance operation and charging for a few miles. Therefore, the ship is additionally equipped with two diesel generators of 2 × 500 kW power for a speed of 13.5 knots. Only these allow the ship to make the sailing to its home port, regularly taking 2 weeks, but extended to more than a month due to the winter conditions. In mid-January, GLOPPEFJORD carries out its first test run off Anda. At this time, the sister ship EIDSFJORD is on transfer voyage from the builder and waiting in Vigo for favourable weather conditions to cross the Bay of Biscay.

The 106 m length and 16.8 m breadth hybrid ferries are equipped with the Siemens BlueDrive PlusC system of 2 × 520 kWh output. The electrical energy drives two azimuth propellers from Rolls Royce giving the ship a speed of 13.5 knots. The batteries are recharged after each crossing on the 2.4 km long route by an automated system of the German company Stemmann. At 590 V/1650 A it has a power of 1.5 MVA. The ferries can carry 349 passengers, 120 cars and 12 trailers. Their mooring is done automatically by a vacuum system, as well as the navigation of the ferries is automated. The optimal crossing speed is based on wind and current conditions and the loading state of the ship (Fig. 5.26).

Even before the first battery-powered ferries were put into service, Norwegian companies, which are particularly environmentally conscious in the narrow fjord geography, relied on LNG as fuel, especially because the

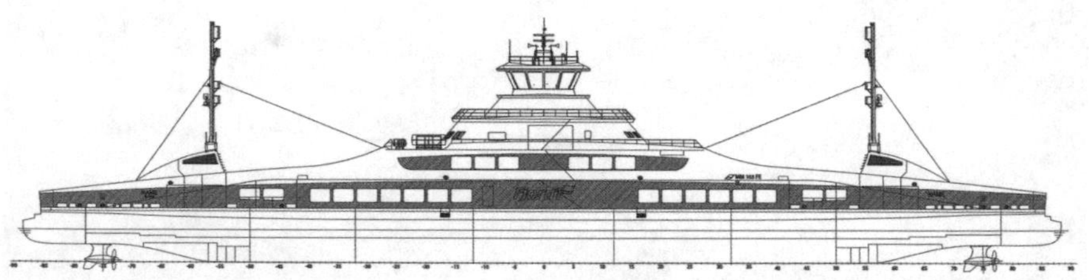

Fig. 5.26 The battery driven fjord ferry GLOPPEFJORD, built in Turkey according to a Norwegian design, achieves a speed of 13.5 knots (Tersan)

natural gas is produced in the country and well available. In 2013, two gas-powered newbuildings were delivered by Poland's Remontowa Shipbuilding, well experienced in the construction of LNG burning, diesel-electrically driven ferries, to Norwegian ferry operator Norled for employment on the Stavanger-Tau line. Realized according a design of the Norwegian company LMG Marin the 123.2 m length, 17.65 m breadth double-ended ferries RYFYLKE and HARDANGER are laid out for 550 passengers and 165 cars or alternatively 18 trucks. Propulsion energy of the 11 knots achieving ship is generated by four Mitsubishi generator sets. Two of them burn LNG, two CNG.

Remontowa shipyard in 2015 as well built an LNG-powered, diesel-electrically driven ferry, named SAMSØ, for a line from Saelvig on the isle of Samsø to Hou in Jutland. Renamed PRINSESSE ISABELLA shortly after delivery, the first such Danish LNG newbuilding, operated by the Samsø Kommune, is accelerated to 14 knots by four Wärtsilä 6L20DF engines acting on four azimuth propulsion units of German manufacturer Schottel. The 40 m^3 gas and 80 m^3 diesel oil tanks of the 99.9 m length, 18.5 m breadth double-ended ferry, which operational concept includes the carrying of the Samsø fire brigade, are bunkered in the port of Hou. The LNG is delivered by trucks from Rotterdam. There are yet mid-term plans to switch to biogas fuel produced on the isle.

5.2.6 Electrically Driven Ferries AR VAG TREDAN and ELLEN

In 2012, STX France in Lorient as part of Korea's STX group developed, in cooperation with Stirling Design International (SDI), the first electrically driven port ferry. Powered by supercapacitors AR VAG TREDAN was to be employed in the port of Lorient to carry up to 150 passengers. The EUR 3.4 million costing ferry evolved from the Ecorizone research programme. Supercapacitors supply the energy required for propulsion and on-board operation. They are charged through a shore connection after a relatively short period. Unlike batteries, the lighter capacitors supply the energy without an electrochemical process. They can store the energy much faster and are able to repeat this many times without requiring maintenance. Their French developers assume more than 1 million possible charging processes and a service life of the capacitors of 15 years. The contract was awarded by Lorient Agglomération, based on the decision to use only zero-emission ferries with the aim to protect residents, ferry passengers and the environment. A total of

128 Batscap supercapacitors were installed on the ferry. Their aluminum cathodes have pores of nanometric size with a surface area of 3000 m^2 per gram of material. They store the motive energy for one round trip. The 400-volt connection at the bow of the ship is made manually via a two-pole plug with mechanical support from a crane within 4 min. During the shuttle trips, which are carried out every 15 min, 28 times a day, the super capacitors feed two Masson Marine azimuth propellers of 2 × 75 kW output, giving the 22.1 m length, 7.2 m breadth hull of the electric ferry a speed of 10 knots. Technical problems which developed in 2019 could not be solved since, and the AR VAG TREDAN was lying idle in Lorient in 2022.

In July 2019 Søby Vaerft at Søby/Denmark delivered the purely electrically driven ferry ELLEN to the Æro Kummune at Ærosköbing/Denmark. The 996 gt ferry was built on base of a 57 m length, 12.8 m breadth hull supplied by Turkey's Kedat shipyard. The ship runs on power from batteries charged in harbour and sufficient to cover a 22 nm long, 55 min taking route from Søby on the island of Ærø to Fynshav in Denmark. ELLEN is laid out to carry 31 cars or 5 trucks and 198 passengers. The EUR 21.3 million costing so called E-Ferry project was funded by the EC and the industry as part of the Horizon 2020 research and innovation programme.

5.3 Railway Ferries

As a type of ship almost as old as railways themselves, rail-bound ferries have lost their importance due to the construction of fixed crossings, for example through the English Channel, across the Great Belt or the Øresund. The increase of individual car traffic and the strongly expanded use of trucks has also reduced the volume of passengers and cargo that could be transported by this particular mode of transport.

5.3.1 Swedish SKANE from Trelleborg to Rostock

Captain and officers of the railway ferry SKANE coming in from Trelleborg are highly concentrated while conducting the entry manoeuvres into the Rostock sea channel. The ferry COPENHAGEN, which arrives from Gedser at the same time off the Warnow estuary, is given priority, because its timetable is much tighter on the by far shorter route. Oncoming is a Swedish tanker, while an incoming bulk carrier stays behind. Fortunately, none of the cruise liners berthed at Warnemünde is departing in this moment, so that the access route to the Rostock overseas port

Fig. 5.27 The Swedish rail ferry SKANE started its service between Trelleborg and Rostock in 1998 (Ralf Witthohn)

would be blocked. Thus, the command of the rail ferry can initiate the turn of the ship in the Breitling bay without getting under pressure. The officers then quickly change to the aft wheelhouse and from there carry out the berthing manoeuvre at Skandinavienkai. Only minutes later cars, trucks and rail vehicles roll over the ship's special ramps (Fig. 5.27).

Built in 1998 as the world's largest rail ferry by Astilleros Espanoles in Puerto Real/Spain the 42,705 gt, 9314 dwt SKANE of Sweden's Stena Line provides passenger, car, truck and rail wagon transport services between Rostock and Trelleborg. The 199 m length, 29.6 m breadth SKANE is accelerated by a four-engine propulsion system via two propellers to a speed of 21 knots. The vessel can carry 600 passengers and on 1110 m of track 55 rail vehicles or cars. SKANE is operated together with the similar rail ferry MECKLENBURG-VORPOMMERN, which was in 1996 built by Schichau Seebeckwerft in Bremerhaven on order of Deutsche Fährgesellschaft Ostsee (DFO). Since 2013, this ferry is as well operated by Stena. The 199 m length, 28.8 m breadth MECKLENBURG-VORPOMMERN has capacities for 600 passengers, 440 cars on 2140 m and wagons on 920 m of track. The 37,987 gt, 7520 dwt ship is accelerated by two propellers to 18 knots. Due to the long harbour entrance to Trelleborg, which has to be approached astern, it was retroactively fitted with a bow rudder in addition to two transversal thrusters forward, to improve manoeuvring during unfavourable weather conditions (Fig. 5.28).

At the Rostock terminal for combined freight traffic, 34 trains a week were in 2015 running to Verona, Hamburg, Karlsruhe, Brno, Novara, Duisburg, Domodossola and Wels. In 2019 there were still 30 trains maintaining weekly intermodal cargo transports, 15 to Verona/Italy, 4 to Lovosice/Czech Republic, 3 each to Cervignano/Italy, Brünn/Czech Republic, Wuppertal/Germany and 1 each to Halle/Germany and Curtici/Romania. Just as the division of

Fig. 5.28 Passengers, cars, trucks and railway carriages are transported on board the ferries MECKLENBURG-VORPOMMERN and SKANE between Trelleborg and Rostock (Ralf Witthohn)

Germany once led to the construction of the Rostock overseas port, the numerous liner services of the Deutsche Seereederei soon fell victim to the end of the GDR. After drastic losses in cargo handling from over 20 million to only 8 million tonnes, the end of the Eastern Bloc, however, opened up alternatives in ferry and passenger shipping for the traditional Baltic Sea port with new connections to Scandinavia and Eastern Europe. In 2015, 6100 of the port's total of 7900 ship arrivals had been made by ferry and ro ro vessels. They carried 342,000 trucks and 21,700 rail wagons with a total of 14.7 million tonnes of freight. In 2019, ro ro cargo amounted to 16.2 million tonnes, a share of 63% of Rostock's total handling volume. 379,800 trucks rolled on and from board of ro ro ships, the number of trailers achieved 125,300 units, of rail vehicles 18,388, that of private cars 554,000. 2.5 million passengers boarded the ferries.

In 2021, Rostock handled of 27,100 railway wagons carried to and from Trelleborg by SKANE and MECKLENBURG-VORPOMMERN, an increase by 41% from 19,200. In total, 18 million tonnes of wheeled cargo from ferries and ro ro ships passed the harbour, a 2.7 million tonnes plus compared to the pre-year. The number of truck units increased from 367,000 in 2020 to 407,000, the number of trailers fom 126,000 to 161,000 in 2021 (Fig. 5.29).

In April 2020, Stena Line announced plans, to terminate the ferry service from Saßnitz to Trelleborg, maintained by the 1989-built combined passenger/railway ferry SASSNITZ. The ferry had been laid up in March due to decreasing bookings in course of the COVID-19 pandemic and sailed for

Fig. 5.29 On main deck of the rail ferry MECKLENBURG-VORPOMMERN there is space for either wagons or trailers (Ralf Witthohn)

Udevalla, which port it left to be scrapped in Aliaga/Turkey. In 2019 about 300,000 passengers had been carried on the tradition-rich railway ferry connection. Freight and rail cargo volumes as well had decreased after the shifting of rail traffic routes. Due to the pandemic, Stena line suspended the car and passenger ferry service between Frederikshavn and Oslo in May 2021 (Fig. 5.30).

5.3.2 Denmark's PRINSESSE BENEDIKTE on *Fugleflugtslinien*

Following the transfer of five ferry lines to Sweden's Stena Line in 2012, Scandlines, once formed by the merger from the national ferry companies DFO from Germany and Scandlines Danmark, concentrated on the traditional routes between Rostock and Gedser, Puttgarden and Rødby, as well as from Helsingør to Helsingborg. On the ferry route between Puttgarden/Germany to Rødby/Denmark, named *Vogelfluglinie* in German and *Fugleflugslinjen* in Danish, the double-ended, 15,187 gt rail ferries DEUTSCHLAND and SCHLESWIG-HOLSTEIN had been operating under the German flag since 1997. For the last time in December 2019 railway wagons were transported on the line. Due to maintenance work on the Copenhagen—Lolland high-speed line in preparation for the fixed Fehmarn

Fig. 5.30 The railway ferry SASSNITZ was one of the ferries handed over from Danish/German Scandlines to Sweden's Stena Line in 2012, but terminated its service on the old-established Saßnitz—Trelleborg route in March 2020 (Ralf Witthohn)

Belt crossing, the wagons have since been rolling via Odense, Fredericia and Flensburg to Hamburg.

DEUTSCHLAND and SCHLESWIG-HOLSTEIN were built at van der Giessen-de Noord in the Netherlands, while the Danish-flagged PRINSESSE BENEDIKTE and PRINS RICHARD are buildings of Ørskov Christensen Staalskibsvaerft in Frederikshavn/Denmark. The 142 m length, 24.8 m breadth German ships can accommodate 1200 passengers, 364 cars, have 625 m of track for trucks and 118 m for a passenger train. The Danish ships offer trucks 580 lane metres. From 2013 on, the 14,822 gt PRINSESSE BENEDIKTE and its sister vessels wer fitted with a Siemens hybrid system for storing energy in batteries of 2600 kW output, which saves one in five diesel generators. The diesel-electric propulsion system comprises four generators of 3500 kW each feeding four electrical motors of 3200 kW output (Fig. 5.31).

As well fitted with a hybrid system are the double-deck ferries BERLIN and COPENHAGEN contracted by German-Danish Scandlines at Germany's Volkwerft Stralsund in 2010 as replacement for PRINS JOACHIM and KRONPRINS FREDERIK on the Rostock-Gedser route, built as rail ferries at Nakskov in 1980/1981. However, the ferry operator refused to take over the ships, which had become too heavy. Financing of EUR 184 million for a total investment volume of EUR 230 million, including the adaptation of the port facilities to the two-storey handling system, was granted over 75% by

Fig. 5.31 A passenger train rolling ashore through the ramp of PRINSESSE BENEDIKTE in Rødby belongs to history since December 2019 (Ralf Witthohn)

Germany's state-owned KfW Ipex Bank and covered by a state export credit insurance. The design called for newbuildings with a deadweight capacity of 2900 tonnes on a draught of 5.5 m and of 4200 tonnes on 6 m. On 1600 lane metres there should be space for 96 trucks or 460 cars on the two vehicle decks that were loaded simultaneously on two ramp levels. Due to the shallow water conditions with corresponding draught limitation and scheduled departures every 2 h, a special propulsion system was designed with three propellers, two outer diesel-electrically driven azimuth propellers and a central controllable pitch propeller rotated through a reduction gear. The good manoeuvrability, which is given anyway by three propellers, was to be further improved by two transversal thrusters forward. The initial intention to create the conditions for LNG propulsion was abandoned in the course of construction in favour of exhaust gas scrubbers. In addition, the 21 knots fast ferries are equipped with an energy storage system from Siemens based on batteries that can store unused energy for the on-board system or for propulsion (Fig. 5.32).

Following the insolvency of Volkswerft Stralsund, Scandlines acquired the newbuildings, which were unfinished, at a fraction of the originally agreed

Fig. 5.32 Six years elapsed between the order and completion of the two-deck ferries BERLIN and COPENHAGEN (Ralf Witthohn)

price. Hamburg-based Blohm+Voss developed a modification concept implemented at Denmark's Fayards in Munkebo, according to which the ship's weight was reduced by lowering the passenger decks and using aluminium. Equipped with four 18,000 kW MaK diesel engines, the 169.5 m length, over fenders 25.4 m breadth BERLIN was finally commissioned under the German flag in 2016, COPENHAGEN under the Danish. The ships' tonnage amounts to 22,319 gt. The replaced KRONPRINS FREDERIK remained in reserve and had to stand in for the BERLIN, which returned to the shipyard in January 2017, because the exhaust gas scrubbers of the ferry did not work, so that the ship had to burn marine gas oil instead of the planned heavy fuel oil. In 2021, KRONPRINS FREDERIK temporarily served the Puttgarden—Rødby connection, before returning to its lay-up berth in Rostock. The sister ship PRINS JOACHIM was acquired by the Greek shipping company European Seaways for operation on the Italy-Albania route from May 2016 under the name PRINCE. In December 2016 it was resold to Africa Morocco Links in Athens and renamed MOROCCO STAR for service on the Algeciras—Tanger route, on which the ferry still sailed in 2022 (Fig. 5.33).

Fig. 5.33 After being replaced by a newbuilding, PRINS JOACHIM moved to Italy and ran on a service to Albania (Ralf Witthohn)

5.4 Combined Cargo and Passenger Carrying Ships

While ships were designed for centuries to carry passengers and freight simultaneously, this changed fundamentally with the success of aviation. From that time on, people and goods were transported separately on specially designed ships. After the Second World War, European shipping companies still commissioned a small number of combined cargo and passenger ships, like for colonial service, but the attempt to reconcile cargo transport with passenger transport and to have special cargo and passenger ships equipped with appropriate cabins and lounges for this purpose, ended even earlier than pure liner shipping in the 1960s. At the same time, combined transport received a new impetus with the introduction of roll on roll off transport in ferry shipping. In some remote areas, however, the simultaneous transport of people and freight on conventional ships is still carried out, often supplemented by carrying animals. In the Chilean fjords of Tierra del Fuego, in shipping to Mauritius, to St. Helena, the Falkland Isles and the South Sea islands, combined vessels, some of the latest design, are still in operation.

5.4.1 Black Pearls from Bora Bora

At noon time the combined cargo and passenger ship ARANUI 5 has left Papetee on Tahiti. Next stop is Fakarava, an atoll of the Tuamotu archipelago, which is reached 25 h later. After a day at sea, ARANUI 5, which name means "Great Way" in the indigenous language, reaches Hiva Oa on Puamau on the fourth day and Ua Huka on Vaipaee-Hane-Hokatu the next day. From there the ship sails to Nuku Hiva on Taiohae and Ua Pou and after another day at sea to the Tuamotu archipelago. The penultimate stop of the two-week journey is Bora Bora, before the ship moors again at the starting point and home port of Papeete in the morning time. Six days pass before ARANUI 5 starts a new round trip with the same destinations.

ARANUI 5 plays an important role in the French Polynesian economy. As its main task, the ship carries out seven to 14-day supply voyages between the islets of the South Seas, provides the inhabitants with the things of everyday needs and participates in the transport of the country's few export goods. These include cultured pearls, such as those that have been growing for 40 years on a farm in Vaitape on Bora Bora, processed into valuable jewellery and marketed internationally, but also offered to tourists for home sale. Among the agricultural products of the islands belonging to France are exotic fruits such as vanilla beans, some of which are very rare and used not only as foodstuff, but also in cosmetics. At the same time the ARANUI 5 sailings are marketed worldwide under the slogan "Freighter to paradise". In this way, the ship gives tourists the opportunity to get to know some of the 118 islands and atolls of the archipelago, which economy is largely dependent on tourism.

In 2017 ARANUI 5 carried out 19 trips on the described route, in 2020 a total of 20 sailings were intended to be carried out, but all cruises were stopped, when the Haut-Commissariat de la République en Polynésie Francaise in the wake of the COVID-19 pandemic issued a travel ban to French Polynesia in March 2020. ARANUI 5 is equipped with 103 cabins and common dormitories, each for up to eight people, so that space for a total of 254 passengers is provided. In addition to a restaurant, an outdoor swimming pool, massage room, boutique and bars are available to them. Delivered by China's Huanghai Shipbuilding to SNC Marquises Investissement in 2015, the multi-purpose ship was specially designed for operation in the South Seas. At a length of 126.1 m, deadweight is figured at 3300 tonnes. Four cargo holds of 6400 m^3 capacity are served by two revolving cranes, which can also launch a tender boat for the transfer of passengers and cargoes. Two propellers, driven by two main engines with a total output of 8000 kW, provide a cruising speed of 15 knots and high manoeuvrability in the narrow island waters. A new ship to be named ARAMANA and intended for passenger transport only has been designed by the Shanghai Ship Research and Design Institute (SDARI). Its commissioning was first planned in 2022, but due to the COVID-19 pandemic is expected not before 2024. At 139 m length, 22 m breadth and laid out for 280 passengers, the 14,500 dwt ship is to be built by Huanghai Shipbuilding in Rongcheng/China (Fig. 5.34).

Fig. 5.34 The Chinese-built 11,468 gt ARANUI 5 in the roads of Hiva Oa (CPTM)

The French-flagged, 11,468 gt ARANUI 5 replaced the 116 m length ARANUI 3 built in 2002 by Severnav shipyard in Turnu-Severin/Romania at costs of EUR 16 million. The ship offered space for 208 passengers and 3800 tonnes of cargo. The accommodation offered rooms for 12 passengers and three different cabin categories comprising ten suites, 12 luxury and 63 standard cabins. A floating pontoon could be lowered at the stern for water sports activities, and a swimming pool on the aft deck was another leisure facility. Besides the ship's regular crew of 33, 25 crew members took care of the passengers. ARANUI 3, for its part, had replaced the former German general cargo vessel BREMER HORST BISCHOFF, built in Bremen in 1971 for liner services from Germany to Norway and in 1985 fitted with an additional superstructure for passenger transport, before being used as an island supplier in the southern hemisphere.

Owner of the ARANUI vessel generations is Papetee-based Compagnie Polynésienne de Transport Maritime. It has been managed by the Wong family since founding in 1954. Although only about 12% of the population is of Chinese origin, they play an important role in the economy of French Polynesia. This facilitates relations with China, which wants to develop its tourism to the isles. Chinese engineers have been sent to the South Seas for infrastructure projects such as the expansion of Tahiti airport and a highway from Papetee to Taravao. At the end of 2017, the Papetee port company, which is active in the handling of local ferries, cargo ships, naval vessels, but also cruise liners of over 360 m length, 4500 teu container ships and oil tankers, commissioned the tug newbuilding AITO NUI 2. The 60 tonnes bollard pull tug of type ASD 2810 joined the smaller tugs AITO NUI 1 of 40 tonnes bollard pull and AUTE, after undertaking a 6635 nm trip from the Vietnamese builder Damen Shipyard Song Cam to Tahiti on ist own keel. In order to ensure the maintenance of the tugboat on the distant island, a local

Fig. 5.35 The passenger and cargo ship ARANUI 3 built in Southeast Europe replaced the former general cargo vessel BREMER HORST BISCHOFF in 2003 (Gina Bara)

engineering company was contracted to cooperate with a Damen service station in Brisbane, which is 3220 nm away (Fig. 5.35).

French Polynesia's foreign trade links are maintained mainly by container lines. A 16-day service named Greater Bali Hai was in 2022 maintained by the South Pacific Liner Consortium of Japan's NYK Bulk & Procect Carriers and Kyowa Shipping. The ships employed were the 900-teu vessels CORAL ISLANDER II, PACIFIC ISLANDER II, SOUTH ISLANDER and TROPICAL ISLANDER. They provide shipment options to the Far East. The port rotation of the 63-day round trips is Busan, Kobe, Nagoya, Yokohama, Honiara/Solomon Islands, Port Vila/Vanuatu, Luganville, Noumea/New Caledonia, Lautoka/Fiji, Suva/Fiji, Nuku'alofa/Tonga, Apia/Samoa, Pago-Pago/American Samoa, Papeete/French Polynesia, Busan.

Sailing to the east, to the US ports of Long Beach and Oakland, from Papetee, Apia, Pago-Pago and Nuki'Alofa there is a joint service of Hamburg Süd and San Francisco-based Polynesia Line, in March 2020 employing the container ships FESCO ASKOLD, CAP PAPATELE and POLYNESIA of 1100/1700 teu capacity. In the following year, the name Polynesia Line consolidated under Swire Shipping, the service was 2022 maintained by Maersk and Swire by employing the 1300/1700 teu ships CAP SALIA and MOUNT CAMERON.

Every 14 days Pacific Direct Line (PIL) operated a cargo ship between Papeete and Auckland/New Zealand, from where onward services to South East Asia are possible. In March 2020 this was the long-term employed 777 teu carrying SOUTHERN TRADER, which is equipped with two container cranes and for which break bulk cargo is also accepted. In November 2017 French operators CMA CGM and Marfet established a container link to Europe via Panama canal using four 2260 teu reefer container vessels of type SEATRADE ORANGE of the before participating dutch operator Seatrade. Every week one of a total of 13 ships of 2200/2500 teu ships started the 91 days lasting voyage from Papetee, Noumea, Brisbane, Sydney, Melbourne and Tauranga. In 2022 was still one of the Seatrade reefer container vessels, the SEATRADE GREEN, participating in the service, meanwhile maintained every 14 days and affording 89 days lasting voyages (Fig. 5.36).

Tasks similar to ARANUI were performed by MAURITIUS PRIDE, built in 1990 at Husumer Werft in Germany for Mauritius Shipping, between the Mascarenes. The 5234 gt ship was equipped to carry 268 passengers in cabins and public rooms, but also cargo and livestock. In 2014 the combi carrier was sold to India and renamed MALDIVES PRIDE, later SOBAT, under which name it was broken up in 2017. The 5492 gt combined passenger and cargo vessel MAURITIUS TROCHETIA, built by Hudong shipyard in Shanghai in 2001, and the German 5030 dwt cargo vessel ANNA, which is suitable for container transport, continued the transports between the islands of the Indian Ocean.

Fig. 5.36 Four units of the reefer container type SEATRADE BLUE were in 2017 in the 13 ships comprising fleet connecting South Sea ports with Europe (Ralf Witthohn)

Fig. 5.37 The superstructure of XIANG XUE LAN offers space for 392 passengers, in the hold and on the hatches can be stowed 293 container teu, (Ralf Witthohn)

5.4.2 XIANG XUE LAN from China to Korea

In 1995/96, MTW shipyard in Wismar built the 16,071 gt passenger and cargo ships ZI YU LAN and XIANG XUE LAN for China. The 150.5 m length, 24 m breadth ships have 122 cabins for 392 passengers and a deadweight capacity of 6500 tonnes. Three cargo holds are served by a container crane, and for 293 teu there is space in the holds and on the hatches. Two MaK diesel engines with a combined power of 15,000 kW act on one shaft to give the ships a speed of 20 knots. In February 2022 ZI YU LAN sailed from Lianyungang/China to Pyeongtaek/South Korea, XIANG XUE LAN between Yantai and Incheon (Fig. 5.37).

5.5 High-Speed Ferries

While the speed requirements of 27 to 28 knots set for a whole series of newbuildings from the 1990s onwards could still be met by conventionally driven twin-screw ferries, there were developed designs that went beyond this speed demand with an innovative approach, such as that of the FINNJET, which was accelerated by gas turbines to 33 knots as early as 1977. Yet, striving for speed produced a long list of failures in the construction and operation of high-speed ferries. One of the most glaring examples of a failed project is the Japanese ferry SUPER LINER OGASAWARA, completed by Mitsui in Tamano in 2005. The ferry was accelerated to 45 knots by the most powerful

Fig. 5.38 The Japanese ferry SUPER LINER OGASAWARA, completed by Mitsui in Tamano in 2005, reached 45 knots only during trial runs, as it was subsequently decommissioned (Mitsui)

gas-turbine-water jet propulsion system available at the time. But, for economic reasons, the ship did not enter a commercial service. It was yet activated in May 2011 after an earthquake in Japan to support relief measures.

In other cases as well the attempt failed to create a high attractiveness in passenger traffic by fast crossings. The reasons for such problems are manifold. On one hand, high speeds require the provision of large engine powers, which have to be provided by special propulsion systems that are complicated and susceptible during operation. At the same time, the high fuel consumption of these systems often leads to higher than calculated costs. Among the various technological approaches used in the design of high-speed vehicles, there have been no further developments in civil air-cushion and hydrofoil vehicles (Fig. 5.38).

5.5.1 LEONORA CHRISTINA in Bornholm Service

One month after delivery in June 2011 and after a voyage halfway around the world, the ferry newbuilding LEONORA CHRISTINA reached its operational area in the Baltic Sea, a ferry route of Bornholmer Faergen from Rønne on Bornholm/Denmark to Ystad in southern Sweden. The Rønne-based company belongs to the Danske Faerger shipping group, which also operates ferry lines on short routes from Esbjerg to Nordby, between Ballen and Kalundborg, Bøjden and Fynshav and from Spodsbjerg to Tars. From Bornholm, the company operated two connections to Køge on Sjaelland, a five and a half hour voyage, and from March to October to Mukran on Rügen, reached in four and three and a half hours, resp., on the conventional car and passenger ferries

POVL ANKER and HAMMERODDE. However, the shipping company lost the tender for the new, 10-year Bornholm ferry licence valid from September 2018 on to Denmark's Molslinjen. The new licensee ordered the 40 knots fast catamaran EXPRESS 4 at a cost of 100 million Australian dollars from Austal shipyard for delivery in 2019. The catamaran has a capacity of 1006 passengers, 425 cars and 610 truck metres. Another ferry newbuilding of 115 m length has been ordered by Molslinjen from Austal for delivery in spring 2022 for the Rønne-Ystad line. The 158 m long ro pax ferry newbuilding HAMMERSHUS was built at costs of EUR 68 million by Rauma Marine Constructions (RMC) in Finland. The ferry operator announced to undertake sailings from Saßnitz/Rügen to Bornholm between January and March 2019.

LEONORA CHRISTINA was built by Austal shipyard in Henderson/Australia. Constructed from aluminium alloy, the 112.6 m length, 26.2 m breadth, 8.5 m depth catamaran carries 1400 passengers and 357 cars, which are loaded through bow and stern ramps. Due to its light weight and special hull shape, the 10,371 gt ship, powered by four MAN engines of 9100 kW output each, reaches a speed of 40 knots by a water jet propulsion system. This allowed a crossing time of 80 min. LEONORA CHRISTINA was end of 2016 sold to Ferry Gomera and in 2022 operated under the Spanish flag and the name BETANCURIA EXPRESS by Norway-based company Fred. Olsen between Morro del Jable in Spain and Las Palmas/Canary Islands.

LEONORA CHRISTINA had replaced another Austal building on the Bornholm route, the 2000-delivered, gas-turbine driven VILLUM CLAUSEN. The water-jet propelled catamaran of 86.6 m length and 24 m breadth was in 2017 sold to Greece operator Seajet and renamed WORLDCHAMPION JET. In March 2020 the ship was laid up at Piraeus since October, but during the 2021 season carried out sailings to various Greek destinations (Fig. 5.39).

Fig. 5.39 The catamaran ferry LEONORA CHRISTINA built in Australia maintained a fast connection between Bornholm and Sjaelland in 2011 before switching to Spain (Ralf Witthohn)

5.5.2 Unsuccessful SPIRIT OF ONTARIO 1

Of the same type as VILLUM CLAUSEN is the ferry SPIRIT OF ONTARIO 1, built in 2004 by Austal for the line Toronto-Rochester on Lake Ontario. After experiencing various technical problems, the 42 knots ship switched to the Strait of Gibraltar in 2007. Renamed TANGER JET II the ferry was operated by the Flensburg-based shipping company FRS, and from 2012 for a short time as DOLPHIN JET in the Kattegat between Aarhus and Kalundborg. One year later it changed name to VIRGEN DE COROMOTO and operated the route from Isla Margarita to Puerto La Cruz on behalf of Venezuela's Conferry, until it was in July 2016 laid up in Punta de Piedras, later in Puerto Cabello. Another newbuilding of Austal shipyard is the 82.3 m length catamaran BOOMERANG, which was delivered to Poland's Polferries in 1997. From 2001 the 38 knots fast ship was used as TALLINK AUTOEXPRESS 2 for the Estonian shipping company Tallink between Tallinn and Helsinki. In 2007 the ship switched to Conferry as well, but sank in the port of Guanta/ Venezuela in 2018 (Figs. 5.40 and 5.41).

5.5.3 Italian Monohull Ferry SUPERSEACAT FOUR

On the Tallinn-Helsinki route, the BOOMERANG competed, among others, with the monohull ships SUPERSEACAT FOUR, SUPERSEACAT THREE SUPERSEACAT ONE of Estonia's shipping company Superseacat, a joint venture between the Bermuda shipping company Sea Containers and the Greek company Aegean Speed Lines, which existed as the successor of a

Fig. 5.40 After failing on a US-Canada route, SPIRIT OF ONTARIO I connected the European and African continents under the name TANGER JET II (Ralf Witthohn)

Fig. 5.41 After an assignment in the Baltic Sea, BOOMERANG was employed in South America (Ralf Witthohn)

Fig. 5.42 SUPERSEACAT FOUR operated between Estonia and Finland until 2008 (Ralf Witthohn)

Silja Line subsidiary from 2006 until bankruptcy in 2008. Built in 1998/99 by the Fincantieri shipyards in Riva Trigoso and Muggiano, the 100 m length, 17.1 m breadth ships for 800 passengers are accelerated to 34 knots by water jet propulsion with a total output of 27,500 kW (Fig. 5.42).

In 2016 SUPERSEACAT FOUR was employed as SPEEDRUNNER IV for the Aegean Speed Lines between Piraeus and Serifos, in 2017 as SUPERRUNNER between Piraeus and Rafina. The SUPERSEACAT THREE operated as SPEEDRUNNER III between Piraeus, Seriphos and Milos. The sister ship SUPERSEACAT ONE, built in 1997, sailed as ALMUDAINA DOS from Malaga to Melilla in 2016 and was laid up in Almeria at the end of 2017, where it was still reported as berthed in 2022. After initial operations between Dover and Calais as well as between Liverpool

Fig. 5.43 The Italian-built GOTLANDIA II travels the Gotland routes from Swedish mainland (Ralf Witthohn)

and Dublin, SUPERSEACAT TWO sailed as HELLENIC HIGHSPEED on the Rafina-Cyclades route in 2016 and from Ios to Piraeus the following year. In January 2022, the former SUPERSEACAT FOUR ran on Greek routes with calls at Sitia, Kasos, Karpathos, Diafani, Chalki and Agios Nikolaos under the name SUPERRUNNER JET. At this time, the SUPERSEACAT THREE called at Aegean Sea ports including Syros, Serifos, Sifnos, Kimolos, Milos as HELLENIC HIGHSPEED.

Riva Trigoso was also the place of build of the larger GOTLANDIA II, which was built to the same basic concept for the Swedish shipping company Destination Gotland in 2006. The 122 m length, 16.7 m breadth fast ferry reaches 36 knots with the help of four water jet propulsion systems. On the two routes from Nynäshamn and Oskarshamn to Visby the ship can carry 780 passengers and 160 cars. The GOTLANDIA II being laid-up in Visby during the winter 2021/22, was the traffic to the isle maintained by the ro pax ferries VISBY and GOTLAND (Fig. 5.43).

5.5.4 Wavepiercer SICILIA JET from Incat

Although the rest of Australia's commercial shipbuilding industry is of less importantance internationally, another fast ship supplier, Hobart/Tasmania-based Incat supplies the world with high-speed ferries. The builder has developed a fast ferry type which is known as wavepiercer, a designation derived from the special double hull shape with a keel-like centre section. Italy's SNAV, which had already been operating hydrofoils on short routes before, started the first long-distance service Naples-Palermo in 1997 by commissioning the Incat-built SICILIA JET, a 40 knots achieving wavepiercer carrying 800 passengers and 200 cars. The 86.6 m length, 26 m breadth 5007 gt ship changed in 2004 as SARDINIA JET to the Naples-Golfo Aranci route, 1 year

Fig. 5.44 By introducing the Incat-built SICILIA JET Italy's ferry company SNAV maintained a fast ferry connection with Palermo (Ralf Witthohn)

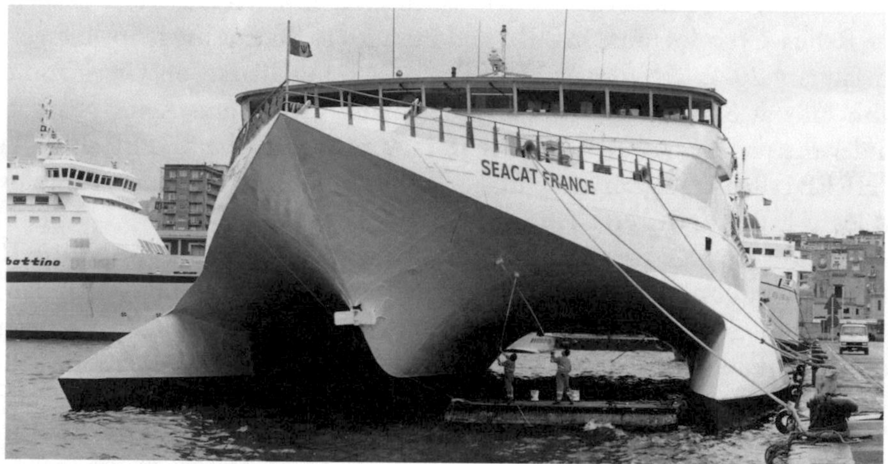

Fig. 5.45 Wavepiercer SEACAT FRANCE, built by Australian Incat shipyard as SEACAT CALAIS, returned to the canal route after an assignment in the Adriatic Sea under the name CROAZIA JET and was in 2022 registered as HSC CAT (Ralf Witthohn)

later as PESCARA JET to the Pescara-Hvar-Split line. After sale to Germany's FRS in 2006, the ferry took up service from Algeciras to Tangier as TARIFA JET, which was interrupted during the winter season 2021/2022 (Fig. 5.44).

A smaller Incat type is represented by the ferry SEACAT TASMANIA delivered in 1990. The 74 m length, 26 m breadth twin hull ship operated under the name SEACAT CALAIS for UK-based shipping company Hoverspeed on the Channel route, from 2000 for SNAV as CROAZIA JET, subsequently on the Channel as SEACAT FRANCE, was then chartered out as EMERAUDE FRANCE and came into Greek hands in 2014 as SUPERFAST CAT. In 2022, the vessel was still registered in Greece as HSC CAT.

In 2013, the Incat shipyard completed an order from Denmark's Mols Linjen for the construction of the 112 m length, 30.2 m breadth catamaran KATEXPRESS, which, with an engine output of 36,400 kW, reaches 37 knots and at maximum 42 knots. The 10,500 gt vessel, renamed EXPRESS 2 in 2017, still sailed between Aarhus on Jylland and Sjaellands Odde on Sjaelland in February 2022. At 3000 dwt, the ferry can carry 1000 passengers and 411 cars on 567 m track length. EXPRESS 2 sailed along with the 10,842 gt Incat newbuilding EXPRESS 3, while MAX MOLS, built in 1998 and of half the size by measuring 5617 gt, was in 2021 employed as MAX between Rønne and Ystad and in Februray 2022 between Aarhus and Sjaellands Odde (Fig. 5.45).

Part II

Work at Sea

6

Offshore Work

The activities at sea, which do not involve the transport of people and animals or the exchange of goods, are many and diverse. In order to maintain shipping movements, a wide range of work is done by dredgers of various types, harbour, deep-sea and emergency tugs, buoy layers and icebreakers. Numerous special ships, including supply vessels, anchor handling tugs, pipe transporters, accommodation and guard vessels are employed in the offshore industry. Floating cranes, crane ships, cable and pipe layers as well as research vessels perform other exceptional tasks (Figs. 6.1 and 6.2).

Subsea pipelines are laid for the transport of oil and gas, for which highly sophisticated ships with special laying equipment of various technologies are designed. Due to the high standard, such high-tech vessels are among the few types of ships for which European shipyards still received orders. At the end of 2014, Lloyd Werft in Bremerhaven/Germany delivered the 199.3 m length, 32.3 m breadth construction and pipe-laying ship CEONA AMAZON to London-based shipping company Ceona. Due to a decline in offshore production, however, the 27,450 dwt newbuilding remained without employment for months, so that the shipping company went bankrupt. In 2017, US group McDermott, which is active in oil and gas production technology, took over the ship. Renamed AMAZON the Gibraltar-registered vessel was converted, and from the end of the year deployed in the Zuluf field off the Gulf Coast of Saudi Arabia on behalf of the world's largest oil company Saudi Aramco based at Dharan/Saudi Arabia. In February 2018 the AMAZON was sighted in Jebel Ali/United Arab Emirates, 2 years later in Rotterdam, where it was still being berthed in the beginning of 2022 (Fig. 6.3).

Numerous novel special ship types have also been designed and deployed for the new offshore wind power sector. The construction of offshore wind

R. Witthohn, *International Shipping*, https://doi.org/10.1007/978-3-658-34273-9_6

Fig. 6.1 The 16,188 m³ trailing suction hopper dredger BONNY RIVER, in 2019 delivered to Belgian DEME group by China's COSCO Shipping Heavy Industry (Guangdong), is powered by two four-stroke dual fuel diesel engines and thus prepared to burn gas fuel (Ralf Witthohn)

Fig. 6.2 The offshore support vessel EDT HERCULES has in 2014 been built by Spain's Construccuiones Navales Del Norte according to an Ulstein X-bow PX 105 design for operation by Cyprus-based EDT Shipmanagement (Ralf Witthohn)

Fig. 6.3 Before its first job on an oil field off the Saudi coast, CEONA AMAZON was laid up for several months, so that grass grew through its mooring lines (Ralf Witthohn)

farms requires extensive preparatory work, such as ground and environmental exploration, for which auxiliary vessels, often converted from other ship types, are used. Especially due to adverse weather conditions the construction of the power plants by installation vessels is technically demanding. This refers as well to maintenance and control tasks or the repairs and replacement of faulty rotors and generators, to be carried out by fast personnel transporters, so-called crewboats. In the German Bight area alone, two dozen such special ships were stationed in the summer of 2017, most of them in Cuxhaven and on the isles of Borkum and Helgoland. Three years later, the number had become considerably smaller due to the downturn of the offshore wind industry. Speed and high seaworthiness are the top priorities for the design of the crewboats, characteristics that are particularly met by twin-hull ships.

6.1 Pipe Layers SEVEN SEAS and SEVEN OCEANS

From 2007 onwards the Dutch shipyard IHC Offshore & Marine in Krimpen on the river Ijssel built a series of pipe and cable laying ships of similar design. The series was initiated by SEVEN SEAS and SEVEN OCEANS for the UK-based shipping company Subsea 7, in 2008 followed by HOS ACHIEVER and HOS IRON HORSE for Texan Hornbeck Offshore Services and in 2014 by SAPURA JADE, SAPURA DIAMANTE, SAPURA TOPZIO, SAPURA ONIX, SAPURA JADE and SAPURA RUBI for Sapura Navegacao Maritima, a joint venture of Malaysia's oil company SapuraKencana and London-based drilling company Seadrill (Fig. 6.4).

SEVEN SEAS and SEVEN OCEANS differ in their eye-catching installation devices. SEVEN OCEANS was designed for the laying of rigid pipes as a so-called rigid reeled pipe lay-construction vessel and, for this purpose, at the stern equipped with an adjustable laying ramp achieving a maximum holding force of 600 tonnes. The SEVEN SEAS, on the other hand, was planned as a flexible pipe-lay-construction vessel with a laying tower arranged at half the length of the ship for laying elastic pipes through a moon pool. On the SEVEN OCEANS, rigid pipes with a diameter of up to 406 mm (16 inches) can be deployed with the help of the ramp, but also flexible pipes or so-called umbilicals. During the laying process, the pipe is unwound from the main storage and laying drum and led into the water via a second alignment drum on the laying tower at the stern of the ship. The laying system also

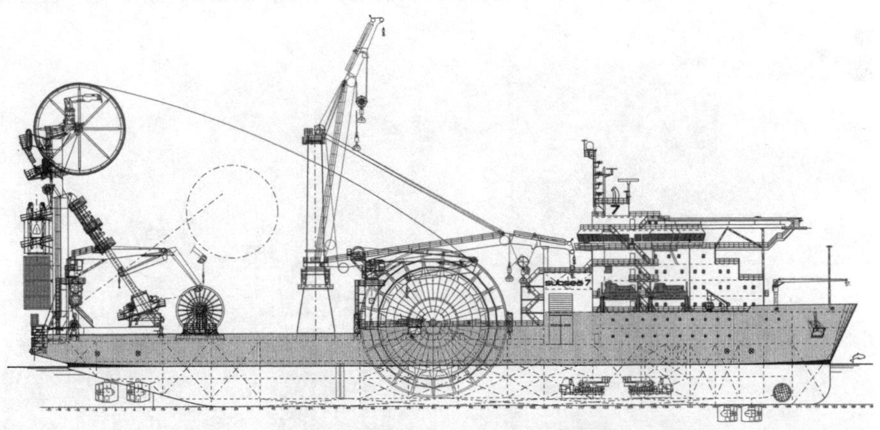

Fig. 6.4 At almost the same main dimensions and propulsion systems, IHC built two different types of pipe-laying and installation ships, of which the SEVEN OCEANS has a laying device at the stern (IHC)

includes pipe aligner, roller drum and two ramp elevators. The A&R (Abandonment and Recovery) winch system has pulling powers of 80 tonnes and 450 tonnes, resp. The main laying reel has a storage capacity of 3800 tonnes. Its hub diameter is 18 m, the largest diameter 28 m, the distance between the flanges 10 m. The centre of the drum is 15.4 m above the base of the ship, so that its radius extends to just above the double bottom of the ship. Pipes with diameters from 152 to 406 mm (6–16 inches) can be stored and unwound on the reel. A smaller piggyback reel has a hub diameter of 4.4 m, a largest diameter of 7.8 m and a flange distance of 5 m. The heave compensated Huisman crane on SEVEN OCEANS has an SWL of 350 tonnes in double fall operation at 13 m outreach in offshore use. In port the lifting capacity increases to 400 tonnes at 16.5 m (double fall) and 200 tonnes at 32 m (single fall) (Fig. 6.5).

In contrast to SEVEN OCEANS with its stern ramp equipment for rigid pipes, SEVEN SEAS, also known as flexlay or J-lay ship, has a vertical system for laying flexible pipes through a 7 × 7.5 m moon pool in the ship's bottom. The laying tower, set up above the moonpool, is equipped with two tensioning devices arranged one behind the other with a tractive force of over 400 tonnes. The A&R capacities are 125 tonnes and 360 tonnes, resp., and can be increased to 450 tonnes for the laying of flexible pipes. On SEVEN SEAS the pipes are stored below deck in two horizontal carousels, each with a capacity of 1250 tonnes. There is the option to install one carousel of 3000 tonnes on the aft deck. In the J-lay mode a 400 tonnes clamping device is used in combination with an adjustable ramp. The ramp, which laying tower can be inclined up to 15°, has facilities for storage, welding, coating and inspection

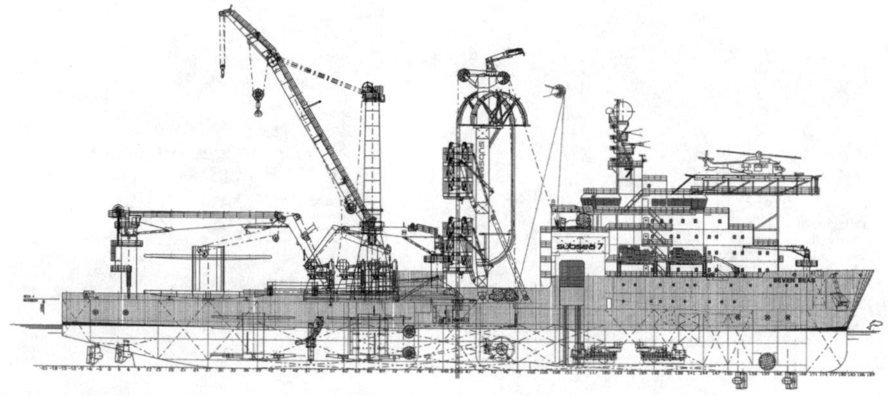

Fig. 6.5 SEVEN SEAS is equipped with a laying device at half of the ship's length (IHC)

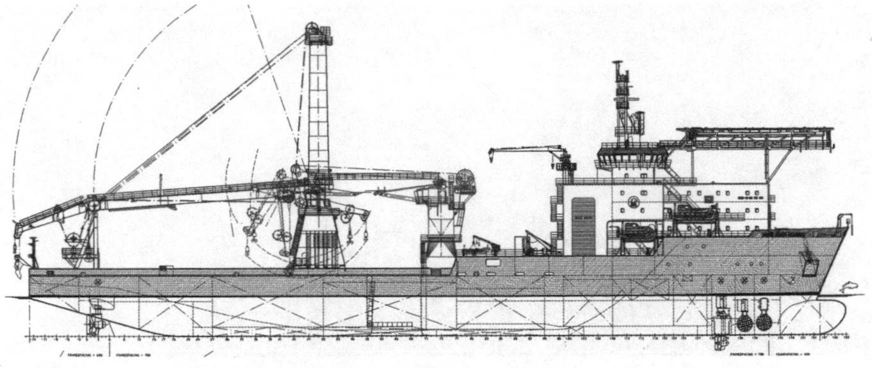

Fig. 6.6 In 2009 Merwede shipyard delivered the offshore construction ship HOS IRON HORSE based on the previous SEVEN SEAS (Merwede)

of the pipes. In both the flexible and J-lay configurations, the maximum pipe width is 24 inches (610 mm) (Fig. 6.6).

On both ships, a dynamic positioning system of type Kongsberg SDP-22, indispensable for the demanding offshore work, enables the precise determination of the correct location and work at the predetermined position. The system comprises two HiPAP (High Precision Acoustic Positioning) units, three DGPS, taut wire and a fanbeam. To monitor the work, the vessels also have an extensive surveying system. The vessels are permanently equipped with two work-class remotely operated vehicles (ROV). They can be launched from the ship's side to a diving depth of maximum 3000 m. Pre-installed foundations also allow the optional carrying of a third ROV.

Due to the extended stern for the ramp construction, the length of SEVEN OCEANS of 157.3 m exceeds that of the 151 m length SEVEN SEAS. At the same time, three frame spaces have been inserted in SEVEN SEAS to accommodate the moonpool. All other main dimensions of the two ships are identical. The energy for propulsion and electrical supply is generated by six main engines of type Wärtsilä 12V26 suitable for HFO and MDO operation and putting out 6 × 3360 kW. Three azimuth stern propellers have an output of 3 × 2950 kW, two retractable azimuth thrusters at the bow have an output of 2 × 2400 kW. The attainable speed of the SEVEN OCEANS is laid out for more than 13.5 knots, that of the SEVEN SEAS specified as 13 knots. A transversal thruster forward develops 2200 kW, so that the total thruster power of the vessels amounts to 15,850 kW.

Since delivery, SEVEN OCEANS worked in Europe, both Americas, Africa and Australasia in water depths from 19 to 2300 m. In the summer of 2017, SEVEN OCEANS completed the laying of the production pipelines of the

Fig. 6.7 2015-built SAPURA JADE represents a further development of the SEVEN SEAS type (Ralf Witthohn)

Maria field off the Norwegian coast, exploited by German oil companyWintershall. A total of 100 km of pipelines were laid on the Haltenbank to three platforms as part of the project. In August 2019 the pipe layer arrived in the Wiltonhaven of Rotterdam, where it was still moored in March 2020. In 2021 the ship was involved in the Mad Dog 2 project in the Gulf of Mexico together with the companie's SEVEN PACIFIC. SEVEN SEAS was stationed in the Gulf of Mexico at the end of 2019, and still in 2021. In January 2022, the Subsea 7 company was awarded an offshore installation contract by US-based Beacon Offshore Energy concerning the tie-back of four subsea wells to the Shenandoah host facility in the Gulf of Mexiko, where SEVEN SEAS and SEVEN OCEANS were anchoring off Corpus Christi in February 2022. In December 2021 had been contracted a tie-back to the Ivar Aasen platform in the North Sea Hanz field with Aker BP (Fig. 6.7).

6.2 Diving Support Ship TOISA PEGASUS

A construction ship with special equipment to support divers was completed by Merwede shipyard in Hardinxfeld-Giessendam/Netherlands in 2008 under the name TOISA PEGASUS on behalf of Toisa Ltd. in Piraeus. The 131.7 m length, 22 m breadth, 9.5 m depth ship can accommodate 18 divers who work in two saturation diving bells to a depth of 300 m. The offshore crane installed on the aft deck lifts 400 tonnes at an outreach of 16 m. The

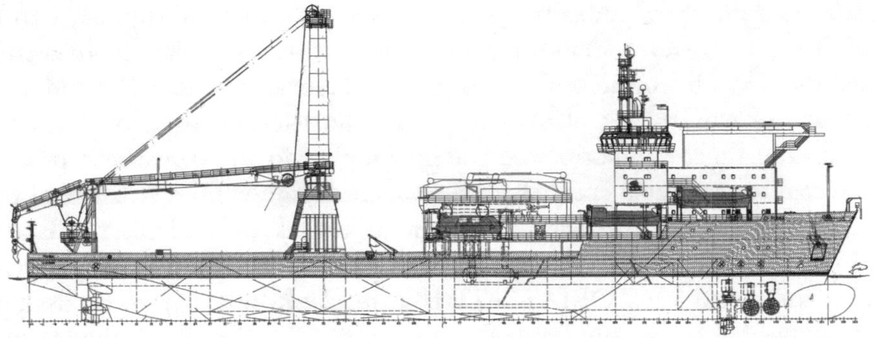

Fig. 6.8 SEVEN PEGASUS ex TOISA PEGASUS is the combination of a diving support and crane ship (Merwede)

diesel-electric propulsion system of TOISA PEGASUS powers two azimuth propellers under the stern of 3000 kW output each for a speed of 13 knots. Three transversal thrusters, one of which is retractable, are installed forward to support the dynamic positioning system. A total of 64 cabins accommodates 99 persons. The maximum deadweight capacity amounts to 9440 tonnes. After operating off the North American west coast in spring 2017, TOISA PEGASUS anchored in early 2018 in the bay of Eleusis, Greece, which the ship had entered in August of the previous year. In January 2019 Subsea 7 acquired the TOISA PEGASUS, which was subsequently renamed SEVEN PEGASUS. In February 2019, the ship left Deer River in the US for the Tarantula platform, for which Subsea 7 had In October 2018 been awarded a contract from US company Fieldwood Energy for work in the deepwater Katmai field in the US Gulf, including the provision of umbilicals, riser and flowline systems. In May 2021 SEVEN PEGASUS left the waters bound for Rotterdam, and in January 2022, the ship completed a voyage from Peterhead/ UK to Amsterdam (Fig. 6.8).

6.3 Wind Farms Replacing Coal Mines

Following the decision by the German government to shut down nuclear power plants and facing increasing critics about the use of fossil fuels, the traditional energy companies were faced with a restructure of their activities. Essen-based RWE Innogy saw offshore wind farms as an alternative for the production of renewable energy. In doing so, the group wanted to avoid the high charter costs for installation ships and opted for building up its own special ship fleet. After all, there had already been strong commitments to the maritime industry by industrial groups in the past, such as Thyssen, Krupp and Stinnes. However, the operation of two jack-up installation ships supplied from South Korean builders for employment in the transport and construction of wind power plants remained a short-term activity of the RWE group. The inner conflict of the energy supplier in the conversion process from coal to renewable energy became evident from the nomenclature of the two newbuildings, which were given the names of disused coal mines in Essen.

In February 2012, the first of the each EUR 100 million costing special ships, christened VICTORIA MATHIAS, reached Bremerhaven on deck of the Norwegian semi-submersible transport vessel EAGLE. RWE Innogy had selected Daewoo Shipbuilding and Marine Engineering (DSME) in Korea in a complex global tendering process involving 40 shipyards. German suppliers provided equipment worth EUR 40 million each for the newbuildings. The order for two ships plus one—not exercised—option was awarded in December 2009. Steel cutting of the first ship was carried out in September 2010, keel laying in January 2011, floating out in March and delivery in December 2011. The first steel plates for the sister ship FRIEDRICH ERNESTINE were cut in October 2010, keel laying took place in April 2011, floating in May and delivery in December 2011.

The actual production of the ships in less than 12 months was only made possible by the construction method as commonly realised in Korean shipbuilding. The hulls were built in three large sections, which were joined on a building pontoon. The superstructure was delivered from a distant construction site in a widely equipped condition on a transport pontoon, discharged with help of a heavy load system and placed on the hull. The production of the main crane and four jack-up rigs, each weighing 650 tonnes, in Europe required delivery by a heavy-lift carrier. Because the jacking system failed during the test, the legs had to be dismantled and transported back to Europe separately by ship. After a 55 days lasting voyage of 14,600 nm via South Africa and Tenerife on deck of the Norwegian transport vessel, VICTORIA MATHIAS reached its homeport Bremerhaven in February 2012, where the ship was unloaded at the river quay of the container terminal. The sister ship FRIEDRICH ERNESTINE left South Korea on deck of the transporter FALCON on 26 December 2011 and arrived in Bremerhaven shortly after the first ship, in February 2012 (Fig. 6.9).

Following the mobilisation phase at Lloyd Werft in summer 2012, VICTORIA MATHIAS started the installation of 48 Senvion turbines of 6.15 MW output each at the Nordsee-Ost wind farm 35 km north of Helgoland. In order to ensure the safe standing of the installation ship at the loading quay, the southern end of the Bremerhaven container terminal, steel

Fig. 6.9 Mobilisation of the installation ship VICTORIA MATHIAS took place in the Kaiserhafen of Bremerhaven, from which container quay jacket foundations where subsequently transported to the Nordsee Ost wind farm (Ralf Witthohn)

foundations were sunk there and backfilled. In the wind farm field were initially rammed foundation pipes of 100 tonnes weight and 2.5 m diameter for the jacket foundations. They were positioned in a water depth of 22–26 m over an area of 36 km^2. After laying the cables between the plants and erecting the converter platform, the wind farm went into operation in May 2015. It yet had to be temporarily shut down the following month following a rotor failure. FRIEDRICH ERNESTINE erected the 576 MW Gwynt y Môr wind farm off the North Welsh coast from a base in Wales.

The two ships were initially operated by RWE Offshore Logistic Company (OLC), founded in Hamburg as a subsidiary of RWE Innogy. NSB Niederelbe Schiffahrtsges. from Buxtehude/Germany supported the supervision during the construction period and was entrusted with the nautical and technical management of the ships for 5 years. When the special technical challenges became apparent and state subsidies were reduced, RWE Innogy decided to discontinue the operation of offshore installation ships. VICTORIA MATHIAS was sold to MPI Offshore of the Dutch Vroon group at the beginning of 2015, while retaining the charter contract. The ship was renamed MPI ENTERPRISE and initially continued to operate in RWE's

Nordsee-Ost wind farm. In 2020 it was registered for the Danish wind farm maintenance company Ziton and renamed WIND ENTERPRISE. In February 2022, the WIND ENTERPRISE worked in the Amrumbank windfarm off Germany's North Frisian coast.

The sister ship FRIEDRICH FERNESTINE was in 2015 chartered out for 5 years to ZPMC Profunda Wind Energy in Shanghai, a joint venture of Zhenhua Shipping and Profunda Offshore Contractor in Shenzhen, under the name TORBEN. In the following year, TORBEN was used within the first phase of the Formosa 1 project in a charter to A2SEA for initially carrying two 4 MW Siemens turbines and ZPMC-built foundations, but ended the employment before completion. In the beginning of 2019, the ship was reported as being renamed TUO PENG by Shanghai Zhenhua Shipping, its owner since 2016, but register entries reveal the name GUO DIAN TOU 001 from 2020 on, its owner described as Shanghai Geejia Offshore. The ship had been reported anchoring off China's coast in April 2020, in February 2022 as being berthed in Weihai/China.

The procurement of the newbuildings was based on the intention of RWE to be able to independently carry out the installation of offshore wind farms under its own management. For this purpose, a ship was designed to take over the functions of transporting the wind turbines as well as their installation. Important advantages of this mode of operation, known as the mono-vessel concept, are the reduction of waiting times and the avoidance of shuttle traffic between the base port and the jack-up platform. Previously implemented installation concepts provided for a division of the functions, such as the transport of the components on barges and the erection by floating cranes, crane ships or work platforms. Under the difficult sea conditions to be expected, however, it seemed advisable to make use of the advantages of a fixed platform. Due to the short distances from the bases in Germany and England, a relatively low speed was considered to be sufficient. In this respect, the design represents a compromise between a transport ship and a construction ship, primarily focussing on the installation capabilities.

On the basis of these requirements, the design office Wärtsilä Ship Design in Hamburg, in cooperation with IMS Ingenieurgesellschaft, drew a ship on a hull similar to a barge, in which little importance was attached to ship lines that would reduce resistance. The low speed requirement of 7.5 knots and the need for a dynamic positioning system led to the choice of a diesel-electric propulsion system with six retractable azimuth propellers. The pontoon-shaped hull was to meet the conditions of sufficient transport capacity both in terms of deck area and load capacity. On the other hand, the weight of the hull and the payload should not exceed the capacity of the jacking system figured at a maximum of about 17,000 tonnes. The deadweight capacity of the ship at the largest draught of 5 m is stated as 6315 tonnes, the maximum payload as 4500 tonnes (Fig. 6.10).

These basic conditions resulted in a 100 m length, 40 m breadth, 8 m depth pontoon hull, the deck area of which is restricted by the forward placed superstructure, the crane foundation and the four jack-up houses at the deck corners. This created transport capacities for four 6.3 MW wind turbines of Repower or six 3.6 MW Siemens turbines. At the time of the ships' deliveries, the classification of the jack-up cylinder components was not yet included, because material defects had been discovered during the jack-up tests in South Korea. They occurred on the eyes, by which the legs are connected to the collar plates. This delayed the commissioning of the ships and made the retrofitting with a total of 96 new eyes in Germany necessary. The 78 m long jack-up rigs allow the ship to work as a platform up to a water depth of 45 m. The legs, each weighing approx. 650 tonnes, were rolled out of an S460 special steel of 100 mm thickness from Dillinger Hütte and in South Korea installed by a 3600 tonnes lifting floating crane. Essential elements of the hydraulic jacking system supplied by the Dutch company Muns are two collar plates, one upper

Fig. 6.10 The deck of the installation vessel VICTORIA MATHIAS offers space for 4500 tonnes of wind power plant components (Ralf Witthohn)

and one lower unit, which enclose the platform's rig cylinders. They alternately take up the load by engaging and disengaging in recesses in the rigs. 12 hydraulic cylinders each exert the upward or downward force. The working pressure of the 12,35 tonnes weighing, 8655 mm long cylinders of Muns und Montanhydraulik is 370 bar, the clamping force 9200 kN, their stroke 3000 mm, their outer diameter is 795 mm. The safe working load (SWL) is figured at 4300 kN per cylinder, the lifting speed 0.7 m/min.

A foot extension with a footprint of 10 × 8.74 m and a maximum height of 3 m can be fitted under each of the jack-up rigs in dock or by crane, even in floating condition. The 130 tonnes weighing spud cans, also referred to as elephant feets, were manufactured according to a design of the US classification society ABS. They are mounted with the help of 120 mm thick bolts. The lead throughs for the lifting legs at each corner of the ship are surrounded by a jack house in which the hydraulic drives are placed.

The main working tool of the installation vessel is its starboard side positioned, electro-hydraulically operated revolving crane designated BOS35000 by its supplier Liebherr. The boom of the crane projects far beyond the stern, where it is supported. The crane can be configured in different ways. Fitted with a 78 m jib, the lifting capacity is 1000 tonnes at a radius of 25 m and 644 tonnes at 38 m. A 102 m long jib lifts 800 tonnes. The smallest working radius is 12.2 m. The main deck of the ship, which is laid out for a load of 15 t/m^2, serves for cargo transport. Motions caused by loading operations can be compensated by an anti-heeling system, which electrically driven pump delivers 1500 m^3/h. The ship has between the main crane and the superstructure been equipped with an articulated crane of company Lintec, which can lift 6 tonnes at 15 m outreach (Fig. 6.11).

The different energy requirements for the ship's four operational modes, such as sailing, dynamic positioning, jacking and crane operation, occur largely independently of each other in terms of time. They make plausible the use of a diesel-electric system with its special ability to adapt the energy generation to the actual demand. Initially projected as power source were five MTU four-stroke MTU 16V4000M43 engines, each with an output of 2240 kW at 1800 rpm and mounted on the double bottom. Additional power requirements due to the increase of the dynamic positioning status to DP2 required the equipment of a further MTU type 12V4000P83 diesel engine of 1680 kW output on the main deck. Leroy Somer generators feed ABB electric motors to drive the six 1600 kW azimuth propellers of Rolls-Royce. At a diameter of 2150 mm the propellers have four non-variable blades. The propellers are hydraulically retractable into the hull. They allow a sailing speed of 7.5 knots on the design draught, and in combination with the dynamic

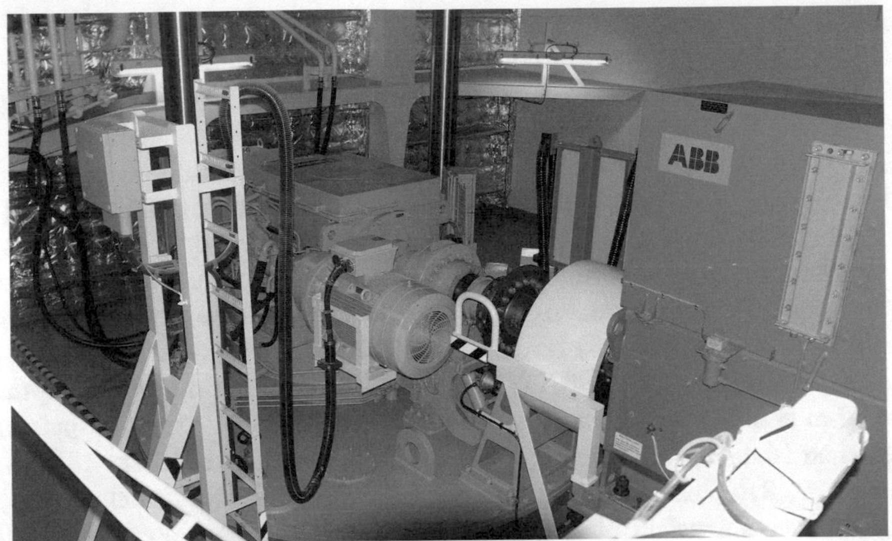

Fig. 6.11 Six azimuth propellers of VICTORIA MATHIAS can hydraulically be retracted (Ralf Witthohn)

positioning system of manufacturer L-3, can maintain an exact position or a transverse speed of 3 knots. A total of two reserve propulsion units were provided for both ships. A particular design challenge was the supply of cooling water to the propulsion system, when the platform is raised. In order to extract sea cooling water, the ship is equipped with a pump supplying a 40 m long pipe which is lowered into the sea.

The six decks high accommodation block on the forecastle offers room for 60 people. The crew concept was initially laid out to comprise 25 nautical and technical personnel in the common three-watch system. On the basis of a crew factor of 2.5, a total of around five crews were available for both ships. The installation team works in a two-shift scheme. As an extension of the ISM Code, the head of the construction team belonged to the management team, together with the captain and the first technical and nautical officer. Rescue means include a free fall lifeboat for 60 persons at the stern next to the main crane support and a conventionally launched closed lifeboat for 80 persons near the main crane. Crew changes can be undertaken by helicopter. For this purpose there is a 16.7 m diameter landing deck for Agusta 139 class helicopters in front of the superstructure, offset to portside.

6.3.1 Installation Ship MPI DISCOVERY on Amrumbank

Following the sale of the third, 2012-built German jack-up installation ship INNOVATION by the Hochtief group to Belgium's Deme company in 2014, foreign shipping firms took over work in the German offshore wind farms, including the installation vessel MPI DISCOVERY. End of 2013, the ship began transporting the foundation components for 80 wind turbines, weighing a maximum of 670 tonnes, from Cuxhaven to the Amrumbank-West wind farm. There they were placed in the predetermined positions with the help of the ship's heavy-duty crane. On six 72.8 m long jack-up rigs instead of the usual four, the MPI DISCOVERY is able to operate at maximum depths of 40 m as long as the wave heights do not exceed 2.8 m, the wind speed is less than 14 m/s and the current does not stronger than 2 knots. The 138.6 m length, 40.8 m breadth ship has a deadweight of 13,704 tonnes, including the rigs it reaches a weight of 17,560 tonnes. The jack-up system can lift 6000 tonnes of cargo at a speed of 1 m/s. The maximum jacking load per rig is 3750 tonnes, the holding load 7500 tonnes. The 1000 tonnes lifting main crane is ready for operation up to wind speeds of 21 m/s. The power supply for the propulsion system, the jacking system, the crane and on-board operation is provided by six Bergen C25:33L-8 diesel engines. Three Rolls Royce azimuth propellers serve as main propulsion units, while three transversal thrusters support dynamic manoeuvring. The accommodation of the Dutch-flagged MPI DISCOVERY, built by COSCO (Nantong) shipyard under contract of UK's MPI Offshore, can accommodate 112 people. Having changed hands to company Jan de Nul Luxembourg in 2018 and renamed TAILLEVENT the installation vessel in 2020 left Germany's Trianel Windpark Borkum II for Taiwan, of which coast 21 turbines for the Changhua Phase 1 project of Taiwan Power Company were built up in cooperation with Japan's Hitachi group. In August 2021, Jan de Nul resold the TAILLEVENT to China's Huihai No 1 Wind Power company, which renamed the vessel HUI HAI YI HAO for operation off the Chinese coast (Fig. 6.12).

The ownership of the sister vessel MPI ADVENTURE switched to the Rotterdam-based Van Oord group in 2018. Van Oord's subsidiary MPI Offshore operated the vessel along with the MPI RESOLUTION and Van Oord's AEOLUS in the North Sea region in 2021/22. In early 2021, the company was awarded a four-year contract by Sweden's Vattenfall concerning planned and unplanned maintenance work at about 600 offshore wind turbine plants. In April Van Oord was contracted to install 100 monopile and 350 km of array cables in the 1.4 GW Sofia offshore wind farm on Dogger

Fig. 6.12 The jack-up vessel MPI DISCOVERY carried the foundations and turbines from Cuxhaven to the Amrumbank-West wind farm (Ralf Witthohn)

Bank in the central North Sea using the installation vessel AEOLUS. Later in the year, the MPI RESOLUTION commenced work on Italy's first offshore wind farm located off Taranto comprising ten turbines. To meet the expected growing market demands, Van Oord in October 2021 contracted another jack-up construction vessel with Yantai CIMC Raffles shipyard in China delivery in 2024, including an option for a second unit. Designed by Danish bureau Knud E. Hansen, the 175 m length newbuilding will feature a Huisman heavy-lift crane, a 5000 kW battery pack and a propulsion and energy supply system able to burn methanol.

6.3.2 BRAVE TERN and BOLD TERN to Build Global Tech 1

Shortly after the MPI DISCOVERY started operations from Cuxhaven, the installation ship BRAVE TERN in 2014 began transporting towers and nacelles loaded in Bremerhaven with help of its own crane for the Global Tech 1 wind farm north of Helgoland, a project of Germany's Hochtief building concern. Delivered in September 2012 by Lamprell Energy shipyard in Dubai, BRAVE TERN and its sister vessel BOLD TERN, completed in February 2013, are operated by company Fred. Olsen Windcarrier (FOWIC) from

Fredericia/Denmark. The ships were financed by a consortium of the German bank KfW IPEX and Danish Ship Finance, which contributed EUR 85 million and EUR 60 million, resp., to the in total EUR 145 million demanding project. State-owned KfW reasonned its financial commitment by the use of the ships for more environment-friendly energy generation and the high proportion of German suppliers by naming the complete propulsion system from Voith Turbo, the electronic control elements from Siemens and generators from ABB, which is yet a Zürich/Svitzerland-headquartered company.

The 131.7 m length and 39 m breadth vessels are fitted with a diesel-electric propulsion system based of Wärtsilä engines of the types 12V32, 6L32 and 9L32 with a total output of 17,100 kW. They generate the power for three aft-mounted Voith-Schneider propellers of 1750 kW output each allowing the ships a speed of 12 knots, as well as for three Wärtsilä type transversal thrusters of 1750 kW, which operation is coupled to a DP2 class dynamic positioning system. The jacking system of Dutch company GustoMSC works on four legs of 81.5 m length and gives the ship a firm footing on water depths up to 40 m. The slewing crane also comes from Gusto. It lifts 800 tonnes at 24 m outreach. The deck area of 3200 m^2 is, for example, sufficient to transport the topsides of five 6 MW wind turbines or three jacket foundations. The Det Norske Veritas classified ships have room for 80 crew members in 56 cabins. In the fourth quarter of 2016, BOLD TERN started the installation of five Alstom 6 MW turbines at Block Island wind farm, the first such offshore project off Rhode Island in the United States. In 2021, BOLD TERN and BLUE TERN were together engaged in the installation of 100 Vestas power plants in Scotland's Moray East farm, BOLD TERN was as well busy in the Dutch Borssele windfarm, while BRAVE TERN worked at Taiwan's Yunlin project. BRAVE TERN was planned to be equipped in spring 2022 with a 1600 tonnes Huisman offshore crane able to lift 400 tonnes 165 m above deck and 256 m above the sea bottom (Fig. 6.13).

6.3.3 AEOLUS in Dutch Gemini Park

The first wind power installation ship built in Germany was delivered by Sietas shipyard in Hamburg to the Rotterdam-based shipping company Van Oord. In February 2014 the 139.4 m length, 38 m breadth AEOLUS was towed to Lloyd Werft at Bremerhaven to be equipped with its four jack-up rigs. The newbuilding, which came into service in June, was fitted with a diesel-electric propulsion system comprising of four MaK main diesels of 4320 kW output each, four Siemens propulsion engines of 2500 kW each,

Fig. 6.13 In April 2014, BRAVE TERN transported Areva windpower plant components from Bremerhaven to the Global Tech 1 wind field in the North Sea (Ralf Witthohn)

two conventional propellers aft and two each Wärtsilä transversal thrusters arranged forward and aft. The 5.7 m draught vessel thus reaches a speed of 12 knots and meets the requirements for a DP2 class dynamic positioning system of Kongsberg. The main working equipment of the construction vessel, which can be deployed to a water depth of 45 m, is a 900 tonnes slewing crane from NMF/TTS, which is mounted around the aft port jack-up rig. The jacking system from IMS/Muns lifts the ship at 0.67 m/min, with full load at 0.4 m/min. The deck area covers 3300 m^2. Accommodation is available for 74 persons. In May 2021, AEOLUS started the installation of 62 jacket foundations of the Saint-Brieuc wind farm off the French coast (Fig. 6.14).

Van Oord did not exercise the option for an AEOLUS sister ship, but ordered the cable layer NEXUS from Dutch Damen Group in October 2013. The keel of the NEXUS was laid at Damen Shipyards Galati in Romania and the ship was delivered at the end of 2014. When the order was placed, the ship's first employment was already secured, the laying of power cables for the Gemini wind farm 60 km north of Schiermonnikoog. Designed as a multi-purpose vessel based on the Damen Offshore Carrier 7500 design, the 120.7 m length, 27.5 m breadth newbuilding is equipped with a heavy-lift crane and a carousel capable of carrying cables weighing more than 5000 tonnes. The vessel, which is equipped with a dynamic positioning system of class DP2, can accommodate 90 persons persons. NEXUS was as well engaged in an alternative power project off Taiwan in 2021 and was still operating from Taichung in early 2022 (Fig. 6.15).

Fig. 6.14 German-built and Dutch-operated AEOLUS is equipped with a 900 tonnes lifting crane (Ralf Witthohn)

Fig. 6.15 Instead of a second wind farm installation vessel of the AEOLUS type, van Oord received the cable layer NEXUS, which became engaged in the laying of power cables between the windmills (Ralf Witthohn)

6.3.4 Bridge-Building Crane SVANEN in Offshore Employment

In August 2014, one of the world's strongest floating cranes participates in the construction of a German wind farm. In Cuxhaven, the special crane SVANEN is being prepared for work on the Amrumbank-West wind project. There, the crane rams single-pipe foundations, so-called monopiles, into the seabed. The monopiles are up to 70 m long, have a diameter of 5 m and weigh up to 800 tonnes. The pieces are transported to the floating crane's position in the wind farm. In order to be able to execute the contract awarded by the German energy company E.ON, the Dutch company Ballast Nedam installs a new hydro-hammer from IHC Merwede between the catamaran hulls of the SVANEN and transports the old unit to the Netherlands for repair.

The 103 m length, 72 m wide SVANEN can lift 8700 tonnes weighing objects up to a height of 76 m. Named SVANEN because of the shape of its design the 1991-built crane was originally laid out for the assembly of the road and rail bridges across the Belt and Öresund. The self-propelled crane is able to achieve a speed of 7 knots. Operated by Dutch company Van Oord Later the heavy-lifter worked in Baltic Sea wind farms, where it supported the erection of several hundred foundations. In April 2016 Van Oord was awarded an E.ON contract for the installing of 60 monopiles and transisition components of the Arkona project, which was completed by SVANEN in November 2017. As an intermediate job, SVANEN, together with AEOLUS, was engaged in the wind farm Walney Extension, finished in August 2017 (Fig. 6.16).

6.3.5 Routes for Wind Energy Cables

In April 2015, the offshore vessel HAVILA PHOENIX for the first time uses its special equipment for laying power cables on the German wind field Global Tech I, located 55 nm northwest of Helgoland in the southern North Sea. The diesel-electrically powered offshore construction vessel was in 2009 delivered by Havyard in Leirvik/Norway to domestic Havila Shipping in Fosnavaag. In spring 2014 the ship has been converted for a long-term charter by the Norwegian offshore company Deepocean to carry out a new special task. To create room for the additional installations, the vessel was lengthened by 17.4–127.4 m. The new equipment includes a 250-tonne A-frame supplied by the company MacGregor at the stern. In conjunction with the existing offshore crane, the frame can lower into the water what is described as the world's largest self-propelled trench cutter for digging cable

Fig. 6.16 The floating crane SVANEN received a new hammer for erecting the foundations of the Amrumbank West wind power plants (Ralf Witthohn)

routes. The 2350 kW machine uses a water jet system. Another jet digging system, lowered at the ship's side, uses an ROV, of which the HAVILA PHOENIX has two on board. Conversion work as well included the fitting of a cable laying device with a horizontally positioned cable drum of 2000 tonnes capacity (Fig. 6.17).

In December 2020, Havila Shipping terminated the Deepocean charter, which should have lasted until 2023, when Deepocean informed the owner, that it would seek protection under British law for winding up the UK-based operations. HAVILA PHOENIX found new employment, when Nexans Norway contracted the ship initially for 90 days plus a 90 days option. The charter was again prolonged in August 2021 for another 100 days until February 2002. At this time, the vessel was carrying out subsea operations from Piraeus.

6.3.6 Cables for the Walney Wind Farm

For 10 days, the Dutch shallow-water barge STEMAT SPIRIT of Rotterdam-based Stemat company is loading power cables intended for the Walney wind farm, located off the British coast, at the quay of Rhenus-Midgard in Nordenham/Germany. A harbour crane is needed to position the cable in the carousel on the main deck of the cable laying vessel. Once loading is complete, the special ship sets

Fig. 6.17 The offshore vessel HAVILA PHOENIX was converted for laying energy cables (Ralf Witthohn)

course for the Irish Sea, where it arrives 7 days later and begins laying the cable at a speed of 0.3–0.5 knots.

Built in 2010, STEMA SPIRIT had already had its first employment at the Walney wind farm right after commissioning. A special design enables the ship to fall dry off the coast during ebb tide and to connect the submarine cable with the landside part of the cable. For this purpose, the ship's hull was constructed with a flat bottom. The ship is also fitted with special manoeuvring aids. Four azimuth thrusters, two of them at the stern and two retractable units forward as well as a transversal thruster are components of a dynamic positioning system of class DP2 ensuring precise locating. During high tide, the cable-laying vessel moves as close to shore as possible and is kept in the planned position by anchors of a six-point mooring system. At low water, the vessel falls dry at a minimum draught of 1.9 m and is thus able to serve as a fixed working platform. The cable is pulled on rollers ashore by a winch mounted at the coastline. After afloating during the tidal period, the ship can start its actual laying operations in the sea. A special plough is used to dig the cable into the sea bed.

In June 2010, STEMAT SPIRIT used this system to create a connection from Heysham in the Morecambe Bay in eastern England to the Walney wind farm, located 27 nm west of Heysham. The project was realised by Dong Energy together with Scottish and Southern Energy. The 90 m length, 28 m breadth, 6.5 m depth cable ship was built by Taizhou Xingang shipyard in China. It is driven by two Caterpillar engines with a total output of 2610 kW, allowing a speed of 10 knots. The superstructure at the front of the vessel can accommodate 60 people. Owned by Netherlands-based BW Marine, the STEMAT SPIRIT was operated by the Stemat company in Rotterdam, which became part of the Dutch Boskalis concern in 2016, including six barges of up to 80 m length and 14 work vessels used for a wide range of tasks (Fig. 6.18).

The loading of high-quality communication or power cables in Nordenham takes place at the quay of the private port operator Rhenus Midgard quay as well as directly at the quay of Norddeutsche Seekabelwerke (NSW), an 1899 founded cable producer. As early as 1903, the company's own cable layers STEPHAN and VON PODBIELSKI loaded intercontinental cables from Nordenham. At that time, the company caused a sensation when it laid a telecommunication line from Borkum to New York.

6.3.7 Nordenham Submarine Cable for Baltic II

In October 2012, the Finnish cable layer AURA of Meriaura shipping company from Turku for the first time takes over power cables at Norddeutsche Seekabelwerke in Nordenham/Germany. The cable is to connect the Baltic Sea wind farm Baltic II with the German mainland. Especially for this purpose, the AURA has within 3 months specially been converted from a pure deck transporter for wind turbine components to a cable laying vessel at Finland's Rauma shipyard, then still part of

Fig. 6.18 Due to its special shallow-water design the Dutch shallow-water cable layer STEMAT SPIRIT can lay cables right up to the shore (Ralf Witthohn)

Korean STX group. The work included widening the ship's hull over a length of 70 m to increase its stability and to accommodate a carousel capable of carrying all types of cables. After conversion, the deck of AURA is a sufficiently large platform to host the auxiliary power supply and the cable installation equipment. This includes the laying device, a heavy-duty-trencher of 1600 kW output, an actively heave compensating crane and a turntable. In order to perform the laying tasks precisely, the ship has been retrofitted with a dynamic positioning system of class DP2. For its operation, AURA was equipped with a second and third transversal thruster, which are electrically driven by power from two additionally installed diesel engines. An accommodation area located behind the original superstructure, offers space for 36 members of the laying crew. The wheelhouse was enlarged as well and two ballast tanks were converted into fuel tanks.

The conversion was based on a long-term charter contract that Norddeutsche Seekabelwerke had concluded with Finland's Meriaura company in March 2012 about an employment for laying power cables in the North Sea and Baltic Sea. AURA is flying the Finnish flag and manned by a Finnish crew. The contract provided the cable works with a second installation vessel in addition to the laying barge NOSTAG 10, which had entered service in 2009. Operated by the companies Ludwig Freytag/Oldenburg, NSW and Hans Schramm/Brunsbüttel, NOSTAG 10 shows a length of 92.9 m, 27.5 m breadth and achieves a deadweight of 6364 tonnes. The shallow-water cable-layer is only equipped with a manoeuvring unit and has otherwise to be moved by tugboats. The pontoon was produced in China, while its forward placed superstructure was built by BVT Brenn- und Verformtechnik in Bremen.

After commissioning, NOSTAG 10 laid a 53 km long power cable between St. Peter-Ording/Germany and the isle of Helgoland, which until then had been dependent on diesel generators. Further deployments of NOSTAG 10, which was subsequently given a raised forecastle deck to improve seaworthiness, included laying work to the Alpha Ventus wind farm off Borkum and to the Danish Rodsand II farm. In December 2017, the cable layer, which can work to depths of up to 50 m, was loaded onto the semi-submersible heavy-duty vessel TRIUMPH of Dutch company Dockwise on Wilhelmshaven roads and transported to Amamapare/Indonesia. There, about 11,000 km of fibre optic cable were to be laid as part of the Palapa ring project, which was completed in October 2019. For this project NOSTAG 10 had previously loaded 850 km of cable at Norddeutsche Seekabelwerke. In 2018 the cable layer was registered for Nostag Nusantara Mardika in Jakarta and hoisted the Indonesian flag (Fig. 6.19).

Fig. 6.19 The non-propelled cable layer NOSTAG 10 is especially suitable for coastal water projects, but dependent on tug assistance (Ralf Witthohn)

6.3.8 Replacement Cable for Borwin 1

In autumn 2012, the cable layer ATALANTI is employed to create a replacement connection between the Globaltech 1 offshore wind farm and the Borwin Alpha converter platform of the Borwin 1 field. Actually, the Globaltech 1 wind farm should be connected to the Borwin Beta platform. However, as the completion of this platform under construction at Nordic Yards in Wismar is delayed, ATALANTI has to create a 30 km long replacement cable connection. The 97 m length ATALANTI of a Greek shipping company gets along with a small propulsion power when laying cables. To increase speed during transiting ATALANTI accepts tug assistance, which is often provided by the tug ANTEUS (Fig. 6.20).

ATALANTI was previously involved in the network connection of the Borwin 2 offshore wind farm. Of the work, which included laying a 125 km cable on the seabed via Norderney to Hilgenriedersiel in Ostfriesland/Germany and 75 km on land, the cable layer took over the seaward part. The cable-laying vessel, built in China in 2008 and completed in Greece in 2010, was chartered by Italy's Prysmian group, which together with Siemens was awarded the contract for the connection by Dutch network operator Tennet. The latter, state-owned company is responsible for laying power cables between the newly emerging offshore wind farms and the German mainland and awards companies the contracts for the technical execution of the projects. The laying of the cables was yet delayed due to unresolved liability and financing issues.

Fig. 6.20 The Greek-owned cable layer ATALANTI laid a replacement line between the wind farm and the converter platform (Ralf Witthohn)

6.3.9 Converter Platform for Meerwind Ost

Coming in from Freeport in the Bahamas at a speed of 11 knots, the crane ship OLEG STRASHNOV arrives in the German Bight in April 2014 and sets course for the Meerwind Ost wind farm. There is to be erected a converter platform built in Bremerhaven in the previous months. The day before, the structure has been carried to the site 23 km north of Helgoland on a pontoon towed by the Italian deep-sea tug CARLO MAGNO. Bremerhaven tugs have towed the transformer station lying on the transport pontoon WAGENBORG BARGE 7 from the fishing port through the lock, after which CARLO MAGNO took over the towing chain of the pontoon. OLEG STRASHNOV is needed to place the around 3500 tonnes weighing platform on the prepared foundation. For this action, ideal conditions with low wind forces and little swell were awaited for. The transformer station is positioned in the middle between the Meerwind Süd/Ost wind farms to serve both (Figs. 6.21 and 6.22).

The contract was at the beginning of 2012 placed by Bremerhaven/Germany-based company WindMW with a consortium of the steel construction company Weserwind and the French plant constructor Alstom Grid and included the steel structure of the platform. It houses the complete transformer station. The 12 m high topside, painted in warning yellow and measuring 31 × 46 m, has three decks accommodating a substation and is fitted a helicopter landing deck above. The dimensions of the 950 tonnes jacket foundation for the platform, which has already been placed before, are 28 × 28 × 47 m. For the two 288 MW wind farms with 80 turbines, components with a total weight of around 90,000 tonnes have been processed. Dutch company Tennet is laying the power cables for the parks, which extend over an area of 42 m² and cost a total of EUR 1.2 billion. The wind

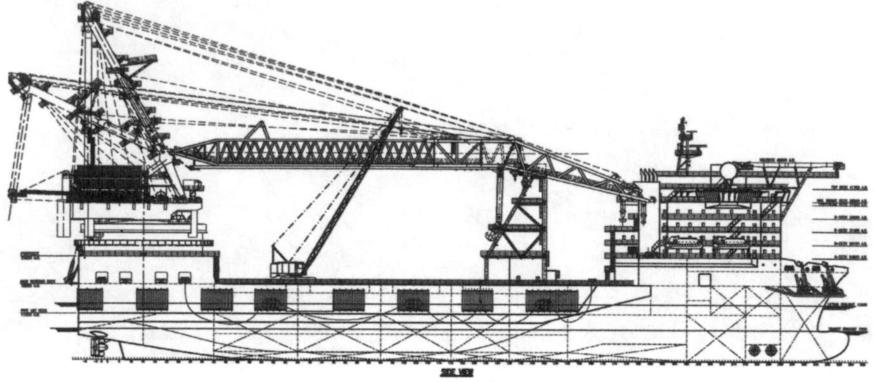

Fig. 6.21 The crane of OLEG STRASHNOV can lift 5000 tonnes (GustoMSC)

Fig. 6.22 The immense crane capacity of OLEG STRASHNOV is in demand for offshore work of all kinds, in the pictured occasion for the erection of transition pieces for wind turbines (Ralf Witthohn)

farm is financed by a EUR 820 million share from a consortium of banks and EUR 360 million from investment capital. The bank consortium is led by state-owned German KfW-Ipex bank and includes Hessische Landesbank, Lloyds and Santander. Shareholders of WindMW and the wind farm are the investors Blackstone and Windland Energieerzeugung with shares of 80 and 20%, resp.

The crane ship OLEG STRASHNOV, built in the Netherlands by IHC Merwede in Krimpen aan den Ijssel in 2011 and registered in Cyprus on behalf of the company Seaway Heavy Lifting based in Zoetermeer/Netherlands, has been equipped with a 5000 tonnes lifting crane. With its help, the crew of up to 395 people carries out the installation of the heaviest components on and under water. The basic concept of the ship was developed by the customer, founded under Russian influence, together with the Dutch company GustoMSC, on its turn supplying the offshore crane under sub-contract of the shipyard. In order to achieve the comparatively high speed of 14 knots for such a sophisticated ship, a special hull form was designed and patented. The vessel achieves its stability required for crane operation by a hull widened to 47 m above the design draught waterline. Ballast is taken by eight pumps with a total hourly capacity of 20,000 m^3 in four tanks. During crane operation it is pumped against the heeling motion into the side bulges, so-called sponsons. Welcome effect of the upward widening of the hull is the creation of a large working deck with a surface of 4000 m^2. Exact positioning at the operation site can be achieved by eight retaining anchors in combination with a mooring and dynamic positioning system on the base of eight winches. Six Wärtsilä diesel engines of the main generators provide a total output of 27,000 kW. They supply the power for two aft-installed azimuth propellers, two retractable azimuth units arranged at half the ship's length, two transversal thrusters forward, as well as for the ballast pumps and for crane operation. The crane, which rotates through 360 degrees, has a hook that lifts 5000 tonnes at 32 m outreach up to 102 m above the waterline and 740 tonnes at the maximum outreach of 84.5 m. An auxiliary hook lifts 800 tonnes to 72 m, another can handle 200 tonnes. The supporting frame of the main crane, which consumes up to 6000 kW of electrical energy, can be laid down, so that passages otherwise restricted by bridges, such as those of the Suez canal or the Bosporus, are made possible.

After commissioning, OLEG STRASHNOV carried out the first installation work on the offshore wind field Sheringham Shoal off the English east coast. During this work, 66 monopiles, 71 transition pieces and the upper parts of two high-voltage distribution systems were installed. In February 2012, the ship completed the installation of a 4000 tonnes offshore structure in India. In June 2016, work was carried out off Sheringham on the east coast of England again. In 2017, Luxembourg-based Subsea 7 acquired a 50% share in Seaway Heavy Lifting from K&S Baltic Offshore (Cyprus) and became the sole owner of the company, now named Seaway 7. The company was awarded the contract for works related to the inner array grid and export cable system of the Yunlin wind farm in the Taiwan Strait. OLEG STRASHNOV was in 2019 renamed SEAWAY STRASHNOV. The heavy-lift vessel was in May

Fig. 6.23 Using its own 2500 tonnes SWL crane STANISLAW YUDIN loaded tripod foundations for Global Tech 1 in Bremerhaven in March 2013 (Ralf Witthohn)

2021 working at the Southwark platform off the Englisch East Coast. In February 2022, the ship was moored in a Maasvlakte dock at Rotterdam.

Seaway Heavy Lifting operates another crane ship of much older origin, named SEAWAY YUDIN ex STANISLAV YUDIN and built by Wärtsilä in Turku in 1985. Of 183 m length and 36 m breadth, the ship uses its 2500 tonnes lifting revolving crane to build up wind farms, like on the Formosa 2 field off Taiwan in 2021. In early 2022 the SEAWAY YUDIN was on its way from the Indonesian port Batu Ampar to Anping in Taiwan (Fig. 6.23).

6.3.10 Working and Living in a Wind Farm

Offshore work requires new technical solutions in the transfer of personnel to their place of work and in the provision of accommodation. For this purpose, a large number of special ships, mainly crew boats, were put into operation. Particular attention was paid to equipping them with complex gangway systems for a safe transfer from the tender or the accommodation vessel to wind power plants. One of the first newbuildings to combine the tasks of accommodation and transport of workers was the accommodation vessel BIBBY WAVEMASTER 1, delivered in August 2017 by the Dutch Damen group to UK-based Bibby Marine Services under a long-term charter contract with Siemens Gamesa and EnBW. A year later, Damen Shipyards Galati in Romania began constructing the sister vessel BIBBY WAVEMASTER HORIZON, which was delivered in October 2019 and entered a 10-year charter of the same German companies to serve as hotel ship in the German North Sea wind parks Hohe See and Albatros for periods of 4 weeks, interrupted by replenishment at Emden/Germany.

The outstanding component of the ship's special equipment is a transfer system installed on the aft deck. A 2-tonne lift that can be held on six levels allows the personnel to get access to the offshore facility over a variable gangway with a height-adjustable platform and a three-dimensional motion compensation system. The transfer can also take place in an 11 m long work boat. A dynamic positioning system of class DP2 in combination with two Schottel propulsion units of 2150 kW each under the stern and three transversal thrusters of 860 kW each forward—one of which is retractable—ensure exact positioning possibilities of the 13 knots fast BIBBY WAVEMATER 1. Its equipment also comprises an articulated crane with a maximum of 5 tonnes SWL, a landing deck for helicopters of up to 12 tonnes weight and a bunker station for crew botas, also called CTV (Crew Transfer Vessel). The ship's accommodations offer space for 20 crew members and 45 workers, with 30 additional berths at double occupancy. There are also six offices, five conference rooms and three recreation rooms with fitness and sauna facilities. In October 2017, BIBBY WAVEMASTER 1 was operating off the east coast of England, in March 2020 in the Dan Tysk offshore windfarm off Esbjerg (Fig. 6.24).

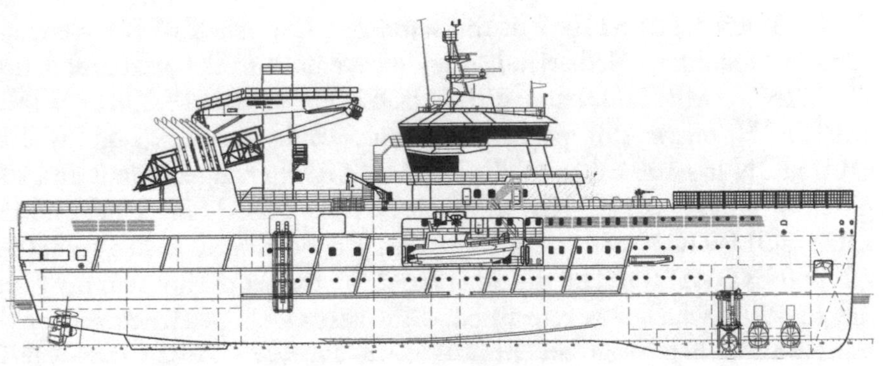

Fig. 6.24 The most important working tool of BIBBY WAVEMASTER 1 and BIBBY WAVEMASTER HORIZON is their high-tech gangway system (Damen)

6.3.11 Ferries to Floating Hotels

In addition to highly specialized newbuildings a number of older ferries are used as hotel ships for workers in the offshore industry. One of the largest of such floating hotels is the former Baltic Sea ferry REGINA BALTICA of 145 m length and 25 m breadth. Built in Turku/Finland in 1980 as VIKING SONG and later renamed BRAEMAR, BALTIKA, ANNA KARENINA and ANNA K., the 18,345 gt ship switched in 2012 to the SWE Offshore company in Vastra Frolunda/Sweden for operation under the Latvian flag as REGINA BALTICA. Still classified as a ro ro passenger ship by Italy's RINA society, the vessel was from 2012 to 2014 the prepared for the new task by fitting 300 single cabins, a bunker station for crew boats and a helicopter landing deck above the forecastle for rapid personnel changes. In 2015, the company also purchased the ferry MOBY CORSE, built in 1978 in Aalborg as DANA ANGLIA and fitted the ship with 250 single cabins, and the PATRIA SEAWAYS, built in 1991 as the passenger ferry STENA TRAVELLER in Landskrona, which was equipped with rooms for 100 workers. SWE Offshore provided floating accommodation for workers employed in the wind farms Sheringham Shoal, BARD Offshore 1, Borkum Riffgrund, Butendiek and Gode Wind as well as for the Scottish gas plant project Petrofac Lerwick. In 2017, REGINA BALTICA was purchased by Spain's ferry company Baléaria Eurolineas and again employed as a ferry between Algeciras and Tanger (Fig. 6.25).

A similar change of use took place in 2010 and 2008 on the passenger and car ferries PRINSESSAN BIRGITTA, built in 1969 by Aalborg Vaerft, and PRINSESSAN CHRISTINA of the same age. On behalf of the company C-bed in Hoofddorp/Netherlands, they were rebuilt to the accommodation ships WIND AMBITION and WIND SOLUTION. The 13,336 gt WIND AMBITION offers 150 passenger berths, while the 8893 gt WIND SOLUTION has 100 cabins. WIND AMBITION changed hands and was sighted in Greece as PRINCE in 2018, WIND SOLUTION became the AQUA SOLUTION of Piraeus-based Sea Jets Maritime. C-bed's 7880 gt WIND INNOVATION, on the other hand, is a former offshore supply vessel built in 1999, which was retrofitted with passenger superstructures for 80 cabins and a heli deck aft in 2016. In February 2022, the WIND INNOVATION proceeded from Den Helder to the Hornsea Two wind farm off the Englisch east coast (Fig. 6.26).

Fig. 6.25 The former Baltic Sea ferry REGINA BALTICA has been carrying out personnel change in port, but could also be reached by helicopter (Ralf Witthohn)

Fig. 6.26 Christened in 1969 as ferry PRINSESSAN CHRISTINA, the WIND SOLUTION served as a floating hotel for offshore workers from 2008 on (Ralf Witthohn)

6.3.12 Freight to Passenger Transporters

In 2015, the multipurpose cargo vessel AMSTELGRACHT of Dutch shipping company Spliethoff was equipped with accommodation facilities for workers, under a contract of Switzerland's Allseas group, which is involved in offshore work including pipe-laying and subsea construction. In a similar way, the Antarctic supply ship ICEBIRD in 1984 delivered by Heinrich Brand

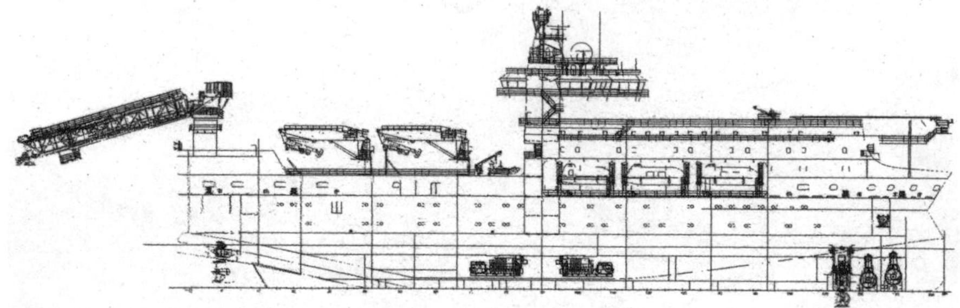

Fig. 6.27 The Spanish-built accommodation ship REFORMA PEMEX can host 699 persons (Barreras)

Schiffswerft in Oldenburg to Hamburg-based shipping company Günter Schulz and chartered out to the Australian government was designed from the outset with two-deck high living quarters for 90 scientists. Another special example of an accommodation ship, in this case used in oil and gas production, is the 21,383 gt vessel REFORMA PEMEX of Mexico's oil company PMI Norteamerica, delivered in September 2016 by Hijos de Barreras in Vigo/Spain. The ship has facilities for 699 people and is equipped with appropriate lifesaving equipment, a total of eight lifeboats and two evacuation systems. In early 2002, REFORMA PEMEX was positioned off the Mexican coast east of Cotzacoalcos (Fig. 6.27).

6.3.13 Hybrid Propulsion on GEO FOCUS

In addition to the large installation vessels, a whole range of sophisticated ships perform a wide range of special functions. These include scientific exploration of the seabed, research into ecological issues, rapid personnel transport, watchkeeping and towing tasks. The tasks are partly carried out by existing offshore or by converted ships, but also by newly designed, innovative ship types. One of the vessels involved in preparatory work for offshore wind farms is the surveying vessel GEO FOCUS of the Dutch shipping company Geo Plus. Among the international offshore surveys undertaken by the 34.5 m length special survey vessel was work for the Swedish wind farm Karehamn and for the installation of the converter platforms Dolwin, Helwin and Sylwin in the German North Sea sector. On behalf of the Belgian company G-tec, GEO FOCUS investigated the planned cable route connecting the wind farms of the ELIA group off the Belgian coast with a shore station near Zeebrugge (Fig. 6.28).

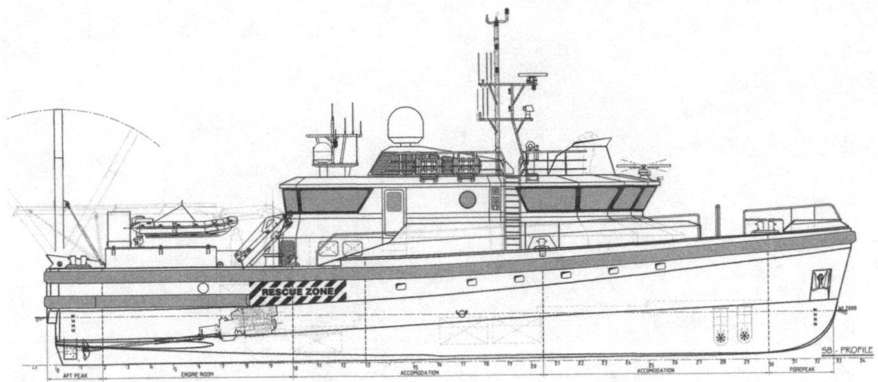

Fig. 6.28 On the high-tech ship GEO FOCUS the working deck can easily be watched through rear wheelhouse windows (De Haas)

GEO FOCUS was in July 2012 delivered by shipyard De Haas in Maassluis. According to its owner, the ship achieves fuel savings of 75% thanks to a hybrid propulsion installed. Two Caterpillar C-18 type diesel engines with a combined output of 1300 kW act through a Reintjes 344 RHS gearbox on two propellers ensuring a maximum speed of 15 knots. In the electric mode, the power of two John Deere diesel engines of 209 kW output each feeds through a PTI an electric motor giving the ship a speed of 9 knots. Alternatively, the adverse transmission fulfils the function of a PTO. The ship, which is classified as a workboat by Lloyd's Register, is equipped with two Voith inline transverse beams and a DP1 class dynamic positioning system. A hydraulically operated stern frame allows the use of remote-controlled underwater vehicles. In order to carry out the various researches, GEO FOCUS has extensive scientific equipment, such as a multibeam echosounder and a HiPAP. There are cabins for 18 persons, the crew numbers five persons (Fig. 6.29).

6.3.14 Crewboat GESA for Personnel Transfer from Helgoland

After 24 offshore workers have boarded, the crew Transport Vessel (CTV) GESA sets sail from the quay of the Cuxhaven fishing port, passes the lock and, reaching the Elbe fairway, quickly increases its speed to over 20 knots. The personnel carrier is bound for Helgoland, which has developed into a logistics centre for the construction of offshore wind farms off the German North Sea coast. From here, the crewboat takes workers to the 80 wind turbines of the offshore wind farms Meerwind Süd and Meerwind Ost, which are located about 40 minutes sailing

Fig. 6.29 By the stern frame GEO FOCUS can launch remotely operated underwater vehicles (Ralf Witthohn)

time from Helgoland. GESA has for this purpose been chartered by Bremerhaven-based company WindMW, the realisation and operating company for the construction and commissioning of the windpower plants (Fig. 6.30).

The 29.5 m length, 9.5 m breadth double-hulled vessel GESA has been specially designed by Maløy Verft in Deknepollen/Norway to carry out such personnel transports. Designated S-Cat, the fast catamaran was developed according to the SWATH principle (Small Waterplane Area Twin Hull). The design of the newbuilding, which hull material consists of glass-reinforced plastic (GRP) and thus helps to reduce the ship's weight of 160 tonnes, was based on two ship-theoretical considerations. Firstly, it was to benefit from the stability advantage inherent in ships with two hulls that are relatively far apart, and, secondly, from the design of the two torpedo-like hulls. These allow to adapt the hydrodynamic properties of the ship to the operating conditions by respective ballasting with corresponding changes of draught. In calm seas and without ballast, the submergence and thus the resistance is reduced, the power requirement is lower. In heavy seas, the draught is increased and provides a higher seaworthiness, by especially reducing slamming. Under these conditions the S-Cat can cope with wave heights of up to 4.7 m. In each of the two hulls there are two ballast tanks with a total capacity of 60 tonnes. The ballast water system, which is remotely controlled from the bridge,

Fig. 6.30 The requirements of transporting personnel in adverse weather conditions were thought to be best met by the choice of the SWATH design principle for the crew-boat GESA (Ralf Witthohn)

comprises four pumps, each of 90 m^3/h capacity. The dynamic stabilisation system includes four Humphree interceptors and two Servo Gear wings. In transit, a draft of 1.6 m with correspondingly lower resistance is aimed for. Then the ship will reach a service speed of 22.5 knots. During the test runs a maximum of 24 knots were achieved. In the rear end of each of the two hulls there is only little space for a four-stroke propulsion engine of type MTU10V2000M72, which was selected by the customer. The common rail diesel engines each produce 900 kW at 2250 rpm. The engines act on controllable pitch propellers via reduction gears. Their consumption is stated as 4 tonnes in 12 hours, the bunker capacity amounts to 10 tonnes.

Cargo of up to 10 tonnes, mainly consisting of spare parts, can be transported on the forward and aft deck offering a total area of 70 m^2. At a draught of 2.2 m, GESA achieves a deadweight capacity of 55 tonnes, the gross measurement is figured at 194 gt. Navigation takes place in a cockpit-like wheelhouse with two adjacent seats. For the offshore personnel there are 24 seats at tables in the rear area of the superstructure. A hydraulically operated platform is used for the transfer of personnel at the bow or stern of the ship, which is classified as passenger ship by Det Norske Veritas. Specially manufactured rubber fenders forward and a standard fender at the sides cushion the docking manoeuvres. In February 2022, GESA was still in crew transfer off the German coast, sailings including trips from Büsum and Cuxhaven to Helgoland (Fig. 6.31).

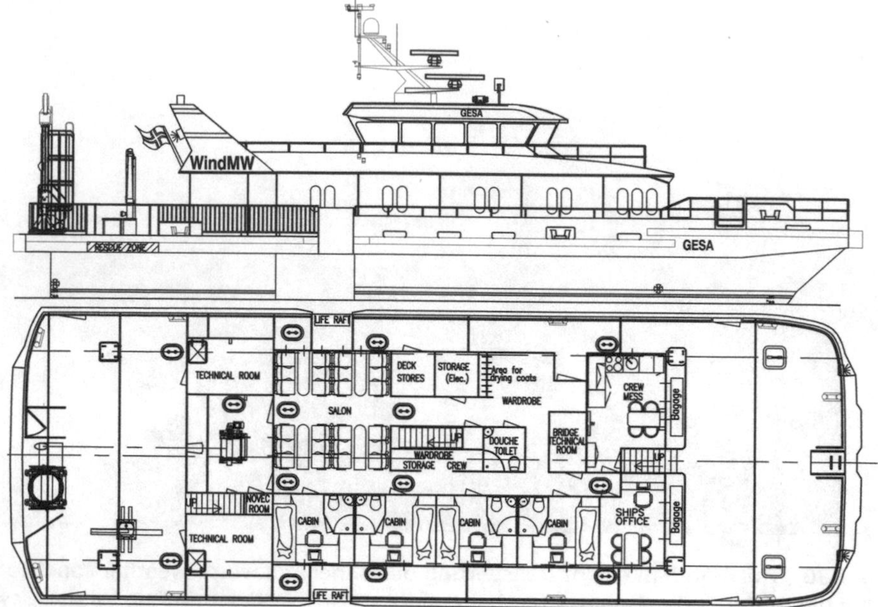

Fig. 6.31 The main deck of the crewboat GESA accommodates the rooms for the crew and the personnel to be carried (Maløy)

6.3.15 Fast Ship with Soft Bow

With the exception of the German-owned catamaran OFFSHORE TAXI ONE built in Cuxhaven in 2016, most of the fast transporters specially developed for the wind power industry were built at Dutch, British and Scandinavian shipyards. The shipping company World Marine Offshore in Esbjerg commissioned four fast personnel and material transporters of the WORLD SCIROCCO type from Norway's Fjellstrand shipyard in 2013. The ships combine the characteristics of a trimaran with those of a SWATH ship. With the help of ballast water, the regular transit draught of 2.2 m can be increased to 2.6 m to improve seaworthiness, when weather conditions should impede the personnel transfer. A hydrofoil at the bow and bilge keels help ensuring a smoother behaviour in rough seas. A "soft" bow in the form of an automatically operating hydraulic fender as known from railways, is integrated in the ship's stem. This allows collision forces, which are recorded for control, to be reduced by 50%.

The 23.2 m length crewboats are capable of carrying 12 people in a lounge and in three single cabins as well as a 20-ft container on the forecastle deck. The ships are powered by four Scania DI 13 main engines for a speed of 20 knots, so that operations can be continued even if one engine should fail. The sea endurance is 4 days. Besides transport tasks, which include the carrying of hazardous cargo, the vessels can serve for inspection, rescue and diving operations. Larger are the 30 m length, 13 m breadth WORLD BORA and WORLD CALIMA. In 2014 World Marine Offshore took over the 40 knots achieving air cushion tenders UMOE FIRMUS and UMOE RAPID for 24 passengers from Norway's Umoe shipyard as well as the catamarans WORLD SEAGULL and SEA COMFORT (Fig. 6.32).

Another high-tech crewboat under the Danish flag is the SWATH-type ship SEA GALE, which was deployed from Cuxhaven to the Meerwind field, but had to interrupt its service in February 2014 after both shaft systems were damaged, so that the ship had to be lifted onto a quay of Cuxhaven's fishing port for repairs. Operated by A2SEA the vessel is one of a series of four newbuildings made of carbon fibre which keels were laid at Danish Yards in Skagen. SEA GALE can accommodate 24 passengers and has a sea endurance of several days. Two MTU 10V2000 common rail engines acting on controllable pitch propellers through reduction gears provide a speed of 22 knots. A total of four Sleipner transversal thrusters at the bows and sterns of the two hulls increase manoeuvrability. In February 2018 the SEA GALE was engaged in the Irish Sea from Barrow-in-Furness as base port, but was back in German waters in 2021 as FOB SWATH 3 of Denmark's company Offshore Windservice.

Fig. 6.32 The soft-bow of the trimaran WORLD SCIROCCO facilitates mooring manoeuvres at the wind power towers (Ralf Witthohn)

6.3.16 Olsen's High-Speed Fleet

Since 2011 the shipping company Fred. Olsen Windcarrier (FOWIC), based in Fredericia/Denmark, has built up a fleet of catamarans named BAYARD 1 to BAYARD 7. They were developed in cooperation with shipyard Batservice Mandal/Norway. These ships also have ballast tanks, by which filling the draught can be increases, so that the sea-keeping qualitites are improved. A special feature is the stern configuration, which is designed to reduce movements and drifting. The BAYARD type ships are accelerated to a speed of 25 knots by two MAN engines of 749 kW each. At an economical speed of 22 knots and in laden state, the ships consume 186 l/h, at a maximum speed of 30 knots 350 l. The deck area of 51 m^2 offers storage space for three 20-ft containers. In addition to crew transport, as carried out by BAYARD 3, BAYARD 4 and BAYARD 6 to the Meerwind Süd/Ost wind farm, BAYARD 3 also assisted the cable layer NORMAND FLOWER in its work. In 2018 the crewboats joined the fleet of Sweden-based Northern Offshore Services (NOS) and were renamed BUILDER, BRINGER, BOARDER, BOOSTER, BACKER, BOLDER and BRAVER, resp. In 2022, NOS operated 42 special offshore crafts (Fig. 6.33).

Fig. 6.33 The special stern shape of BAYARD 6 is to improve seaworthiness and manoeuvring characteristics (Erich Müller)

6.3.17 Waterjet Propulsion for High Speeds

In September 2013, the new catamaran WATERLINES of Dutch shipping company Royal Wagenborg, coming from its homeport Delfzijl, arrived Bremerhaven Nordhafen in a charter of Germany's energy supplier RWE Innogy to transport passengers to the Nordsee Ost wind farm. RWE Innogy set up a service base on Helgoland from which the wind farm can be called at. The crewboat, initially completed under the name WHALE OF THE WAVES, was the first in a series of newbuildings of the Dutch VEKA Group and handed over to the Wagenborg shipping company in April 2013. Operated by a crew of three persons the 19.4 m length and 7 m breadth ship can accommodate 12 passengers. The aft deck offers space for three 10-ft containers. Powered by two Hamilton water jets the WATERLINES reaches a maximum speed of 27 knots, the cruising speed is 23 knots. The water jet pumps are turned by two MTU 8V2000 M72 diesel engines putting out 720 kW each. Renamed FOSTABORG the boat was employed by Dutch Wagenborg group in a ferry service between Holwerd and Ameland in 2021.

The most frequently realized design of a crew transporter is represented by a type from the Dutch Damen group, designated as Fast Crewboat Supplier FCS 2610. Examples are the newbuildings SURE STAR and SURE SWIFT, which were handed over to UK's Sure Wind Marine at the beginning of 2014, after contract signing had only taken place in September 2013. Among the nine ships of the shipping company at the beginning of 2018, many of which operate for German charterers, there is a third FCS 2610 unit, the prototype SHAMAL of the series, which was named SURE SHAMAL. Made of aluminium alloy the crewboats were built at Damen Shipyards Singapore, where they were built in large numbers for stock from June 2010 on and subsequently sold. The 26 knots reaching catamarans with a range of 1200 nm have a so-called twin axe bow and operate under the Workboat Code of Category 1. At 26 m length and 10 m breadth they offer facilities for four crew members and 12 passengers. A deck crane of 20 tm SWL is available for handling two 10-ft or one 20-ft container. Damen also developed the smaller twin axe type FCS 2008.

In contrast to most crew tenders, Dutch-flagged LIZ V, delivered in May 2013 from Neptune Shipyards in Aalst/Netherlands to shipping company Stemat in Rotterdam, represents a monohull design. Subsequent to commissioning the vessel was used for crew transports from Cuxhaven and Helgoland to the Meerwind Süd/Ost wind farm. Three Rolls Royce water jets of 970 kW each accelerate the 31.3 m length ship to a speed of 32 knots. Classified by Bureau Veritas, LIZ V is allowed to operate at a distance of 4 h sailing time from a safe haven. The crewboat is equipped with a special fender on its

Fig. 6.34 The monohull crewboat LIZ V is powered by three water jets (Ralf Witthohn)

vertical stem for mooring at offshore structures. The tender has room for 24 passengers, on deck can be stowed a 20-ft container. In 2021, operation of LIZ V switched to Dutch company Sima Charters, which renamed the ship SC ROTTERDAM and employed it in a fleet of 11 crew transfer boats, besides the tug ANTEOS (Fig. 6.34).

6.4 Six WG COLUMBUS Type Seismic Vessels

Due to the complex technology involved, the intensity of offshore oil production is highly dependent on energy prices. At times of high prices, large fleets of special ships were built, which suffered from low rates and underemployment in a weaker developing market. For the search for oil and gas, modern seismographic research vessels have been developed in addition to numerous conversions from stern trawlers. Based on the offshore type SX124 of Norway's Ulstein group, shipyard Hijos de J. Barreras in Vigo/Spain in 2009/2010 built the seismic vessels WG COLUMBUS, WG MAGELLAN, WG AMUNDSEN and WG TASMAN under contract of the British company Western Geco, a company of the US-based Schlumberger group, while Drydocks World Dubai contributed the same-type WG VESPUCCI and WG COOK. The ships were in 2020 registered for Shearwater Geoservices, which changed the ships' first names from WG to SW. Shearwater is a 2016 founding of Rasmussengruppen and Rieber Shipping and in 2019 entered an agreement with Schlumberger to purchase the assets and operations of Western Geco. In 2022, Shearwater Geoservices employed a fleet of 25 seismic research vessels.

These special ships, constructed with a so-called X-bow, are equipped with three high-pressure compressors that trigger sound waves on the seabed with the help of airguns. Hydrophones towed behind the ship on long cables record the reflections, from which conclusions can be drawn about the geology of the seabed and the presence of hydrocarbons. At the stern of the WG COLUMBUS, 12 reels are installed, from which the cables are let into the water via the stern. The necessary high towing power is provided by two diesel-electric azimuth propellers of 2 × 3000 kW output, which are supplied by six main generators of 1800 kW each. A retractable azimuth propeller under the forecastle puts out 850 kW. The 88.8 m length, 19 m breadth vessel, equipped with a DP2 class dynamic positioning system, reaches a speed of 15 knots. It can accommodate 69 crew members and the seismic crew. Its first mission took WG COLUMBUS to the US Gulf (Fig. 6.35).

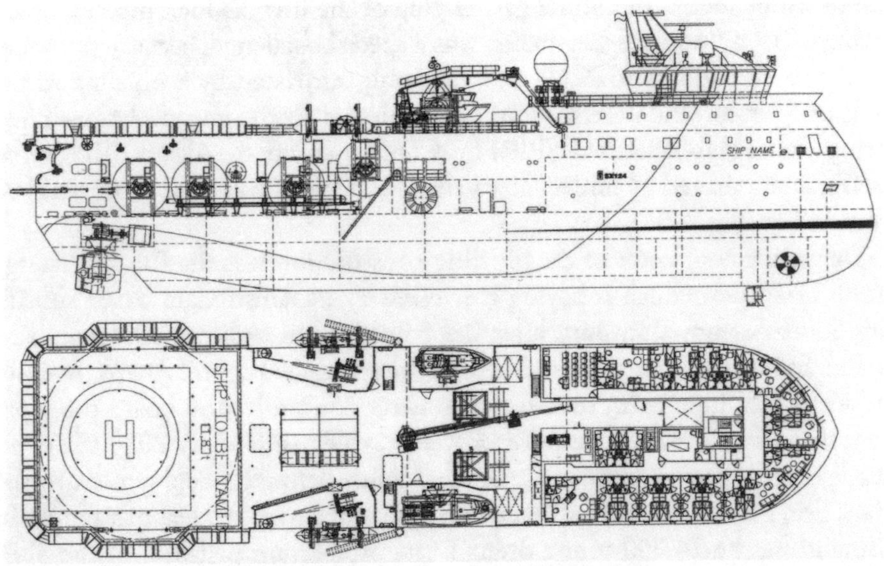

Fig. 6.35 The seismographic research vessel WG COLUMBUS, later named SW COLUMBUS, was in 2009 built by Hijos de J. Barreras in Vigo for the British shipping company Western Geco on the base of an Ulstein design (Barreras)

6.4.1 Russian Gas Through the Black Sea

For one of the longest trips that Hamburg harbour tugs have ever made, the assistant tugs BUGSIER 20 and BUGSIER 21 take plenty of provisions for a one-month voyage to Novorossiysk in early April 2017. After 14 days the tugs make a stopover in Valetta on Malta. There, the crews, which have been increased from three to six men for two-watch operation, are being exchanged. Their first location of work is situated off the Black Sea coast near Anapa. Hamburg-based tug operator Bugsier, which is tradition-rich in towing, salvage and offshore work, is subcontractor for assistance services during pipeline laying work. BUGSIER 20 and BUGSIER 21 are, thanks to their two Voith-Schneider type propellers, particularly manoeuvrable and thus well suitable for this kind of work done in co-operation with the Norwegian offshore vessel SIEM SPEARFISH, the Dutch tugboat BARRACUDA and the Dutch general cargo vessel EGMONDGRACHT.

At Anapa, the pipes of the TurkStream pipeline leave mainland and enter the Black Sea. The two parallel pipe strings of the offshore section of the pipeline cross the Black Sea over a distance of 930 km and reach land again 100 km west of Istanbul near the town of Kiyikoy. The pipeline feeds Russian gas into the Turkish gas network and is connected to the Luleburgaz network in south-eastern Europe. After the outbreak of the Russia-EU gas dispute in 2022, Bulgaria was cut off from deliveries. Its annual capacity is estimated at 31.5 trillion cubic metres of natural gas. Laying of the first 32-inch pipeline, consisting of 12 m long pipe segments, was a special challenge, because the water depths reached down to 2200 m, so far only surpassed by a pipeline in the Gulf of Mexico. The pipe sections were shipped to the world's largest pipe-laying ship PIONEERING SPIRIT of Dutch company Allseas. The laying itself was monitored by underwater vehicles, the welding of the pipes checked from inside (Fig. 6.36).

The prime contractor of the pipeline construction was the Russian energy group Gazprom, which is having it realized by its Amsterdam-based subsidiary South Stream Transport, after the SouthStream project had been abandoned. Seven weeks after the arrival of the two Bugsier tugs off Anapa, Russian President Vladimir Putin started the construction project on board the pipe-layer that had passed the Bosporus 3 weeks earlier. In January 2020 the pipeline laying work was completed. In the month before its deployment to the Black Sea, PIONEERING SPIRIT had performed a special feat of strength in dismantling the 24,000-tonne Brent Delta production platform of the Shell group in the North Sea. Over the next decades, the dismantling of several hundred offshore structures in the North Sea will have to be undertaken.

Fig. 6.36 The world's largest pipe-laying ship PIONEERING SPIRIT passed the Bosporus on its way to the Russian Black Sea coast (TurkStream)

6.4.2 Higher Yield Through Stimulation

In order to better exploit the yield of oil and gas wells, special ships are in use, known as well stimulation vessels. They are equipped with highly sophisticated equipment that uses various methods, for example by pumping acidic liquids into a blocked well. During a two-year conversion in 2013/2014, Lloyd Werft in Bremerhaven converted the offshore supply vessels ISLAND CENTURION and ISLAND CAPTAIN for such special tasks requiring great experience and special knowledge of the crew.

Under contract of the Norwegian shipping company Island Offshore, the ships had in 2012 been built to the standard design UT 776 CD of domestic Ulstein group. In their basic version, the ships are used to ransport cargo for platforms on the 1000 m^2 aft deck as well as liquid cargo in tanks. However, they were also equipped with a DP2 class dynamic positioning system for standby tasks at platforms. During the conversion, ISLAND CENTURION and ISLAND CAPTAIN were extensively redesigned and refit with special equipment. This included high-pressure pumps with a total output of 20,000 hhp (hydraulic horsepower) as well as large cable drums at the stern and on deck. After the conversion, the 93 m length ISLAND CENTURION and ISLAND CAPTAIN took up a 7 years charter of US-based Schlumberger group for work on behalf of an oil company in the North Sea (Fig. 6.37).

Fig. 6.37 The offshore supply vessel ISLAND CENTURION of Ulstein's standard type UT 776 CD after conversion into a well stimulation vessel (Ralf Witthohn)

6.4.3 ISLAND CONSTRUCTOR in Oil Production Operations

One of the most unusual vessels for assistance tasks in the search for and exploitation of hydrocarbon deposits is the ISLAND CONSTRUCTOR, built in Poland and Norway under contract of the Norwegian company Island Offshore. Built according to a design of the Ulstein engineering company with the designation SX121 the ship has been constructed with a patented X-bow. The vessel can perform various special offshore tasks worldwide, primarily working on oil or gas wells as so-called well intervention vessel. The 120.2 m long, 25 m wide and 10 m high hull of the ISLAND CONSTRUCTOR was built by Maritim shipyard in Gdansk and towed to Ulsteinvik for fitting out. After completion of work at the shipyard, the special ship started service in 2008 on the sealing of oil wells on behalf of the British Acteon group. Afterwards the vessel, which is entered in the Norwegian International Register and homeported in Aalesund, took up a long-term BP charter.

The diverse requirements profile includes the construction and equipping of underwater installations, their inspection, maintenance and repair, as well as work with remote-controlled underwater vehicles, known as ROVs (Remotely Operated Vehicles). The newbuilding is also capable of transporting liquid cargo such as fuel, fresh water, brine, sludge and chemicals in the ship's own tanks as well as up to 5600 tonnes of cargo on the 800 m^2 large aft deck. In the forward part of the ship is provided accommodation for a maximum of 90 persons. The Ulstein X-bow configuration of the stem promises better sea-keeping characteristics with lower fuel consumption. Main equipment components of this high-tech ship are its extensive working tools.

These include an 8 × 8 m moon pool, above which a 32.3 m high tower, designated module handling tower (MHT) for well intervention tasks is mounted on main deck. The tower is made of aluminium and equipped for the coil tubing laying method. The main hoisting wire and hook are designed for well intervention work at a safe working load of 100 tonnes, for drilling work via two halliards for a maximum of 200 tonnes. The 100-tonne active heave compensation (AHC) winch system has a control system that allows four different operating modes. The winch's 1100 m long, 66 mm thick wire is pulled by a hydraulic motor in AHC mode at a maximum speed of 165 m/min, in normal mode at up to 30 m/min. The working tower is designed for an SWL of 300 tonnes, but the static of the AHC winch only allows a maximum SWL of 200 tonnes. The pressurized tower cabin offers three seats for operation of the tower functions.

On the aft deck, which has a total area of 1470 m^2 and which is designed for a maximum distributed load of 10 tonnes/m^2, a special sliding system is installed for the safe and efficient transport of the underwater equipment across the deck of the ship to and from the moonpool. In addition to sliding beams welded onto the deck, it consists of two push-pull units, six equipment pallets totalling 60 tonnes SWL behind the moonpool, two equipment pallets totalling 100 tonnes SWL in front of the moonpool and two control stations. They handle the horizontal movement of the equipment, while the tower itself is used for vertical movement to the seabed. A special halliard and winch system is available for this process. The purpose of the system is to ensure safe working with the equipment even in rough seas. A steering system consisting of wires, which is used to guide the underwater equipment between the seabed and the ship, but not for hoisting, comprises four electrically operated winches of 5 tonnes SWL. They allow speeds of up to 186 m/min in AHC mode, while a maximum of 75 m/min is possible in normal operation. A wire-supported holding device consisting of four arms with an SWL of 0.7 tonnes serves to engage or disengage the four winches in the steering system of the underwater modules. It operates in a horizontal sector of 70° and in a vertical sector of 60° in conjunction with the steering winches, the working tower and the moonpool hatch. The umbilical compensator with a lifting force of 10 tonnes and a holding force of 17 tonnes is designed to compensate for movements of ±3.5 m at an amplitude of 2 m/s, while the wire compensator integrated in the tower with its top and bottom rollers equalizes cycles of up to 4 m and rotations of 80°. The static load for the hoisting operation is 10 tonnes.

Four passively acting tanks arranged below main deck—one each in the fore and aft ship and two in the middle ship area—serve to improve the rolling behaviour of the ship. In addition, an active anti heeling system is used, which has an interface with the ship's main crane. It operates automatically or can be manually remote controlled from the crane or bridge. The system consists of four 190 m^3 wing tanks and two electrically driven pumps. The main crane, which serves both cargo handling and offshore operations, is an articulated crane and specially equipped for ship operation. It has an SWL of 250 tonnes at 12 m outreach. Its maximum SWL down to a water depth of 500 m is 140 tonnes at 10 m radius. Up to 2500 m it carries a load of 94 tonnes.

The double hangar for the remotely operated underwater work vehicles (ROVs) is integrated in the rear part of the superstructure. A part of the middle section is designed as a separate room. On each side of the ship, the hangar has a 5 × 10 m gate in the shell plating. A gate of 5 × 8 m also provides rear access. The ROVs are launched into the water and recovered by means of two gantry-type launching systems (LARS), one on each side of the vessel. The helicopter landing deck above the wheelhouse with access to D-deck has been designed and equipped according to the regulations for helicopters of type EH101.

In accordance with the requirements of dynamic positioning class DP 3, the main engines, propulsion units and cross radiators are located in separate rooms. Their operation is secured by separation into two independently operating systems. The power for the diesel-electric propulsion is supplied by a total of four diesel engines, two of type B32:40L9 and two of type B32:40L6 from Rolls Royce Bergen. At a speed of 720 rpm, the nine-cylinder engines generate an output of 2 × 4145 ekW, the six-cylinder engines of 2 × 2765 ekW. This results in a total output of the four diesel generators of 13,820 ekW. The power is primarily used to supply two electrical motors with an output of 2 × 3500 ekW at 1200 revolutions. They turn two azimuth propellers of type Rolls-Royce AZP 120360°, which are mounted under the stern of the vessel and act as main propulsion units.

The main generators also provide power for a total of four manoeuvring aids, three of which are located in the fore and one in the aft. Two Rolls-Royce transversal thrusters of type TT 2650 DPN CP with an output of 2 × 1800 kW at 900 rpm are installed in the foreship. Their speed and pitch are variable, the propeller diameter is 2650 mm. A retractable manoeuvring unit of type TCNS 92/62-220 with a propeller diameter of 2200 mm and with variable speed and pitch is mounted forward as well. Its output is 1500 kW at 1800 1/min. A manoeuvring unit of the same Rolls-Royce type is installed under the stern. As auxiliary diesel acts an MTU 12 V2000 engine delivering 550 kW at 1800 rpm, while a Scania DI12 emergency diesel delivers 318 kW at 1800 rpm. The DYNPOS-AUTRO IMO Class III dynamic positioning system has three operator stations. Its function is based on the evaluation of data collected by three wind and motion sensors, two hydro-acoustic systems that operate according to the water depth, a radar scan system and two DGPS systems. Interfaces exist with the gyro compass as well as the two main propulsion and four manoeuvring units of the ship. The status of the DP system is displayed on the navigational bridge, engine control room, ROV control room, ROV hangar, main deck, captain's chamber and working tower. At a draught of 7.5 m, the vessel complies with the classification requirements for maintaining position according to an ERN (Environmental Regularity Number) of 99,99,99. The vessel is equipped with a joystick system that integrates the operation of the two propulsion and four manoeuvring propellers in one control lever.

6.4.4 German Oil from Mittelplate

Shortly after midnight, the supply vessel COASTAL LIBERTY departs from Cuxhaven for Germany's only offshore oil field, located off the coast of Schleswig-Holstein. The tide determines the supply schedule of the artificial island located near the Kaiser-Wilhelm-Koog coast line in the sensitive Nationalpark Wattenmeer. About 4 h later, before daybreak, the COASTAL LIBERTY is back at the Helgoland quay in Cuxhaven. The 233 dwt supply ship is part of the fleet of the Dutch Acta Marine, which has also stationed its sister ship COASTAL LEGEND and the COASTAL EXPLORER in Cuxhaven, while SARA MAATJE IV and SARA MAATJE VII are transporting passengers from Cuxhaven to Mittelplate.

The small fleet of Den Helder-based Acta shipping company is carrying out its supply duites on behalf of RWE Dea, the operator and, together with its partner Wintershall, shareholder of Germany's largest oil field with an annual production capacity of about 1.4 million tonnes. The Mittelplate production island was built for the exploitation of this field starting in 1987. Until 2005, the crude oil was transported to the Brunsbüttel oil port by tugboat-pushed barges. Since then, the oil has been transported through a 10 km pipeline to a production facility near Friedrichskoog and then pumped on to Brunsbüttel or the oil refinery in Hemmingstedt (Fig. 6.38).

Fig. 6.38 The Dutch deck transporters COASTAL LIBERTY and COASTAL LEGEND supply the only German offshore oil exploitation site Mittelplate from Cuxhaven (Ralf Witthohn)

Fig. 6.39 The floating production and offloading unit PETROJARL I can produce 30,000 tonnes of oil per day (Ralf Witthohn)

6.4.5 Oil Production Vessel PETROJARL I

The oil production vessel PETROJARL I of America's Teekay group, listed at the New York Stock Exchange and operating from Vancouver, was in 2016 prepared for a five-year charter by the Brazilian oil and gas production company Queiroz Galvao Exploracao e Producao (QGEP) at the Damen ship repair yard in Rotterdam. Built in 1986 by Japan's shipyard Nippon Kokan in Tsurumi, the floating production storage and offloading unit (FPSO) can produce 30,000 tonnes of oil per day. The ship is equipped with a turret mooring system that supports up to nine production lines anchored to the seabed, independent of the ship's movements. The 215.3 m length, 32 m breadth 31,473 dwt vessel is electrically driven by power from diesel and gas turbines on two propeller shafts for a speed of 10 knots. In the forward positioned superstructure there is accommodation for 68 crew members. At the end of 2017, the ship was on its way from Haugesund to the Atlanta field in the Santos basin 185 km off the Brazilian coast, where its position was still reported in 2020. The QGEP estimates the deposit to be developed in 1500 m water depth at 260 million barrels of oil. In early 2022, PETROJARL I was reported to be anchoring in the river Parana/Brazil (Fig. 6.39).

7
Dredging

The rapid growth particularly in the size of container ships has given rise to demands for the deepening of river fairways, which, once implemented, require permanent maintenance measures. The construction of new ports is also dependent on modern dredging technologies. Depending on the procurement of the subsoil, dredgers of various types are used in the creation and maintenance of waterways and ports. In the case of firmer subsoils such as marl, chalk or hardened sand, these are usually non-propelled bucket dredgers operating along land-connected ropes with the aid of winches and emptying the buckets into barges. Bucket dredgers are also employed for softer grounds such as sand or mud, which are otherwise worked by suction dredgers, in case they are to ensure a more even removal.

Cutter suction dredgers are used for work in particularly hard ground such as even rocks. While they lower the cutting head to the ground on ropes, the dredger itself is anchored for the dredging operation and swung around a fixed pile with the help of winches. Trailing suction dredgers, which usually dump the dredged material picked up by the drag head through bottom doors and are therefore described as hopper dredger, are designed in such a way that they can alternatively pump the dredged material through pipelines or spray it ashore for land reclamation. Other types of dredgers include stilt dredgers, backhoe and grab dredgers, which all deliver the dredged material into barges. Dredgers are also designed and employed to extract sand from the sea for use as building material.

Although the fees levied on shipping for navigating canals are generally lower than the cost savings from the reduction of the sailing time, such charges are not common practice for waterways of natural origin, so that general

R. Witthohn, *International Shipping*, https://doi.org/10.1007/978-3-658-34273-9_7

public has to bear the costs for maintaining their navigability. Many rivers that have lost their original shape and won a canal-like character as a result of being made navigable have to be permanently maintained at costs that far exceed those of canals. The maintenance and particularly the deepening of fairways, which is often necessary to meet the rapid growth in the size of ships, represents a significant intervention in nature and a threat to the coastal population by floods, which is already affected by higher water levels due to climate change. The reason for the failure to take advantage from the commercially orientated main users is to be found in the traditional competitive situation of ports and their unwillingness to cooperate. While emerging countries, in particular, are attempting to improve the conditions of navigability for constantly growing ship units by means of large-scale infrastructure measures, legal actions in developed countries are often leading to years of delays in the further deepening of rivers.

At a cargo capacity of 46,000 m^3 described as the world's largest trailing suction hopper dredgers are CRISTÓBAL COLÓN and LEIV EIRIKSSON, built in 2009 and 2010 by Construcciones Navales del Norte in Sestao/Spain for the Belgian contractor Jan de Nul. The 213 m length, 41 m breadth vessels show a draught of 15.2 m, resulting in a deadweight capacity of 78,386 tonnes. The dredgers, equipped with a twin propulsion system consisting of two 19,200 kW main engines turning two propellers, can dredge to a depth of 155 m. Their two suction pipes have a diameter of 1.3 m. The dredgers can dump their cargo through nine bottom doors or discharge it ashore via a pressure pipe (Fig. 7.1).

While undertaking dredging work in the Elbe river in January 2010, CRISTÓBAL COLÓN did not dump the dredging material at deeper points of the river as usually is the case, but pumped the sand through a pipeline on to a projected harbour area east of Cuxhaven/Germany. Though being assisted by three tugboats, the ice flow of the Elbe repeatedly proved to be too strong and interrupted the pipeline. The dredger had to be coupled to the floating

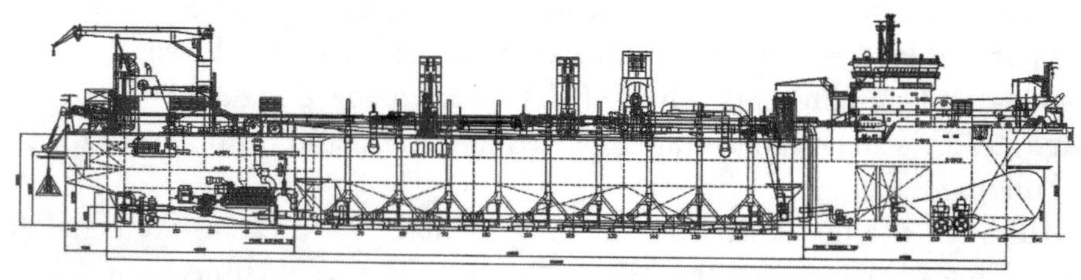

Fig. 7.1 Built in Spain and operated by a Belgian owner, the suction dredgers CRISTÓBAL COLÓN and LEIV EIRIKSSON represented a new size class (CN del Norte)

Fig. 7.2 In January 2010, CRISTÓBAL COLÓN pumped dredged Elbe sand through a pipeline on to a projected harbour area east of Cuxhaven/Germany (Ralf Witthohn)

pipeline up to three times a day, which proved to be an often difficult manoeuvre in the prevailing ice conditions. For 6 weeks, 2 million cubic metres of sand were transported ashore, until the area had reached the necessary height of the quay to be built afterwards and to take over the dyke protection. In the beginning of 2020 CRISTÓBAL COLÓN proceeded from Mailiao/Taiwan to Galang anchorage/Indonesia, where the ship still anchored in May 2021 with de Nul's VASCO DA GAMA, GALILEO GALILEI and J. F. J. DE NUL. In early 2022, CRISTÓBAL COLÓN was undertaking dredging work at Manila with LEIV EIRIKSSON, which had in 2020 been sighted off Takoradi/Ghana and in 2021 reported as berthed in Pula/Croatia along with the companie's IBN BATTUTA and L'ETOILE (Fig. 7.2).

7.1 Dredging in Bronka

In April 2015, the Chinese suction dredgers TONG YUAN, coming from Jeddah, and TONG XU bound in from Qinhuangdao shortly after each other arrive in St. Petersburg. The trailer suction hopper dredgers of the China Communication Construction Company (CCCC) have made the long voyage to continue the construction of the new Bronka multi-purpose terminal on the south coast of the Gulf of Finland west of St. Petersburg, which will finally be completed in December 2015. To create a water depth of 14.4 m in the 6 km long entry to the port, in the turning basin and at the berths, a dredging volume of more than 13 million cubic *metres will be moved. In the previous year, the Dutch company Boskalis had started the EUR 130 million project by deploying its dredgers. The 155 m length, 27 m breadth TONG XU has a capacity of 13,000 m*^{3}*. The suction dredger is a 2008 newbuilding of China's Guangzhou Wenchong shipyard, which equipped the ship with a dual fuel Wärtsilä propulsion system. The two 17,400 kW Wärtsilä main engines of type 12V38 comprising propulsion system turns two controllable pitch propellers accelerating the ship to a speed 15.5 knots, and, through a reduction gear and also drives the dredging pump. In April 2020 the dredgers were sighted in the waters of their homeland, TONG YUAN was anchored off Caofaidian, TONG XU off Weifang. In early TONG XU was engaged in dredging work at Dar es Salaam/Tanzania.*

In August 2017, BLG Logistics Group from Bremen and the owner and terminal operator LLC Fenix decided on the joint business development of the port of Bronka, aiming to replace road transports to and from central and western Europe. By then, 18,000 new vehicles had been handled since the port had been inaugurated in December 2016. The storage zone within a total port area of 120 ha is planned to be extended to 12 ha. Russia has started to develop several more port projects. In the beginning of 2020, the Russian Far East Investment Promotion Agency finished a feasibility study on the Slavyanka port expansion project, in which the South Korean infrastructure and engineering company Hyein E&C was reported to plan investments supported by Korea's Hyundai Engineering and Construction company with up to USD 540 million. The project is to include the construction of a multipurpose port with terminals for the handling of containers, grain and the export of LNG. The Slavyanka port project shall become part of the Primorye-2 transport corridor for the transport of cargo between Russia, China, Korea and Japan (Fig. 7.3).

7.2 Deployment to Port Harcourt

A high level of dredging technology is represented by the suction hopper dredger BREYDEL. The dredger was in 2008 delivered by Dutch shipyard IHC to the Belgian dredging company DEME in Zwijndrecht for worldwide operation by its subsidiary Dredging International within a fleet of over 80 dredgers and 200 auxiliary vessels. The dredger has specially been designed for work in shallow waters. By choosing a large ship breadth of 28 m and keeping the ship's weight as low as possible, the draught was limited to a maximum of 9.1 m. This enables dredging to be carried out in fairways that would not allow the work of commonly designed dredgers of the same capacity class. As

Fig. 7.3 The trailer suction hopper dredger TONG XU, built in China according to European design, left its home country in 2015 to carry out work for the new Bronka terminal near St. Petersburg (Ralf Witthohn)

a so-called trailing suction hopper dredger, the ship loads the dredged material from the seabed into its holds by its dredging equipment and disposes of it by opening the ship's bottom doors. When working in shallow waters, four pre-disposal doors are used to dump a significant part of the cargo in advance in order to reduce the draught of the vessel, before it reaches the actual dumping position. The doors are installed in a way, that they do not protrude below the keel when opened. As an alternative to the conventional form of dumping, the ship can also spray its cargo through a nozzle or, after a docking manoeuvre at a floating pipe, pump it ashore for the purpose of land reclamation or bank construction. The hydraulically remote-controlled suction pipe of the dredger can operate in water depth of up to 43 m. It is lowered or raised by a gantry crane portside aft by hydraulically operated winches.

The design incorporates the technical advances achieved with the previous dredger types PALLIETER and BRABO of 5400 and 10,650 m^3 capacity, resp. At 121.5 m length, 28 m breadth and 9.8 m depth, the capacity of the BREYDEL is 9000 m^3. In its parallel central hull part the dredger has a cargo hold which is divided into five compartments by transverse bulkheads. The maximum deadweight at a draught of 9.1 m reaches 18,710 tonnes. On this draught, which corresponds to a freeboard of 0.7 m, the dredger is allowed to operate up to 8 nm off the coast. On a draught of 8.15 m, which correlates to a freeboard of 1.65 m, offshore operation up to 15 nm is permitted. The bottom of the dredger is formed by ten double bottom doors arranged in two rows. They can be opened or closed hydraulically by remote control. At the front end of the hopper a telescopic overflow device is installed which is adjustable in height. A jet system on both sides of the double bottom ensures that the dredged material liquefies when being pumped ashore or dumped. It also supplies water at a pressure of 16 bar to the suction head, to ejectors, and is used for spilling the cargo hold and filling the forepeak with ballast water. Start and stop of the two jet pumps as well as the control of the valves is done by an automated remote control in the wheelhouse (Fig. 7.4).

The ship's twin propulsion system is designed to maintain dredging and pumping operations by means of specially developed reduction gears. A distinct feature of the optimized underwater hull is a pronounced bulbous bow, which increases the ship's deadweight, but still enables a sufficient ship speed at relatively low engine power. The power for the double-walled centrifugal dredging pump, installed directly in front of the forward engine room bulkhead, is provided by the ship's portside main engine. This is connected through a reduction gear which allows three gear ratios, one for dredging operations and two for unloading ashore. Measuring devices and sensors are used to monitor the dredging process, the draught of the ship, the cargo and the position of the suction pipe. They supply data to a dredge control system (DCS), which is operated by only one specially trained officer, responsible for both navigation of the ship and dredging operations. By combining the activities

Fig. 7.4 A comparatively large breadth of 28 m makes the 9000 m³ trailing suction hopper dredger BREYDEL better suitable for work in shallow waters (Ralf Witthohn)

on an integrated navigation and dredging console, dredging is expected to be carried out more economically and with less production losses, misunderstandings in ship management and near-accidents. While safe navigation of the vessel is looked upon by the shipping company as a primary goal, the actual dredging work is seen as a more intuitive task.

In the aft arranged engine room, two Wärtsilä type four-stroke irreversible main engines are acting on two controllable pitch propellers. The port engine of type 12V32 puts out 6000 kW at 775 rpm, the starboard engine of type 8L32 produces 4000 kW at the same revolution. Both engines are reduced by gears to a propeller speed of 140 rpm. They accelerate the loaded vessel to a maximum speed of 14.8 knots. The starboard engine is equipped with a power-take-off (PTO) for a 2125 kVA shaft generator. Diesel generators with outputs of 905 and 207 kVA are operating as auxiliary generators. The high maneuverability required when dredging is improved by two spade type flap rudders. In addition, the ship is equipped with two transversal thrusters forward.

BREYDEL was deployed to undertake dredging work off Port Harcourt in the Niger estuary delta at the beginning of 2018. In March 2020, the dredger was engaged in the port of Salerno, together with the DEME owned dredger SCHELDT RIVER. In April 2019, DEME group took over the 16,188 m^3 trailing suction hopper dredger BONNY RIVER from China's COSCO Shipping Heavy Industry (Guangdong). The 158.2 m length, 138.8 m breadth, 13.7 m depth, 10.0 m depth vessel is powered by two four-stroke dual fuel diesel engines type Wärtsilä of 2 × 8000 kW output and thus prepared for burning gas fuel. After jurisdictional approval of the Elbe deepening, BONNY RIVER became in 2020 engaged in the deepening of the river. In February 2022 BONNY RIVER proceeded from Abu Qir/Egypt to Ravenna/Italy. DEME participates in a project intending to build a plant in the Ostend port area that is to produce green hydrogen from wind farm power by 2025 (Fig. 7.5).

7.3 Dredging off Yuzhny

In May 2013 the German trailing suction hopper dredger EKE MÖBIUS started deepening of the fairway to the Ukrainian port of Yuzhny. Delivered by J. J. Sietas to Hamburg-based company Josef Möbius Bau in August 2012 the dredger is designed for fairway deepening and maintenance work. In view of the fact that the energy requirements during the different modes of operation are timely independent of each other, a diesel-electric supply system was

Fig. 7.5 Fitted with dual fuel main engines, the 2019 built BONNY RIVER is reducing the environmental impact when using gas as fuel during dredging operations, which are mostly undertaken near the shore and ports (Ralf Witthohn)

chosen which supplies the energy for both the propulsion of the ship and for dredging or unloading. The generators and electric motors for the propulsion or the dredging pump drive are controlled by a power management system, which is to minimize fuel consumption, wear and emissions. The energy is supplied by four Caterpillar diesel engines, each producing 1630 kW and connected to generators. They supply two electrical motors of 1700 kW output each. The Caterpillar diesels also cover the power requirements of the dredging pump, which delivers 900 kW for dredging and 2000 kW for transfer of the dredging material ashore.

At a ship's speed of about two knots the cargo hold is loaded in 62 min through a suction pipe of 1 m diameter reaching down to 30 m. The dredged material is transported at a speed of 13 knots and can within 66 min be discharged over the bow station or dumped through eight bottom doors. The classification rules allow the 120.4 m length, 21 m breadth hopper dredger to work at a draught of 6.8 m and to navigate without restrictions on 5.7 m draft. The cargo hold has a volume of 7350 m^3 up to the overflow and of 8000 m^3 up to the coaming. The deadweight capacity is stated as 9900 tonnes. EKE MÖBIUS and sister ship WERNER MÖBIUS had originally been built to participate in the deepening of the Elbe, but due to the delay of the project EKE MÖBIUS carried out dredging in the Black Sea. After take-over of the Möbius company as part of Austrian Strabag group by the Dutch Boskalis Westminster building concern EKE MÖBIUS joined their fleet in 2016 and was renamed MEDWAY. In March 2020, the dredger operated on the Weser and at Rotterdam. The former WERNER MÖBIUS, renamed BEACHWAY, reached Singapore on a voyage from Vancouver and was in 2021 engaged in Taiwan's Yunlin offshore wind farm project, from where the dredger proceeded to Singapore, where it was moored in the beginning of 2022 (Fig. 7.6).

7.4 Belgian Cutter Head Dredger VESALIUS

While trailing suction dredgers are self-propelled vessels with their own cargo hold and bucket-chain dredgers traverse the ground with the aid of ropes, cutter suction dredgers are stationary dredgers, which are equipped with a rotating cutter head for cutting and fragmenting harder soils and rock. The material is sucked up by dredge pumps and either discharged through a floating pipeline ashore, to a deposit area or into split hopper barges moored alongside the dredger. While operating the dredger is secured by a spud lowered into the seabed. With the help of winches and anchors the dredger swings round. Smaller cutterhead excavators are often towed to the site by tugboats, while larger ones can be self-propelled. One of the largest fleets of cutterhead dredgers is owned by Belgium's hydraulic engineering company Jan de Nul, which employed 14 of these special vessels worldwide in 2020. The largest of these is the 151 m length, 36 m breadth, 5.75 m draught newbuilding WILLEM VAN RUBROECK, which was commissioned in 2017. Its cutter head has a power of 8500 kW. The total installed engine power amounts to 41,000 kW, supplying an 8500 kW submersible pump, two equally powered on-board

Fig. 7.6 Originally built to take part in the Elbe deepening project, WERNER MÖBIUS had after its delay to be occupied otherwise and became part of the Boskalis group in 2017 as BEACHWAY (Ralf Witthohn)

Fig. 7.7 The non-propelled Belgian cutter suction dredger VESALIUS can lower its cutter head to a depth of 25.5 m, which ability was in no way needed during its 2021/22 work in the shallow Caspian Sea (Ralf Witthohn)

pumps and the propulsion power of 2 × 3000 kW. The maximum dredging depth is 45 m, the suction pipe has a diameter of 1.1 m.

Smaller is the 9260 kW non-propelled cutter dredger VESALIUS, built in 1980 at Sliedrecht/Netherlands-based Scheepswerf C. M. van Rees. Its cutter head has a capacity of 1325 kW, the total engine power is figured at 9260 kW. The submersible pump power amounts to 1100 kW, the two on-board pumps have an output of 2 × 2650 kW. The 94 m length, 17 m breadth VESALIUS has a draught of 3.5 m and can operate at a maximum water depth of 25.5 m. Reflagged to Kazakhstan, the dredger was in May 2021 berthed in the Caspian Sea port of Bautino/Kazakhstan and in 2022 in Qaradag/Azerbaijan (Fig. 7.7).

7.5 Backhoe Dredger PETER THE GREAT

With a view to the Russian hydraulic engineering market, in 2012, the backhoe dredger PETER THE GREAT was built under a contract awarded by a Belgian-Russian joint venture of Zwijndrecht based company DEME and the Northern Dredging Company in St. Petersburg. The newbuilding from Dutch shipyard Ravestein is intended for worldwide operation under the Cyprus flag. The Bureau Veritas classification society allows dredging operations up to 15 nm from the coast and a maximum of 20 nm from the nearest port. The dredger was built on the basis of a 60 m long, 17.2 m wide, 4 m deep pontoon with seven watertight compartments. The deck is designed for

Fig. 7.8 The backhoe dredger PETER THE GREAT does its work with help of a grab-fitted crane (Ralf Witthohn)

a distributed load of 10 t/m^2. At the stern a Liebherr revolving crane with an 11 m^3 bucket can work in a water depth of up to 20 m. During dredging operations, the dredger is fixed in its working position with the help of two spuds led through the ship's hull. Energy supply is provided by two diesel generators of 250 and 98 kW output. Accommodation for ten crew members is created in the raised forecastle. Due to its lack of propulsion, the dredger has to be transported to its place of operation by tug or piggyback on a transport vessel. In October 2016, the semi-submersible heavy-lift carrier COMBI DOCK III transported the dredger from Zeebrugge to Kaliningrad after having floated PETER THE GREAT into its hold. In May 2021, the dredger was stationed at Barry near Cardiff/UK, in February 2022 it was moored in Antwerp (Fig. 7.8).

8

Salvage and Towage

By about 17,400 units, tugs are the most frequently represented type of ship, but due to their type they only contribute 0.4% to the world fleet's tonnage. Together with those of trawlers, the designs of tugs differ fundamentally from all other types of vessels, as their most important performance requirement does not relate to a high cargo capacity or an optimal economical sailing speed, but to the power achievable on the towing hook or the towing winch, known as bollard pull. This is of even higher priority for harbour tugs than for high-sea salvage tugs, which at the same time require a high maximum speed in emergency operations. However, the number of salvage operations has decreased due to the improved technical reliability of ships. Furthermore, governments have stationed emergency tugs off their coasts, after serious accidents with major environmental damage had occurred. At the same time, anchor-handling tugs of the offshore business used as an alternative in salvage and deep-sea tugboats have reduced the employment opportunities of classic salvage tugs, so that their once flourishing business has lost much of its importance and only few newbuildings for these special purposes have been commissioned.

The opposite is true for harbour tugs with the task to assist cargo and passenger ships. Due to the considerable growth in ship sizes, especially of container ships and car carriers, their bollard pulls increased from around 30 tonnes in the 1980s to up to 90 tonnes after the turn of the millennium. At the same time, in addition to the innovations that have always been sought for tugs such as the Voith-Schneider or Schottel propulsion systems, new ideas have been implemented to combine the high maneuverability required for assistance work with maximum towing power. Recently developed propulsion

R. Witthohn, *International Shipping*, https://doi.org/10.1007/978-3-658-34273-9_8

systems have taken into account aspects of increased economy and lower environmental pollution, since tugs in particular offer great potential for fuel savings and emission avoidance. They have yet so far only been realized in isolated cases, for example by the installation of hybrid battery drives.

The compulsion to invest as a result from technical developments coincided with a revolution in economic conditions. In many ports, only one company or a tugboat pool of monopoly character had over decades offered their services. Often, large liner shipping companies were involved by subsidiaries, such as Hapag in Hamburg through Lütgens & Reimers or Norddeutscher Lloyd in Bremen and Bremerhaven. But when in liner shipping a strong concentration process took place, this also affected the harbour towage business. Globalisation initated the development of companies operating several hundred tugs, such as the Danish shipping company Svitzer or Spain's Boluda (Fig. 8.1).

The turning point had been marked by the market entry of the Kooren family in Rotterdam, who had from 1987 onwards operated six second-hand tugs acquired in the United States in competition with the established tug operator Smit. Under the name of Kotug, the shipping company expanded its activities to Hamburg in 1996 and to Bremerhaven in 1999, after having concluded long-term contracts with major container shipping companies, for

Fig. 8.1 Denmark's Maersk Line was a Kotug customer in Bremerhaven, until Maersk's sister company Svitzer displaced the Dutch company, which finally sold its share in the northern European business, meanwhile marketed as Kotug Smit, to Spain's Boluda group (Ralf Witthohn)

which towage duties had already been performed in Rotterdam. Prior to this, Germany's container carrier Hapag-Lloyd had terminated its towage activities and sold its Hamburg tug company Lütgens & Reimers to Bremer Unterweser-Reederei (URAG), while the subsidiary Transport & Service, which was active on the rivers Weser and Jade, was taken over by Hamburg-based Bugsier-Reederei. An attempt by Kotug to gain a foothold in Le Havre since 2006 was abandoned at the beginning of 2011, after several years of legal disputes with French authorities and trade union defence. In contrast, expansion efforts in London, Port Hedland, Tanjung Pelepas, Cameroon and off the coast of Brunei were more successful. Out of a total of more than 40 Kotug units in May 2016, RT AMBITION and SD SHARK operated in London, while RT FORCE, RT INSPIRATION, RT ATLANTIS, RT ROTATION, RT ENTERPRISE, RT EDUARD, RT DISCOVERY, RT TOUGH and RT ENDEAVOUR, RT DARWIN and RT SENSATION worked in Port Hedland/Australia, RT STÉPHANIE and RT CLAIRE were engaged in Tanjung Pelepas/Malaysia, RT MARGO in Cameroon for the Limbé oil refinery. RT TASMAN, RT CHAMPION, RT LEADER and RT ZOÉ assisted at Shell's Brunei LNG terminal, in Mozambique were stationed RT SPIRIT and RT MAGIC. In Rotterdam, Bremerhaven and Hamburg were about 19 tugboats stationed, and three in the offshore business.

In London and Bremerhaven, however, the shipping company came into competition with the globally active Danish tugboat company Svitzer, which was in many cases able to rely on towage contracts with Maersk Line of the joint parent company A. P. Møller. In 2019, the joint venture Kotug Smit, founded in 2016 by Kotug and the Boskalis group and responsible for the northern Europe business, was sold to Spain's Boluda Towage (Fig. 8.2).

The increased competitive pressure had already forced the established tug shipping companies to rationalise, lay off and reduce their fleets or to close down their businesses, when they were no longer competitive with the large tugboat operators due to the small size of their fleet. In 2014, Unterweser Reederei (URAG) in Bremerhaven had temporarily entered a joint venture with the Svitzer shipping company, which chartered two tugs from URAG. In Bremerhaven, the pool of URAG and Bugsier fell apart, and in Hamburg, too, the URAG subsidiary Lütgens & Reimers left the towage community, while in 2016 URAG tried for the first time to gain a foothold in the Cuxhaven market, which until then had been covered by local tug operator Wulff, by stationing two tugs there. Such efforts proved unsuccessful and after high operating losses URAG was in 2017 integrated in the Boluda company, which then operated 230 tugs. Shortly afterwards, the Schuchmann family as owners announced their withdrawal from Bugsier Reederei and the companie's

Fig. 8.2 The hard daily round-the-clock towage work, as experienced by the crew of the harbour tug SVITZER MARKEN when taking the line of the container vessel EVELYN MAERSK in a hurricane, was made even more difficult, after a war of tug broke out that created high losses, a number of company sales and the loss of jobs (Ralf Witthohn)

sale to Hamburg-based Fairplay company. In December 2017 German government's Bundeskartellamt imposed total fines of around EUR 13 million on the shipping companies Fairplay, Bugsier and Petersen & Alpers and their managers for splitting up the towage contracts according to fixed quotas in several German ports (Fig. 8.3).

In a counter-action to the aggressive market performance of the big players Svitzer and Boluda, Hamburg-based Fairplay announced in the beginning of 2020 to operate four Spanish-flagged tugs in ports of the Canarian islands, which had until then been a monopoly of Boluda, but is offering only a limited number of tug jobs, e. g. at the local container terminal. In March 2020, the tugs FAIRPLAY-29 (FAIRPLAY NUBLO) ex MULTRATUG 14, BUGSIER 19 (FAIRPLAY BENTAYGA), BUGSIER 20 (FAIRPLAY BANDAMA) and FAIRPLAY TAMADABA ex CLAUS were assembled in Las Palmas. The latter tug was chartered from Lübeck-based tug company Johannsen, as was the tug CARL for work in the Weser region at the same

Fig. 8.3 Tradition-rich Unterweser-Reederei from Bremen was in 2017 purchased by Spain's Boluda Towage, including the harbour tug VB RÖNNEBECK, which had in 1977 been built in Wilhelmshaven/Germany for the Hapag-Lloyd towage branch Transport & Service as HERKULES and in 2016 been purchased from Dutch company Iskes by Unterweser (Ralf Witthohn)

time. Later in 2020 Fairplay gave up its Canarian islands due to regulatories imposed by the Spanish authorities. The Fairplay Towage group fleet list actually named over 70 operated units. Already in 1997 the company had persued a counter-strategy and stationed large sea-going tug newbuildings of the FAIRPLAY-21 type at Rotterdam, in reaction of the Kotug activities in Hamburg and Bremerhaven (Fig. 8.4).

8.1 Grounded with Dragging Anchors

On October 13 2017, the G.T.H. grain terminal Hamburg expects the panamax bulk carrier GLORY AMSTERDAM, built in Japan in 2004. However, the ship is delayed and does not dock at the silo in Kuhwerder Hafen before October 20. Discharge is completed on 26 October, the ship leaves the Hanseatic city and anchors on 7 m ballast draught in the German Bight to wait for new orders. At this time, dozens of container ships, car carriers and bulk carriers are assembled on

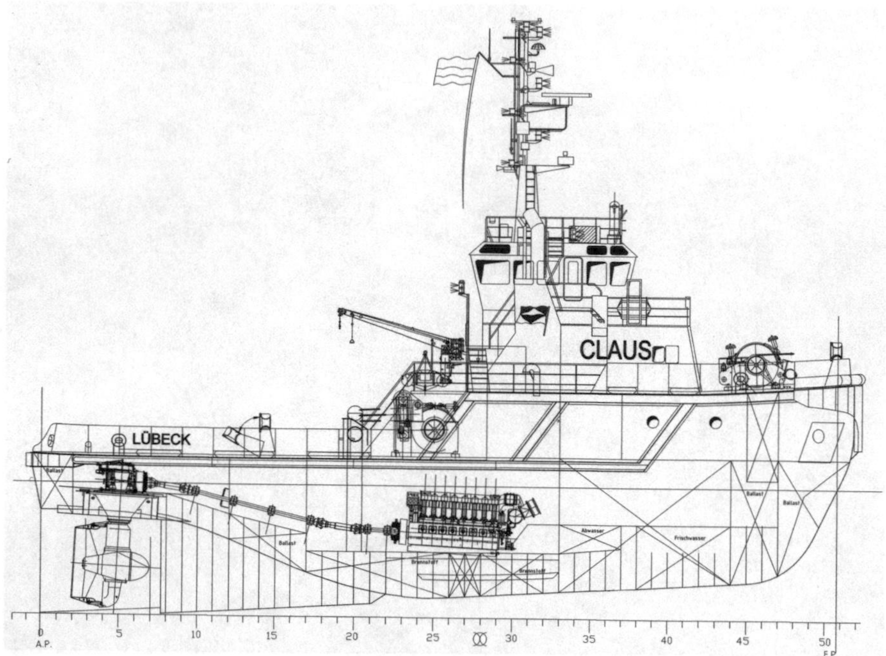

Fig. 8.4 Hamburg tug operator Fairplay chartered the 64-tonnes bollard pull ASD tug CLAUS built at Schiffswerft Lindenau in Kiel in 2006 from Lübeck operator Johannsen and stationed it under the name FAIRPLAY TAMADABA at Las Palmas to extend competition with Spain's Boluda group into their homeland (Lindenau)

the roads. The hurricane "Herwart" is announced to cross the southern North Sea 28 October, and as the wind force increases causing a rising swell, the leave the anchorage for the open sea, because the anchors would no longer be able to hold the ship in position at the announced wind forces and swell heights. This applies in particular to ships in ballast, which, due to their larger freeboard, offer the wind a greater surface area to attack and, because of the smaller displacement, become the plaything of the waves more quickly.

Although the main engine of the 77,171 dwt, 225 m length GLORY AMSTERDAM is running, the crew is unable to weigh the anchors in time. The storm grows to hurricane force and the waves pile up to 7 m high. On the morning of 29 October, wind and waves begin to drive the bulk carrier into the direction of the coast. The emergency tug NORDIC, which is constantly in position off the coast near the islof of Borkum on behalf of the German government, is alarmed and makes its way to the ship in distress. Upon arrival, the distance to the East Frisian Islands is still more than 30 nm. The emergency tug manages to connect a towing line, but the rope breaks. Attempts of a boarding team dropped by helicopter to establish a line connection fail as well. Driven by the hurricane, the GLORY AMSTERDAM drifts further south and at about 6.30 p.m. touches the ground, 1.6 nm off the island of Langeoog, at the most unfavourable time, namely at high tide. The storm has caused the water level to rise more than 2 m above normal tide. This means that the vessel is stranded high on the sands. As the ship is empty, there is fortunately less danger, that cargo would exert pressure on the ship's structure and force it to break up. But 1800 tonnes of heavy fuel oil and 140 tonnes of diesel oil in the fuel tanks create the potential of an environmental disaster (Fig. 8.5).

Apart from NORDIC, the government owned emergency response vessel MELLUM, the tugs JADE and VB WESER of Unterweser-Reederei and two tugs of Bugsier, BUGSIER 9 and BUGSIER 10, have been deployed to the scene of the accident. However, the four tugs of private shipping companies are being withdrawn in the course of the day. The stranded vessel lies much too high on the sand to be pulled free. Having suffered from seasickness, the 22-strong Chinese crew is taken care of by a team of doctors, who have landed on the ship by helicopter (Fig. 8.6).

The ship's owner, Glory Ships in Singapore, commissions the Dutch company Smit Salvage to salvage the GLORY AMSTERDAM. Smit deploys the offshore tugs FAIRMOUNT SUMMIT and UNION MANTA. They sail from Esbjerg/Denmark and Aberdeen/Great Britain, resp., for Wilhelmshaven, in order to be prepared there for the actual salvage work. Three days after the grounding the auxiliary vessel HURRICANE succeeds in connecting the first over 1 km long tow line between FAIRMOUNT SUMMIT and the casualty, later the second line is being carried to UNION MANTA. Afterwards 16,000 m^3 *of water ballast are*

Fig. 8.5 Despite having dropped both anchors, the bulk carrier GLORY AMSTERDAM stranded off the isle of Langeoog/Germany in October 2017 (Havariekommando)

Fig. 8.6 The highsea salvage tug NORDIC, chartered by the German government especially for such emergency cases, was not able to protect the damaged bulk carrier GLORY AMSTERDAM from grounding near Langeoog (Ralf Witthohn)

pumped out of the bulk carrier's tanks. At the morning tide, 4 days after the stranding, the ship is being towed free. The large tugs pull it towards the fairway, from where the harbour tugs MULTRATUG 4, JADE and BUGSIER 11 take over. GLORY AMSTERDAM reaches Wilhelmshaven in the course of the day for damage assessment. Eight days later, three tugs tow the "dead" ship from Wilhelmshaven to Bremerhaven for docking. There, the repairs take several months and until the next year. It was discovered, that the rudder had broken in course of the grounding and had to be newly constructed (Fig. 8.7).

An investigation carried out by Germany's Bundesstelle für Seeunfalluntersuchung (Federal Bureau of Maritime Casualty Investigation, BSU) found out, that the accident was favoured by a series of striking omissions of Germany's Havariekommando, which is based in Cuxhaven since establishment in 2003. The command had been founded, in order to prevent an accident such as the casualty of the timber carrier PALLAS, which had under similar conditions caused severe oil pollutions at the German coast. BSU stated, that the Chinese captain of AMSTERDAM GLORY did not realize, that NORDIC was a government-chartered emergency tug, so that the salvage action was delayed. Havariekommando did not dispose of an ECDIS chart to get direct informations on AIS data and radar informations, nor was it able to directly use the relevant radio frequencies. The boarding team was not sufficently equipped with VHF radio communication means to

Fig. 8.7 As a new rudder had to be manufactured to replace the one broken during the grounding of the bulk carrier GLORY AMSTERDAM, the ship was out of service for over 3 months (Ralf Witthohn)

communicate appropriately during the salvage action. Havariekommando was not able to brief the captain of AMSTERDAM GLORY about legal and technical measures in connection with the intended towage action, nor did it brief the boarding team adequatly or was able to maintain sufficient communication with the team.

Two and a half year after the AMSTERDAM GLORY casualty another, similar incident in the German Bight was managed in a better way. The 24,460 dwt general cargo vessel SANTORINI of the Geek subsidiary of German company Aug. Bolten lost its rudder in a heavy storm in February 2020, when on voyage from Abidjan to the Baltic in ballast. NORDIC achieved to connect a tow line and to hold the disabled ship on a position, from where it was towed to Bremerhaven by two local tugs to be fit with a new rudder in a repair dock. In February 2022, the container vessel MUMBAI MAERSK had again to be towed free after grounding in the Weser estuary by a salvage tug, which had to be deployed from Rotterdam, after the NORDIC was undergoing propeller repair in drydock at the time.

8.2 Boskalis Tugs Operated Worldwide

The highsea tugs FAIRMOUNT SUMMIT and UNION MANTA, engaged in the salvage of GLORY AMSTERDAM, were operated by Fairmount Marine in Rotterdam, a subsidiary of the Dutch Boskalis company and meanwhile integrated in that firm, which is mainly engaged in the dredging business. Among the 80 offshore vessels in the Fairmount fleet at the end of 2017 there were three cable-laying tugs, four diving support vessels, two dumping vessels, seven heavy-lift carriers, 19 semi-submersible transport vessels and 17 deep-sea and anchor-handling tugs. In early 2022, Boskalis operated a combined fleet of over 200 vessels assting ships in Asian-Pacific ports and at remote offshore terminal, in a joint venture with Keppel Smit Towage and Smit Lamnalco. Furtermore was provided emergency response, salvalge and wreck removal work through Smit Salvage employing a salvage fleet including tugs with bollard pulls exceeding 120 tonnes (Fig. 8.8).

The largest of these are named FAIRMOUNT EXPEDITION, FAIRMOUNT SHERPA, FAIRMOUNT SUMMIT, FAIRMOUNT ALPINE and FAIRMOUNT GLACIER, built by Japan's Niigata Shipbuilding from 2005 to 2007. They feature a bollard pull of 205 tonnes, which is achieved by four Wärtsilä 6L32 diesel engines of 3000 kW output each, allowing a free sailing speed of 15 knots. Through double reduction gears the main engines act on two controllable pitch propellers running in nozzles. The individual drums of the three-part electro-hydraulically operated towing and

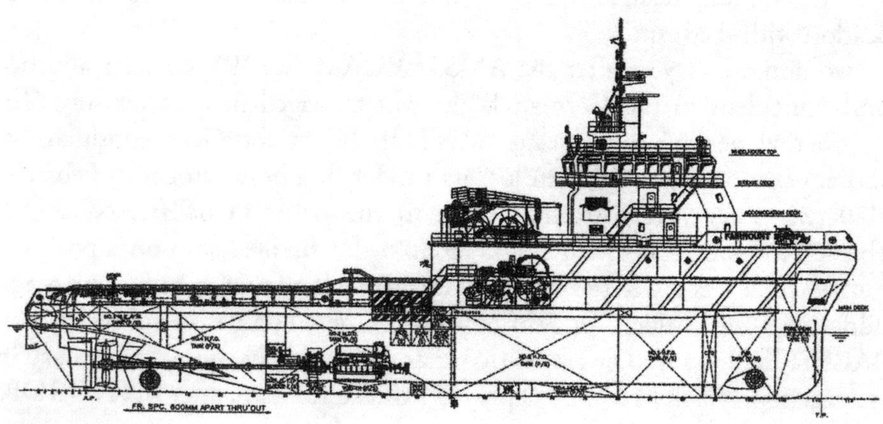

Fig. 8.8 A towing winch installed in the forward half of the vessel and two propellers running in knozzles under the stern of the FAIRMOUNT tugs form the force vectors that must be brought into an optimized relation of their positions in any deep-sea tug (Fairmount)

anchor winch can be belayed with two 1500 m long towing wires of 76 mm diameter and one equally strong working wire of 300 m, or with chains of maximum 100 m length. The holding force of the brake is 400 tonnes. At the stern there is a 5.4 m long roller with a diameter of 2.5 m and an SWL of 300 tonnes, to reduce the resistance of the towing wire. The 75 m length, 18 m breadth and 8 m depth tugs have a ballast draught of 4 m and a maximum draft of 6.8 m. Their accommodation provides space for 36 persons, 12 of them for the ship's crew and 24 for the project specialists. In September 2017, FAIRMOUNT EXPEDITION and FAIRMOUNT SUMMIT had arrived in Esbjerg from Port Said. During the same year, the FAIRMOUNT SHERPA sailed from Singapore via Cape Town to Rio de Janeiro and Paramaribo. FAIRMOUNT ALPINE was off the South Korean coast in November, and the FAIRMOUNT GLACIER was moored in its home port Rotterdam on a voyage from Larnaca, in March 2020 the tug undertook a towage from Port Louis/Mauritius to Chittagong. FAIRMOUNT ALPINE, renamed BOKA ALPINE in 2018, completed a voyage from Singapore to Port Nelson/New Zealand. In February 2022, the tug was on voyage from Rio de Janeiro via Durban to Dubai, where also the BOKA EXPEDITION had arrived from Singpore (Fig. 8.9), while BOKA GLACIER was underway in South African waters and BOKA SUMMIT had arrived in Rio de Janeiro from the Indian Ocean.

Fig. 8.9 205 tonnes bollard pull give BOKA ALPINE ex FAIRMOUNT ALPINE a top rank among the world's strongest salvage tugs (Ralf Witthohn)

8.3 Ten Tugs in Harriersand Salvage Action

At the end of September 2009, Bremen-based Unterweser Reederei and Bugsier Reederei from Hamburg call in the strongest tugboat fleet that has ever been deployed in German waters. Partly in rotation a total of 17 tugs, mainly from Bremerhaven, are trying to refloat the bulk carrier ALGOMA DISCOVERY, which has run aground on the Weser river isle of Harriersand opposite the port of Brake/Germany. The bulk carrier had completed the loading of steel coils with a total weight of 27,000 tonnes in the Bremen iron and steel port 22 September. In the evening of the same day, the ship was on way downriver to Ravenna, when a defect in the electrical system cripples the function of the controllable pitch propeller. After passing the light buoy 96, ALGOMA DISCOVERY left the fairway to starboard and ran aground on Harriersand. The grounding took place at the time of the highest tide, so that the bow of the ship is still only 150 m away from the bank of the isle. Two tugs of Unterweser-Reederei, ROTESAND and BREMERHAVEN, which are undertaking assistance work in Brake, quickly make their way to the grounded vessel. But they have to give up their refloating attempts after just half an hour. For another try at high tide in the next morning, eight tugs, named JADE, ROTESAND, ELBE, BLUMENTHAL, BUGSIER 1, BUGSIER 6, BUGSIER 20 and STIER, are deployed from Bremerhaven to the scene of accident. However, their salvage attempts fail. Immediately afterwards, the oil pollution fighting vessel LUNEPLATE begins to discharge the fuel oil from the ALGOMA DISCOVERY.

In the early afternoon, the floating cranes ATHLET I and ATHLET II, deployed from the port of Bremen, reach the position and begin to discharge steel coils from the holds and to load them onto a pontoon and into three barges. By the evening the cranes have managed to lighten 2060 tonnes of the cargo, LUNEPLATE takes over 138 m^3 of heavy fuel oil. The water injection ship AKKE and the dredger HARRIERSAND simultaneously remove sand and silt from below the foreship of the ALGOMA DISCOVERY. However, the preparatory works prove to be insufficient for the next salvage attempt during the evening flood—1 day after the grounding has taken place. After floating cranes and oil-fighters have cleared the field, even ten tugs are now being tied up. Moored at the stern and sides of the ship, they develop a concentrated power of over 36,000 kW for a combined bollard pull of more than 500 tonnes.

Such an action is only possible due to the large number of tugs stationed in near-by ports, which strongest units have more than twice the towing power that was common in recent years, in order to meet the demands of assisting ships rapidly grown in size. The main engine of ALGOMA DISCOVERY is supporting the salvage manoeuvre, but the ship has sunk even deeper into the sand of the Weser and does not move from its position. In the next morning, the lightering work is therefore being continued by the floating cranes. In the afternoon a total of ten tugs are sent to make another attempt. 6000 tonnes of cargo less and the tide support the manoeuvre, so that the ship can be towed into the direction of the fairway already by half of the deployed tug fleet and 1 before the tide reaches it peak. Under assistance of the tug ROTESAND, ALGOMA DISCOVERY proceeds to Brake Niedersachenkai for inspection (Fig. 8.10).

8.4 From Tampa to Bremerhaven for Repair

Due to a fire in the auxiliary engine room on voyage from Cartagena/Colombia to Veracruz/Mexico, the container ship MONTEVIDEO EXPRESS of Germany's E. R. Schiffahrt gets lost of all propulsion and electrical power on 21 January 2015. Although the crew is able to extinguish the fire, a tug is needed to take the vessel to Tampa, Florida, where 1039 containers are unloaded. The shipping company declares general average and it is decided, to have the ship repaired in Germany. This means that a transoceanic towage is necessary, starting from Tampa on 24 March by the Dutch anchor-handling tug BOULDER. Via Ponta Delgada/Azores the MONTEVIDEO EXPRESS arrives at Bremerhaven/Germany on 8 May.

Fig. 8.10 Up to ten tugs were deployed to refloat the grounded bulk carrier ALGOMA DISCOVERY, but only lightering of the steel cargo and fuel oil let the salvage become a success (Ralf Witthohn)

Ten months after the towing assignment for MONTEVIDEO EXPRESS from Tampa to Bremerhaven, BOULDER returns to Bremerhaven to tow the fire-damaged US car carrier COURAGE, to Aliaga/Turkey for scrapping. After a major fire in its cargo of vehicles, the ship of American Roll-on Roll-off Carrier (ARC) has spent 9 months in Bremerhaven's Kaiserhafen. On 6 March 2016, BOULDER arrives inbound from Rotterdam to prepare the towing connection with the help of a mobile crane. Two anchor chains are led through Panama chocks at the bow of COURAGE and connected with the actual towing bridle to form a connection, with the help of which the tug can exert a steering effect on its appendage. Along the entire starboard side of the transporter a reserve towline is installed which could be picked up by the tug in case of a break of the first line.

In the morning of 8 March, four Bremerhaven harbour tugs tow the disabled car carrier through the Kaiserhafen lock onto the river Weser. The tug BOULDER accompanies the tugs and takes over the car carrier off the Columbuskaje. One of the tugs, BUGSIER 3, is acting as steering stern tug, until the tow reaches the pilot vessel on the outer Weser after several hours at 10 knots maximum speed. End of the month the tow reaches the demolition site at Aliaga/Turkey. Still in 2017 Tschudi sells BOULDER to India. Later in the same year the tug is renamed HURRICANE-I and broken up at Chittagong/Bangladesh. The sister ship HURRICANE-II ex BLUSTER is likewise scrapped in 2017, in Alang/India.

The offshore tug BOULDER had in 1988 been delivered under the name MAERSK LIFTER by Scheepswerf Waterhuizen in Pattje/Netherlands in a series of four such newbuildings for Denmark's Maersk group and later joined the fleet of Norwegian owner Tschudi. In July 2014 the sister ship BLUSTER played an important role in an even more spectacular towing operation. The tug towed the wreck of the Italian cruise liner COSTA CONCORDIA, which had before been salvaged from the rocks under highly difficult circumstances and was hanging on buoyancy boxes, from the island of Giglio/Italy to Genoa for scrapping (Fig. 8.11).

8.5 Nine High-Sea Tugs for Harms from Mützelfeldt

The abandonment of the traditional deep-sea towing and salvage business by Hamburg-based Bugsier-Reederei, which had for decades been leading in this branch together with the Dutch shipping company Smit, meant a longer interruption on this business field in Germany. A renaissance took place in Germany, when the Hamburg-based shipping company Harms Bergung,

Fig. 8.11 The former anchor-handling tug BOULDER carried out several long-distance towages before being scrapped in Bangladesh (Ralf Witthohn)

Transport and Heavylift ordered a series of nine deep-sea tugs of different sizes and performances from Mützelfeldtwerft in Cuxhaven. The first of the conventionally twin-propelled newbuildings was commissioned as PRIMUS in 2004, followed by MAGNUS and TAURUS in 2006/2007, JANUS and URSUS in 2007/08, URANUS, PEGASUS and CENTAURUS in 2009 and ORCUS in 2010. Ferrostaal in Essen, a subsidiary of the MAN group with a stake in Mützelfeldtwerft, acted as contractor for the tugboats, which were financed by funds at inflated prices. Like the two shipping company directors, the shipbuilding managers of Ferrostaal used millions of euros from the fund capital. An Augsburg court judged this to be embezzlement and in 2011/2012 imposed prison sentences of several years, which were reduced on appeal (Fig. 8.12).

In 2011, the first of the newbuildings, named PRIMUS, switched to the management of Bremen-based Unterweser-Rederei tug and was renamed BREMEN FIGHTER. In March 2021, Boluda Deutschland won a two-year plus three-year option contract for the stationing of a government-chartered emergeny tug in Sassnitz to protect the eastern part of Germany's Baltic Sea region as replacement of the tug FAIRPLAY-25 of Arbeitsgemeinschaft Küstenschutz. Of the further eight tugs with bollard pulls of 100 tonnes to 300 tonnes, the Hellespont company in Hamburg took over the management of PEGASUS and CENTAURUS in 2014. The remaining six were acquired by Dutch ALP Maritime Services and renamed ALP ACE, ALP IPPON, ALP WINGER, ALP FORWARD, ALP GUARD and ALP CENTRE. ALP from 2016 to 2018 also received the anchor-handling and salvage tug newbuildings ALP STRIKER, ALP DEFENDER, ALP SWEEPER and ALP KEEPER with a bollard pull of over 300 tonnes and designed with a patented X-bow from Japanese Niigata shipyard.

Fig. 8.12 The 218 tonne bollard pull tug ALP FORWARD, built as URSUS in 2008, resumed towage operations in June 2019 after a several months lasting lay-up period at Bremerhaven due to poor market conditions (Ralf Witthohn)

8.6 Rapid Emergency Help for SPLITTNES

Only by the courageous efforts of two tug crews it is avoided, that the consequences of a collision between a container ship and a bulk carrier remain relatively small. In the evening of 22 November 2011 the Japanese container ship MOL EFFICIENCY approaches the Bremerhaven container quay, off which a turning manoeuvre has to be initiated before berthing. But the manoeuvre is delayed and it is to fail, because the container ship has to stop and reverse at low speed, until the oncoming bulk carrier OCEAN PEGASUS has passed. The German bulk carrier SPLITTNES is following in the course line of MOL EFFICIENCY under pilot assistance and, and due to the prevailing fog, under radar advice as well. It has been agreed, that the SPLITTNES should remain in a waiting position, until the turning of MOL EFFICIENCY is completed.

However, the agreement is changed in favour of an overtaking action. The 2 knots running tide yet pushes the SPLITTNES under the stern of MOL EFFICIENCY. Both ships are severely damaged. The SPLITTNES starboard side is holed and the ingress of water causes a list of about 20°. The crew gets panic, and half of the 18 members leave their ship in a freefall boat, only the ship's officers remain on the bridge. The deckman of the harbour tug MONTALI has to climb onto the bulk carrrier to be able to establish a line connection. With the support of a second tug, RT PIONEER, SPLITTNES is towed to Columbuskaje, where the ship is prevented from sinking by the use of pumps. Germany's Bundesstelle für Seeunfalluntersuchung (Federal Bureau of Maritime Casualty Investigation, BSU) in Hamburg, which in contrast to the acting of the dissolved German Sea Courts (Seeamt) basically does not blame the pariticipants of an acident any more, does not find a single cause for the severe collision, but brings the idea into play, that the OCEAN PEGASUS sailing downstream would have, due to the more favourable tide conditions, found it easier to assume a waiting position. The traffic control centre, which would have been responsible for the organisation of the shipping traffic, is merely accused of not having pointed out the reverse course of the MOL EFFICIENCY (Fig. 8.13).

Fig. 8.13 Poor visibility and inadequate traffic planning caused the near sinking of the bulk carrier SPLITTNES (Ralf Witthohn)

8.7 Five Floating Cranes Raise the SEKI ROLETTE Wreck

Months of preparations have to be made to remove a wreck lying in the main shipping lane off the German coast as a result of a fatal collision in April 1992. Five floating cranes from the Netherlands and Germany are being deployed by the companies Bugsier and Harms from Hamburg to salvage the ro ro vessel SEKI ROLETTE of a Japanese shipping company. The ship sank in a collision with the Russian container ship newbuilding CHOYANG SENATOR. The container carrier had only 2 h before left Bremerhaven for a positioning trip to the first port of loading and just dropped the Weser pilot at the outer position, before it crossed the southern route of the traffic separation area, when SEKI ROLETTE, bound for Hamburg, approached from the west. North of the island of Wangerooge, the bulbous bow of CHOYANG MOSCOW hit the starboard side of SEKI ROLETTE and tore a large hole, through which large amounts of water immediately flow into the ship's cargo hold. On SEKI ROLETTE, the doors in the transverse bulkheads for rolling cargo were not closed, so that the freighter sank within minutes. Not even all of the crew managed to get outside, five of them died, three crew members saved themselves on the bulbous stem of the container ship. Among the dead, who could only be recovered from the wreck after months, was also the captain of SEKI ROLETTE. The investigating authority (Seeamt) in Bremerhaven stated in its hearing, that the command of SEKI ROLETTE did not sufficiently evaluate the radar informations, did not obey the warnings of the Wilhelmshaven traffic control and did not give fog signals (Fig. 8.14).

Six months after the SEKI ROLETTE sinking the wreck is righted, lifted into a floating state by five floating cranes and carried to Hamburg in the ropes of the cranes ROLAND and THOR of Bugsier and under assistance of nine tugs. There, the wreck is loaded on the semi-submersible heavy-lift carrier GIANT 3 and transported to Aliaga/Turkey for demolition, which takes place in November 1992 (Fig. 8.15).

8.8 New Tug Designs and Techniques

One of the two tugs used in the salvage of the self-unloading bulk carrier SPLITTNES represents the prototype of a newly developed type designated Rotortug by the Dutch Kooren group. Built by shipyards in Spain and the Netherlands, RT PIONEER, RT INNOVATION, RT MAGIC and RT SPIRIT were put into service in 1999 for mainly use as harbour tugs but also

Fig. 8.14 Five floating cranes worked off Wangerooge to salvage the ro ro vessel SEKI ROLETTE (Ralf Witthohn)

Fig. 8.15 The semi-submersible carrier GIANT 3 carried the holed SEKI ROLETTE from Hamburg to Turkey for demolition (Ralf Witthohn)

for shorter deep-sea towing operations Instead of the usual two, the 80-tonnes bollard pull tugs were equipped with three Schottel propellers, two under the foreship and one under the aft ship. This patented propulsion system made it possible to carry out the berthing of car carriers in Bremerhaven with only two instead of three tugs as practised before. The first tugs of this type were the reason for the rise of the Kotug fleet to one of the largest in northern Europe. In January 2018 Kotug took delivery of the strongest Rotortug, named RT RAVEN, from the Albwardy Damen shipyard in Sharjah/UAE. Designed by Robert Allan and Damen Hardinxveld, the 100-tonnes bollard pull tug started offshore operation on behalf of KT-Maritime Services Australia in the ConocoPhillips Bayu Undan field in the Timor Sea off

Australia's north coast. Originally a 50/50 joint venture between Kotug International and Teekay Australia, Kotug became the sole owner of the shipping operator in April 2018 (Fig. 8.16).

Among the Rotortug newbuildings with the patented propulsion system were the 85-tonne bollard pull harbour tugs RT DARWIN and RT TASMAN delivered by ASL Shipyard in Singapore and commissioned in Bremerhaven in the beginning of 2012. The tugs are equipped with three Niigata Z-Peller propellers, powered by three diesel engines of 1654 kW output each. From October 2014, RT TASMAN was together with Kotug's RT CHAMPION and RT LEADER deployed to Shell's LNG terminal in Brunei. The German tug operators Bugsier, URAG and the Dutch company Smit as well received Rotortugs, which entered service under the names BUGSIER 4, BUGSIER 5, BUGSIER 6, GEESTE, HUNTE, SMIT KIWI and SMIT EMU. The 32 m length, 12 m breadth, 6.45 m draught tugs, designed for three-man operation and equipped with a towing winch fore and aft, are among the most powerful harbour tugs (Fig. 8.17).

Fig. 8.16 After starting with second-hand tugs like the former US-owned ZP CONDON (left) Dutch operator Kotug invested in Rotortugs of the RT PIONEER type (Ralf Witthohn)

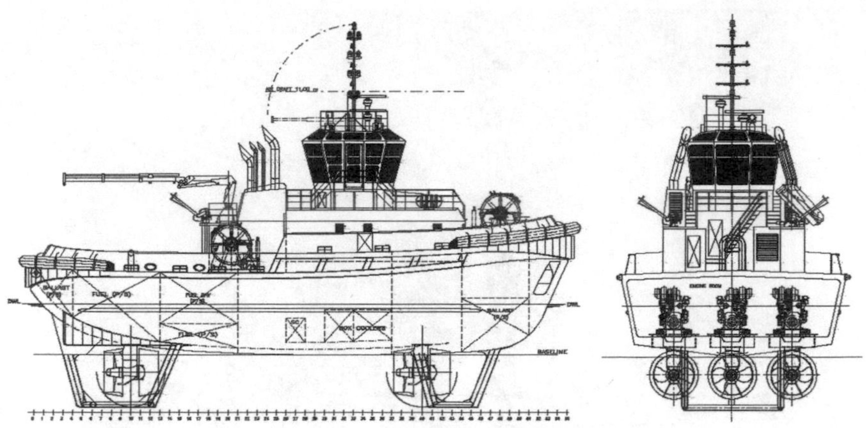

Fig. 8.17 On the 60-tonne bollard pull tug RT CLAIRE, the Rotortug principle is realised by arranging two azimuth propellers forward and one aft (Kotug)

8.9 Large ASD Tug Series from Damen

In September 2013, Hamburg-based tug operator Lütgens & Reimers commissioned its first tug newbuilding of the ASD type from the Damen shipyards group head-quartered in the Netherlands. Named PROMPT, the tug designated ASD 2411 (Azimuth Stern Drive of 24 m length and 11 m breadth) was built at Song Thu Shipyard in Da Nang, Damen's Vietnamese production site. The tug's two propellers arranged under the stern of the ship are rotating in knozzles. They are driven by two Caterpillar main engines of type 3516C, each with an output of 2100 kW. Gearboxes reduce the engine speed from a maximum of 1600 rpm to 235 rpm of the five-bladed rudder propellers. The 24.5 m length, 11.3 m breadth tug achieves a free sailing speed of 13 knots and a bollard pull of 71 tonnes. To make room for the propulsion propellers the tug was designed with a steeply rising rake of keel. Counter. As a result, the main engines rest on forward inclined foundations and thus provide the necessary space for the gearboxes in the stern. In spite of this technical design measure the propellers still protrude below the base, so that in order to allow a docking of the tug, an extensive support structure is fitted under the actual keel. This is also inclined 13° to the stern. The towing winch with a split drum is located on the foredeck. The tug is designed with the today usual all-round vision bridge and facilities for four crew members, including a double cabin (Fig. 8.18).

On the base of the experiences made with the 2008-built ASD 2411 tug PETER tradition-rich Hamburg tug company Petersen & Alpers (P&A)

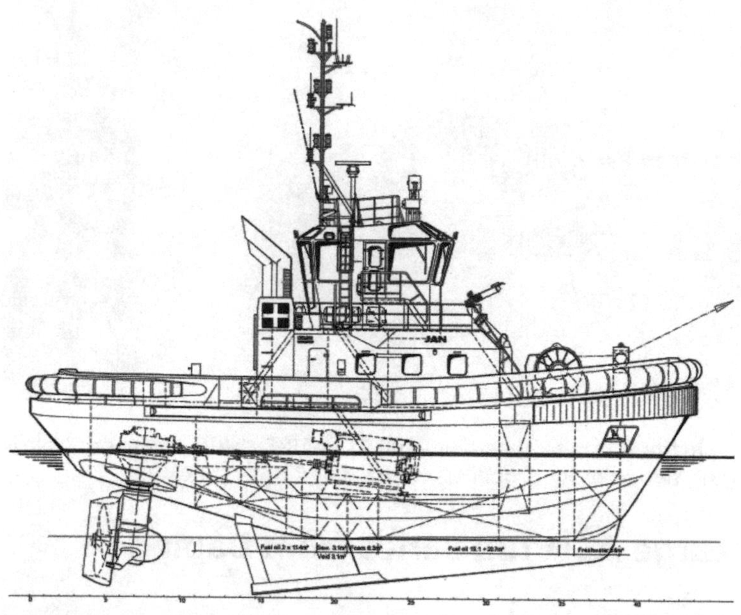

Fig. 8.18 The 2019-delivered Hamburg tug JAN of Damen type ASD 2411 is operated by Fairplay-Reederei under the name FAIRPLAY XV (Damen)

ordered a larger and more powerful tug of type ASD 2913 named MICHEL and delivered at the end of 2014 by Damen Shipyards Galati in Romania. In order to meet the increasing requirements to be met by growing ship sizes, MICHEL has a bollard pull of 80 tonnes. The 28.9 m length, 13.2 m breadth tug is fitted with two Caterpillar diesel engines of type 3516C, each with an output of 2525 kW at 1800 rpm. The four-stroke engines turn two Rolls Royce US255 propellers with a diameter of 2.8 m. To increase leakage safety the hull of the tug, which is equipped with extended fenders, is designed according to the double hull concept. In 2019, Neue Schleppdampfschiffsreederei Louis Meyer commissioned another Damen ASD 2411 type tug of 71 tonnes bollard pull named JAN for management and operation in Hamburg by Fairplay Schleppdampfschiffs Reederei Richard Borchard as FAIRPLAY XV. Another Hamburg harbour tug designed according to the ASD principle is the VB RESOLUT of Spain's Boluda group (Figs. 8.18 and 8.19).

The relatively new ASD concept has quickly gained worldwide acceptance. The Damen yard list includes a long series of ASD 2810 tugs. Among them are ARGUS for the shipping company Iskes in Ijmuiden, SD REBEL, SD RANGER and SD ROVER for the Rotterdam Kotug shipping company,

Fig. 8.19 2014-built ASD tug VB RESOLUT, owned by URAG-Boluda since the purchase of Unterweser-Reederei by Spain's Boluda group, is mostly working in the port of Hamburg (Ralf Witthohn)

ARCANGEL SAN MIGUEL, VIRGEN DEL VALLE for the Meyer group in Panama, RUA CAP. LUCIO R for Rua Remolcadores Unidos Argentinos and VICE ADMIRAL OWUSU-ANSAH for Ghana. Other versions of the ASD type, which vary in dimensions and engine power, include SPRIGTHLY of ASD type 2009 and TANKRED (ASD 2310) for DMS Maritime in Australia, the ASD 2411 type HAURAKI for Auckland, an ASD 3111 type named CAPO NOLI for Scafi in Italy, the ASDs 3212 MULTRATUG 19 for Terneuzen, KIN for International Tug in Mexico, SD DOLPHIN for Kotug, BALADHA and BULGU for BM Alliance Coal Operations in Australia, SL SERVAL for Smit Lamnalco and MERCURIUS for Iskes and—in the largest version—the 87 tonnes towing ASD 3213 tug TANOK for Intertug in Panama. In May 2014 the Danish shipping company Svitzer received the ASD 3212 type SVITZER LONDON with a bollard pull of 81 tonnes. The wide use of this type was also shown by a major order from state-owned Petroleos de Venezuela about ten ASD 2810 tugs for operation in the country's oil ports. Due to the series production for stock, seven of the tugs could be delivered by the Romanian and three by the Vietnamese shipyard of the Damen Group shortly after the order was placed until the end of July 2014.

Special design versions of Damen's ASD tugs meet the requirements of high ice classes, such as VOLODYMYR IVANOV and KAPITAN MARKIN of ASD 3212 type, 2012 delivered to the authorities of the Ukrainian ports of Yuzhny and Mariupol. The 32.7 m length, 12.8 m breadth tugs were designed according to the ice notation ARC4 of the Russian Maritime Register of Shipping. The power of two Caterpillar main engines is brought into the water by two Rolls Royce azimuth drives, which generate a bollard pull of 69 tonnes. A hydraulically operated double-drum winch on the forecastle achieves a force of 100 tonnes. The equipment includes two fire fighting monitors with an hourly capacity of 2 × 1200 m^3 of water or 2 × 300 m^3 of foam.

The Russian newbuildings BULAT and SIVER are smaller ice-class tugs of type ASD 2810, which are operated in the Baltic ports of Vyssozk and Vyborg, while CHARA and KUZBASS of the same type are stationed at the Far Eastern ports of Vostochny and Vanino. BULAT, which was built in Vietnam, achieves a bollard pull of 51 tonnes when sailing ahead and of 48 tonnes when steering astern. The corresponding free sailing speeds are indicated as 13.5 knots and 13 knots. They are achieved by a total engine power of 3130 kW, which is transmitted to two Rolls Royce US255 propellers. In December 2016 the Dutch shipyard handed over the ASD 2810 type CAPO BOEO built in Song Cam to Italy's Rimorchiatori Riuniti group for work in the Sicilian port of Augusta. In June 2017, the keel of SST-ARUÁ out of a new series of four ASD 2411 tugs for SAAM Smit Towage Brazil was laid at Wilson Sons shipyard in Guarujà/Brazil. In the same month, another ASD 2810 tugboat for Russian customers, named TIS, was delivered to the Russian company RPK Nord for operation at the transshipment tanker UMBA anchored in Kola Bay since 2016, as well as the ASD 2913 type DANIMARCA and ASD 2411 COLUMBIA for Rimorchiatori Riuniti. In October, a unit of this type was handed over to the Port Autonome de Papetee, christened AITO NUI II (Fig. 8.20).

In August 2017, the heavy-lift carrier LONE transported a dozen "stock-built" Damen tugs from the Vietnamese shipyard to Rotterdam, including nine ASD tugs. In January 2018, a shallow water version of an ASD 2310 SD (Shallow Draught) modified for special requirements under the name PAPILLON was delivered to De Boer Remorquage, a subsidiary of the Dutch companies Dutch Dredging and Iskes Towage & Salvage, for a 12-year assignment in the ports of Cayenne and Kourou of French Guyana. In the following month the Dutch tug builders completed the 80 tonnes towing ASD 2913 newbuilding ST. ANGELO for shipping company Tug Malta in Valetta, a subsidiary of the Rimorchiatori Riuniti Group.

Fig. 8.20 Thanks to their high ice class, the Russian tugs BULAT und SIVER can operate in Wiborg und Wyssozk (Ralf Witthohn)

Of the hybrid version of the ASD 2810 tugs, BERNARDUS and MULTRATUG 28 were put into service in 2014/15 for Dutch shipping companies. They can operate under different modes, either powered by diesel engines, diesel-electrically or electrically thus having the potential to save 30%, with emissions reduced by up to half. In February 2018, Ijmuiden-based Iskes Towage & Salvage commissioned the conventional ASD 2810 newbuildings LYNX and PHOENIX, which were built for stock by Damen's Vietnamese shipyard and which are sister ships of ARGUS and ARASHI already in service for Iskes for assistance services in Delfzijl and Eemshaven. By commissioning the 60-tonne bollard pull tugs the share of ASD tugs in the companie's fleet increased to half of the 18 tugs. In the beginning of 2021, Spain's Boluda group acquired the Iskes company, which tugs retained their names with the prefix VB (Fig. 8.21).

From 2016 to 2019, Damen's ASD building programme as well included the ASD 2009 type CMS WRESTLER for Clyde Marine to serve UK ports, the ASD 2411 tugs S. E. BARKAT GOURAT HAMADOU for Djibouti, SAMSON and ASD NANDA for operation in Nassau, the 2412 units BUFFALO and BEAGLE for Boluda to serve at Rotterdam, the 2609 version PETER for Nakhodka, the ASD 2810 types ROSE for Tug Services in Nassai, VYAZ (ELM) for RPK Nord, MASSABI for Port Pointe Noire, the 2811 unit PIONEER for the UK, the ASD 2813 versions VINCENZINO O. for Moby Lines in Sarroch/Italy, OCEAN GRANVILLE and OCEAN KITSILANO for Quebec, the 2913 design OSPREY for Mina Saqr Port, 3010 AYA I for Vanino/Russia, the 3212 types SL LULU and SL MURJAN for Smit

Fig. 8.21 The 2014-Romanian built Damen ASD 2810 type BERNARDUS of Ijmuiden/ Netherlands-based tug operator Iskes, in 2017 purchased by Dutch operator Multraship and renamed ADVENTURE, represents a hybrid type able to operate under electric power during low power demands (Ralf Witthohn)

Lamnalco, SOCHI and TUAPSE for Taman/Russia, MULTRATUG 29 for Terneuzen, MARS and MERCURIUS for Iskes, JAWAR ANBAR, JAWAR DHI QAR, JAWAR DIYALA and JAWAR MUTHANNA of different sizes for Dubai.

8.10 Bugsier Tugs from Fassmer Yard

In 2014, Fr. Fassmer shipyard in Berne/Motzen on the Lower Weser ended a 7-year interruption in German tug construction by delivering two 31.5 m length, 11.3 m breadth, 6.2 m draught harbour tugs for Hamburg-based Bugsier-Reederei. BUGSIER 7 and BUGSIER 8 represent a tug type which is fundamentally different from Damen's ASD types. Two high-scew Schottel propellers running in nozzles are installed forward. They are driven by two ABC type main engines of 2 × 2250 kW output installed in the aft part of the ship. Designed for 12 knots and a bollard pull of 72 tonnes, the tugs have an aluminium alloy wheelhouse and no raised forecastle, as they are intended for operation in port (Fig. 8.22).

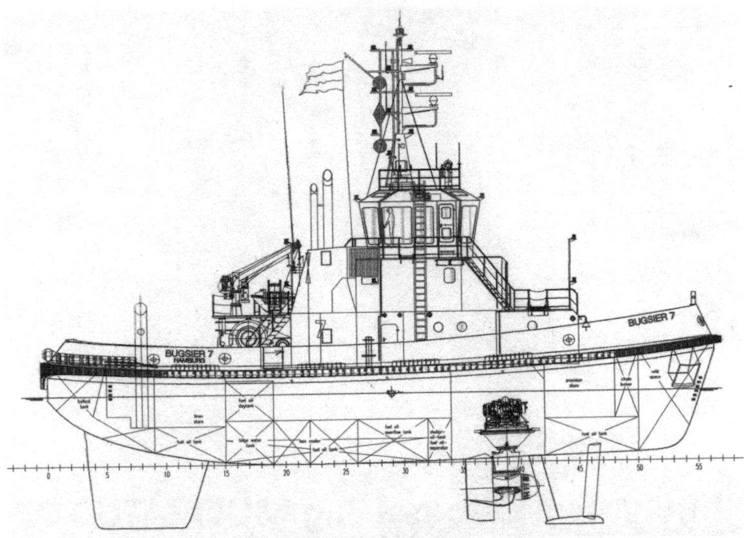

Fig. 8.22 Two rudder propellers under the foreship characterize the propulsion concept of the Fassmer buildings BUGSIER 7 and BUGSIER 8 (Fassmer)

8.11 The EDDY TUG Project

A hybrid tug project of Hollands Shipyard in Hardinxveld-Giessendam extended the choice of tug operators between completely different propelling systems for specific local navigational conditions and task spectra. The first newbuilding of the Dutch type, the double-ended tug EDDY 1, came into service in June 2014. In order to achieve a 360° all-round performance spectrum, the tug was equipped with two newly developed azimuthing propellers of Schottel type, which rotate 290 times per minute and are installed forward and aft in the midship line. The power take-in (PTI) is a PEM electric motor mounted on the reduction gear and powering the tug, when during the approach trip the main engine is switched off. The towing winch is located about one third of the shop's length of the vessel from the bow. Due to the special propeller arrangement, the same pulling and pushing forces of about 65 tonnes are developed in almost every direction. Main engines are two Mitsubishi motors, each producing 1610 kW. Two diesel generators, each with an output of 568 kW, ensure economical operation without the use of the main engines. They accelerate the free sailing tug to 11 knots, one even to 9.2 knots. EDDY 1 became temporarily owned by Panamanian interests in 2016 and in 2020 operated by Norway's FFS Marine as FFS ATHOS in Norwegian and Danish waters. In early 2022, FFS ATHOS was sighted in Stavanger (Fig. 8.23).

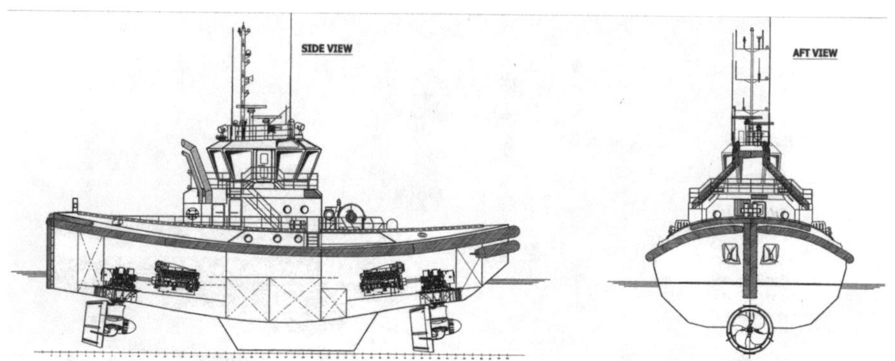

Fig. 8.23 The innovative hybrid tug EDDY 1 is driven by one propeller each under the fore and aft hull (Holland Shipyard)

8.12 Innovative Carousel Tug MULTRATUG 32

At the beginning of 2018 Damen Shipyards handed over the tug MULTRATUG 32, known as the Carrousel Rave Tug, to Dutch shipping company Multratug on the basis of a leasing contract as the first of two newbuildings. The tug, which is stationed in Terneuzen on the Scheldt, is the first to combine special towing equipment with the new type of propeller arrangement as chosen on the tug EDDY 1. Like EDDY 1, MULTRATUG 32 is equipped with two propulsion units, in this case of type Voith, arranged one behind the other in the midship line of the ship. In addition, the tug is fitted with a novel, patented towing system. This includes a steel ring installed around the superstructure, on which a towing winch is running at an inclination of up to 45°. Due to the flexible installation, the winch remains in the line of force of the tug line during all manoeuvres the tug is performing. This should enable the tug to manoeuvre and work much faster than conventional tugs without compromising its stability. This does not only reduce the risk of capsizing during towing operations, but also makes a better use of the pulling power power, thus hoped to increase the efficiency of assistance work by about 25%.

The so-called RAVE concept of a carousel tug is based on a design by the Canadian engineering bureau Robert Allan in cooperation with the supplier of the propellers, the German company Voith. The special winch was manufactured by Dutch company Machinefabriek Luyt in Den Oever, the two

Fig. 8.24 A towing winch running around the deckshouse is the special feature of MULTRATUG 32, which was stationed in Antwerp in 2022 (Ralf Witthohn)

main engines with an output of 2650 kW are from Belgian engine builder ABC. They provide the tug with a bollard pull of 77 tonnes. The building contract was signed in November 2015 between Novatug, a company of the Dutch Multratug shipping company, and the Damen Shipyards subsidiary Van der Velden Barkemeyer. The hull of the 32 m long tug was built by Theodor Buschmann in Hamburg, like Multratug a company of the Fairplay group. Completion of the tug took place at Damen Maaskant shipyard in Stellendam/Netherlands (Fig. 8.24).

9 Icebreaking

Nowhere are ships subjected to greater stress than while navigating in ice. Cargo ransports in ice-covered waters, but also the exploitation of resources in polar zones and research expeditions require the assistance of large icebreakers such as those operated in Russia, Canada and the United States. In contrast to the northern Polar Sea, the Antarctic has by international treaty been declared a research area with a temporary ban on mining. Many countries including South Africa, Argentine, China, South Korea, Australia and Germany deploy specially built research icebreakers and supply ships to this area (Fig. 9.1).

9.1 Russian Nuclear Icebreaker YAMAL

The development of nuclear-powered merchant ships in the 1960s led to a dead end. The Japanese research vessel MUTSU as well as the cargo ships SAVANNAH of the United States and OTTO HAHN of the Federal Republic of Germany were decommissioned when a high risk, geographically limited operational capability and the high costs of fuel exchange became apparent. Further projects, such as the construction of large container ships to be powered by nuclear energy, therefore stayed in the drawer. The only reactor-powered civilian ships are Russian icebreakers.

Between 1974 and 2007, six nuclear-powered icebreakers of the ARKTIKA class were delivered by Baltic shipyard at St. Petersburg/Russia. Three of them are still in service, named SOVETSKIY SOYUZ, YAMAL and 50 LET POBEDY, completed in 1989, 1992 and 2007, resp. The construction of the 150 m length, 30 m breadth YAMAL drew out for 6 years, not at least due to the political collapse of the Soviet Union. The icebreaker is equipped with two

R. Witthohn, *International Shipping*, https://doi.org/10.1007/978-3-658-34273-9_9

Fig. 9.1 One of the largest icebreaker fleets, unemployed in the summer, is stationed in Finland, which also has a high level of internationally demanded expertise in the development of technologically sophisticated ships for navigation in ice (Ralf Witthohn)

nuclear reactors of 171,000 kW output each, of which only one is in operation at the time. The reactors supply the energy for three electrical motors with a total output of 52,800 kW to drive three propellers. Breaking ice of up to 9 m thickness, the ship was able to reach the North Pole. The 50 LET POBEDY is used for cruise activities during the summer by Murmansk-based Atomflot. In 2017, two 14-day voyages to the North Pole were offered in June and July at prices starting at USD 28,000. In October 2020 a new nuclear-powered icebreaker named ARKTIKA was commissioned, the first of four LK-60Ja class units driven by two electrical motors of 20,000 kW each. The sister ships under construction are to be named SIBIR, URAL and JAKUTIA (Fig. 9.2).

9.2 Diesel-Electrically Driven Ice Breaker MOSKVA

The icebreakers MOSKVA and SANKT-PETERBURG, delivered in 2008 and 2009 by Baltiyskiy Zavod in St. Petersburg to state-owned Rosmorport FSUE in Moscow, have been equipped with a diesel-electric propulsion system. Four diesel generators with a total output of 21,000 kW supply the energy for two azimuth propellers running in nozzles. The 114 m length, 27.5 m breadth ships are able to pass through 1 m thick ice with a snow layer of 20 cm at a speed of 3 knots forward or astern. In ice-free waters a speed of 17 knots is achieved. The icebreakers have been designed with a gap in the stern to accommodate the stem of vessels, which are being moored and towed by means of a double-drum winch of 2 × 250 tonnes pull. The icebreakers

Fig. 9.2 The Russian icebreaker YAMAL was one of only four nuclear-powered civilian ships in 2022 (Ralf Witthohn)

have as well been equipped with powerful fire-fighting equipment, including two monitors with a total capacity of 5000 m^3/h. The basic design of the MOSKVA and SANKT-PETERBURG was as well realised in the slightly modified icebreakers VLADIVOSTOK, MURMANSK and NOVOROSSIYSK built in 2015/2016.

Rosmorport carries out icebreaking services in 15 Russian ports with a fleet of eight large icebreakers, six icebreakers operating in harbours, 14 shallow water icebreakers and three auxiliary icebreakers, assisting more than 3000 vessels in the winter months. The icebreakers are stationed in Azov, Rostov-on-Don, Taganrog, Yeysk, Arkhangelsk, Kandalaksha, Astrakhan, Olya, Magadan, Murmansk, St. Petersburg, Vyborg, Vyssozk and Vanino. In January 2018, MOSKVA was deployed at the Sabetta LNG terminal. The Rosmorport icebreakers had already provided icebreaking services to ships when the port was built in 2015. SANKT-PETERBURG was berthed in the port of the name-giving city, together with the older icebreakers FORT, TOR, MURMANSK, CAPTAIN NIKOLAEV and SEMYON DEZHNEV. Despite increasing obstructions by ice, the busy shipping traffic ensured that the shipping channel to St. Petersburg was passable without the need of icebreakers (Fig. 9.3).

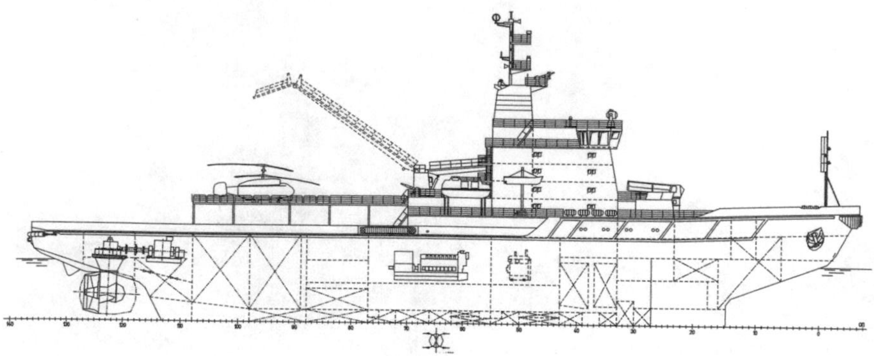

Fig. 9.3 The final construction of the Russian icebreaker MOSKVA differed from this earlier design state of 2006 (Baltiyskiy Zavod)

9.3 Icebreaking Offshore Vessels for Kazakhstan

The 3600 km long transfer voyage from their builders in the Netherlands via the inner Russian canal system with 14 locks, 36 bridges and 11 high-voltage lines between St. Petersburg and Astrakhan set essential parameters for the construction of two ice-breaking supply ships, which Royal Niestern-Sander shipyard in Delfzijl built for the Dutch Wagenborg group in 2012. Designed for year-round use in the Caspian Sea, SANABORG and SERKEBORG were designed and built according to the regulations of ice class 1A Super of the French classification society Bureau Veritas. The 68.2 m length, 14 m breadth newbuildings were as well laid out as distinctly shallow-water vessels with a draught of only 3.15 m to meet the shallow water conditions of the Caspian Sea. SERKEBORG had also to undertake the long and difficult voyage, which is only possible from May to October, in the opposite direction, before the ship was in 2017 converted into a diving supportship at its builders in Delfzijl. For the new employment, the ship was equipped with a decompression chamber, moon pool, dynamic positioning system of class DP2 and larger accommodations for 24 instead of 12 persons. To fulfill a 10-year contract for subsea work, the ship, renamed REDSBORG, sailed to the Dutch Caribbean island of St. Eustatius in June 2017. Renamed STATIABORG the ship was still stationed there in 2022.

The diesel-electric propulsion system of the two offshore vessels consists of two generator sets with Caterpillar diesel engines of type 3516C-HD, each producing 2000 kW. They provide the energy for two icepods specially developed by supplier Wärtsilä from Finnland and providing a speed of 12 knots and a bollard pull of 34 tonnes. By these two newbuildings Wagenborg strengthened its commitment in Kazakhstan, which goes back to 1997, when seven companies active in the oil industry (AgipKCO) awarded the contract to participate in the exploration of an area of 300 × 300 km in the northern part of the Caspian Sea with average water depths of only 3.5 m and ice covers up to 0.8 m thickness. In October 1998 Wagenborg had initially deployed the icebreaking supply vessels ARKTICABORG and ANTARCTICABORG with main engines of 2 × 1950 kW output. In 2017, Wagenborg Offshore operated a fleet of icebreakers, tugs, barges and accommodation ships in the Caspian Sea (Fig. 9.4).

9.4 Gas Turbine-Powered US Icebreaker POLAR SEA

POLAR STAR and POLAR SEA, built in 1976/77 by Lockheed shipyard in Seattle for the US Coast Guard, are regarded to be the most powerful conventionally powered icebreakers. The 121.6 m length, 25.5 m breadth ships were equipped with a hybrid propulsion system, which gives the ship's command

Fig. 9.4 After operating in the oil production sector in the Caspian Sea, SERKEBORG was in Delfzijl converted into a diving support ship for employment in the Caribbean (Ralf Witthohn)

the choice between diesel-electric and gas turbine propulsion for particularly high requirements. The ships are diesel-electrically driven by three shaft motors with a combined output of 13,000 kW via three propellers. Three gas turbines produce 43,500 kW, at maximum 54,350 kW. This allows breaking of 1.8 m thick ice at 3 knots and the navigation in up to 7 m thick crushed ice. Together with Canada's Coast Guard icebreaker LOUIS S. ST-LAURENT, the POLAR SEA was the first American surface vessel to reach the North Pole in 1994. The two ships also completed the first west-east crossing of the Arctic Sea. POLAR SEA undertook trips to Alaska and Greenland, but was was as well used for Antarctic expeditions and got to the southernmost point reachable by a ship in 1994. Including nine pilots for the helicopters, the crew numbers 155, plus 32 scientists. Due to damage to five of six diesel engines, the POLAR SEA was laid up in its home port of Seattle in 2010 and was in 2022 still used as spare parts store for the sister ship (Fig. 9.5).

9.5 Helicopter Transfer from Arctic Supply Vessel DP POLAR

Seven cargo vessels of type VITUS BERING built between 1986 and 1992 by Kherson shipyard were designed for flexible use in Polar zones. Another unit, China's icebreaker SNOW DRAGON, was developed from this ice-breaking vessel type in 1993. Although described by the Ukrainian shipyard as an Arctic Supply Ship, the vessel type has as well been used for Antarctic supply

Fig. 9.5 Until it was decommissioned due to engine damage, POLAR SEA, together with its sister ship POLAR STAR, was the most powerful non-nuclear-powered icebreaker (Ralf Witthohn)

operations, such as IVAN PAPANIN in the Antarctic summers of 2013 and 2015. A special procedure using helicopters and hovercrafts was devised for handling cargo off the Siberian coasts. For this purpose, the ships were equipped with a hangar and a landing deck aft, from which KA-32 helicopters can lift up to 5 tonnes of cargo. Cargo transfer from the anchorage to port can also be carried out by feeder vehicles of various types.

The 167 m length, 22.6 m breadth, 13.5 m depth, 9 m draught vessels achieve a deadweight capacity of 10,125 tonnes and offer 340 slots for containers. Five cargo holds have a combined volume of 16,900 m^3. Cargo handling is also undertaken by two twin cranes of 2 × 25 tonnes SWL. An ice ram is mounted behind the rudder, and the propeller is running in a nozzle. The high propulsive power required during navigating in ice, is generated by a two-stroke MAN B&W main engine of type 8L60MC manufactured under licence and putting out 17,920 kW for a speed of 17 knots of the laden ship in open water. Taking into account the consumption required for handling operations, the range is 8000 and 14,000 nm if this is neglected (Fig. 9.6).

The prototype VITUS BERING was already scrapped in 2003 under the name DEEPWATER 1 in Alang/India. The former ALEKSEY CHIRIKOV was converted to the Mexican offshore vessel CABALLO AZTECA and operated in the Gulf of Mexico until about 2017. VLADIMIR ARSENYEV was as NEVIS PEARL scrapped in Alang/India in 2016. VASILIY GOLOVNIN of Russia's Far Eastern Shipping Co. (FESCO) was the only unit of the building series still engaged in Polar operations, when in March 2020 it proceeded from the Antarctic via Cape Town to Maiori/Italy. In the first months of 2021 the ship was berthed in Cape Town, in February 2022 in Murmansk. STEPAN KRASHENINNIKOV, converted into a research vessel in Gdansk in 2002, was scrapped in Turkey in 2013. In March 2018, IVAN PAPANIN suffered a leak from a damaged hull when leaving India's Antarctic Bharati station. 100

Fig. 9.6 In its original architecture, the polar supply vessel type from Kherson looked like IVAN PAPANIN (Ralf Witthohn)

expedition members had to be disembarked, the ship sailed for Cape Town and was broken up in Chattogram (Chittagong)/Bangladesh later in the year. YUVENT was dismantled after operations in the Antarctic under the name ICE MAIDEN I in Wallsend in 2013 (Fig. 9.7).

9.6 Supplies for the Neumayr Station

Even before the German polar research vessel POLARSTERN was commissioned, the Alfred Wegener Institut in Bremerhaven chartered the Norwegian special vessel POLARSIRKEL, classified as research vessel, icebreaker and seal hunter. The 590 dwt ship of 50 m length and 11.5 m breadth, built in 1976 under high ice class regulations by Hoivolds shipyard in Kristiansand/Norway under a contract awarded by R. C. Griber shipping company, supplied the first German Antarctic station Neumayr. In 1981, after changing to the Canadian flag, the ship was renamed POLAR CIRCLE and in 1989 POLARSYSSEL, before it came into Spanish hands in 2007 to be operated as research vessel and from 2011 on as a small cruise liner.

The sister ship POLARBJORN, built the year before, undertook campaigns in polar waters, but also on the Amazon and Congo under the name ARCTIC SUNRISE for the Greenpeace organization. In 1995, Griber shipping commissioned a larger research vessel of a different design, laid out to undertake logistic tasks as well, from Kvaerner Kleven shipyard in Leirvik/Norway. The 80 m length, 17 m breadth POLAR QUEEN is twin-propelled, built with a double hull and features a deadweight of 2134 tonnes. As early as 1999, Griber sold the ship to the British Antarctic Survey to supply the British Antarctic stations under the name ERNEST SHACKLETON. In August 2016, the ship was chartered out to guide the cruise liner CRYSTAL

Fig. 9.7 The research vessel DP POLAR, built as Arctic supply vessel STEPAN KRASHENINNIKOV, was scrapped in Aliaga after operations in the Antarctic in 2013 (Ralf Witthohn)

Fig. 9.8 Built in Norway in 1995, POLAR QUEEN entered British service in 1999 as ERNEST SHACKLETON (Ralf Witthohn)

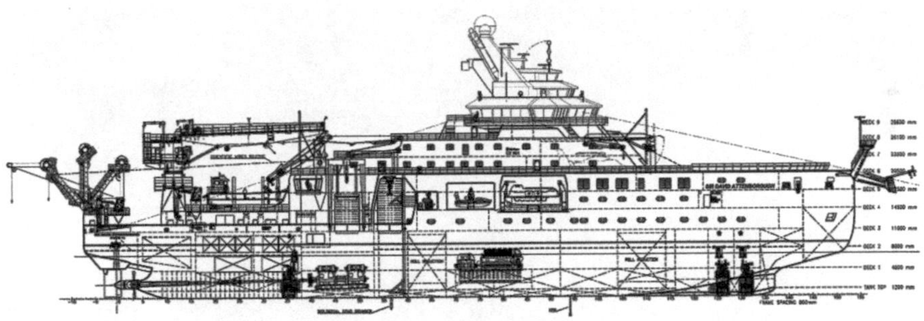

Fig. 9.9 The British icebreaker newbuilding SIR DAVID ATTENBOROUGH started its first voyage to the Antarctic in 2021 (Cammell Laird)

SERENITY through the Northwest Passage. In October 2017, the ship loaded supplies bound for Port Stanley on the Falkland Islands in Bremerhaven.

In 2014, the British government announced the building of a new polar research vessel to be named SIR DAVID ATTENBOROUGH, which in 2021 replaced both ERNEST SHACKLETON and the 5732 gt vessel JAMES CLARK ROSS, built in 1990. Designed by Cammell Laird shipyard in cooperation with Rolls Royce in accordance with the specifications of the British Natural Environmental Research Council (NERC), the diesel-electrically driven twin-screw ship is fitted with a dynamic positioning system and has space for 30 crew members and 60 scientists. The ship is to break ice up to 1.5 m thick and shall be able to survive in ice for 60 days. Its cargo capacity is 4000 tonnes is transported over a range of 19,000 nm at 13 knots. The 128 m length, 24 m breadth, 11 m depth, 7.5 m draught vessel is equipped with helicopters, a tender boat, a remotely operated underwater vehicle (ROV), moon pool and winches, in addition to 12 km long cables for acoustic and seismic underwater research (Figs. 9.8 and 9.9).

Part III

Recreation, Travel and Rescue at Sea

10

Cruise Shipping

For the first time in modern history, a complete shipping branch came to a total standstill in March 2020, when the COVID-19 pandemic had reached the cruise industry and paralyzed it. All operators, including the big US-based concerns Carnival and Royal Caribbean, but also the inland cruise operators nearly simultaneously announced the suspending of all their cruise activities, by this following the recommendation of the Cruise Lines International Association (CLIA). When port authorities had announced, that they would not grant the ships permission to enter, the cruise liners had to be retracted from service, assembled at the world's main cruise hotspots or returned to their homelands, where they were laid up. This meant the abrupt end of the cruise business. Germany's Aida Cruises, a subsidiary of Italy's Costa Crociere and thus part of the US Carnival group, immediately ended the actual cruises of all its 14 liners (Fig. 10.1).

Exemplary was the sudden end of the world cruise of the 44,697 gt liner ARTANIA operated by Phoenix Reisen from Bonn/Germany. The cruise was stopped at Sydney, but the captain had to give up the plan, to directly bring back his ship and 832 of its passengers still on board to Bremerhaven/Germany, when seven of them were positively tested for the virus. Therefore all passengers, of whom one person meanwhile had died, were to be taken home by plane after quarantine in a hotel. Even harder hit were the passengers of Holland America Line's ZAANDAM, a Carnival subsidiary, on which at least two persons died off the Panama coast and numerous got ill, while South American countries did not allow the ship to enter their ports. In a desperate move, yet welcomed by US president Donald Trump, the world's leading, US-based cruise concern Carnival had before offered their liners to be used as hospital ships for COVID-19 patients. Facing the nearly total loss of income

R. Witthohn, *International Shipping*, https://doi.org/10.1007/978-3-658-34273-9_10

Fig. 10.1 Costa Crociere's Italian-flagged 2680-berth cruise liner COSTA MEDITERRANEA was at Port Louis/Mauritius, when the COVID-19 pandemic let collapse the international cruise business in the mid of March 2020. Homeward-bound, the ship passed the Suez canal end of the month. The ship's succeeding itinerary included Mediterranean cruises from Venice, but the operator cancelled the voyages to start in April. The ship proceeded to Sembawang/Indonesia and spent the winter 2020/2021 at anchor, before proceeding to Dubai in August 2021 (Ralf Witthohn)

Fig. 10.2 Particularly hit by the termination of the whole world's cruise activities were the passengers, crew and operator of the 1984-built, 1200 berth ARTANIA, which had to make a long trip home from Sydney to Germany (Ralf Witthohn)

and high expenses for the laid-up fleet and their crews, Carnival offered to pay up to 11.5% rates to finance new debts in a volume of USD 6 billion (Fig. 10.2).

The way particularly cruise shipping was hit put up the question, whether such maritime leisure activities, which total output worldwide is figured by CLIA at USD 150 billion, would ever return to their old state. Up to this fiasco, cruise shipping had shown strong growth rates and forecast a worldwide increase of ocean passengers from 30 million in 2019 to 32 millions in 2020. In fact, not more than about 200 cruises were carried out from July to December 2020, the number of passengers dropped to 5.8 millions. In May 2021 Germany's TUI Cruises re-started *Blaue Reisen* in Germany from the port of Kiel, from Mallorca in June 2021 besides cruises from the Canary Islands and from Greece. At falling infection rates, Hapag-Lloyd and MSC from Kiel and Aida Cruises from Rostock resumed operations shortly afterwards.

The cruise business, yet, remained constantly affected by the pandemic, particularly, when the infection rates sharply increased again from autum 2021 on. End of the year, Hapag-Lloyd Cruises' EUROPA 2 had to suspend a cruise from Dubai, while the EUROPA had to change its itinerary. In January 2022, about 3000 passengers had to leave the AIDANOVA in Lisbon, because 60 crew members and eight passengers had been infected. The ship proceeded to Gran Canaria, to resume sailings there. At the same time, a cruise of MEIN SCHIFF 6 from Dubai had to be suspended, the passengers had to return by plane. In January 2022 MEIN SCHIFF 1 returned with a number of infected passengers and crew members from a cruise to the Caribbean to Bremerhaven. During a second such cruise 88 passengers and 115 crew members had to leave ship due to Corona infections. Six infected passengers and 53 crew members were still on board, when the ship arrived in February several days earlier than scheduled. The re-start concept to concentrate on shorter cruises, in order to avoid the uncertainties bound to long-term sailings, yet affected the Baltic Sea itineraries, when the operators cancelled St. Petersburg after the Russian attack against Ukraine (Fig. 10.3).

The cruise liner builders, too, had to interrupt or to reduce their building programmes. Germany's MV Werften group in Wismar, Warnemünde and Stralsund were promised financial help from the government of Mecklenburg-Vorpommern, after work on cruise liners for the Hong Kong-based Genting group had been stopped in March 2020 and payments to suppliers were not anymore secured. In January 2022, MV Werften filed for insolvency proceedings, and so did the yard group's sole customer Genting, leaving the 201,000 gt newbuilding GLOBAL DREAM uncompleted in the Wismar building dock.

In Italy, the Fincantieri concern suspended work for an initial, later extended 2 weeks period in the mid of March 2020. As there would be a risk that orders for cruise vessel newbuildings could be cancelled and investments in newbuildings postponed with fatal consequences for thousands of employees in Europe's shipbuilding industry, the governments of Germany, France,

Fig. 10.3 In the wake of the COVID-19 pandemic, Hapag-Lloyd Cruises' EUROPA and EUROPA 2 hat to terminate their scheduled cruises and assembled at Bremerhaven Columbuskaje in November 2020 (Ralf Witthohn)

Finland, Italy and Norway in April 2020 agreed, that cruise shipping companies would, due to their sudden deterioration in earnings, be allowed to suspend for 1 year the repayment of debts of newbuildings that were financed by state export credit guarantees. In December 2020, the deferral was extended until March 2022. Germany alone had covered payment obligations for the financing of cruise ships built in Germany at a total volume of around EUR 25 billion. In February 2022, Meyer Werft was granted a EUR 14 million aid equally shared by the of Lower Saxon and German governments, to secure the companie's 3500 jobs.

Meyer Werft in Papenburg/Germany decided to bring forward the transfer of the nearly completed 184,700 gt newbuilding IONA for UK-based P&O Cruises, a Carnival subsidiary, from the builder's site to the North Sea in March 2020. But the planned work programme on the liner type had to be reduced during its stay at Bremerhaven. Meyer Werft boss Bernard Meyer expected the conditions of the cruise market not to be reinstated before 2030. In February 2021, yet, the shipyard started the building of the IONA sister vessel due for delivery to P&O in 2022. In March, the builder delivered the 169,000 gt ODYSSEY OF THE SEAS to Royal Caribbean and announced the contract signing of a 52,000 gt cruise liner to be delivered to Japan's NYK in 2025. Despite some corona cases that had previously occurred on board, the Meyer Werft newbuilding AIDACOSMA left Eemshaven at the end of October 2021. Afterwards, the ship was berthed at the Columbuskaje in Bremerhaven to be fitted out. After further sea trials off the Norwegian coast lasting several days the 183,774 gt AIDACOSMA was delivered to Italy's Aida Cruises in December, but started its cruise programme to western European ports from Hamburg not before February 2022 (Fig. 10.4).

Fig. 10.4 Delivered in December 2021, the AIDACOSMA started a first cruise from Hamburg in February 2022 (Ralf Witthohn)

The Meyer Werft newbuilding IONA was one of 19 cruise liner newbuildings that were to be commissioned in 2020. Several of the newbuildings did yet not start passenger operations right after delivery due to the pandemic situation. In February 2022, Carnival's MARDI GRAS, which delayed maiden voyage had taken place a year before, completed a cruise at Cape Canaveral/US from Roatan/Honduras. At the same day the port was called by the newbuilding 130,818 gt CELEBRITY APEX coming from Costa Maya/Mexico. The 5602 gt CORAL GEOGRAPHER of Coral Expeditions, later than scheduled commissioned in 2021, sailed at this time in Australian waters off Cairns. Another 2020 newbuilding was the 135,156 gt COSTA FIRENZE for Costa, carrying out cruises from Dubai in February 2022, along with MSC VIRTUOSA of MSC. The 20,499 gt Global class CRYSTAL ENDEAVOR of Genting's Dream Cruises was forced at anchor off Montevideo because of non-paid fuel bills. Mystic's 9934 gt WORLD VOYAGER was underway in the US Gulf, Ponant's 9988 gt LE BELLOT was lying idle off Zanzibar, the sister vessel LE JACQUES CARTIER rounded the Arabian peninsular. Princess Cruises' 145,281 gt ENCHANTED PRINCESS and the 56,182 gt SEVEN SEAS SPLENDOR of Regent Seven Seas Cruises sailed the Caribbean. Delivery of the 22,498 gt SCENIC ECLIPSE II to Scenic was post-poned to 2023. Silversea's 40,844 gt SILVER MOON was sighted off the South American north coast, the 6365 gt SILVER ORIGIN anchoring of Baquerizo Moreno/Ecuador, while the 108,192 gt SCARLET LADY of Virgin Voyages, like the ENCHANTED PRINCESS only delivered in 2021, returned to Miami from a Bahamas cruise.

The delivery of 20 newbuildings was scheduled in 2021. Among the completions that took place was P&O's IONA, which in February 2022 sailed for the Canarias Islands from Southampton. ODYSSEY OF THE SEAS was engaged on the Fort Lauderdale-Mexico route. The 108,192 gt VALIANT LADY of Virgin Voyages arrived in Portsmouth from Civitaveccia in March 2022. The 47,842 gt VIKING VENUS of Viking Ocean Cruises sailed along the Norwegian coast, while 181,541 gt MSC VITUOSA operated from Dubai. Delivery of the 25,401 gt EVRIMA to The Ritz-Carlton Yacht Collection at El Astillero/Spain was post-poned to 2022, as was the commissioning of the 23,000 gt SEABOURN VENTURE under construction at Genova/Italy and of the 140,600 gt CELEBRITY BEYOND not completed according to schedule in St. Nazaire/France. The 31,283 gt COMMANDANT CHARCOT was in February 2022 bound for the Antarctic. The 186,364 gt COSTA TOSCANA undertook a cruise in the western Mediterranean, the 40,844 gt SILVER DAWN the eastern Mediterranean, while the 99,935 gt ROTTERDAM sailed the Caribbean from Port Everglades, as did the 170,412 gt MSC SEASHORE from Miami. AIDACOSMA was delivered in December 2021, but did not take up service before February the following year from Hamburg. Delivery of the 145,282 gt DISCOVERY PRINCESS took place in early 2022 in Monfalcone, from where the liner was via Suez and Singapore bound for Los Angeles to undertake a first cruise to Mexico in March 2022. From 2022 to 2027 a total number of only still 25 newbuildings were announced to debut, including the 64,000 gt EXPLORA 1 and EXPLORA II due for delivery to the luxury brand Explora Journeys of MSC group in 2023 and 2024 in a series of four ships from Italy's Fincantieri shipyards in Monfalcone and Castellamare di Stabia.

The decline of employment in the wake of the pandemic, accelerated the demolition of a long series of cruise liners. CARNIVAL FANTASY arrived at the breaker's yard in Aliaga/Turkey in August 2020. The same fate was experienced or faced by the CARNIVAL IMAGINATION, CARNIVAL INSPIRATION, CARNIVAL SENSATION CARNIVAL SENSATION, CARNIVAL ECSTASY and CARNIVAL FASCINATION, the latter having been lying idle off Sri Lanka for months and arrived off Gadani/Pakistan unter the name Y HARMONY in February 2022. The on-going market conditions forced Celestyal Cruises to sell their flagship CELESTIAL EXPERIENCE to Pakistani breakers in October 2021. Among the first scrapped liners was the former German ASTOR, beached in Aliaga/Turkey in November 2020. At the same place broken up was Fred. Olsen's BOUDICCA in May 2021. The COLUMBUS of Greece-based Seajets was in March 2021 sent for demolition. COSTA VICTORIA was towed to Turkey in the beginning of 2021. Cruises were as well terminated by the cruise liners GRAND CELEBRATION, KARNIKA, MAGELLAN, MARCO POLO, MARELLA CELEBRATION, MARELLA DRAM, MONARCH, OCEAN DREAM, SOVEREIGN, SUPERSTAR LIBRA and HORIZON.

A downright cruise industry had from the 1960s onwards, initially in the United States, been emerged and was still growing until the end of 2019, after a temporary downturn as a result of the 9/11 events. In 2018 a number of 7400 passenger ships with a total of 41.8 million gt was recorded. The cruise liners transported about 28.2 million passengers globally in that year. The year before, 11.9 million passengers had come from the United States, 2.4 million from China, 2.2 from Germany and 1.9 from the UK. In 2019 at total of 29.7 million passengers were counted world-wide. A forecast for 2020, which became obsolete due to the coronavirus problem, estimated 14.4 million of passengers were said to come from North America, 6.7 million from Europe, 4.2 million from Asia and 1.5 million from Australia/Pacific. 32% of cruise liner deployments should be made in the Caribbean, 17% in the Mediterranean,

Fig. 10.5 The Caribbean is the prime destination of American passengers, but is also visited by guests from other countries, like on the AIDADIVA berthed here in St. Lucia (Lutz Witthohn)

11% in Europe and 5% each in China, Australia/Pacific and Alaska. But in fact, the pandemic let drop the figure to just 5.8 millions in 2020 (Fig. 10.5).

At the beginning of the modern cruise industry, the available ships were measured under 20,000 gross tons (gt). The economic formula of other shipping branches that larger ships could be operated more economically, was soon taken over by the cruise line operators and led to rapidly increasing newbuilding sizes. In 1996, the 100,000 gross tons mark was exceeded for the first time by the 2600-berth liner CARNIVAL DESTINY, a newbuilding of Italy's Fincantieri shipyard for US-based cruise operator Carnival. In 2016, the USD 1.4 billion costing HARMONY OF THE SEAS, delivered to the likewise American company Royal Caribbean International (RCI) by STX France in St. Nazaire reached a peak by 226,963 gt and a number of 6400 passengers. Only cruise concerns, of which Carnival and RCI are the leading ones, or such of the entertainment industry like US-based Disney or Genting from Hong Kong were still able to finance newbuilding series of the largest ships (Fig. 10.6).

The extent to which the construction of large cruise liners does not only depend on the shipyard's necessary infrastructure, but also on experience and a wide network of capable suppliers is demonstrated by the fact that, in contrast to the rest of international shipbuilding, the construction of large cruise liners is currently only carried out by the European shipyard groups Fincantieri, Meyer and Chantiers de l'Atlantique at their locations in Italy, Germany, Finland and France and just beginning to emerge in China. While the unlimited transfer of know-how has made it possible to relocate the construction of

Fig. 10.6 At 226,963 gt, the 2016-built HARMONY OF THE SEAS is the world's largest cruise liner (Arne Münster)

almost any type of ship worldwide, the design and realisation of large cruise liners has up to now remained a high-tech affair of just two market participants.

At the same time, the dimensions of the largest cruise liners limit the choice of ports to call at, so that smaller ships find niches in remote trade areas such as the Antarctic, the Arctic or on rivers and canals, like the Amazonas and St. Lawrence River. Due to limited growth in the industrialised countries, US companies have been focusing on the Chinese market for some years now, but the companies based there are not willing to give it up without a fight. In order to be able to provide passengers with the joy of travel and recreation at reasonable prices, the service personnel on the ships, mostly originating from cheap wage countries, work under tough conditions at low pay. The organization of cruises requires some logistical effort, as passengers have to be taken care of for days or weeks at a good level and their travel to the port of embarkation, which is more frequently made by plane, is to be planned. Cruises on inland waterways, which, due to their limited navigability, only allow the use of smaller and low-designed ships, are increasingly accepted by the public.

10.1 Caribbean, Alaska, Great Lakes, Hawaii

The United States is by far the largest cruise market. By the destinations Bahamas/Caribbean/Central America, Alaska and Hawaii are being offered three attractive areas that can be reached from domestic ports. Accordingly, US ports also lead the international ranking of passenger numbers. In 2016, 5 million passengers embarked and disembarked in Miami, 4.2 million in Port Canaveral and 3.8 million in Port Everglades. 1.8 million passengers travelled via New York and 1.1 million via New Orleans. On a world scale, the Mexican port of Cozumel was busiest by 4.39 million passengers in 2020, followed by Miami by 3.64 million passengers, Nassau/Bahamas (3.5 million), Barcelona (2.88 million), Port Canaveral (2.59 million) and Civitaveccia, the port of Rome, where 2.55 million passengers were counted. In the wake of the COVID-19 pandemic worldwide port calls fell by about 75% from January 2020 to December 2021.

Carnival Cruise Lines, founded in 1972 and headquartered in Doral/Florida, is by far the market leader with a worldwide revenue of USD 20.8 billion in 2019, which corresponded to a share of almost a quarter of the US market. All companies of the Carnival concern together operated a total of 27 large cruise liners in 2019 and still 25 in 2021. These included the 133,500 gt CARNIVAL HORIZON built by Fincantieri group's Marghera shipyard near Venezia. The ship undertook a number of Mediterranean cruises after delivery in April 2018, subsequently sailed from New York during the summer and began six to eight-day Caribbean voyages in September. In September 2016, 12 years after the construction of the 85,942 gt CARNIVAL MIRACLE, Carnival had returned to Finland, when it ordered two newbuildings of the 180,000 gt class, of which the first was delivered as COSTA SMERALDA in October 2019. Of similar size is the 180,800 gt MARDI GRAS, a 5200-berth vessel to be delivered in October 2020 by Finland's shipyard Meyer Turku, after floating out took place in January 2020 and completion, originally scheduled in August, took place in December. Carnival had received their first cruise liner newbuildings from Kockums shipyard in Malmö and Kvaerner Masa Yards in Turku, from the latter the identical panamax ships CARNIVAL MIRACLE, CARNIVAL PRIDE, CARNIVAL LEGEND, CARNIVAL SPIRIT and for the Italian subsidiary COSTA ATLANTICA and COSTA MEDITERRANEA of around 85,000 gt. With the exception of the Costa ships deployed in Europe and CARNIVAL LEGEND, the ships mainly conducted voyages in America (Figs. 10.7 and 10.8).

Fig. 10.7 Carnival's 181,808 gt flagship MARDI GRAS was floated out at Turku in January 2020 (Meyer Turku)

Fig. 10.8 The Carnival group had six panamax cruise liners of type CARNIVAL LEGEND built in Finland (Ralf Witthohn)

In April 2016, CARNIVAL VISTA, measuring 133,596 gt and designed for 3936 passengers, was delivered by the Fincantieri yard in Monfalcone/Italy. The first ship of a new generation and then the largest in the shipping company's fleet was equipped with a diesel-electric propulsion system of five MAN 48/60CR main engines and a total output of 62,400 kW. The initial cruise programme until September included sailings to the Caribbean from Miami, subsequently until 2019 from Galveston. The sister ship CARNIVAL HORIZON, built at the Marghera site, undertook its first test runs in the Adriatic in November 2017. The first four voyages of the ship started in April 2018 from Barcelona, followed by four-day Bermuda and eight-day Caribbean voyages from New York. In March 2017, in the presence of Italian Prime Minister Paolo Gentiloni, Fincantieri handed over the 144,216 gt cruise liner MAJESTIC PRINCESS, the largest of 77 passenger vessels built since 1990. Destined for the Chinese market, the newbuilding was the third ship in a series for the Carnival company Princess Cruises, which started operation in 2013 by the 142,714 gt ROYAL PRINCESS.

After the suspension of cruises due to the COVID-19 pandemic, CARNIVAL VISTA was anchored off Houston/US, where also the LIBERTY OF THE SEAS had dropped anchor. CARNIVAL HORIZON was lying off Miami, where a dozen large cruise liners had assembled, including the companie's CARNIVAL CONQUEST, CELEBRITY SUMMIT, COSTA MAGICA, and together with MSC SEASIDE, MSC DIVINA, MSC MERAVIGLIA, MSC ARMONA, NORWEGIAN BLISS, NORWEGIAN ENCORE, MARINA, MARINER OF THE SEAS and not far from CARNIVAL SUNRISE, VIKING SKY and INDEPENDENCE OF THE SEAS lying in or off the port of Fort Lauderdale.

10.1.1 Hurricane Irma Destroying Cruise Destinations

On the north side of the island terminal between Miami and Miami Beach, the crews of four large cruise liners are awaiting their guests on October 28, 2017: MSC DIVINA, CARNIVAL SENSATION, CARNIVAL GLORY and the NORWEGIAN ESCAPE. ADONIA has moored on the south side. The cruise liner of UK's P&O Cruises, a subsidiary of the Carnival group, has arrived from Southampton via Bilbao, to take up Caribbean cruises from Barbados in the coming months. A 14-night voyage to Grenada, Mayreau, Pont-à-Pitre, Antigua, St. Kitts, Martinique, St. Lucia, Bequia and Bridgetown will begin on 3 November. 710 passengers have paid between GBP 999 for the inside cabin and GBP 4139 pounds for the suite. It is the last Caribbean season of ADONIA, which had been reserved for adult passengers, under the P&O flag. The shipping company has sold the 30,277 gt ship to Azamara Cruises, a subsidiary of Royal Caribbean Cruises, for delivery in March 2018. Voyages already booked for later trips are cancelled. Until then, same-sex couples can still be married by the captain, after the Supreme Court of Bermuda, where all P&O liners are registered, has allowed marriages to take place.

On MSC DIVINA of Mediterranean Shipping Co., which has arrived from Cozumel/Mexico on 25 October, the crew is preparing for departure at 7 pm. The route to the eastern Caribbean had to be shifted to the west due to the devastation caused by hurricane Irma. Philipsburg on St. Maarten and San Juan on Puerto Rico can no longer be called at. Instead, Montego Bay/Jamaica, George Town/ Cayman Islands, Costa Maya/Mexico and Nassau/Bahamas have been included in the itinerary. The timetable of CARNIVAL SENSATION moored in front of MSC DIVINA has also been disrupted by the hurricane in the previous weeks. The five-night voyage leads to Nassau/Bahamas, the shipping company's own island Half Moon Cay/Bahamas and Freeport/Grand Bahama. Meanwhile, NORWEGIAN ESCAPE is bunkered for a seven-day voyage to Falmouth/ Jamaica, George Town/Cayman Islands, Cozumel/Mexico and the shipping company's own private island Great Stirrup Cay for a price starting at USD 429 (Fig. 10.9).

Built by Meyer Werft in Papenburg under contract of US-based Norwegian Cruise Line (NCL), NORWEGIAN ESCAPE and its three sister ships reflect a tendency towards the return to a supposedly outdated class system of passenger shipping. On the ship, delivered in October 2015, an exclusive area was created for 95 owner's suites of up to 120 m^2 in size, described by the shipping company as "ultra-luxurious", while offering a number of exclusive privileges, including a private restaurant with an outdoor terrace, a lounge, a

Fig. 10.9 As for many other cruise liners, the itinerary of NCL's NORWEGIAN ESCAPE had to be altered due to the devastations caused by hurricane Irma on Caribean isles in September 2017 (Ralf Witthohn)

courtyard area extending over two decks with a retractable canopy, three pools, sauna, spa, an own sun deck, butler and concierge service, preferred embarkation and many other luxuries. By comparison, the smallest interior cabin of 2175 cabins on the NORWEGIAN ESCAPE has a space of 12 m^2, thus, only a tenth of the luxury suites. Just 15 years earlier, NCL had presented a newly developed freestyle cruising concept, according to which traditional constraints were to be replaced by a more open leisure culture.

Each of the five MAN diesel engines of the NORWEGIAN ESCAPE, consisting of three twelve-cylinder and two fourteen-cylinder V48/60CR engines, i.e. common rail engines with a total output of 76,800 kW and the ability to consume heavy fuel oil, is equipped with exhaust gas scrubbing devices. The so-called scrubbers were supplied from the Norwegian company Green Tech Marine. They are installed in the exhaust system instead of silencers, which function they take over as well, so that no additional space is required in the funnel casing. The water using so-called hybrid scrubbers have an open and a closed circuit. While the closed circuit used in port does not require a permanent supply of water, the open circuit used while sailing uses seawater, which is intended to neutralise the acidic exhaust gases due to its alkaline properties. According to the manufacturer, the wash water would be constantly monitored so that its composition would comply with IMO regulations when being let out. In reality, however, this technology only represents a diversion of pollutants previously emitted into the atmosphere into seawater.

Year-round departures of Caribbean cruises from New York by the sister ship of NORWEGIAN ESCAPE, the NORWEGIAN BREAKAWAY, required a special air-conditioning design that had to cope with high below zero temperatures (Fig. 10.10).

Not specified innovations on NORWEGIAN ESCAPE were supported with grants from Germany's Federal Ministry of Economics and the State of Lower Saxony. Between 2010 and 2015, Meyer Werft received innovation subsidies of EUR 72.9 million from the public sector, half of which were provided by the Federal Government and half by the State of Lower Saxony. EUR 5 millions, which the state government of Lower Saxony paid out in 2015 for development work in shipbuilding, went almost exclusively to Meyer Werft. In addition to these project-related subsidies, the state's government invested large sums of millions in infrastructure projects, from which shipbuilding benefited as well.

The design of NORWEGIAN ESCAPE (165,157 gt) represents a further development of the newbuildings NORWEGIAN BREAKAWAY and NORWEGIAN GETAWAY (both 145,655 gt) previously delivered by the builder to Norwegian Cruise Line. Thus the following newbuilding was

Fig. 10.10 On NCL's NORWEGIAN ESCAPE an exclusive area with 95 suites is not accessible for the majority of passengers (Ralf Witthohn)

designated as the lead ship of the NORWEGIAN BREAKAWAY-plus class, which included NORWEGIAN JOY (167,725 gt) NORWEGIAN BLISS (168,028 gt) and NORWEGIAN ENCORE (169,116 gt) with steadily growing gross tonnages. The shipping company's demand for a ship with a larger capacity increased from 4028 to 4248 passengers, served by 1700 crew members, resulted in a raising of the already high-rise superstructure by one to 20 decks at the same ship length. To compensate for the reduction in stability, the ship was widened from 39.7 to 41.4 m. Due to the resulting deterioration of hydrodynamic properties, this in turn necessitated an increase in engine power from 62,400 to 76,800 kW by means of an additional MAN main diesel engine of type 12V48/60CR generating 14,400 kW. The engine power of 39,000 kW supplying two ABB Azipods accelerated NORWEGIAN ESCAPE during the trial runs to a maximum speed of 21.8 knots. This meant that the contract speed of 21.5 knots was slightly superseded. In early 2022, NORWEGIAN ESCAPE and NORWEGIAN ENCORE undertook sailings from Cape Canaveral/US to Caribbean destinations, NORWEGIAN JOY from Miami to Mexico, while NORWEGIAN BLISS departed to there from Los Angeles (Fig. 10.11).

When delivered by Meyer Werft to NCL in April 2013 after an 18-month construction time NORWEGIAN BREAKAWAY was hitherto the largest passenger ship built in Germany. The Nassau/Bahamas-registered cruise liner and its sister ship NORWEGIAN GETAWAY were financed by a European consortium of five banks led by Germany's state-owned KfW IPEX-Bank in Frankfurt. It ensured that the export credit with Hermes cover and ship CIRR (Commercial Interest Reference Rate) was made available in conformity with OECD rules and included financing during the construction period. The newbuildings already had a 56% higher tonnage than the 93,502 gt measuring NORWEGIAN GEM delivered by Meyer Werft in 2007. The configuration of the main section, which has been retained in the enlarged NORWEGIAN ESCAPE version, shows some special features resulting from different deck widths and a setback of the superstructure block with the passenger decks by 7 m. The boats hang outside half the breadth of the ship, resulting in a maximum ship breadth of 52.7 m including boats, and a moulded breadth of 39.7 m. In February 2022, NORWEGIAN BREAKAWAY undertook cruises from New Orleans to Cozumel/Mexiko and NORWEGAN GETAWAY from New York to the Bahamas.

Port and Weser pilots have been ordered for 6 p.m., but the schedule for testing the cruise liner NORWEGIAN ESCAPE will not fully be kept to. It is not until an hour later that the newbuilding detaches from Bremerhaven Columbuskaje under its own power and without tug assistance at calm sea. The ship's small fast

Fig. 10.11 As on NORWEGIAN BREAKAWAY, many of the large cruise liners feature an atrium extending over several decks (Ralf Witthohn)

rescue boat, observed from the unfolded platforms at the bow of the ship, circumnavigates the colourful stem paint showing a swordfish and is lifted on board by crane. Only now—the sun has already set an hour earlier in the blue and red shining sky over the coastline of Butjadingen—does the ship slowly pick up speed and glide into the dark and cold North Sea night. This third test run of the NORWEGIAN ESCAPE is said to have become necessary because of problems with the ship's sea chests.

After undocking from the covered building site of Meyer Werft on 15 August 2015, the ship was transferred from Papenburg to Eemshaven on 18 September on the river Ems, which had to be dammed for this purpose. A week later, the ship made a first short test trip into the North Sea, ending in Bremerhaven. There the fitting out of the ship with the shipping company's equipment began, while final work is carried out in the cabins and public areas. For a nightly black-out exercise during which the ship loses all its power supply, two tugs are ordered for holding it at the quay. Subsequently, the actual six-days lasting trial to the Norwegian south coast starts. The tests include all manoeuvres like mile trip and crash stops, the most important measurement referring to the maximum speed.

To continue its equipment, NORWEGIAN ESCAPE subsequently returns to Bremerhaven. However, during the trial trip it has become apparent, that further tests and a docking stay at the shipyard of Blohm + Voss in Hamburg will be necessary. The ship is docked for 1 day in the repair dock Elbe 17 and returns to Bremerhaven for completion after further test runs in the German Bight. Only 2 h after the handover on 22 October 2015, NORWEGIAN ESCAPE sails, for the first time under the flag of Bahamas and that of the RCI shipping company, to Hamburg. After a short presentation trip with travel agents on 24 October, the maiden voyage via Southampton to Miami begins, where the christening of the ship takes place on 9 November 2015 (Fig. 10.13).

Only a few buildings built for a public audience are less attractive in terms of exterior design than a number of the cruise ships that have come into service in recent years. For many decades, the harmony between technical necessities and the pleasing exterior of a passenger ship was regarded as a prerequisite for a successful design, not least to serve the image of the shipping company. Nowadays, elegant lines are often sacrificed to the goal of high capacities and interior design specifications. The cruise liner types of the NORWEGIAN BREAKAWAY, QUANTUM OF THE SEAS and GENTING DREAM classes delivered by Meyer Werft to Norwegian Cruise Line and Royal Caribbean from 2013 onwards are clear examples of this (Figs. 10.12, 10.13, and 10.14).

10.1.2 From New York to Bermuda

The fact that the high-rise architecture of the recent large cruise liners is not only an aesthetic shortcoming was demonstrated by the accident of the ANTHEM OF THE SEAS delivered by Meyer Werft to Royal Caribbean International in April 2015. The ship was caught in a storm off the US East

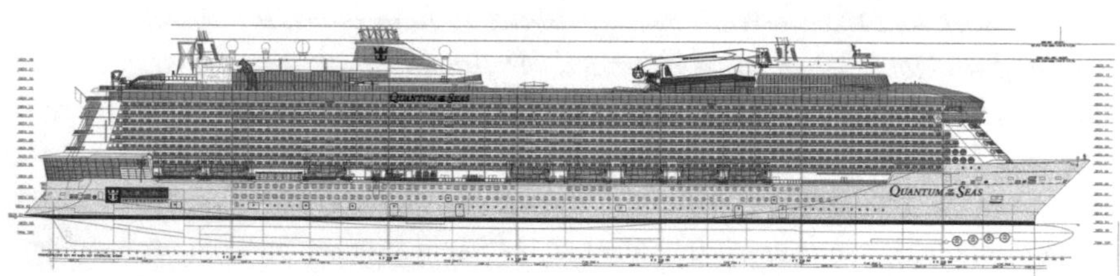

Fig. 10.12 The outer architecture of QUANTUM OF THE SEAS is dominated by a total of 1571 balconies (Meyer Werft)

Fig. 10.13 The interior lay-out of the NORWEGIAN ESCAPE, including the pictured reception, distinguishes the ship's outer architecture nearly without compromise (Ralf Witthohn)

Fig. 10.14 A series of trial runs and the subsequent exchange of wheelhouse windows were to be carried out, before the newbuilding NORWEGIAN ENCORE as fourth unit of the NORWEGIAN BREAKAWAY-plus class was delivered by Meyer Werft to NCL in October 2019 (Ralf Witthohn)

Coast in February 2016, injuring several passengers due to severe listings and causing damage to the interior fittings, so that the voyage had to be terminated. The ship's itinerary, which was to be completed by spring 2019, provided for 8- to 13-day voyages of the ship from New York to Florida, Bermuda, the Bahamas and the Caribbean. At a tonnage of 168,666 gt, the 347 m length, 49.4 m breadth ANTHEM OF THE SEAS can accommodate up to 4819 passengers (Fig. 10.15).

The third newbuilding of this type, OVATION OF THE SEAS, undertook sea trials off the southern Norwegian coast and in the German Bight in March 2016. During the over 2 weeks lasting tests top speeds of just under 24 knots were reached and cornering at high speeds of more than 20 knots carried out. During a second dock stay at Blohm + Voss, the propulsion propellers were subsequently replaced by a pair of different design. The voyage planning of the OVATION OF THE SEAS included voyages from Australia and East Asia in 2018/2019. During a cruise in December 2019, passengers

Fig. 10.15 The design of the ANTHEM OF THE SEAS incorporates, in addition to the architectural flaw, a shortcoming in seaworthiness with high acceleration forces even at small angles of inclination and the associated dangers for passengers and crew (Ralf Witthohn)

were killed or injured by the outbreak of a volcano while visiting White Island/ New Zealand (Whakatane), after the ship had anchored off the coast en route from Sydney to Wellington.

Another Royal Caribbean Cruise Line newbuilding for the Asian market and fourth unit of the class was after delivery in April 2019 christened SPECTRUM OF THE SEAS. Among the special features of the 169,379 gt ship is a glass capsule, which can be raised 90 m above deck. The fifth and last ship of the class was delivered as ODYSSEY OF THE SEAS in April 2021 5 months later than scheduled and was subsequently planned to take up two- to seven-night trips from Haifa/Israel to Cyprus and Greece with passengers vaccinated against the corona virus. The plans were cancelled after the break-out of the Israeli-Palestinian conflict in May 2021, and the ship sailed to the US, where it undertook sailings from Port Everglades to Mexiko in February 2022. At the same time, ANTHEM OF THE SEAS still continued its Bahamas sailings from New York, but started trip as well from Cape Canaveral. The OVATION OF THE SEAS cruised off the West Coast from Seattle. SPECTRUM OF THE SEAS anchored off Hong Kong (Fig. 10.16).

For two further newbuildings to be supplied by the Meyer Werft group to RCCL at costs of over EUR 2.5 billions, a banking consortium led by

Fig. 10.16 Following four units of the SPECTRUM OF THE SEAS class, the Meyer group was to build two 5500-berth liners for Royal Caribbean at its Finnish production place (Ralf Witthohn)

Germany's KfW IPEX-Bank provided credits. The consortium comprises BNP Paribas, HSBC, Commerzbank, Santander, BBVA, Bayern/LB, DZ Bank, JPMorgan and SMBC. KfW is participating by a share of EUR 686 million in the largest syndicated financing of the bank with loans granted over 12 years. In addition to Hermes cover, the investment is secured by the Finnish Export Credit Agency (ECA) Finnvera. The newbuildings, designed to carry more than 5500 passengers, are being built at Meyer Turku shipyard in Finland and scheduled to be commissioned in 2022 and 2024. Their engine room sections, including the main engines, will be supplied by Neptun Werft in Rostock/Germany and are be towed to Turku. The ships will be equipped with dual-fuel propulsion systems and run on LNG in regular operation, otherwise on marine diesel oil. On one of their cruise liners, RCCL in 2016 tested a total of 100 kW generating fuel cells from Ballard Power Systems for the energy supply of the hotel operation. The fuel cells are as well to be used on the newbuildings from Finnland.

10.1.3 Mickey Mouse on Caribbean Course

A small fleet of large southbound cruise liners is heading for the port of Nassau/Bahamas.

CARNIVAL ELATION has departed from Freeport, CARNIVAL VICTORY from Miami and CARNIVAL SUNSHINE from Charleston. Another big liner, MAJESTY OF THE SEAS of Royal Caribbean Cruise Line, has left Cape Canaveral just half an hour before DISNEY DREAM. The operator of the latter liner, Disney Cruise Line, has chosen Port Canaveral, located on Florida's east coast 160 nautical miles north of Miami, as home port of its cruise ships.

Founded by the Walt Disney Company in 1995, Disney Cruise Line was to extend the concept of family holidays to the high seas from the group's adventure parks, which had initially been established in Florida in 1971. For this purpose, the 83,000 gt measuring DISNEY MAGIC and DISNEY WONDER were in 1998/99 built by the Fincantieri shipyard site in Monfalcone/Italy. The architecture of the second newbuilding duo in 2010 and 2012 from Meyer Werft, the 129,690 gt ships DISNEY DREAM and DISNEY FANTASY, was based on the Italian buildings in their two-funnel look, despite a considerable increase in size. In March 2016, the Walt Disney company announced, that it intends to commission two more LNG-powered cruise liners, slightly larger at around 139,300 gt and offering 1250 cabins for a total of about 4000 passengers. Meyer Werft plans to deliver the ships in 2021 and 2023. In July 2017 Disney ordered a third ship, which is

Fig. 10.17 Two funnels give the DISNEY FANTASY a more classic liner architecture (Ralf Witthohn)

scheduled to be completed in 2022. The 144,000 gt, LNG burning newbuilding DISNEY WISH has been floated out from Meyer's building dock in February 2022 and was delivered in June 2022, delayed by several weeks (Fig. 10.17).

The Bahamas-flagged Disney ships of German origin were primarily designed for three- to five-night cruises from Port Canaveral, but the two older liners also undertake trips in Europe, from California, to Alaska and Mexico. In 2018, the route planning for DISNEY MAGIC envisaged 4- to 13-night voyages from Miami to the Bahamas, from New York, from Barcelona, San Juan, Dover, Quebec, Civitaveccia, Copenhagen and Dover with prices ranging from USD 1300 to USD 12,000. DISNEY WONDER simultaneously undertook 2- to 14-night trips from San Diego, Port Canaveral, Vancouver, San Juan and Galveston. The DISNEY DREAM and DISNEY FANTASY departed from Port Canaveral. In the wake of the COVID-19 crisis, DISNEY MAGIC, DISNEY FANTASY and DISNEY DREAM had suspended their cruises in April 2020 and were anchored off Port Canaveral, next to the cruise liners CARNIVAL FANTASY, CARNIVAL LIBERTY, CARNIVAL BREEZE and CARNIVAL MAGIC. In February 2022, DISNEY DREAM and DISNEY FANTASY undertook cruises from Cape Canaveral to the Bahamas, DISNEY MAGIC from Miami and DISNEY WONDER from New Orleans to Cozumel and Costa Maya in Mexico.

While the newbuildings from Italy at that time were still panamax ships of 294 m length and 32.2 m breadth, the German-built liners could only pass the enlarged Panama canal. On delivery in December 2010 in Bremerhaven, DISNEY DREAM was the largest and, at a price of around USD 900 million, also the most expensive passenger ship hitherto built in Germany. Despite the substantial enlargement to 339.5 m length, 36.9 m breadth and an increase in the number of decks by two to 14, the proportions and essential architectural features such as the shape of the two funnels, an invading stern and a strongly raking stem were, in the sense of a uniform brand, retained. A significant difference to the first newbuilding pair, however, is evident in the midship section. While the passenger decks on the Italian-built ships are extended to the entire breadth of the ship, the cabin decks including the balconies on the Meyer ships reach only 85% of the total width in favour of a larger outdoor space.

The second Disney duo was designed for 4000 passengers, who are accommodated in 1100 outside and 150 inside cabins. Of the outside cabins, 901, including 41 suites, have a balcony, the remaining ones round windows. The maximum number of beds is 5007, the number of crew members 1453. The ship's health centre is located on the lowest deck 1 accessible to passengers. On deck 2, 58 outside and 27 inside cabins are arranged next to the Enchanted Garden restaurant. The family-friendly Deluxe Oceanview staterooms offer three to four berths and are 19 m^2 in size. They have either one large or two smaller portholes. These rooms, equipped with TV, refrigerator and safe, are also located in two different categories on decks 5, 6, 7 and 8. Deck 3 houses the lobby with atrium and bar, and in the front the two-story Walt Disney theatre, restaurants, lobbies, shops and guest information, but no cabins. The Buena Vista theatre on deck 4 shows films, and there is also an area reserved exclusively for adults with five nightclubs on this deck. On deck 5 above there are facilities for children and infants, in addition to 88 outside and four inside cabins, which are distributed along the deck's length (Fig. 10.18).

The sixth deck is a pure cabin deck, with 194 outside and six inside cabins. The layout of decks 7 to 10 is similar, but deck 11 offers public spaces with a partly open air restaurant at the stern, followed by a pool for adults and children. On the forward starboard side are fitness and wellness facilities, while port side a dozen of partly spacious cabins are arranged. The Sun deck 12 has two restaurants, lounge, bar and the entrance to the water slide aft. At the front, three of the 29 outside cabins offer a view over the bow of the ship. In the aft section of deck 13, mini golf and basketball can be played at Goofy's Sports. The dummy of the forward funnel allowed the installation of a lounge with video wall and computers.

When choosing the propulsion system, the shipping company stuck to the more conservative diesel-electric option with conventional propellers, as it had already been equipped to the forerunners, despite the proliferation of—not always trouble-free—pod systems. The system comprises five main diesel

Fig. 10.18 The Mickey Mouse figure plays an important marketing role on the Disney ships (Ralf Witthohn)

gensets to supply the energy required to operate and propel the vessel. The portion required for propulsion is transmitted to two fixed pitch propellers via two electrical motors. In contrast to the first two ships equipped with Sulzer engines, five MAN four-stroke diesel engines of type V48/60CR were installed, two in the 14-cylinder version with 2 × 16,800 kW and three in the 12-cylinder version with 3 × 14,400 kW. This results in a total output of 76,800 kW. The two Converteam electrical motors each have an output of 21,000 kW. The ship is manoeuvred by means of two steering gears. Its manoeuvrability in port is increased by three transversal thrusters forward and two aft. A pair of stabilizers improves the seaworthiness.

On deck 4, eight closed motor rescue boats of the Drochtersen/Germany-based company Hatecke are placed on each side of the ship. Their total height including the launching systems requires two deck heights. The boats have two side doors on each side, which can be closed with tarpaulins, and an aft steering position. Forward of the boats, there is fitted a fast rescue boat on each side of the ship, and in front of and behind the boats there are life rafts stored, which can be lowered into the water by cranes (Fig. 10.19).

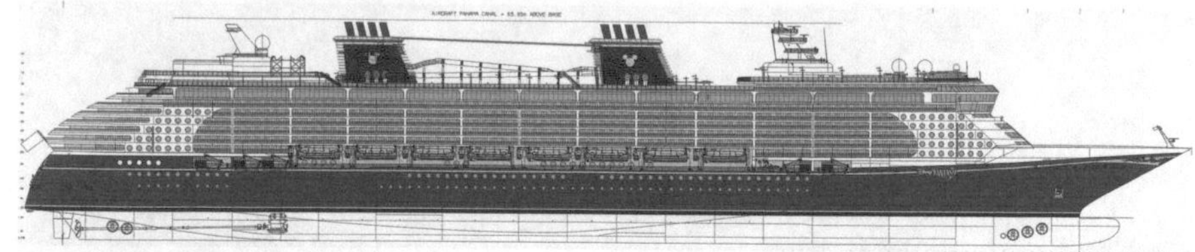

Fig. 10.19 The look of the first Disney ships DISNEY MAGIC and DISNEY WONDER has been retained on the DISNEY DREAM and DISNEY FANTASY despite a considerable increase in size (Meyer Werft)

10.1.4 PEARL MIST on the Great Lakes

The high challenges of cruise liner construction are not only evident from the still low number of only a handful of shipyards established in it. The difficulties occasionally encountered in the implementation of such projects also show the complexity of the task. Irvine Shipbuilding in Halifax, well experienced in the building of special ships, worked for several years on the PEARL MIST newbuilding project in Halifax, for which a contract had been signed with Pearl Seas Cruises in Guilford/USA already in 2006. The ship was only commissioned in June 2014 after outfitting work by Chesapeake Shipbuilding in Salisbury/Maryland. The building had been delayed due to legal disputes between the Canadian shipyard and the US shipowner about the performance of the construction work, which were finally settled in April 2013. The shipping company, a sister firm of American Cruise Lines operating river and smaller ships in North America, had complained, that the ship did not meet the regulations of the classification society Lloyd's Register and those of the flag state Marshall Islands. PEARL MIST finally came into service under supervision of DNV GL and the Marshall Islands flag, a register often chosen by US shipping companies.

The 99 m length, 16.8 m breadth, 4.8 m depth and 3.5 m draught cruise liner of 5109 gt is a novelty in that it is primarily used on the Great Lakes and the St. Lawrence River during summer, but seasonally also for sailings along the North American Atlantic coast and to the Caribbean, e.g. from Fort Lauderdale to Cuba. The ship's outside architecture features a balcony structure as typical of today's passenger ships. On four cabin decks are fitted 108 outside cabins ranging in size from 28 to 54 m^2 for a maximum of 210 passengers. The lowest of the six passenger decks, which is also the main deck, with the restaurant aft and a promenade on both sides, is only at short distance from the waterline. Main means of rescue are two closed lifeboats, one of which is placed on the uppermost cabin deck. Two diesel engines of type Caterpillar 3516C HD drive the ship to a speed of 17 knots via two propellers.

In February 2017, American Cruise Lines put into service a similar newbuilding, measured at 4057 gt and capable of carrying 175 passengers under the name AMERICAN CONSTELLATION. Specially designed to navigate the inland waterways of the Pacific Northwest, the ship started cruises from Seattle to Puget Sound and Alaska. The sister ship AMERICAN CONSTITUTION took up voyages on the East Coast, including trips on the Hudson river, Chesapeake Bay and to New England destinations after delivery by Chesapeake Shipbuilding in spring 2018.

10.1.5 Celebrity Cruise Liners to Alaska and Hawaii

The United States is the prime market for the 14 ships fleet of Celebrity Cruises, to be extended by the newbuildings CELEBRITY APEX in 2020 and CELEBRITY BEYOND in 2021, although other continents such as Australia and Europe are as well among the destinations. The companie's fleet includes five 121,000 gt postpanamax vessels built from 2008 to 2012 by Meyer Werft under the names CELEBRITY SOLSTICE, CELEBRITY EQUINOX, CELEBRITY ECLIPSE, CELEBRITY SILHOUETTE and CELEBRITY REFLECTION. The 2850 to 3420 passengers accommodating CELEBRITY SOLSTICE undertook Pacific voyages in 2018, from Seattle along the US Pacific coast, from Vancouver to Alaska, from Sydney to the Great Barrier Reef or from Honolulu to Vancouver. In 2020 the ship visited, among others, Australien ports. The 2590 berth ship CELEBRITY SUMMIT spent most time in American waters, making trips to the Bahamas, Bermuda, the Caribbean, New England and Canada. CELEBRITY ECLIPSE in 2018 undertook trips in the Mediterranean, the Baltic Sea and Scandinavia, which included transatlantic positioning trips. In 2020 the liner was sighted along the west coasts of North and South America, while CELEBRITY REFLECTION, which undertook sailings in the Mediterranean as well as in the Caribbean 2 years before, set in 2020 course to Caribbean destinations. End of March 2020, CELEBRITY REFLECTION had interrupted its cruises and anchored on the Cococay roads north of Bahamas, together with CELEBRITY EDGE, and next to nine other large liners, such as GRANDEUR OF THE SEAS, HARMONY OF THE SEAS, OASIS OF THE SEAS, ANTHEM OF THE SEAS, EXPLORER OF THE SEAS, NIEUW STATENDAM, VEENDAM, VOLENDAM and NIEUW AMSTERDAM.

Having been active in passenger shipping since 1922, Celebrity Cruises was founded in 1989 as a subsidiary of the Greek Chandris group. In the same year, the 27,907 gt liner MERIDIAN entered service, built in 1963 in Italy by the Riuniti Adriatico shipyard as GALILEO and modernized by Lloyd Werft in Bremerhaven/Germany. From 1990 to 1997, the 46,811 gt to 76,522 gt measuring newbuildings HORIZON, ZENITH, CENTURY, GALAXY and MERCURY from Meyer Werft joined the Celebrity fleet. After the takeover of the shipping company by Royal Caribbean Cruise Line in 1997, the Celebrity brand was retained. Between 2000 and 2002 the French shipyard Chantiers de l'Atlantique delivered the 89,900 gt panamax ships MILLENNIUM, INFINITY, SUMMIT and CONSTELLATION. In autumn 2018 was commissioned the 3373 berth newbuilding CELEBRITY EDGE for Caribbean and European cruises, as the first of four sister vessels and in 2020/21 to be followed by CELEBRITY APEX and CELEBRITY BEYOND from Chantiers de l'Atlantique. In 2019, the small, 120 passengers carrying CELEBRITY FLORA, built at Dutch shipyard De Hoop, took up 7-day trips round the Galapagos Islands as the fourth of such so-called expedition vessels operated by Celebrity in this region (Fig. 10.20).

Fig. 10.20 The postpanamax cruise liner CELEBRITY REFLECTION was towed backwards from its builder Meyer Werft over the Ems river to the open sea (Ralf Witthohn)

10.1.6 PRIDE OF AMERICA Built Under Special Permit

Only a politically granted exception from the regulations of the US Jones Act, which requires the building of ships operating solely between US ports to be carried out in the US, allowed to continue the construction of a cruise liner hull at a foreign shipyard. The newbuilding had, due to the bankruptcy of American Classic Voyages in 2001, remained unfinished at Ingalls shipyard. Germany's repair and conversion yard Lloyd Werft, which is principally dependent on the supply of hulls due to a lack of shipbuilding capacity of its own, seized the opportunity together with the US shipping company Norwegian Cruise Line (NCL) to have the unfinished hull towed to Bremerhaven for completion under the name PRIDE OF AMERICA. However, the equipping and lengthening of ship proved to be complex and resulted in one of the most spectacular accidents in shipbuilding history (Fig. 10.21).

In January 2004, when the ship's draught was increased due to the progressing equipment state, assembly holes provisionally cut for supply lines in the portside shell plating went under water, so that 45,000 tonnes of water flooded the ship, which lost stability and was only prevented from complete capsizing by lying on the quay wall. The accident of PRIDE OF AMERICA, which was in a delayed state of construction, remained without loss of life. In a complicated action the ship was salvaged in February and March 2004 with

Fig. 10.21 The hull of the PRIDE OF AMERICA as towed from the US to Germany was of such minor state, that the completion took much more efforts than expected (Ralf Witthohn)

the aid of a pontoon of Dutch company Smit, and its construction continued after dismantling the water-damaged equipment components. As the accesses to the two electric pod propulsion units had been open during the accident, these too were damaged and had to be replaced by two new propulsion units originally intended to be fitted to the sister ship PRIDE OF HAWAI'I. However, the diesel generators, which had also been submerged, could be restarted.

In the wake of the accident, Lloyd Werft declared its insolvency in February 2004, but was able to complete the ship in July 2005 after compensation of the damage costs by the insurers. Altogether, construction of the 80,439 gt/2186 berth PRIDE OF AMERICA took 7 years since the original order had been placed in America. The first passenger ship newbuilding under the US flag after almost 50 years could begin to undertake seven-day voyages all year round from Honolulu to Kahului, Hilo, Kona and Nawiliwili. The steel, which had also already been shipped to Bremerhaven for the construction of the sister ship, was transported further on to Meyer Werft and in 2006 used in the building of the 93,558 gt PRIDE OF HAWAI'I, later to become NORWEGIAN JADE. Like all other cruise liners, PRIDE OF AMERICA interrupted its Hawaiian cruises due to the COVID-19 pandemic and was tied up at Honolulu, near to the companie's NORWEGIAN JEWEL. In February 2022, PRIDE OF AMERICA was berthed in Portland/Oregon, NORWEGIAN JEWEL was cruising the western Mediterranean along with the NORWEGIAN JADE (Fig. 10.22).

Fig. 10.22 Simple misconduct caused the sinking of the cruise liner PRIDE OF AMERICA, one of the most spectacular accidents in shipbuilding history (Ralf Witthohn)

Fig. 10.23 As the Hawaii market did not meet NCL's expectations, PRIDE OF HAWAI'I switched to Europe under the name NORWEGIAN JADE (Ralf Witthohn)

10.1.7 PRIDE OF HAWAI'I in Europe

Meyer Werft building PRIDE OF HAWAI'I, which had originally also been planned to sail round Hawaii, suffered a case of Novovirus in November 2007 and was the following year withdrawn from Hawaii to become the first NCL ship mainly deployed in Europe under the name NORWEGIAN JADE. During the Olympic Games 2014, the ship served as a hotel ship in Sochi/Russia (Fig. 10.23).

10.2 Europe

Unlike in the Caribbean, which is sailed by cruise liners all year round, voyages in European waters are limited to the warm season from spring to late summer. Exceptions are short-term Christmas trips by English operators or occasional world cruises starting in Europe at the end of the year. Aida Cruises made an exception, when its newbuilding AIDAPRIMA undertook voyages only from Hamburg to Western Europe in the winter season 2016/2017 (Fig. 10.24).

At the same time, Europe was for a long time the sole continent on which large cruise liners were built. Only in November 2019, China's Shanghai Waigaoquiao Shipbuilding started steel cutting of the first of two 135,500 gt cruise liners, built according to the Vista class design of Italy's Fincantieri group and to be delivered in 2023 and 2024. State-owned shipyard group Fincantieri builds passenger ships at three of its nine shipyard sites, in

Fig. 10.24 Langelinie in Copenhagen is on the itinerary of almost every Baltic Sea cruise (CMP)

Monfalcone near Trieste, Marghera near Venice and Sestri Ponente near Genoa. In January 2013, Fincantieri, which had for several years also held a share in Germany's Lloyd Werft, extended its activities to Norway, where it purchased the Vard shipyard. The Scandianvian builder was awarded a contract from Australia's Coral Expeditions to build a series of expedition cruise liners at Vard Vung Tau shipyard in Vietnam.

In September 2017, the French government approved the 50% takeover of STX France in St. Nazaire by Fincantieri, previously majority-owned by Korea's STX group, by. But the deal fell through after the outbreak of the COVID-19 pandemic. The builder was subsequently given its old name Chantiers de l'Atlantique and became majority-owned by the French state holding 84.3%. The second large group, Germany's Meyer Werft, is constructing ships at three locations, in addition to the main facility in Papenburg, this is Meyer Turku in Finland and the Neptun Werft facility in Rostock-Warnemünde, where a large number of inland liners and, more recently, large sections for the cruise liner newbuildings in Papenburg and Turku are being built. In contrast, the second attempt of Japan's Mitsubishi shipyard in Nagasaki to succeed in this sector failed. The completion of the 116,000 gt ship DIAMOND PRINCESS for Princess Cruises was, due to a major fire, delayed by more than half a year in 2004, the construction of the AIDAPRIMA ordered from Germany's Adia Cruises by almost a year. The delivery, scheduled for March 2015, was not finally completed until February 2016, so that the shipyard decided to abandon cruise liner construction (Fig. 10.25).

Fig. 10.25 The delayed and loss-making building of AIDAPRIMA in Japan strengthened the exclusive position of two European shipyard groups in the construction of large cruise liners (Ralf Witthohn)

10.2.1 Carnival's European Subsidiaries Costa, Aida and P&O

Fincantieri had started to supply the Carnival group with postpanamax cruise liners in 1996, by delivering CARNIVAL DESTINY, later named CARNIVAL SUNSHINE. From 2006 to 2009 the 114,000 gt ships COSTA CONCORDIA, COSTA SERENA, CARNIVAL SPLENDOR and COSTA PACIFICA came from the Sestri shipyard in Genoa. The first ship of the series, COSTA CONCORDIA, suffered the loss of 32 lives in January 2012 in a severe casualty following irresponsible behaviour of the captain, when the ship grounded on the Italian island of Gigliu in the Mediterranean. The cruise liner was scrapped in Genoa after a difficult and costly salvage operation a year and a half later. The Fincantieri building programme for the Carnival group as well included the constantly in size growing CARNIVAL MAGICA, CARNIVAL TRIUMPH, CARNIVAL VICTORY, CARNIVAL CONQUEST, CARNIVAL GLORY, CARNIVAL VALOR, CARNIVAL LIBERTY, CARNIVAL DREAM, CARNIVAL MAGIC, CARNIVAL BREEZE, CARNIVAL FREEDOM, COSTA SERENA, CARNIVAL SPLENDOR, COSTA PACIFICA, COSTA LUMINOSA, COSTA DELIZIOSA, COSTA FAVOLOSA, COSTA FASCINOSA and COSTA DIADEMA (Fig. 10.26).

Fig. 10.26 COSTA CONCORDIA, built in Genoa, was scrapped at the same location only 6 years later after one of the most hair-raising cruise liner accidents (Ralf Witthohn)

The type of Carnival's MARDI GRAS is as well being realized at Meyer's Papenburg site for Carnival's European subsidiaries Costa, P&O and Aida. The first, 183,858 gt measuring newbuildings of the lng burning so-called Helios class class was delivered by Meyer Werft in Papenburg in November 2018 under the name AIDANOVA (Fig. 10.27). Scheduled to undertake trips from Las Palmas, the liner was in March 2020 lying idle in the port of Santa Cruz de Tenerife/Spain due to the COVID-19 outbrek.

Succeeding IONA for P&O, initially scheduled for delivery in May 2020, a second liner for P&O is under construction at the Papenburg site, originally scheduled for completion in 2022. As for Royal Caribbean newbuildings built by Meyer, Germany's state-owned KfW IPEX bank provides Carnival with a loan of approximately EUR 786 million for the newbuilding. The bank took over the complete structuring of the loan and acts as bookrunner, Initial Mandated Lead Arranger (MLA), Facility Agent and ECA Agent. Intending to syndicate up to 80% of the financing sum, the financing has a term of 12 years from delivery. Financining is insured by an export credit insurance, the so-called Hermes cover, and subject to OECD's ship CIRR (Commercial Interest Reference Rate).

In 2021 Meyer Papenburg planned to deliver the similarly sized AIDACOSMA to Aida Cruises and a sister ship to the same operator 2 years later. Also of this type class are the Finnish-built COSTA SMERALDA, delivered end of 2019, and COSTA TOSCANA, intended to be constructed at

Fig. 10.27 AIDANOVA represents the prototype of the 180,000-gt class built at Meyer Papenburg and Meyer Turku for four different Carnival group cruise operators (Ralf Witthohn)

Turku until 2021. The engine rooms of the LNG burning, 17 knots reaching ships, come from by Neptun Werft in Rostock-Warnemünde. After delivery, the 337 m length, 42 m breadth COSTA SMERALDA cruised the western Mediterranean from Italian ports, but had to suspend its sailings and was sighted laying in Marseille after the COVID-19 outbreak.

10.2.2 COSTA VICTORIA on Mediterranean Routes

Joseph Lambert Meyer, who had founded the iron ship building yard Barth & Meyer in Papenburg in 1872, 21 years later participated by 6% shares in the foundation of Bremer Vulkan, Schiffbau & Maschinenfabrik. Victor Nawatzki, a former employee of Meyer, was appointed as the Vulkan's first and long-standing director. Meyer himself turned down an offer from Bremen to take over the shipyard. After the interests of the Bremen and Papenburg shipbuilders had, due to Meyer's concentration on passenger and special shipbuilding, for a long time hardly overlapped, this changed, when Bremer Vulkan shipyard group drew up a concept for entering the cruise liner construction sector, in which the three shipyards Schichau-Seebeck, Bremer Vulkan and Lloyd Werft were to share the work. However, during the construction of the first ship according to this scheme, the 75,166 gt COSTA VICTORIA, for Italy's Costa Crociere, the shipyard group fell into insolvency in 1996. As result the builders Bremer Vulkan and Schichau-Seebeck went bankrupt, while Lloyd Werft continued to equip the newbuilding according to the original design in the course of insolvency proceedings. In continuation of the concept, Lloyd Werft also completed the sister ship of COSTA VICTORIA, laid down in 1999 under the name COSTA OLYMPIA, but completed as NORWEGIAN SKY on behalf of Norwegian Cruise Line (NCL).

After two Far East voyages from Singapore at the beginning of 2018, COSTA VICTORIA embarked on a 21-day positioning voyage to Savona in February 2018, from where the ship undertook seven- and nine-day voyages in the western Mediterranean in the summer with calls at Olbia, Port Mahon, Ibiza, Palma de Mallorca and Tarragona. In autumn, 2 weeks trips in the Indian Ocean from Port Louis/Mauritius followed, until the return to the Mediterranean in spring. In March 2020, COSTA VICTORIA was moored at Civitaveccia/Italy, the cruise port for Rome visitors, together with MSC GRACIOSA. But only several months later, the liner was sold for scrap after less than 25 years of service (Fig. 10.28).

Fig. 10.28 During the summer months, COSTA VICTORIA called at Mediterranean ports, such as pictured Malta's capital Valetta, although concerns are growing about thousands of passengers over-flooding the narrow streets of the small city cevery day (Ralf Witthohn)

10.2.3 AIDAPRIMA Sailing from Las Palmas

Established in 1993 from the origins of state-owned Deutsche Seereederei (DSR) in Rostock after its privatization, Aida Cruises became part of UK-based P&O in 1999. By sale of the British operator to the Carnival group in 2002, the German operator was integrated into the US-based concern as a subsidiary of Italy's Costa Crociere. Aida primarily seeks its customers in German-speaking countries, but also offers them trips outside Europe such as in the Caribbean or the Middle East. The newbuildings AIDAPRIMA and AIDAPERLA initially undertook seven-day round trips from Las Palmas and Palma de Mallorca, resp., during the winter season 2017/18. In spring 2018, AIDAPRIMA switched to the western Mediterranean for voyages from Mallorca and Barcelona, followed by trips in the Persian Gulf on a fixed seven-day route from Abu Dhabi to Khalifa Bin Salman, Dubai, Muscat and Abu Dhabi until February 2019. The trips were offered despite the political tensions and military conflicts in the region (Fig. 10.29).

In 2020 AIDAPRIMA was to continue weekly round-trips from Dubai, which had yet to be suspended due to the COVID-19 pandemic keeping the ship in Port Rashid/Dubai. End of March 2020 a fleet of cruise liners had assembled in the Arabian port, including the companie's AIDAVITA, NCL's NORWEGIAN JADE, MSC LIRICA, MSC BELISSIMA, HORIZON and NAUTICA. AIDAPERLA had initially chosen Hamburg as its port of

Fig. 10.29 After undertaking winter voyages from Hamburg in the first year, AIDAPRIMA switched to Las Palmas in 2017 as port of departure for a week lasting round trips to Madeira (Lutz Witthohn)

departure for seven-day tours of Western Europe from spring 2018 until December, when two-week voyages from Barbados were scheduled. In March 2020, the ship was caught there, together with AIDALUNA, MSC PRECIOSA, TUI's MEIN SCHIFF 1, MEIN SCHIFF 2, RCI's FREEDOM OF THE SEAS, SERENADE OF THE SEAS, VISION OF THE SEAS as well as SEABOURN ODYSSEY and SEA CLOUD. At the end of 2017, Dubai set a target of 1 million passengers in 2020, which became obsolete due to the COVID-19 pandemic. The capacity of Mina Rashid was to be expanded from six to seven mega ships. Dubai's neighbouring countries are as well investing in cruise tourism. Abu Dhabi received 315,000 cruise guests in 2017, Bahrain 76,000.

Italian-flagged AIDAPRIMA, the eleventh newbuilding of Aida Cruises, and its sister AIDAPERLA, delivered in 2017, differ fundamentally from the shipping company's previous cruise liners in terms of size, design and propulsion technology. A tonnage of 125,572 gt made them the largest cruise liner newbuildings of a shipping company operating in Germany, about 75% larger than the last four Aida ships built in Germany between 2010 and 2013 and featuring 71,304 gt. The external characteristics and lines of AIDAPRIMA and AIDAPERLA were developed in cooperation between the shipyard and Hamburgische Schiffbau-Versuchsanstalt (HSVA) and resulted in a turning away from the classic bulbous bow configuration in favour of a vertical stem with only a slight bulb. Instead of a diesel-electric propulsion system with two fixed-pitch propellers as fitted to the recently delivered AIDASTELLA, AIDAPRIMA was equipped with two pod drives. One of the diesel generators installed alongside three conventional four-stroke MaK M43C engines, an MaK M46DF engine, offers the opportunity of using LNG as fuel.

A special innovation installed on AIDAPRIMA and AIDAPERLA refers to the Air Lubrication System (MALS). Developed by the shipyard the system had already been successfully tested in other types of ships. Air bubbles escaping through openings in the bottom of the ship provide a resistance-reducing film between the ship and the water. At 300 m length and 37.6 m breadth, the ships offer space for about 3300 passengers in 1643 cabins. The passengers should be able to carry out their activities regardless of weather conditions, as the AIDAPRIMA initially operated all year round from Hamburg. In January 2017, yet, several people were injured on the ship, which ran into a North Sea storm, so that the concept was not pursued further.

After the delivery delays caused by the Japanese builders, the Rostock-based shipping company returned to domestic Meyer Werft and in 2015 ordered the 5200-berth newbuildings AIDANOVA and AIDACOSMA for delivery in autumn 2018 and spring 2021. These newbuildings are characterised by the first-time possible use of LNG as fuel in four dual fuel engines from Caterpillar supplying the entire ship and hotel operation. The first liner's keel was keel laid in Papenburg in September 2017. In February 2018, Aida ordered a third newbuilding of this type to be delivered in 2023. The Aida fleet is employed according to uniform route schemes with round trips of 4 to 14 days duration or longer positioning voyages. For the oldest liner named AIDACARA, built in Turku in 1996, the timetable after wintering with voyages between Gran Canaria and Mallorca and a world cruise from Singapore to Hamburg from April 2018 onwards specified Scandinavia, the Baltic Sea and Greenland/Iceland with departures from Kiel and Hamburg as well as the Mediterranean, Cape Verde Islands, Canaries and Azores as destinations in autumn. The AIDAVITA, which was built in Wismar in 2002/03, set off for East Asia in autumn 2018 after a summer in Europe, visiting Vietnam and China. The sister ship AIDAVITA remained mainly in Europe after voyages from Mauritius at the beginning of 2018 (Fig. 10.30).

Built in Papenburg in 2007, AIDADIVA returned from the Caribbean to the Baltic Sea in spring 2018, only to return to Central America via Canada in autumn. Trips to Scandinavia and the Baltic Sea were on the AIDABELLA's programme, which had Thailand, Malaysia and Singapore on its sailing schedule at the beginning of 2019. AIDALUNA's programme for 2018 and 2019 included the Caribbean, Baltic Sea and Caribbean again, while AIDABLU started with a trip from Portugal to West Africa, then switched to the Adriatic Sea and at the end of 2018 visited the Seychelles, Madagascar and Mauritius. AIDASOL took its passengers to Las Palmas for one-week voyages at the

Fig. 10.30 After the first call at Malta, the battery saluted the 69,203 gt panamax ship AIDADIVA built by Meyer Werft in 2007 (Ralf Witthohn)

beginning of 2018, made four-day trips from Hamburg to Amsterdam and London in the summer and 10-day sailings to Norway, before heading for the western Mediterranean from Mallorca. AIDAMAR sailed in northern and southern Europe and in the Caribbean. In February 2022, the ship made 7 days round-trips round the Canary Islands along with AIDANOVA. AIDADIVA and AIDAPERLA undertook 14 day cruises from the Dominican Republic to Mexico. AIDABELLA round-trips from Dubai, AIDABLU was on way from Palma de Mallorca to Civitaveccia, AIDASOL made a fortnight trip from Hamburg to Norwegian destinations, AIDASTELLA a round-trip from Mallorca in the western Mediterranean. No sailings were announced of AIDAPRIMA, AIDALUNA, AIDAAURA and AIDAVITA.

In 2019, Aida Cruises took over a second-hand ship from its parent company Costa Crociere. The 48,200 gt COSTA NEORIVIERA, built in 1999 by Chantiers de l'Atlantique at Nazaire as MISTRAL for Festival Cruises, was renamed AIDAMIRA and renovated at a Genovan repair yard, including the fitting of an exhaust gas cleaning system. An initial cruise scheduled to depart from Palma de Mallorca in December 2019 and the transition sailing to South Africa had yet to be cancelled due to uncompleted renovation work. In April 2020 AIDAMIRA was waiting for order at Tenerife/Spain. In February 2022, the ship was berthed at Pula/Croatia (Fig. 10.31).

10.2.4 Scrubber-Fitted MEIN SCHIFF 3, 4, 5, 6, 1, 2 and 7

TUI Cruises, founded in Hamburg in 2008 as a joint venture between the Hannover/Germany-based tour operator TUI and Royal Caribbean Cruises from Miami, is as well focussing on the German-speaking market. Their ships

Fig. 10.31 AIDASOL, built in 2011 by Meyer Werft, is measured at 71,304 gross tons (Ralf Witthohn)

cover the cruise regions of Europe, North and Central America and Asia. As their first ships, the company had the Meyer Werft buildings CELEBRITY GALAXY and CELEBRITY MERCURY of 1996/1997 converted to European standards, to put them into service under the names MEIN SCHIFF, later MEIN SCHIFF 1, and MEIN SCHIFF 2. MEIN SCHIFF 2 left TUI Cruises in 2015 and was renamed SKYSEA GOLDEN ERA, later MARELLA EXPLORER 2. In April 2018, MEIN SCHIFF 1 was handed to the buyer Marella Cruises and renamed MARELLA EXPLORER (Fig. 10.32).

The operator's first newbuildings were MEIN SCHIFF 3, MEIN SCHIFF 4, MEIN SCHIFF 5 and MEIN SCHIFF 6, delivered in May 2014, May 2015, June 2016 and May 2017, resp., from STX Finland, later named Meyer Turku. The 295 m length, 35.8 m breadth ships with tonnages of 99,526 gt and 98,785 gt, resp., have diesel-electric propulsion systems enabling a speed of 21.8 knots via two fixed propellers. The main engines are four Wärtsilä four-stroke motos of two different types with a total output of 44,000 kW. The two electrical motors from ABB have an output of 14,000 kW each. A scrubber system from American Electric Power (AEP) is to almost completely eliminate sulphur emissions and reduce particle emissions by 60%. Of the in total five transversal thrusters three of 3000 kW output are installed forward and two of 1000 kW aft. At two-bed occupancy, the ships can accommodate 2506 passengers served by a crew of about 1000. Accordingly to the sale of the second-hand ships, the fifth and sixth ship of the newbuilding series entered service in May 2018 and May 2019 by taking over the names MEIN SCHIFF 1 and MEIN SCHIFF 2 from the first ships. They are intended to be followed by MEIN SCHIFF 7 in 2023 (Fig. 10.33).

Fig. 10.32 After renovation the Celebrity liners GALAXY and MERCURY in 2009 became the first TUI Cruises liners MEIN SCHIFF and MEIN SCHIFF 2 under the Maltese flag (Ralf Witthohn)

Fig. 10.33 In February 2022, TUI Cruises' MEIN SCHIFF 1 had to end a cruise from Bremerhaven to the Caribbean several days earlier and was moored in the German port (Ralf Witthohn)

10.2.5 Reservation for 90 British Pounds

Shortly after the announcement of contract signing with Meyer Werft in September 2015, Saga Cruises in the UK offered potential cruise customers a reservation on their first newbuilding to be commissioned in summer 2019 against payment of a refundable fee of GBP 90. The five-day maiden voyage of the 58,250 gt SPIRIT OF DISCOVERY was scheduled to go to the Channel Islands and Brest in July, followed by voyages to Norway, Spain/Portugal, the Baltic Sea and a 36-day excursion to the Caribbean in December. The 236 m length, 31.2 m breadth new building was laid out for a passenger capacity of 999 persons. An option for a sister ship to be realised under the name SPIRIT OF ADVENTURE by summer 2020 was converted into a firm order in September 2017.

Saga Cruises belongs to a group of companies, which is specialized in services for older people and also active in the insurance and health sectors. Over the years, the company has built up a solid base of senior customers by offering them cruises on second-hand ships. At the time of the first newbuilding order, these were SAGA SAPPHIRE, built in 1981 as EUROPA for Hapag-Lloyd, and SAGA PEARL II, which had started first voyages in that year under the name ASTOR for Hamburg's HADAG and switched to GDR ownership as ARKONA in 1985. In keeping with the architecture of these ships and the target group envisaged, the newbuildings have a more traditional outer design than other Meyer buildings. For their construction, Meyer Werft reactivated an elder building dock und thus created the conditions for the production of three cruise liner newbuildings per year instead of two, a production target that had already been issued 10 years before. Saga Cruises suspended the cruises of their actually operated ships on 15 March 2020, particularly following the UK government announced changes to cruise travel advice for those over 70. On that day, SAGA SAPPHIRE proceeded from Dover to Tilbury docks, SPIRIT OF DISCOVERY to there from Southampton. In July 2020, SAGA SAPPHIRE was purchased by the Anex Tourism Group and renamed BLUE SAPPHIRE under the brand Selectum Blu Cruises for a first cruise with visits at Alexandria, Kusadasi, Marmaris and Antalya. In February 2022, the ship was berthed at Antalya (Fig. 10.34).

Fig. 10.34 After purchasing ships such as Hapag-Lloyd's former EUROPA, operated as SAGA SAPPHIRE, Saga Cruises ordered their first two newbuildings from Meyer Werft (Ralf Witthohn)

10.2.6 EUROPA 2 on World Tours

In contrast to the ships of the large fleets operating cruise concerns often deploying their ships on repeated round trips, Germany's Hapag-Lloyd Cruises works out part of the itineraries of their cruise liners in such a way, that passengers can book sections of varying lengths of global tours. From October 2017 to January 2018, EUROPA 2 travelled in North and South America and the Caribbean before crossing the South Atlantic from Buenos Aires to Cape Town. This was followed by voyages to Walfish Bay, to Mauritius, Colombo, Bali, Laem Chabang, Hong Kong, Singapore, Colombo, Abu Dhabi, Dubai, Limassol, Venice, Palma de Mallorca, Monte Carlo, Palma de Mallorca, Lisbon and Hamburg with arrival in July 2018. From October 2018 the ship's itinerary included calls at Gran Canaria and Cape Town, before the voyage was continued via the Far East to Nouméa, Tahiti, Auckland to Sydney, and from April 2019 to Bali, Hong Kong, Tokyo, Singapore and Colombo.

Due to the spreading of the COVID-19 respiratory disease, the EUROPA 2 cruise scheduled to start from Hamburg in May 2020 was cancelled, as were the voyages of EUROPA from Hamburg, BREMEN from Otaru, HANSEATIC NATURE from Porto and HANSEATIC INSPIRATION from Boston. In March, EUROPA 2 had arrived at Marseille from Palma de Mallorca. EUROPA headed for the Panama canal from Puerto Vallarta/ Mexico, bound for Hamburg. HANSEATIC NATURE set course for Hamburg from Colon, while HANSEATIC INSPIRATION was on its way from Rio de Janeiro to Bridgetown/Barbados. The passengers were carried back by plane. Shortly before the overall cruise shipping shut-down Hapag-Lloyd Cruises, which was since the TUI touristic concern's divestment in Hapag-Lloyd in 2008 a direct subsidiary of TUI, had in February 2020 announced, that by summer 2020 a 50/50 joint venture with Royal Caribbean Cruises was to be established. In the wake of the COVID-19 crisis TUI was end of March 2020 granted a EUR 1.8 billion credit by the German government to keep its business going (Fig. 10.35).

Hapag-Lloyd had taken over the Malta-flagged EUROPA 2 from STX France shipyard in St. Nazaire in 2013, which at that time belonged to Korea's STX group. The luxury ship is an enhanced and enlarged version of EUROPA built at Kvaerner Masa-Yards in Helsinki in 1999. This 28,890 gt, 198.6 m length, 24 m breadth ship was designed for 516 passengers accommodated in 251 suites and served by 370 crew members. The new 42,830 gt EUROPA 2 is of 225.4 m length and 26.7 m breadth. Both ships are equipped with a

Fig. 10.35 The France-built luxury liner EUROPA 2 is a considerably enlarged version of the EUROPA built in Finland in 1999 (Ralf Witthohn)

diesel-electric propulsion system with an unchanged Azipod output of twice 6650 kW for 21 knots. For the first time on a cruise liner, SCR catalytic converters reducing nitrogen oxide emissions were installed on EUROPA 2. Scrubbers to reduce sulphur oxides were yet not being used (Fig. 10.36).

The Bahamas-flagged EUROPA built in 1999 began the year 2018 departing from Acapulco to Tahiti, Melbourne, Bali, Rangoon, Dubai, Limassol, Piraeus, Venice, Nice, Bilbao and Hamburg. From September onwards, northern European trips were followed by trips to Nice, Monte Carlo, Barcelona, Piraeus, Dubai, Mauritius, Cape Town, Buenos Aires, Valparaiso, Tahiti, Auckland, Melbourne, Hong Kong, Singapore and Dubai. The Hapag-Lloyd fleet was in 2019 expanded by the luxury expedition liners HANSEATIC NATURE and HANSEATIC INSPIRATION capable to sail the Antarctic. The ships are laid out for 199 to 230 passengers. Their 138 m long hulls were delivered by the Vard's Romanian shipyard site in Tulcea. When commissioning the third expedition liner to be named HANSEATIC SPIRIT in 2021, Hapag-Lloyd Cruises plans to hand over its elder cruise ship BREMEN to a buyer. The 1993-built expedition cruise liner HANSEATIC was in 2018 sold to One Ocean Expeditions and renamed RCGS RESOLUTE. In April 2020, the ship was in collision with a Venezolan patrol boat, which sank (Fig. 10.37).

Fig. 10.36 After a first cruise liner newbuilding named EUROPA from Bremer Vulkan, Hapag-Lloyd contracted a second luxury ship of this name in Finland (Ralf Witthohn)

Fig. 10.37 Delivered in 2019, HANSEATIC INSPIRATION was the second of three planned Hapag-Lloyd Cruises expedition liner newbuildings (Ralf Witthohn)

10.2.7 Four QUEENs from France and Italy

In contrast to the relatively young American companies newly founded for the cruise business, the traditional European liner companies, which are actually nearly all US-owned, but operate independently on the market, can rely on well-known brands and names for image building. In this way, the 1936 and 1940-built QUEEN MARY and QUEEN ELIZABETH, by tonnages of 81,237 gt and 83,673 gt the largest passenger ships when built in 1936 and 1940, and QUEEN ELIZABETH 2, launched in 1969 and of 65,863 gt, experienced a renaissance when UK-based Cunard Line commissioned the 148,528 gt QUEEN MARY 2 from Chantiers de l'Atlantique in 2003 and the 90,901 gt QUEEN ELIZABETH from Fincantieri in 2010. The latter is a sister ship to the 2007-built QUEEN VICTORIA. In September 2017, Cunard Line, a Carnival Group company, ordered another newbuilding of 113,000 gt and laid out for 3000 passengers from Fincantieri for delivery in 2022 (Figs. 10.38 and 10.39).

The 2620-berth vessel QUEEN MARY 2 was as one of the first large cruise liners equipped with a podded propulsion system, which repeatedly needed repair. The ship required a great deal of design and development work contributung to the total costs of around USD 800 million. In contrast, the 2000-berth ships QUEEN VICTORIA and QUEEN ELIZABETH, which were built at costs of around USD 500 million in 2007 and 2010, are versions of the shipyard's Vista class panamax standard design. This was first realized in 2002 by ZUIDERDAM for the Carnival subsidiary Holland America Line, followed by OOSTERDAM in 2003 and WESTERDAM in 2004. The fourth ship, initially planned as the NOORDAM, was delivered to P&O Cruises as ARKADIA after modification. Further ships of the type for Holland America Line, enlarged by lengthening, were commissioned in 2006 under

Fig. 10.38 QUEEN MARY 2 is an expensive and impressive single design that continues the tradition of the Cunard liners (Ralf Witthohn)

Fig. 10.39 In summer 2019, QUEEN ELIZABETH made several round trips in Europe, including calls at Le Havre as port of disembarkation for visits of the passengers to Paris and the Normandy (Ralf Witthohn)

the name NOORDAM, in 2008 as EURODAM and in 2009 as NIEUW AMSTERDAM. Costa Crociere's COSTA LUMINOSA and COSTA DELIZIOSA are as well Vista class newbuildings of the years 2009 and 2010.

The ship's passengers having been flown home except those unable to fly due to medical conditions, QUEEN MARY 2 made a stopover end of March 2020 almost empty on Durban anchorage/South Africa on her return trip from Fremantle/Australia to Southampton, which was reached in April. Australia being a favourite cruise destinations of British passengers during the north hemisphere winter season, QUEEN ELIZABETH, too, was cruising off Australia during the COVID-19 outbreak and had to proceed from Sydney to Gladstone anchorage about 600 nm north, waiting there for order and not being expected to return home soon. At the same time passengers and crew of QUEEN VICTORIA were luckier, when they arrived at their homeport Southampton from Port Everglades at the end of March. Southampton was also the port of destination for other British cruise liners, including P&O's BRITANNIA and VENTURA, while AURORA was berthed at Dover. In February 2022 QUEEN MARY 2 crossed the Atlantic from Hamburg to New York, from where it headed for the Caribbean. QUEEN ELIZABETH undertook cruises from Southampton to western European destinations including Gran Canaria, while QUEEN VICTORIA was underway in the Caribbean (Fig. 10.40).

Fig. 10.40 QUEEN VICTORIA was the first of three Cunard liners to reach its homeport Southampton, after all of the companie's cruises had to be suspended due to the COVID-19 crisis (Ralf Witthohn)

10.2.8 NORWEGIAN SKY from Lloyd Werft

Lloyd Werft in Bremerhaven/Germany has had a long business relationship with Norwegian Cruise Line (NCL), since the shipyard converted Cunard's 14,110 gt newbuilding CUNARD ADVENTURER into the SUNWARD II in 1977 and the former French Atlantic liner FRANCE into the cruise liner NORWAY in 1979/80 on behalf of the Norwegian shipowner Knut Kloster. The 28,221 gt cruise liners ROYAL VIKING STAR, ROYAL VIKING SKY and ROYAL VIKING SEA, which were taken over by Kloster shortly afterwards, had been lengthened by Bremerhaven-based Seebeckwerft in the Lloyd Werft drydock. Seebeckwerft had already built Klosters' first cruise liners under the names STARWARD and SKYWARD in 1968/1969 and the luxury cruise liner ROYAL VIKING QUEEN in 1990 as the last newbuilding commissioned by Kloster's Royal Viking Line. NCL was in 1966 founded in Oslo as one of the first cruise lines and taken over by Star Cruises of the Asian Genting group in 2000. In 1999, Norwegian Cruise Line took over the 77,104 gt panamax cruise liner NORWEGIAN SKY, planned as COSTA OLYMPIA, from Lloyd Werft, 2 years later the similar 78,309 gt newbuilding NORWEGIAN SUN. The hull of the latter ship was built at Aker MTW shipyard in Wismar in a large building dock once built by the Bremer Vulkan yard group with the help of high subsidies. NORWEGIAN SKY undertook a cruise in Western Europe in January 2022 and returned to the Caribbean subsequently. NORWEGIAN SUN anchored off the western Panama canal entrance (Fig. 10.41).

Fig. 10.41 Only Lloyd Werft survived the cruise liner experiment of the Bremer Vulkan shipyard alliance and, after its bankruptcy, completed the NORWEGIAN SKY ex COSTA OLYMPIA as well as the pictured NORWEGIAN SUN, based on a hull supplied from Wismar, for NORWEGIAN SUN for Norwegian Cruise Line (Ralf Witthohn)

10.3 Asia

Even though the port infrastructure in Asia is often inadequate for cruise liners, so that they sometimes have to be handled in cargo ports, even in Japan, the major cruise companies are not only increasingly offering Asia as an interesting destination for their American and European clientele. Exploiting the great potential of Chinese guests, 2.1 million of whom had already made a cruise until 2016, in a prospering country has become a strategic goal. Especially US shipping companies and Carnival's European subsidiaries regularly deploy some of their ships to Asia.

10.3.1 Jewels of Vietnam: From Hong Kong to Ho Chi Minh City

In the early morning of 26 October 2017, the cruise liner newbuilding WORLD DREAM reaches Bremerhaven Columbuskaje after a short trip from Eemshaven in the Netherlands. The passenger terminal is located in the free port area and thus meets tax-efficient conditions for the handover of the newbuilding by the builders, Meyer Werft in Papenburg, to customer Dream Cruises from Hong Kong. After only 7 h, all formalities are completed, the fuel is supplemented by a bunker boat, and the cruise liner giant sets off again under the three times not quite clean sounding roar of its tyfon, bound for Hong Kong via Gibraltar and Suez. At more than 21 knots, WORLD DREAM ploughs the southern North Sea, the English Channel and the Bay of Biscay. The first voyage from Hong Kong/Guangzhou, lasting just two nights, starts on 17 November. It is followed by a five-night voyage to Manila and Boracay on the Philippines. A trip of the same length to Ho Chi Minh City and Nha Trang is marketed under the title "Jewels of Vietnam" (Fig. 10.42).

As the 44th cruise liner built by Meyer Werft, WORLD DREAM was specially designed for the Asian market, following its sister ship GENTING DREAM delivered in spring 2017. A special feature of the second cruise ship for Dream Cruises, a subsidiary of the Malaysian Genting group, is a submarine that can carry four people to a depth of 200 m. A speed boat bookable for excursions reaches 37 knots. Measured at 151,300 gt, WORLD DREAM can carry over 3376 passengers in 1686 cabins. The 335 m length, 39.7 m breadth ship achieves a speed of more than 23 kn.

To meet the specifications of Dream Cruises not an increase of height or widening of the NORWEGIAN BREAKAWAY type but a lengthening of the ship to 335 m was found to be the best design solution, in contrast to the

Fig. 10.42 The WORLD DREAM's exterior and interior design is orientated towards Chinese guests, but the Dream Cruises' filing for liquidation ended all cruise operations of the Genting subsidiary (Ralf Witthohn)

development of draft chosen for NCL's NORWEGIAN ESCAPE. At that time Malaysia's Genting group still had shares in Norwegian Cruise Line, which it sold 2 years later in December 2018. Delivered in October 2016, GENTING DREAM, initially ordered as GENTING WORLD on behalf of Genting's Star Cruises, features 35 restaurants and bars with a catering concept orientated towards the Chinese taste and a large casino area. In the so-called Genting Club with 142 suites, guests enjoy special privileges. The ship achieves a speed of over 23 knots. Like the WORLD DREAM, the ship has been equipped with two small submarines. A wide range of entertainment activities is offered in a theatre, by virtual reality applications, climbing parks, casino and outdoor activities. 75% of the cabins are outside cabins, most of them fitted with a balcony. The design of the ship was supported with innovation grants from Germany's Federal Ministry of Economics and Energy and the State of Lower Saxony. Since the Dream Cruises' filing for insolveny, the GENTING DREAM is being anchored off Hong Kong, the WORLD DREAM off Singapore (Fig. 10.43).

10.3.2 Genting's GLOBAL DREAM from Wismar

The great demand for cruise liners, resulting in long delivery times, prompted Hong Kong-headquartered Genting in 2015 to involve Lloyd Werft in Bremerhaven in cruise liner projects for its subsidiary Crystal Cruises. The hulls of the 100,000 gt luxury newbuildings to be delivered from 2018 onwards should be built at the East German shipyards in Wismar, Warnemünde and Stralsund, just as for NCL's NORWEGIAN SUN in 2001 from Wismar,

Fig. 10.43 Hundreds of lamps create the companie's name Dream Cruises over the balcony front of GENTING DREAM (Ralf Witthohn)

and equipped by Lloyd Werft. The builders in the State of Mecklenburg-Vorpommern had in the meantime passed into Russian hands, but were unemployed, not least because of the embargo against Russia. In March 2016, Genting took over also those shipyards. In June, in addition to six inland waterway cruise liners, a large yacht of 183 m length for polar operations, a large luxury cruise liner and two 204,000 gt cruise liners for 9500 passengers were reported as contracted, with the number of orders constantly changing. In July 2016, the order bubble burst for Lloyd Werft, when Genting announced that the Bremerhaven shipyard would concentrate on its "core business" conversion and repair and that the cruise liner buildings would be built entirely at the Mecklenburg-Vorpommern shipyards. Design contracts for the two 204,000 GT ships were signed in the same month with the Finnish engineering firms Elomatic and Deltamarin (Fig. 10.44).

After delivery of four inland cruise liners operated by Crystal River Cruises, the MV Werften newbuilding programme in the beginning of 2020 comprised an expedition mega yacht of 200 passengers capacity to be delivered in the same year, two Global Class cruise liners of 208,000 gt for delivery in 2020/2021 and six Universal Class liners of 88,000 gt to accommodate 2000 passengers and to be commissioned until 2024. In November 2019, the 216 m long main section of the first Global class newbuilding assembled of 145 large volume sections, was towed from the Rostock-Warnemünde site of MV Werften over a distance of 43 nm along the Baltic coast to Wismar for

Fig. 10.44 Partly supplied by Volkswerft Stralsund, 145 large volume sections were at Warnemünde assembled to the 216 m long midship section of the cruise liner GLOBAL DREAM, which was in November 2019 towed to Wismar for completion. The large blue sign giving the information, that this would be the "world's largest cruise liner" is yet mis-leading, because, at about 204,000 gt, the ship is considerably smaller than the 226,963 gt HARMONY OF THE SEAS, built in France in 2016 (Arne Münster)

assembling it with the fore and aft ship in the yard's covered building dock. Construction of GLOBAL DREAM was only made possible by distributing work among three MV shipyard sites.

In January 2020 Genting Hong Kong, which was at this time operating the cruise liners GENTING DREAM and EXPLORER DREAM, sold a 33% stake in Dream Cruises to Darting Investment Holdings, thus generating USD 459 million cash. Genting's three German shipyards terminated work on the cruise liner projects end of March 2020, because the precautionary measures implemented in the wake of the COVID-19 pandemic were not able to be met during the construction work. The government of Mecklenburg-Vorpommern signalled financial help, so that suppliers could be paid and the construction of the cruise liners be continued. It came to another stop, when the builder and customer both filed for insolvency in January 2022.

Representing the world's second largest cruise liner type by size, the GLOBAL DREAM was to be operated by 2016-founded Dream Cruises. The ship was scheduled to undertake cruises from Shanghai from December 2020 on and to switch to the US or Australian market during the northern hemisphere winter. The keel of a second Global class newbuilding destined for Genting's Star Cruises and planned to enter service in 2023 was laid in December 2019. Financing of a volume of EUR 2.6 billion for the two newbuildings was arranged by Germany's KfW IPEX within a consortium of international banks. The governments of Germany and of Mecklenburg-Vorpommern granted export credit guarantees for part of the finance volume. End of 2019, Genting presented the design of the 88,000 gt, LNG-powered Universal class for 2000 passengers, of which the first unit was planned to be delivered end of 2022 for operation by Dream Cruises, which yet filed for liquidation in January of that year (Fig. 10.45).

10.3.3 China's First Luxury Liner

Starting in 2013, China's HNA Cruises organized voyages to South Korea, Taiwan and Vietnam on board the 47,678 gt HENNA, the largest and most luxurious cruise liner in the country at the time. The 1960-berth ship was in 1986 built in Malmö as JUBILEE for the Carnival group. In September 2013, the ship was arrested for several weeks on the Korean island of Jeju due to payment demands from a shipping agent. The liner was released, after the operator paid a part of the claims. The majority of passengers had meanwhile returned to China by plane. As no buyer was found for the ship, it was scrapped in Alang/India in 2017. The former owner of JUBILEE, Carnival,

Fig. 10.45 Carnival's former JUBILEE had only a short career as China's first cruise liner HENNA (Lutz Witthohn)

on the other hand, operated a ship in East Asia for several years, named DIAMOND PRINCESS of its subsidiary Princess Cruises. In autumn 2016, the ship undertook cruises from Singapore to Japan, South Korea, Taiwan, Vietnam, Thailand, Malaysia and Indonesia for the first time all year round.

In February 2020, DIAMOND PRINCESS was the first liner hit by the COVID-19 virus, which infected 700 of the ship's passengers. After a passenger who had disembarked in Hong Kong was tested positively, the ship with 3711 passengers and crew members was quarantined in Japanese waters. In July 2021, DIMAON PRINCESS proceeded from Singapore to Augusta/Italy, from where it departed for Limassol/Cyprus in February 2022. In the weeks after the first outbreak, over two dozen cruise liners reported COVID-19 cases. In February 2020, all RCCL sailings in Asia were cancelled due to the outbreaks (Fig. 10.46).

10.3.4 From Shanghai and Hong Kong

US-based concern Royal Caribbean Cruise Line (RCCL), the second largest cruise company to Carnival, welcomed 800,000 Chinese passengers on its cruise liners in 2017. Depending on the season, RCCL ships called at Shanghai, Tianjin, Shenzhen and Hong Kong. QUANTUM OF THE SEAS sailed from Shanghai throughout the year, while OVATION OF THE SEAS started its voyages from Tianjin for 6 months at the end of April 2017, before moving to Hong Kong. VOYAGER OF THE SEAS sailed from Shenzhen, MARINER OF THE SEAS from Shanghai between April and October 2017.

Fig. 10.46 Year-round, DIAMOND PRINCESS undertook cruises in the Far East from 2016 on, until the ship became the first, on which the Corona virus broke out in February 2020 (Lutz Witthohn)

In 2012, Royal Caribbean began design work on the 100,000 gt cruise liner project CHINA XIAMEN, which was to be built by Xiamen Shipbuilding for Xiamen International Cruise by October 2018. The project, estimated at USD 5 billion, should include a new cruise liner port and holiday park in Dongdu Port and was to be realized by the Xiamen Provincial Government and China World Cruises, a subsidiary of Beijing-based project development company Shan-Hai-Shu. The agreements with Royal Caribbean, which has been operating the VOYAGER OF THE SEAS in Far Eastern waters since 2008, also included chartering the LEGEND OF THE SEAS for 21 three- to eight-day voyages from Xiamen, Shanghai, Tianjin and Hong Kong to Taiwan, Vietnam, Japan and Korea. The newbuilding project, yet, did not come to fruition.

Instead, it was announced at the 19th National Congress of the Communist Party in 2017, that the Waigaoqiao shipyard in Shanghai, in cooperation with the Italian shipyard group Fincantieri, was to build the country's first cruise liner for a Hong Kong shipping company and employment on the Chinese market by 2020 at an estimated cost of USD 1 billion.

China's first cruise liner newbuilding was in fact floated out in December 2022 from the building dock at Waigaoquiao Shipbuilding, a member of China State Shipbuilding Corporation (CSSC). The still unnamed 323 m length, 4250 berth ship was at that time scheduled for delivery in September 2023 as the first of two newbuildings. The ships were contracted by CSSC Carnival Cruise Shipping, a 2018-founded joint venture of CSSC and the US-based Carnival group. CSSC as well established a cooperation with Italy's state-owned shipbuilding concern Fincantieri about the transfer of technology (Fig. 10.47).

Fig. 10.47 After the ship's sailings were cancelled, SPECTRUM OF THE SEAS, the sister vessel QUANTUM OF THE SEAS and the companie's VOYAGER OF THE SEAS, together with CARNIVAL SPLENDOR, CELEBRITY SOLSTICE, RUBY PRINCESS, SUN PRINCESS and PACIFIC EXPLORER were end of March 2020 in and off Sydney awaiting order how to further proceed in the wake of the COVID-19 crisis (Ralf Witthohn)

Norwegian Cruise Line, on its turn, strengthened its presence in East Asia from summer 2017 on by deploying the 4200-berth NORWEGIAN JOY, a sister ship of NORWEGIAN ESCAPE, for sailings from Shanghai and Tianjin. Focussing on Chinese customers the ship was given the additional name XI YUE HAO (Inner Joy). Previously, after an interruption since 2002, NCL's NORWEGIAN STAR, another, 91,740 gt Meyer Werft newbuilding of 2001, had been employed from autumn 2016 onwards.

10.4 Expedition Cruises

In addition to the passenger ships for mass tourism, which have been constantly growing in size, a further line of development was the design of smaller, more exclusive cruise liners with more elaborate interior and exterior architecture, as to be watched since the 1960s. The limited ship dimensions not only allow a higher standard of interior design, but, due to their smaller draughts, also voyages on remote routes, including such as the Northwest Passage, in

Fig. 10.48 Smaller cruise liners, like the 34,242 gt BALMORAL, built as CROWN ODYSSEY at Meyer Werft in 1988 and later to become NORWEGIAN CROWN, can easily visit anchorages like that of Sorrento, where calls increased from 50 in 2017 to an announced, but due to the COVID-19 pandemic not anymore realised number of 94 in 2020 (Ralf Witthohn)

Polar zones and on large rivers including the Amazon or St. Lawrence Seaway (Fig. 10.48).

Among the preferred destinations for ships of up to 180 m in length was Rouen as an ideal port for visits of passengers to Paris and locations in the Normandy. In 2020, 22 liners were scheduled to call from February to December for relatively long berthing times of up to four days. Different from many modern cruise terminals, where access is sharply restricted by citing security reasons, the ships can be left easily and offer direct entrance to the old town. One of the 2018 visitors was the THE WORLD, and in 2019 Fred. Olsen's BALMORAL, which was as well intended to visit the port again in September 2020 (Fig. 10.49).

One of the first so-called expedition cruise liner was the 2346 gt LINDBLAD EXPLORER, designed by Copenhagen engineering company Knud E. Hansen and in 1969 built in Uusikaupunki/Finland. Hansen also supplied the general lay-out of the 3153 gt BEWA DISCOVERER, built in Bremerhaven in 1974. Such forerunners inspired the construction of the 4253 gt SEA GODDESS I and SEA GODDESS II in Helsinki in 1984/1985, of the 9975 gt SEABOURN PRIDE, SEABOURN SPIRIT and ROYAL VIKING QUEEN built in Bremerhaven from 1988 to 1992 and the 8378 gt HANSEATIC as SOCIETY ADVENTURER, completed in Rauma in 1991. After nearly 50 years the company of Knud E. Hansen remained active in the design of expedition cruise liners, when in 2015 a drawing of a 140 m length

Fig. 10.49 While the old Rouen port area along the Seine is not anymore used for cargo handling, it is being considered as an ideal place for smaller cruise liners, like Olsen's BALMORAL (Ralf Witthohn)

Fig. 10.50 The first cruise liner with yacht character was in 1974 SEA GODDESS I, later named SEADREAM I (Ralf Witthohn)

ship for 300 passengers with flexible cabin arrangements for up to 50 suites plus 50 cabins was presented (Figs. 10.50 and 10.51).

10.4.1 High Accidents Frequency

A disproportionately high number of medium and small-sized cruise liners were involved in serious accidents or even became a total loss. After LINDBLAD EXPLORER ran aground in the Antarctic in 1972 and had to be towed free by the German deep-sea tug ARCTIC, the ship, then operating

Fig. 10.51 Under the slogan "Semester at Sea", the 1997-built 15,187 gt cruise liner DEUTSCHLAND in the northern winter seasons of 2016, 2017 and 2018 temporarily and in 2019 permanently became the campus WORLD ODYSSEY, which circumnavigated the world with students (Ralf Witthohn)

as EXPLORER, rammed an iceberg off King George Island near South Shetlands in November 2007 and sank. WORLD DISCOVERER ex BEWA DISCOVERER was lost in April 2000 after stranding on a remote Pacific island. Hapag-Lloyd's BREMEN and GRAND VOYAGER, delivered as OLYMPIC VOYAGER by Blohm + Voss in 2000, fell into extreme distress as a result of heavy sea blows and damage to their wheelhouses. After HANSEATIC ran aground in Simpson Strait during an attempt to pass through the Northwest Passage on a voyage from Gjoa Haven to Resolute Bay in August 1996, the passengers had to be evacuated (Fig. 10.52).

While the passengers and crew members of the sunken EXPLORER were rescued by the Hurtigruten vessel NORDNORGE, two cruise Hurtigruten liners themselves suffered severe accidents in 2007, NORDKAPP after touching the ground on Deception Island and FRAM after machinery failure and iceberg collision. In December 2015, the luxury cruise liner STAR PRIDE, built in 1988 as SEABOURN PRIDE, stranded near Isla de Coiba, Panama, and its passengers had to be evacuated. Polar shipping in particular developed unusual business models, for which even Russian icebreakers with conventional or nuclear propulsion were chartered.

In September 2019, the just commissioned 9923 gt expedition cruise liner newbuilding WORLD EXPLORER had to encounter a North Sea storm,

Fig. 10.52 The HANSEATIC passengers had to leave their ship after it got stuck in the Northwest Passage (Ralf Witthohn)

which forced the ship to alter its itinerary. The ship was entered in the Madeira register on behalf of Portuguese Mystic Cruises, which is part of the travel and tourism holding Mysticinvest. WORLD EXPLORER is operated by Stuttgart/Germany-based Nicko Tours and US company Quark Expeditions, which is specialized in cruises to Polar regions. Sister ships of WORLD EXPLORER named WORLD VOYAGER and WORLD NAVIGATOR were delivered in 2020 and 2021. Powered by two Rolls Royce (Bergen) diesel engines of 2 × 2.667 kW acting on two propellers the ice-classed ships are laid out for a speed of 16 knots. Two Schottel pump-hydrojet propulsion units are able to accelerate them to 5 knots. The planned total building programme for Mystic Cruises comprised up to ten more ships, of which the next ships shall be named WORLD TRAVELLER, WORLD SEEKER, WORLD ADVENTURER and WORLD DISCOVERER (Fig. 10.53).

The project for a series of ten 8500 gt expedition ships of 104.4 m length and equipped with 80 cabins is based on a design of Norway's Ulstein group and on European expertise for interior design and equipment. First vessel delivered to Miami/US-based shipping company Sunstone Ships by China Merchants Heavy Industry (CMHI) in Jiangsu in August 2019 was GREG MORTIMER, succeeded by OCEAN VICTORY and OCEAN EXPLORER in 2021 and, as announced, to be subsequently followed by SYLVIA EARLE in March 2022, OCEAN ODYSSEY in April 2022, OCEAN ALBATROS in March 2023 and OCEAN DISCOVERER in 2023. They are constructed with an X-bow, that had so far mostly been realised on offshore ships to

Fig. 10.53 A North Sea storm hit the passengers and crew of the expedition cruise liner newbuilding WORLD EXPLORER, which made an unplanned call at Bremerhaven (Ralf Witthohn)

increase the ship's heavy weather performance. In early 2020, Sunstone Ship had operated a fleet of nine ships, including OCEAN ADVENTURER, OCEAN NOVA, SEA SPIRIT, OCEAN DIAMOND, SEA ENDURANCE, OCEAN ENDEAVOUR and OCEAN ATLANTIC, of which the latter two originally were Soviet ro ro passenger ships.

10.4.2 Yacht-Like LE BORÉAL

Among the outstanding examples of exclusive smaller cruise liners are LE BORÉAL, L'AUSTRAL, LE SOLÉAL and LE LYRIAL, built at Fincantieri's Ancona shipyard from 2010 onwards. The ships were commissioned by Marseille-based Ponant shipping company, founded in 1988 and in 2004 purchased by CMA CGM, one of the major container ship operators. In 2012, Ponant was sold to the financial investor Bridgepoint. At 142.1 m length, 18 m breadth and 4.7 m depth the 10,944 gt LE BORÉAL has a flexible total number of cabins and suites ranging from 112 to 132. The number of suites can be varied from 4 to 24. The maximum number of passengers is 254, the crew comprises 140 members. The ship is powered by two 2300 kW main engines, enabling an average cruising speed of 14 knots. For excursions, the ship, which has been built according to Bureau Veritas ice class IC, is equipped with Zodiacs (Fig. 10.54).

Fig. 10.54 Marseille-based Ponant shipping company received four newbuildings of the LE BORÉAL type from Ancona (Ralf Witthohn)

In November 2015 a fire broke out in the engine room of LE BORÉAL while cruising in the South Atlantic. 347 people and crew members were rescued by helicopters and Royal Navy ships. The damaged liner was transported from Punta Arenas to Europe on deck of the Chinese semi-submersible KANG SHENG KOU and repaired at Fincantieri's shipyard in Genoa until May 2016. An even more yacht-like design is shown by the 92-cabin vessels LE LAPÉROUSE, LE CHAMPLAIN, LE BOUGAINVILLE, LE DUMONT-D'URVILLE, LE BELLOT and LE JACQUES-CARTIER. Their deliveries by Fincantieri's Norwegian shipyard Vard Langsten at Tomrefjord/Norway started in 2018. Like for three Hapag-Lloyd cruise liners from the same shipyard, the hulls were built at Vard Tulcea shipyard in Tulcea/Romania, part of the shipbuilding group. Ponant developed a concept for the lng burning, diesel-electrically driven polar class PC2 expedition ship LE COMMANDANT CHARKOT, delivered in 2021, The 150 m length, 28 m breadth ship of preliminary of 31,283 gt was as well completed and fitted out by Vard Langsten shipyard on base of a hull supplied by Vard Tulcea. The newbuilding was laid out for 270 passengers.

In September 2017, Vard contracted a 120-berth expedition cruise liner to be launched in 2019 for use in the Asia-Pacific region, in particular at the Great Barrier Reef, the Kimberley region, Cape York, Arnhem Land, Papua New Guinea, Indonesia, the South Pacific Islands and Tasmania by the Australian cruise operator Coral Expeditions. The 93.5 m long and 17.2 m wide hull of the CORAL ADVENTURER was built at the Vard Vung Tau shipyard in Vietnam and delivered in 2019, the sister vessel CORAL GEOGRAPHER in March 2021. Vard has also been involved in the development of the 181.6 m long and 22 m wide expedition and research vessel

Fig. 10.55 The 1992 built, diesel-electrically powered CLUB MED 2 and equipped with a schooner rig is one of the most exclusive smaller cruise liners offering space for 368 passengers and 200 crew members (Erich Müller)

project ROSSELLINIS FOUR-10 of investor Kjell Inge Røkke for up to 60 scientists or passengers (Fig. 10.55).

10.4.3 Hurtigruten Liner FRAM

The mail line of the Norwegian shipping group Hurtigruten, having started operation as early as in 1893, developed into a symbiosis of liner and cruise shipping, when a lack of profitability let the original transport tasks along the Norwegian coast be expanded to cruise operations. The latter service is meanwhile taken on by tourists from all over the world. The additional marketing has in turn led to the employment of some of the ships as cruise liners, extending as far as Antarctica. The oldest of the ships, the 1956-built NORDSTJERNEN, which had initially phased out in 2012, was chartered for voyages to Spitsbergen in 2015 and 2016. The shipping company announced in 2016, that the expedition ships SPITSBERGEN and FRAM would be calling at new destinations from 2017, including the Amazon, old Viking settlements on the Canadian coast and the Scoresbysund in Greenland. In the fleet of 13 ships, the oldest ship is LOFOTEN, built in Oslo in 1964. A series of three ships was built in 1993/94 at the Volkswerft Stralsund with the support of the Meyer Werft, named RICHARD WITH, KONG HARALD and POLARLYS.

In 2005 Hurtigruten for the first time chose an Italian shipyard to modernize their fleet by the 11,647 gt measuring FRAM, which keel was laid at the Fincantieri shipyard in Monfalcone, at that time otherwise solely engaged in the construction of postpanamax cruise liners of over 100,000 gt. The vessel is specially designed for expedition cruises and therefore built according to a high ice class, equipped with course tracking systems, iceberg searchlights and a garage for expedition vehicles (Fig. 10.56)

As the first of three newbuildings contracted with Kleven shipyard in Norway, the 2019-commissioned 21,765 gt newbuilding ROALD AMUNDSEN is able to accommodate 530 passengers, 500 on Antarctic cruises. The ship, which delivery was delayed by several months, is diesel-electrically driven via two pods and has additionally been fitted with a battery package of 1360 kWh, which is yet only able to drive the ship over a period of less than one hour. Such a selective measure of green washing does not remove the principal questionability of tourist trips to highly sensitive regions such as Antarctica. In 2020/21, the itinerary for ROALD AMUNDSEN listed 42 voyages to Alaska, Canada, Antarctic, Falkland Islands, Patagonia, Ecudador, Panama, Costa Rica, California, the US East Coast, Northwest Passage, Peru and the Galápagos Isles, but Hurtigruten announced in March 2020, that all sailings would be suspended due to the COVID-19 pandemic. ROALD AMUNDSEN's sister ship FRIDTJOF NANSEN was delivered in 2019.

In 2021, Hurtigruten offered from June on 15 days round cruises from Hamburg on the OTTO SVERDRUP, the companie's former FINNMARKEN, which underwent a modernization programme together with the EIRIK

Fig. 10.56 In addition to Norwegian coastal ports, Antarctica, Alaska, Greenland and Spitsbergen had become destinations of the Hurtigruten vessels, of which FRAM regularly set course for the sixth continent during the Antarctic summer (Ralf Witthohn)

Fig. 10.57 After being fitted out in Bremerhaven, the OTTO SVERDRUP ex FINNMARKEN started cruises from Hamburg to Norwegian ports in 2021 (Ralf Witthohn)

RAUDE ex MIDNATSOL and TROLLFJORD to be employed for cruises, as well as the former MIDNATSOL, which was renamed MAUD and prepared to undertake cruises to the Antarctic, accommodating up to 500 passengers instead of the otherwise regular 528 (Fig 10.57).

After newcomer Havila Kystruten won one of three of the Norwegian government's tenders on the ferry line, the operator contracted four LNG/battery-powered newbuildings. The first one was delivered in November 2021 as HAVILA CAPELLA by Turkey's Tersan shipyard. The sister ships HAVILA CASTOR, HAVILA POLARIS and HAVILA POLLUX are due for delivery from 2022 on.

10.4.4 NATIONAL GEOGRAPHIC EXPLORER

In order to be able to implement the new Hurtigruten cruise concept as quickly as possible, the shipping companies involved had three of their liners transporting freight and passengers, MIDNATSOL, NARVIK and VESTERALEN, enlarged by the MWB shipyard in Bremerhaven. Within a few weeks prefabricated superstructures for additional cabins and a salon were set on the ships. In March 1988, a floating crane hoisted two additional decks in one lift onto the aft deck, which had previously been used for container transport. MIDNATSOL had in 1982 been delivered by Ulstein Hatlo shipyard in Ulsteinvik as the first Hurtigruten vessel with a side loading facility. After the commissioning of a new MIDNATSOL with twice the passenger capacity in 2003, an attempt was made in vain to sell the predecessor, which was then kept as a replacement ship under the name MIDNATSOL II.

From 2008 onwards, after sale to Lindblad Expeditions and conversions in Gothenburg and Las Palmas, MIDNATSOL II was employed as cruise liner NATIONAL GEOGRAPHIC EXPLORER by US-based National Geographic Expeditions, offering cabins for 148 passengers. In 2017, the operator had eight expedition cruise liners in its fleet, including three units operating around the Galapagos Islands, named NATIONAL ENDEAVOUR II, the twin hull ship NATIONAL GEOGRAPHIC ISLANDER and NATIONAL GEOGRAPHIC ENDEAVOUR, which was converted from a factory trawler. NATIONAL GEOGRAPHIC ORION had in 2003 been built by Cassens-Werft in Emden as the mini cruise liner ORION (Fig. 10.58).

The 2906 gt newbuilding NATIONAL GEOGRAPHIC QUEST of Nichols Brothers Boat Builders in Freeland/Washington was carrying up to 100 passengers on seven-day voyages to Alaska and along the coast of British

Fig. 10.58 In 2008 the Hurtigruten liner MIDNATSOL was converted to the cruise liner NATIONAL GEOGRAPHIC EXPLORER (Ralf Witthohn)

Columbia and Central America from June 2018. The sister NATIONAL GEOGRAPHIC VENTURE was commissioned one year later. In November 2017, Lindblad announced a contract with Ulsteinvik shipyard in Norway for the construction of an exclusive expedition cruise liner with 69 cabins and built with an X-bow, that was in March 2020 delivered as NATIONAL GEOGRAPHIC ENDURANCE. Options for two sister ships are planned to be commissioned in the following two years. In the beginning of 2020, the companie's fleet comprised 14 ships, including the sailing vessel SEA CLOUD and four river cruisers. The four-masted barque SEA CLOUD had been built in 1931 at Germania-Werke in Kiel/Germany as private yacht HUSSAR and in 1979 converted to a cruise ship.

After 90 years of service the auxiliary sailing vessel SEA CLOUD had to suspend its cruises due to the COVID-19 pandemic end of March 2020 and was anchored off Bridgetown/Barbados in a fleet of much younger, larger, but as well idle cruise liners. Those included Royal Caribbean's giant cruise liners VISION OF THE SEAS, SERENADE OF THE SEAS, Carnival's AIDAPERLA, AIDALUNA, Hapag-Lloyd's HANSEATIC INSPIRATION, HANSEATIC NATURE as well as MSC PREZIOSA and the 47,861 gt new-building VIKING JUPITER of US operator Viking River Cruises (Fig. 10.59).

In 2002, another Hurtigruten ship, which had in 1960 been built as HARALD JARL in Trondheim, was converted into the small cruise liner

Fig. 10.59 Operator, passengers and crew of the four-masted barque SEA CLOUD 90 became in the same way a victim of the COVID-19 pandemic the newest and largest cruise liners were hit (Ralf Witthohn)

Fig. 10.60 In March 2020, the 60 year old former Hurtigruten vessel HARALD JARL was still underway from Alanya/Turkey to Trogir/Croatia under the name SERENISSIMA, but subsequently entered long-time lay up during the pandemic (Ralf Witthohn)

ANDREA. Ten years later Premier Cruises in Split took over the management of the ship for 100 passengers under the name SERENISSIMA (Fig. 10.60).

10.4.5 High-Class Luxury Vessels

Among the recent newbuildings in the new heights reaching luxury class are the 41,865 gt liners SEABOURN ENCORE and SEABOURN OVATION, ordered by Carnival subsidiary Seabourn for delivery in 2016 and 2018 from Fincantieri's shipyard in Marghera, as well as the 40,791 gt newbuilding SILVER MUSE, completed in April 2017 by Fincantieri's Genoa Sestri Ponente site for Silversea Cruises. The first cruise offered by Seabourn after the break caused by the COVID-19 pandemic was a seven days cruise from Istanbul to Piraeus in May 2020, for which prices from EUR 5800 to EUR 10,500 were published. A succeding fortnight cruise in the same waters was priced at up to EUR 10,500. The sister ships SEABORUN ENCORE had been scheduled for cruises in the western Mediterranean (Fig. 10.61).

Fig. 10.61 SEABOURN OVATION is among the most expensive cruise liners, but got lost of its passengers due to the COVID-19 pandemic like all other cruise ships (Ralf Witthohn)

10.5 Coastal Passenger Shipping

10.5.1 Senegalese ALINE SITOE DIATTA

After one of the most serious accidents in merchant shipping, the loss of the heavily overloaded LE JOOLA in a storm off the Gambian coast with more than 1800 dead in September 2001, the government of the West African state awarded the shipyard Fr. Fassmer in Motzen/Germany the contract to build a replacement ship, which was completed in November 2007 on the base of a hull supplied from a shipyard in Gdansk/Poland. The coastal passenger ship was christened ALINE SITOE DIATTA after the name of a Senegalese resistance fighter against the French colonial rule. Operated by the shipping company COSAMA on behalf of the Gambian Ministry of Transport the 76 m length, 15.5 m breadth twin-screw vessel undertakes sailings twice a week between Ziguinchor on the Casamance River and the capital Dakar. The ship can carry 504 passengers in four classes and over a stern ramp load 18 trailers and 6 trucks on a deck area of 550 m^2 (Fig. 10.62).

10.5.2 LNG-Driven HELGOLAND

The first German newbuilding to be equipped with an LNG burning main engines was the 2256 gt passenger ship HELGOLAND, delivered in December 2015 from Fr. Fassmer shipyard to Cassen Eils in Cuxhaven for the

Fig. 10.62 In 2007, ALINE SITOE DIATTA replaced LE JOOLA, which sinking caused the loss of over 1800 people (Ralf Witthohn)

seaside resort of Helgoland. The realisation of the EUR 30 million project was delayed by 6 months due to technical problems in connection with the new type of propulsion system. The twin-screw vessel was designed for the two-and-a-half-hour lasting trip between Cuxhaven and the southern port of Helgoland. Hullkon shipyard in Szczecin supplied the hull of the ship, which was subsequently equipped with two nine-cylinder dual-fuel Wärtsilä main engines of type 9L20 enabling a speed of 20 knots. The engine equipment includes the fuel preparation system and an LNG tank system of 53 m^3 capacity. The LNG is supplied by company Bomin Linde. At 82.6 m leng, 12.6 m breadth and 3.6 m depth the coastal vessel offers space for almost 1000 passengers, for whose comfort a dynamic stabilisation system has been installed. An atrium extending over several decks with a glass lift, restaurant, sun decks and sky bar is to create a cruise atmosphere for the first time in the Helgoland cruise business. The newbuilding project was based on a transport contract concluded for 15 years between the municipality of Helgoland and the shipping company Eils, which is a subsidiary of Emden-based AG Ems. This includes transports of general cargo and containers in the forward hold of the ship, which is in German designated *Seebäderschiff*. In March 2020 all passenger traffic to the German North Sea isles was temporarily suspended, when, due to the COVID-19 pandemic, no more tourists were allowed to make holiday there. HELGOLAND yet continued supply sailings from Cuxhaven to Helgoland (Fig. 10.63).

Fig. 10.63 The first German LNG burning newbuilding HELGOLAND transports almost 1000 passengers and cargo between Cuxhaven and Helgoland (Ralf Witthohn)

10.6 River Cruises

While the great rivers have always been transport routes for goods and people alike, it was not until the 1980s that a cruise industry on rivers comparable to ocean shipping began to gain importance. Europe in particular, with its major rivers Danube, Rhine, Rhône, Seine, Elbe, Po, Neva, Dnieper or Amur, offers good conditions for this, but inland cruise liners also operate on the Nile, Mississippi, Yangtze or Irrawaddy.

In terms of design, the operated cruise liners pose a particular challenge due to the limited and changing sailing circumstances. Not only are the maximum length and width determined by locks, which often have to be mastered many times, but also limit the fairways of rivers and canals the draught of the ships. At the same time, bridges restrict the maximum height above the waterline, usually described as the air draught of the ship. During the times of high water levels, it can become impossible to pass under the bridges even if the ships are ballasted, just as low water can prevent the voyage. How great the potential danger of accidents due to these restrictions is, was demonstrated in September 2016, when the 135 m long Swiss river cruise liner VIKING

Fig. 10.64 Passau is the port of embarkation for most Danube cruise liners (Ralf Witthohn)

FREYA, built in 2012 by Neptun Werft in Rostock/Germany, rammed a bridge on its way to Budapest while navigating on the Rhine-Main-Danube canal near Erlangen/Germany, killing two crew members (Fig. 10.64).

10.6.1 A-ROSA STELLA on River Rhône

The terrorist threat encouraged investments in the river cruise market, which is considered to be safer. The consequent demand for newbuildings proved to be a stroke of luck for tradition rich Neptun Werft in Rostock. After the cessation of seagoing shipbuilding due to EU restrictions following the German unification and an interlude within the Bremer Vulkan yard group, the builder was taken over by Meyer Werft in 1997 under the name of Neptun Industrie Rostock. Later renamed Neptun Werft, the builders then started a long series of more than 50 river cruise ships in 2002, mainly for Viking River Cruises based in Switzerland. Among the first newbuildings were A`ROSA DONNA and A'ROSA BELLA for the local river cruise company A-Rosa, which had in 2000 been founded as a branch of Deutsche Seereederei in Rostock. The two 124 m ships, each with space for 200 passengers, were transported on deck of the Dutch heavy-lift carrier GIANT 2 via Gibraltar and Bosporus to Constanta, with a stopover in La Coruna/Spain caused by the adverse weather conditions. For the delayed maiden voyages from Passau the temporary built bridges in Novi Sad, Serbia, had to be opened. In 2003 and 2004 succeeded the newbuildings A'ROSA MIA and A'ROSA RIVA, and in 2005 the

Fig. 10.65 A'ROSA DONNA after arrival at Constanta on deck of the heavy-lift carrier GIANT 2 (Gina Bara)

shipyard constructed A-ROSA LUNA and A-ROSA STELLA for sailings on Rhône and Saône (Fig. 10.65).

Under French, Swiss and German flags, a whole fleet of inland cruise liners sailed the 310 km long Rhône river between Lyon and the mouth at Port St. Louis, before they were forced into lay-up due to the COVID-19 pandemic. Among the ships employed were the Rostock-based A-ROSA LUNA and A-ROSA STELLA, which accommodate up to 174 passengers. The breadth of the 125.8 m long ships was designed at 11.4 m according to the dimensions of the locks of the river. A dozen of the locks helps to surmount water level differences of between 6.7 and 23 m. High concentration is required of the French ship's officers when they manoeuvre the inland cruisers into the locks of Beauchastel or Châteauneuf-du-Rhône. Despite all the buildings constructed to make the river navigable, the voyage on the Rhône can have its pitfalls. If the water level is too small, the ships cannot leave Lyon for days. However, it can also happen that the water level is too high to pass under the bridges, even if the wheelhouse, mast and sun roof are lowered. A ballast capacity of 400 m^3 can reduce the height of A-ROSA STELLA and A-ROSA LUNA above the waterline (Fig. 10.66).

At 125.8 m length and 11.4 m breadth the ships have been designed according to the lock dimensions of the Rhône. They were carried in a complicated transport on deck of a lowerable heavy-lift vessel from Neptun Werft to the Mediterranean. In 2017, A-Rosa for the first time sent a ship to the Seine. A-ROSA VIVA sailed from the Rhine to the Seine estuary in January

Fig. 10.66 Lyon is the starting port for the Saône and Rhône voyages of A-ROSA STELLA and A-ROSA LUNA (Ralf Witthohn)

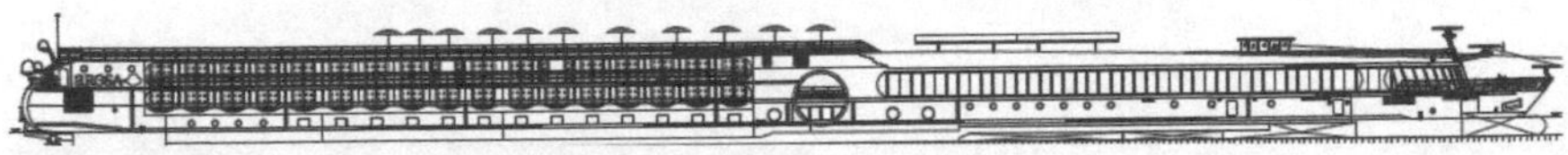

Fig. 10.67 The 135 m long and 11.4 m wide A-ROSA FLORA was launched in 2014 for Rhine cruises (A-Rosa)

on its own keel and with tug assistance during the sea voyage, to undertake river cruises starting and ending in Paris (Fig. 10.67).

In 2009/10, the Neptun-built A-ROSA AQUA and A-ROSA VIVA, classified by Germanischer Lloyd as "passenger ship with cabins" and constructed in accordance with the Rhine Vessel Inspection Regulations, had entered service for voyages on the Rhine. In 2017, the shipping company deployed A-ROSA VIVA to the Seine for sailings from Paris to Rouen or Le Havre. For the Rhine, the same type A-ROSA BRAVA was commissioned. In the first season A-ROSA VIVA undertook 47 river cruises primarily booked by German passengers. The longest of the routes in the summer of 2019 lasted nine days from the Parisian district of Saint Denis via Vernon and Rouen to Le Havre and back to the French capital via Caudebec-en-Caux, Rouen, Les Andelys and Mantes-La-Jolie (Fig. 10.68).

Only in one of the ports called at, in Caudebec-en-Caux, was the energy supply of the ship during the berthing period guaranteed by a compulsory electrical shore connection. Although another berth on the route, the one in Les Andelys, had such a supply facility, the shipping company refused to use it, because the voltage dropped from 390 to 370 V, so that it could cause damage to the ship's equipment. The shipping company calculated EUR 0.39 per kWh for the use of shore-side electricity, while for own generation only 25 cent were to be spent. A-ROSA VIVA can bunker a maximum of 90 m^3 of gas oil, of which up to 5 m^3 cubic were consumed daily, mainly to power the four

Fig. 10.68 Like the high seas cruise industry, inland cruises were hit by the COVID-19 pandemic, including the A-ROSA VIVA, which sailings on the Seine from Paris to Le Havre had to be interrupted (Ralf Witthohn)

331 kW Volvo main engines. They mechanically drive four Schottel twin propellers for a maximum speed of 24 km/h. The ship's manoeuvrability, particularly required in the five locks of the Seine, is improved by a Schottel pumpjet under the foreship, driven by a 404 kW Volvo engine. Particularly during turning and mooring manoeuvres, e.g. during the initial five kilometer long reverse passage from St. Denis along a Seine island, on the Tancarville canal from the Seine to Le Havre and while passing bridges the additional aid is needed. A lowerable wheelhouse reduces the vessel's height above the waterline. The 135 m length, 11.2 m breadth ship has a minimum draught of 1.6 m, which can be increased to 2 m by ballasting. The light ship weight amounts to 1825 tonnes.

That even inland navigation is not free of security problems was demonstrated by the presence of a guard service during the berthing time of A-ROSA VIVA in the St. Denis district of Paris. In March 2020 A-Rosa was forced to suspend all cruises on Danube, Seine and Rhône. Due to the COVID-19 pandemic, entry and hygiene restrictions imposed by the Austrian and French governments made it impossible to carry on cruises (Fig. 10.69).

Fig. 10.69 Only one shore connection at Caudebec-en-Caux, pictured while being created by two crew members of A-ROSA VIVA, among half a dozen of river port berths used by inland cruise vessels on the Seine indicates the inability of the European Commission to protect its populations continent-wide from the harmful exhaust gases generally emitted by ships (Ralf Witthohn)

In 2019, the Portuguese newbuilding A-ROSA ALVA for the first time sailed the Douro. In the same year, A-Rosa, in the previous year acquired by the British finance investor Duke Street, ordered a diesel-electric four-deck ship, to be named A-ROSA SENA, for voyages on the Rhine from the Dutch shipyard Concordia Damen for delivery in 2021. The ship is fitted with a 1.2 MWh battery that can be charged by the ship's generators or by shore connection to provide propulsion power when approaching the cities. At that time, A-Rosa Flusschiff employed 780 people from 25 nations and operated 12 inland cruise liners, which service personnel came mainly from eastern European countries such as Hungary and Bulgaria. In August 2020, A-Rosa announced the shifting of the delivery of the so-called E-Motion Ship A-ROSA SENA to March 2022 due to the COVID-19 pandemic. Cruises in Germany, France, the Netherlands, Belgium, Portugal and on the Danube were then planned to be resumed in June 2021. The first voyage of A-ROSA SENA was to start in May 2022.

10.6.2 Longship VIKING HERMOD from Avignon to Lyon

The take-over of Neptun Werft by Papenburg-based Meyer Werft and its input of deep knowledge in passenger ship construction meant a renaissance of the GDR tradition in the series building of river passenger liners for the Soviet Union. From 2002 to 2015, the shipyard in Warnemünde delivered 56 river cruisers to the shipping companies A-Rosa, Premicon and Viking River Cruises, making the latter Europe's market leader. The US-funded company also carries out voyages in Russia, Asia and Egypt by operating a fleet of 64 ships. By 2016, 34 of them were newbuildings of Neptun Werft. For sailings on Rhine, Rhône and Seine, the shipping company employed more than 40 of its longships accommodating 190 passengers. Four of these, the 2014-built VIKING BURI, VIKING DELLING, VIKING HEIMDAL and VIKING HERMOD sailed on Saône and Rhône. The 135 m length, 11.5 m breadth ships have a 1260 kW diesel-electric propulsion system of four Schottel twin propellers for a speed of 11 knots 11 knots. The in total 65 ships comprising Longship series started in 2010 was ended in March 2021, when VIKING GYMIR and VIKING EGDIR were handed over (Fig. 10.70).

10.6.3 ROSSINI on the Danube

In the city port of Linz, 20 inland cruise liners entered their winter quarters in November 2017. Some of the ships still on cruise pass through the Austrian city, such as VIKING TOR and VIKING EGIL on their way from Vienna to Nuremberg, VIKING VILI on its sailing to Passau or VIKING VE to Budapest. But for the majority of the Danube liners the season is over and their crews are on annual leave after several months of service. BOLERO, BELVEDERE, VICTORIA I, DCS AMETHYST, ADAMANTE, TRAVELMARVEL JEWEL, VERONIKA, THEODOR KÖRNER,

Fig. 10.70 On VIKING HERMOD passengers experience the varied landscapes of the Rhône valley (Ralf Witthohn)

Fig. 10.71 DONAUPRINZESSIN on the Danube (Ralf Witthohn)

WACHAU, VIKING MODI, VIKING BESTIA, CARMEN 1, PRINCESSIN KATHARINA, PRINCESSIN ISABELLA, SCENIC JEWEL and ROSSINI are assembled in the winter berth. All quays at Linz were occupied in April 2020, when not less than 35 river cruisers were moored due to the COVID-19 pandemic as still were one year later in April 2021.

Among them was, still being berthed in February 2022, the ROSSINI, which is the former DONAUPRINZESSIN once owned by Peter Deilmann company in Neustadt/Holstein, a pioneer of German river cruise shipping. After the sale in 2006 to Favorit Reisen in Heilbronn, the ship continued operation on the Danube. The 111 m length, 15.4 m breadth ship, delivered in 1983 by Flensburger Schiffsbau-Gesellschaft, has room for 215 passengers and is powered by two 900 kW engines (Fig. 10.71).

11

Luxury Yachts

When competing for the largest, fastest or most elegant luxury yacht, the greatest possible secrecy applies, contractually regulated between customer and shipyard. The wave of mega-yacht new buildings that has been launched since the 1990s for sheikhs, sultans, monarchs, oligarchs, millionaires and billionaires of every provenance means the repeat of a phenomenon that had already occurred on the eve of the Great Depression in the 1920s.

The largest luxury buildings exceed the dimensions of small cruise liners. While these super yachts are equipped to accommodate a larger number of guests and have to comply with the regulations for passenger ships, the number of passengers on the 80-m types that are particularly in demand is often limited to 12 and thus does not afford to comply with passenger ship safety and manning regulations. A number of the yachts are offered on the charter market at high daily rates. The now outdated rule of thumb from the early days of mega-yacht construction, according to which 1 m ship costs DM 1 million, was soon transferred one-to-one to the euro. And currently the rich of the world have to reckon with 2 or more million euros for every ship metre. Compared with the construction costs of modern cruise liners, which may as well amount to several hundred million, this price level does not even seem excessive given the exclusive equipments. The modelling work regularly carried out by star designers on the exterior and interior design alone swallows up large sums of money.

With the advent of the digital age, yacht builders have gained new customers from the circle of computer magnates. One of the most famous, Larry Ellison, had the 138 m long RISING SUN built at Fr. Lürssen Werft in 2004. The 7600 gt luxury cruiser, which was built for allegedly USD 194 million, achieves 26 knots thanks to its four main engines of a total output of

R. Witthohn, *International Shipping*, https://doi.org/10.1007/978-3-658-34273-9_11

36,000 kW. In view of such data, some of the yachts that have been on test runs in the German Bight in previous years seem rather ordinary, such as the 310 gt BE MINE built in 1991, the 72 m long CORAL ISLAND of 1994, the 60-m CAPRI built in 2003 or the QUEEN M and PHOENIX christened in 2004. This applies less to the military look of the 2000 gt SKAT, which was handed over to software developer Charles Simonyi in 2001. Also standing out from the fleet of the otherwise mostly white or sand-coloured noble runabouts is the deep blue hull make-up of CARINTHIA VII, built in 2002 under the project name FABERGE on behalf of the department store heiress Heidi Horten. In addition to the obligatory power and jet boats, helicopter decks, transparent ship bottoms or submarines, modern shipbuilding techniques are applied to the luxury yachts. These include dynamic positioning systems, azimuth propulsion systems and helipads (Fig. 11.1).

11.1 AZZAM and TOPAZ for the Emirates

The large number of German-built floating luxury palaces is the result of long traditions in building fast ships at shipyards on the Lower Weser. One of the largest mega yachts is the 13,136 gt measuring AZZAM, a product of Fr. Lürssen Werft in Bremen-Vegesack realized under a contract awarded by the President of the United Arab Emirates, Sheikh bin Zayid Al Nahyan. Due to its length of 180 m, the yacht left the former navy building dock of Bremer Vulkan shipyard in April 2013 without its foreship. After adding the missing section in a floating dock, the newbuilding undertook its first trial runs into the German Bight in June 2013 and was handed over by the Fr. Lürssen shipyard in Bremen-Vegesack in October 2014. In the months that followed, the 20.8 m breadth ship carried out numerous further trips into the German Bight, in order to test the four water jet propulsion units accelerating the yacht to over 30 knots (Fig. 11.2).

Another Lürssen newbuilding constructed on behalf of a customer from the United Arab Emirates is the 12,532 gt mega yacht TOPAZ. Following test runs in August, the 147.3 m length, 21.5 m breadth newbuilding was in December 2012 delivered to the country's Deputy Prime Minister. The hull of this ship, too, was constructed in the covered former Bremer Vulkan building dock. As many mega yachts, the ship is classified by Lloyd's Register. Its diesel-electric propulsion system comprises two 6000 kW asynchronous motors turning two propeller shafts. In addition, the yacht has an electrically driven azimuth propeller with an output of 2100 kW. Two transversal thrusters arranged forward have an output of 700 kW each. Energy supply is ensured

Fig. 11.1 As one of the world's largest mega yachts the 147 m long, 12,532 gt mega yacht TOPAZ was in 2012 delivered by Fr. Lürssen shipyard to an Arabian customer (Ralf Witthohn)

by six diesel generators providing a total of 17,300 kW. Equipped with a dynamic positioning system, the ship, managed by Pearl Ships in Dubai, is sailing under the flag of the Cayman Islands. In the TOPAZ build of year, Kröger shipyard in Schacht-Audorf, a construction site belonging to the

Fig. 11.2 AZZAM was the world's longest luxury yacht when delivered in 2014 (Ralf Witthohn)

Fig. 11.3 A sophisticated propulsion and manoeuvring system distinguishes the Lürssen building TOPAZ, built for a VAR customer (Ralf Witthohn)

Lürssen group, delivered the 87 m newbuilding ACE to a Russian client. Other Lürssen projects completed in 2013/2014 were NIKI, a 85-m ship, and the 87-m yacht GLOBAL from Kröger (Fig. 11.3).

11.2 Possible Mission for the Saudi King

As the leading supplier of luxury yachts since their revival, Lürssen shipyard is maintaining traditional contacts with Arab monarchs, such as the Saudi King, for whose navy the shipyard built three fast torpedo boats as early as 1969. In

Fig. 11.4 The 7922 gt measuring KATARA was built for the Emir of Qatar (Ralf Witthohn)

1998 Lürssen fulfilled the monarch's wish for a "Mission Possible" and, under the camouflage name derived from this term, laid the keel of the 12,234 gt MIPOS, put into service in 1999 as AL SALAMAH. On this 139.3 m length, 23.5 m breadth USD 200 million project, a crew of around 100 people takes care of the passengers′ well-being. The 21 knots achieving yacht can be chartered. Another Arab customer of Lürssen is the Emir of Qatar, Hamad ibn Chalifa. In August 2010, he took over the 7922 gt ship KATARA, built under the name CRYSTAL, which 124.4 m long silhouette shows two low-designed funnels (Fig. 11.4).

11.3 AL SAID for the Sultan of Oman

So far the largest Lürssen newbuilding is the 15,850 gt AL SAID ordered by the Sultan of Oman, Qabus Ibn Said. The 155-m vessel delivered by Lürssen in 2008 carried out its trial runs under the name SUNFLOWER. The hull was too large to lay its keel at Lürssen. Therefore it was built on the slipway of SSW Shipyard in Bremerhaven. Only in Lürssen's covered floating dock were fit the superstructure, masts and the two funnels, the forward one of which is a dummy. The ship is powered by two MTU diesels with a combined output of 16,000 kW and equipped with stabilizers that are regularly fit to large yachts. Oman had already received the 10,864 gt state yacht FULK AL SALAMAH from Bremer Vulkan in 1987. The 136.3 m length, 21 m breadth and 19 knots fast double-screw vessel can also be used as a hospital ship. It has 25 cabins for 43 passengers and a cargo hold in the forward. Able to

Fig. 11.5 The largest of all luxury yachts is the Omani AL SAID (Ralf Witthohn)

accommodate 240 soldiers the ship fulfils a double function as a state yacht and warship. After the commissioning of the new AL SAID, the FULK AL SALAMAH, renamed DHAFERAH in 2016, served as its escort ship. The first state yacht AL SAID, in 1970 built at Brooke Marine in Lowestoft as a military hybrid with armament, has been registered as a warship since 1983 (Fig. 11.5).

11.4 157-m Yacht DILBAR for Russia

In February 2016, the third longest Lürssen luxury cruiser in the history of the shipyard undertook its initial test voyage from Vegesack to Helgoland. The 157 m length, 20 m breadth and 6.1 m depth ship undertook the trials, which lasted for days and during which a speed of over 22 knots was reached, still under the project name OMAR. Upon delivery the yacht took over the name from a 110 m yacht that Lürssen had handed over to the Russian entrepreneur and billionaire Alisher Usmanov in 2008 as DILBAR. In May 2016, the new, 15,500 gt DILBAR, left the shipyard bound for Tangier under the Cayman Islands flag (Fig. 11.6).

In March 2018, Lürssen started testing another, 136 m length, 22.5 m breadth measuring megayacht drafted by UK-based Mark Berryman Design. Delivered as FLYING FOX in the following year, the 9022 gt ship offers 11

Fig. 11.6 Lürssen built the 157 m DILBAR for a Russian oligarch in 2016 (Ralf Witthohn)

cabins for 22 guests, who are taken care of by 54 crew members. Propulsion is by means of two MTU diesel engines 16V1163M84 of 2 × 4480 kW output.

In March 2020, the 140 m length, 23.3 m breadth SCHEHERAZADE of 10,167 gt, registered for the Bielor Asset firm at Douglas/Isle of Man and managed by Döhle Private Marine Clients, was completed. The outer architecture of the 10,178 gt mega yacht was designed by Monaco-based Espen Øino, its interior by Zuretti from Nice. It can accommodate up to 18 guests in 9 cabins. Driven by two MTU engines of 2 × 5600 kW the SCHEHERAZADE achieves at least 19 kn (Fig. 11.7).

In July 2020, the 2744 gt AVANTAGE ex HAWAII came from the Rendsburg-based construction site of Lürssen, and in February 2021, the 10,154 gt NORD ex OPUS left its builders under the Cayman Islands flag. The 142 m length, 19.5 m breadth ship, classified as passenger yacht, reaches a speed of 20 kn. In February 2022, the 158 m length BLUE, by the boulevard press reported to have been ordered by the Saudi crown prince Mohammed bin Salman, was floated out from the building dock. Two further Lürssen projects referred to the 146 m OPERA project and the LUMINANCE of similar length (Fig. 11.8).

In October 2021, Lürssen-Kröger shipyard at Rendsburg/Germany delivered the 55 m length, 10.2 m breadth mega yacht MOON SAND reportedly to Hong Kong-based billionaire Charles Ho Tsw-kwok. Designed by the bureau of Bannenberg & Rowell, the 835 gt MOON SAND is laid out for 10–12 guests and accelerated to 17 kn by two a twin-propulsion system. In February 2022, MOON SAND was moored at the builders main yard in Bremen-Vegesack (Fig. 11.9).

Fig. 11.7 The Cayman Islands-registered SCHEHERAZADE is managed from Hamburg (Ralf Witthohn)

Fig. 11.8 The architecture of the 2019-delivered FLYING FOX shows a more filigrane, traditional look (Ralf Witthohn)

Another, considerably larger product of Lürssen's Rendburg yard site was delivered in July 2020 to a Monaco-registered owner. At a length of 87 and 13.8 m breadth, the AVANTAGE measures 2744 gt. Two four-stroke diesel engines of type MTU 16V4000M63 with an output 2 × 1720 kW enable a speed of 17 kn by two propellers. A crew of 18 persons serves up to 14 passengers in seven cabins (Fig. 11.10).

Fig. 11.9 The medium-sized MOON SAND was built at the Rendsburg shipyard of Germany's Lürssen group (Ralf Witthohn)

Fig. 11.10 The Lürssen-built 2744 gt mega yacht AVANTAGE shows elegant lines (Ralf Witthohn)

11.5 AVIVA from Abeking & Rasmussen

Another traditional yacht builder, Abeking & Rasmussen (A&R) in Lemwerder/Germany in the direct vicinity of Lürssen, in 2008 applied the SWATH principle, which had already been realized in pilot vessels and coast guard boats, to the SWATH expedition yacht SILVER CLOUD. In contrast, the 82 m vessel SECRET delivered by A&R in May 2013 is a monohull newbuilding. The 12.4 m breadth, 3.5 m depth yacht measures 2240 gt. After this newbuilding, the Lemwerder shipyard dedicated itself to the construction of the 81.2 m KIBO, which steel hull and aluminium superstructures were built in a hall of the Rönner group's Stahlbau Nord in Bremerhaven, where it was in April 2013 launched from the quay with the help of the two 450-tonne cranes of the heavy-lift carrier PALEMBANG. This unusual way of taking a hull to water required costs of over EUR 100,000.

In April 2016, the shipyard sent the new building CLOUDBREAK on trial runs, during which the approach of the onboard helicopter was also tested. The 72.5 m long steel hull, designed with a straight stem, had been manufactured by the Rönner group and towed to Abeking & Rasmussen for outfitting in June 2014. There, the aluminium superstructures were set on the ship, which equipment included two pairs of stabilizers. The construction of the yacht, which is suitable for worldwide expedition cruises, was supervised by the charterer Super Yachts Monaco on behalf of the customer. Abeking & Rasmussen built further exclusive mega-yachts, the 68 m long AVIVA of 2007, the 78 m EMINENCE, TITAN and AMARYLLIS in 2008, 2010 and 2011, resp., as well as the 60 m ELANDESS in 2009 and EXCELLENCE V in 2012 (Fig. 11.11).

At Abeking & Rasmussen, camouflage names as well serve to disguise projects. ROMEA, delivered in 2015, completed its trial runs as DARTWO. The 98 m long hull for the next A&R newbuilding was launched in the traditional way at Flensburger Schiffbau-Gesellschaft and was towed to Lemwerder for fitting out in May 2015, where the shipyard facilities for the in-house production of hulls of this size were expanded at that time. The largest A&R newbuilding to date was delivered in 2017 as AVIVA (Fig. 11.12).

Fig. 11.11 Its dominating grey colour gives the A&R megayacht CLOUDBREAK a military touch (Ralf Witthohn)

Fig. 11.12 The largest A&R yacht and the second one named AVIVA was first sent on trial runs at the beginning of 2017 and undertook further tests a year later (Ralf Witthohn)

Fig. 11.13 2019-commissioned EXCELLENCE is fitted with a hatch starboard aft, through which tender boats can be launched (Ralf Witthohn)

A complete new architecture developed by London-based designer Winch is presented by the A&R newbuilding EXCELLENCE, which 80 m long and 12.8 m wide hull was built under subcontract in Szczecin/Poland. Delivered in September 2019, the yacht offers room for 14 guests in seven cabins and 23 crew members. Main engines are two MTU four-stroke diesels of 2 × 1,500 kW. Under the difficult conditions of the COVID-19 pandemic, A&R in April 2020 completed the 68-m megayacht SOARING on behalf of Ocean Management from Küsnacht/Switzerland. Entered in the Cayman Islands register, the 1450 gt newbuilding is intended to be available for chartering. Due for delivery in 2022 is a 118 m passenger yacht, which keel of was laid under subcontract of Abeking & Rasmussen at Pella Sietas shipyard in Hamburg in 2019 (Figs. 11.13 and 11.14).

Fig. 11.14 The final fitting-out work on the 68 m SOARING became complicated under the conditions of the COVID-19 pandemic (Ralf Witthohn)

11.6 LE GRAND BLEU, ECLIPSE, LUNA and SOLARIS for Abramovich

More shipbuilders in Germany's Lower Weser area, which can be regarded as the centre for this special shipbuilding branch, have dedicated themselves to luxury yacht building. The first megayacht built in Bremerhaven was floated out at Lloyd Werft in October 2009. The order for the 115 m long and 17 kn fast LUNA was placed by the Russian oligarch Roman Abramovich. The dry dock, otherwise used for repairs, was converted into a building site for 2 years. Under strict secrecy, the hull and superstructure sections of the ship, which were manufactured by a subcontracting Rönner group, were assembled under a temporary scaffolding roof independent from weather influence (Fig. 11.15).

At that time, in order to save the shipyard, the State of Bremen held a 13.16% stake in Lloyd Werft, which carried out the questionable order for an oligarch despite the public shareholding in the company. LUNA's equipment includes the playground equipment usual on this type of luxury ship, such as helicopters and speedboats. The ship, costing over EUR 100 million, was a replacement for the luxury yacht LE GRAND BLEU, which Abramovich sold to an acquaintance and which had been built in 2000 by Bremerhaven-based Rönner group in cooperation with the yacht agency Kusch in the former shipbuilding hall of Bremer Vulkan. LUNA as well did not remain in the possession of Abramovich for long, who sold the yacht to Farkhad Akhmedov in 2012. After the change of ownership, the yacht was rebuilt at MWB

Fig. 11.15 The luxury yacht LUNA, which had been built in Bremerhaven in 2010, underwent an extensive redesign of its interior 5 years later at the same location (Ralf Witthohn)

shipyard in Bremerhaven for over a year without changing the exterior architecture. Abramovich had a number of other luxury yachts built, including the 162.5 m ECLIPSE, measured at 13,564 gt and taken over from Blohm + Voss in Hamburg in October 2010 after a construction period of over 4 years as the world's largest yacht at the time.

In 2017 started the construction of another Abramovich megayacht to be named SOLARIS at Bremerhaven. In April 2018 the foreship section was transported to the floating dock of Lloyd Werft, which had been covered, in order to carry out the construction independent from weather conditions, a condition for building mega yachts. In October and November the middle and aft section of the newbuilding followed. The newbuilding became first visible in March 2021, when it was floated out from the building dock and started trial runs in the German Bight under the name SOLARIS. At 139.7 m length and 21.4 m breadth the 11,247 gt ship offers 18 cabins for 36 passengers and 30 cabins for 60 crew. Eight four-stroke MTU diesel engines of 14,000 kW output drive a twin azipod propeller system of 9000 kW. The ship's equipment includes a dynamic positioning system, helipad, tenders and boats, which are launchable through hydraulically operated side doors, a

Fig. 11.16 The transport of the aft section for Abramovich's Solaris project into the covered floating dock allowed a look on the two azimuth propellers already fitted (Ralf Witthohn)

radar-controlled missile detection system, bulletproof window glasses and a large platform aft for watersport amenities aft (Figs. 11.16 and 11.17).

An unusual luxury yacht conversion project was initiated in January 2022, when the Dutch dockship ROLLDOCK SKY transported the ARGOSY to Bremerhaven, from where the ships was towed to Brake. The ship is a 1931-built product of Germany's Germania Werft, which had delivered it to Charles A. Stone in New York. The 669 gt twin-screw vessel survived the times and was still employed after the turn of the millennium in a passenger service between Sorrento and Capri under the name SANTA MARIA DEL MARE, before conversion back into a yacht was started in 2010 and is now to be completed without further informations having been revealed so far (Fig. 11.18).

Fig. 11.17 Oligarch Abramovich's latest yacht SOLARIS has been fit with bulletproof windows and a missile detection system (Ralf Witthohn)

Fig. 11.18 The 1931-built luxury yacht ARGOSY was in January 2022 towed to Brake/ Germany to complete conversion work (Ralf Witthohn)

11.7 Kiel Stealth Yacht A for Russian Buyers

A visual counterpoint to the generally elegant yacht design is drawn by a 119 m long newbuilding of 2008, developed by the French designer Philippe Starck on the base of more military stealth forms. Originally contracted with Blohm+Voss at Hamburg, the ship was built at HDW shipyard in Kiel for capacity reasons. The extravagant yacht was taken over by the Russian couple Andrey and Aleksandra Melnichenko in 2008 and christened A, the first letter of the owners' first names. A reaches a speed of 23 knots and can accommodate 14 guests. Melnichenko ordered a second exclusive yacht in 2012 from the Kiel shipyard, now operating as German Naval Yards, to replace the first by the second A. The 143 m long three-masted ship was designed for a speed of 21 knots under sails and, after completion in 2016, attracted criticism from a British environmental association for allegedly using protected teak from Myanmar (Fig. 11.19).

Fig. 11.19 The design and choice of name for the A are out of the ordinary for luxury yachts (Ralf Witthohn)

11.8 Expedition Yachts ULYSSES from Kleven Verft

In August 2015, the unfinished yacht ULYSSES, coming from Kleven Verft in Ulsteinvik, reaches the mouth of the Weser, anchors on Blexen roads and the same day enters the fishing port, where the ship is moored at Labradorkai. In the following weeks the ship disappears completely under a tarpaulin-covered scaffold. Warm air is blown under the protective cover via two large hoses attached to the stern to maintain higher temperatures. This system, developed in the absence of a covered dock for the overhaul of warships, allows the newbuilding to be applicated final paints regardless of the weather. In December two boats to be equipped arrive, a 21 m fishing catamaran with water jet propulsion named U-21, which was built in Auckland, New Zealand, and an open speedboat driven by two powerful outboard engines. After the hull cover has been dismantled, the load test of the two 17-tonne cranes mounted on the foredeck of ULYSSES takes place in February 2016. The cranes have sufficient power to lift the catamaran into a sump on the forecastle deck. Two further inflatable boats of different types are positioned on each side of the superstructure. In the same month, the 107.4 m length, 5937 gt vessel undertakes a first test run in the North Sea and returns to port for final fitting out work. In July 2015 the customer, publicity-shy New Zealand entrepreneur and billionaire Graeme Hart, takes over the newbuilding. Fitted with cabins for 60 guests, the vessel is designed for long expedition voyages in rough and icy waters (Fig. 11.20).

Only 2 years after the first ULYSSES, Hart received another luxury yacht from Kleven Verft, which was launched under the project name U116. The designation describes the length of the yacht, which appearance is very similar to the first ship. As with the ULYSSES, the interior and final outfitting works are carried out in Bremerhaven, where the ship arrived in December 2016.

Fig. 11.20 The first ULYSSES arrived in an uncomplete state from Norway for completion and fitting out work at Bremerhaven in August 2015 and was delivered in June 2016 (Ralf Witthohn)

Fig. 11.21 Only 2 years after the first ULYSSES, a New Zealand billionaire took delivery of a second nearly identical megayacht of the same name (Ralf Witthohn)

Delivery of the second ULYSSES took place in June 2018. After sale in 2017 and registry for a Southampton/UK company the first ULYSSES was renamed ANDROMEDA. In February 2022, ANDROMEDA sailed from Port Victoria to Praslin on the Seychelles, while the new ULYSSES cruised off the Bahamas (Fig. 11.21).

Fig. 11.22 In the design of the catamaran-type megayacht ROYAL FALCON ONE, owned by Golden Galaxy Yachts in Labuan/Malaysia, participated the Austrian studio F.A. Porsche and the Australian company Incat Crowther. The 492 gt ship of 4.1 m length and 12.5 m breadth is driven by two MTU diesel engines to 35 knots and was delivered in 2019 by Shipbuilding Haiphong/Vietnam (Ralf Witthohn)

12

Floating Hospitals and Rescue Vessels

A seldom type of passenger ship that is not used for touristic activities is the hospital ship, the vast majority of which are navy units. In order to obtain protection under the Geneva Convention, the outer skin of hospital ships is white with a red stripe and large red crosses on the chimney and hull. Only a few are operated by civilian organizations. The Soviet passenger and ro-ro ship MIKHAIL SUSLOV, in which car deck Lloyd Werft in Bremerhaven installed a modern eye clinic for use off the coasts of developing countries in 1989, was prepared for this task at great financial expense. In Germany a number of civilian hospital ships were operated, including the ferry HELGOLAND and the converted general cargo vessel FLORA, which were operated in Southeast Asia by the German Red Cross. Other passenger ships regularly operated from Hamburg and Bremen in the Helgoland traffic as WAPPEN OF HAMBURG and BREMERHAVEN as well as the coaster PADUA built in 1964 were prepared for short-term deployment in the event of war or disaster, but only performed such tasks during manoeuvres.

12.1 Relief Missions in Africa and South America

Until 2021, the 16,572 gt AFRICA MERCY was the largest civilian hospital ship, a former Danish rail ferry built in 1980 as DRONNING INGRID and converted in 2000 on behalf of Mercy Ships, which is based in Lindale/USA. On the ship up to 400 people from 40 countries volunteer to provide medical assistance and drinking water. From 2012 to 2016, the AFRICA MERCY was in service in Guinea, Congo and Madagascar. The first of the Mercy Ships, which called at more than 500 ports in 70 countries, was the

R. Witthohn, *International Shipping*, https://doi.org/10.1007/978-3-658-34273-9_12

ANASTASIS, an Italian passenger and cargo ship of 11,695 gt, built in 1953 as VICTORIA and scrapped in 2007. The 1035 gt vessel ISLAND MERCY and the 2163 gt CARIBBEAN MERCY ex POLARLYS were deployed for relief operations mainly in Central and South America and in the South Pacific region.

In 2021, Mercy Ships commissioned a new hospital ship, at 37,856 gt by far the largest civilian such ship of this type. The 174 m length and 28.6 m breadth GLOBAL MERCY was built at Tianjin Xingang shipyard in Tianjin and equipped with six operating rooms, a school, fitness rooms, a swimming pool, café, shop and library. The accommodation is laid out for up 600 crew members, comprising surgeons, seamen, kitchen personnel, teachers, technicians etc. The total number of persons on board may reach 950.

12.2 Rescue Ship AQUARIUS of SOS Méditerrannée

In February 2016, the former government-owned German vessel MEERKATZE left Bremerhaven on behalf of the private aid organisation SOS Méditerrannée under the name AQUARIUS for a rescue operation in the area between Sicily, Lampedusa and Libya. For this purpose, the ten persons comprising crew was supplemented by a rescue team of 12 persons of the organisation Médecins du Monde in Marseille. AQUARIUS provided temporary accommodation for up to 500 people. The charter costs for the rescue operation in the Mediterranean Sea were raised for an initial period of 3 months by donations of EUR 750,000. The operations and the distribution of the rescued persons were organised by the Maritime Rescue Coordination Centre (MRCC) in Rome. The ship owned by Hempel Shipping had in 1977 been built by Fr. Lürssen in Bremen for the German government to provide medical and logistical support for the fishing fleet as a so called *Fischereischutzboot*. The MEERKATZE was later chartered out under the name AQUARIUS as a residential ship for wind farm technicians, but taken out of service in 2009. From February 2016 until August 2018, AQUARIUS reported the rescue of over 29,000 refugees from boats. In December SOS Méditerrannée decided to terminate the operation due to the political pressure on the rescue activities. After temporarily named AQUARIUS 2 and AQUARIUS DIGNITUS, the ship is since 2021 sailing as AQUARIUS again and flying the Russian flag. In early 2022, it was stationed at Murmansk being managed by the Murmansk-based Transport Freight Agency, which then

Fig. 12.1 After operations as a fishing protection vessel MEERKATZE and accommodation vessel for the offshore industry, AQUARIUS was in February 2016 deployed to rescue refugees in the Mediterranean (Ralf Witthohn)

operated seven auxiliary vessels of different types. In 2019, SOS Méditerrannée chartered the former offshore supply vessel OCEAN VIKING in partnership with the International Federation of Red Cross and Red Crescent Societies (IFRC) to continue the rescue operations. In February 2022, the ship rescued 247 persons in five separate operations in less and proceeded to Marseille. There, the organization had been attacked by a group of right-wing extremists in 2018 (Fig. 12.1).

Only a few days after the departure of AQUARIUS, another rescue ship left Bremerhaven on the hatch of the Dutch heavy lift carrier ATLANTIC. The former 23.3 m rescue cruiser MINDEN was made available by its private owner to Deutsche Gesellschaft zur Rettung Schiffbrüchiger (DGzRS) in Bremen to undertake rescues of refugees in the Aegean Sea together with other northern European rescue societies under the umbrella of the International Maritime Rescue Federation (IMRF) and under leadership of the Hellenic Coast Guard. During an assignment limited to 3 months the MINDEN reportedly saved over 1100 people from danger. Afterwards, the MINDEN proceeded to Malta. Renamed JANUS, the ship was reported to be berthed in Abidjan/Ivory Coast in 2019 (Fig. 12.2).

Fig. 12.2 In February 2016, the German rescue cruiser MINDEN started its voyage to the refugee mission in the Aegean Sea on the hatch of the Dutch heavy lift carrier ATLANTIC (Ralf Witthohn)

Three large sections of the megayacht project Solaris were separately transported from the covered building hall to the Lloyd Werft floating dock, where they were assembled and fitted out for scheduled delivery in 2020 (Ralf Witthohn)

Glossary

Accommodation vessel Residential ship serving as floating hotel, for example for offshore workers

Aframax Tankers of the 80,000 to 120,000 dwt size class, named after the rate index AFRA (abbreviation for Average Freight Rate Assessment)

Air Draught Height of a ship from the waterline to the topmost structural member, important size indication for the passage of bridges

Alaska pollock Pacific pollock, a fish of the cod family (saithe) living in the northern Pacific Ocean and edible fish frequently processed in the fishing industry

Alliance Cooperation of container shipping companies using their ship in joint liner services

American Bureau of Shipping (ABS) US classification society

Anchorage (Roadstead) Suitable place for anchoring a ship waiting for a free berth, a new order or seeking shelter from storm

Angled stern ramp Rope-operated ramp mounted aft on starboard side on ro ro ships for the rolling transhipment of vehicles up to 400 tonnes weight

Arrest The usual term in the industry for the arrest of a ship by a bailiff for compulsory auction, for example due to the insolvency of the shipping company

Asphalt tanker (bitumen tanker) Special tankers with heatable tanks for the transport of liquid asphalt/bitumen

Asymmetrical aft ship Special fuel-saving lines of a ship to optimise the water flow to the propeller, which has by its rotation a different hydrodynamic effect on the two sides of the ship

Auxiliary diesel engine (generator set, or genset) Air- or water-cooled diesel engine coupled to a generator to produce electrical energy for on-board operation

R. Witthohn, *International Shipping*, https://doi.org/10.1007/978-3-658-34273-9

Azimuth propeller Propulsion and manoeuvring technology initially used primarily on harbour tugs and special offshore ships, but actually also on cruise liners, luxury yachts and cargo ships, comprising a mechanically or electrically driven propeller that can rotate through 360° and thus combines the functions of propulsion and steering of the ship.

Bale Indication of the cargo hold capacity for general cargo, often figured in cbf (1 cubic foot = 0.0283 m^3)

Ballast A weight taken temporarily by seawater or continuously by sludge or solids to increase the stability or to alter the trim of a vessel

Ballast tank Tank in the double bottom, in the side hull, the aft or forepeak or, in the case of ore carriers, a high tank for water ballast

Barge Non-propelled vehicle, moved by tugboats and used for the lightering of ships or—as a sea barge—for transport of cargo over sea and on barge carriers

Barge carrier Cargo ship that can pick up barges by its own crane or by floating them into the cargo hold

Base (line) Upper edge of the keel plate as the most important reference line for the construction of a ship

Bay Container row with counting beginning forward

Bilge Space in the keel of a ship, or the radius between the double bottom and the shell plating

Bilobe Space-saving design of two cylindrical tanks merging into each other, for example for the transport of LPG in specially equipped tankers, named trilobe in the case of three tanks

Block coefficient Ratio of the underwater hull volume to the enclosing cuboid

Bollard pull Tensile force achieved by a tug and determined by means of a measuring device during the bollard pull test in tonnes

Bow Most forward part of the ship, also named stem

Breadth Width of a ship, in shipbuilding terms the distance between the inner edges of the hull (named moulded breadth, in contrast to maximum breadth or breadth overall including the thicknesses of the shell plating, rubrail, fenders etc.)

Breakbulk Dry cargo types transported in larger quantities, such as sawn timber, cellulose or steel, usually by cargo ships equipped with cranes

Breakwater Bulkhead across the forecastle of a ship to protect the ship and deck cargo against heavy seas

Bridge Inhabiting the wheelhouse, usually located on the uppermost superstructure deck, the bridge deck, with all equipments necessary for navigating and manoeuvring the ship

Broker (Ship broker) Term for a person acting as an intermediary for freight contracts, but also for contracts about the construction or sale of ships

BSU (Bundesstelle für Seeunfalluntersuchung) Germany's Federal Bureau of Maritime Casualty Investigation, based in Hamburg

Bulbous bow Special foreship underwater lines to improve the hydrodynamic properties with the side effect of increasing the deadweight capacity

Bulk cargo Uniform cargo such as ore, coal, stones or grain transported in large quantities in bulk carriers

Bulkhead Transverse or longitudinal construction element to divide the ship's hull and increase its strength, of particular importance is the collision bulkhead, for which a minimum distance from the bow is prescribed

Bunker Ship's fuel supply delivered by bunker boats

Cable layer Special ship type equipped with techniques for laying energy or communication cables in the seabed, at the same time often designed as pipe layer

Cabotage In shipping, transport between the ports of a country by a foreign shipping company, often restricted by national rights, but circumvented by the establishment of domestic subsidiaries

Capesize Bulk carrier of about 180,000 tonnes deadweight, which cannot pass the Suez canal in a loaded condition and therefore has to sail round the Cape of Good Hope, the "Cape".

Cargo handling gear Deck equipment for loading and unloading of the ship, in earlier times consisting of a mast, derricks and winches, nowadays mostly largely replaced by electro-hydraulically operated revolving cranes

Casualty Accident of a ship caused by collision, grounding, storm or fire

Catamaran A twin-hull vessel with a low length to width ratio, resulting in high stability and good seakeeping characteristics

Cavitation Cavities occurring especially on the propeller which can lead to material damage

Cell guides Structure of guirders installed in the holds of container ships for guidance of the containers during loading operation and for cargo securing

Charter Hire of a vessel for a limited period of time (time charter) or for a trip (voyage charter)

Class notation Symbols and signs issued by a classification society in accordance with the regulations for a ship and its machinery

Coastal motor vessel (Coaster) Type of ship for near-coastal transport developed to replace sailing ships since the availability of combustion engines

Combi-ship A type of ship for the simultaneous transport of passengers and freight that has become rare, but is still found today particularly on ferries

Con ro vessel Cargo ship for the simultaneous transport of containers and rolling cargoes, e.g. vehicles or cargo loaded on trailers, trucks etc.

Container Standardised box of varying length and height for the continuous transport of general cargo, refrigerated or liquid goods on specially designed ships and after transhipment by truck, rail or inland waterway

Container crane Land-based gantry crane for handling containers by means of a trolley travelling over the hold of the ship on a lowerable jib

Container ship Cargo ship for the transport of standardised boxes (containers), which are secured in the hold in fixed cell guides, on the hatches and on deck by lashings and twist locks

Controllable pitch propeller A propeller equipped with rotatable blades, in contrast to a fixed-pitch propeller, enabling the pitch to be adjusted to the operating conditions, ahead or astern manoeuvring, at zero setting, the propulsive force is cancelled though the rotation speed is maintained

Course Heading of a ship destined by the angle against the north direction

Crane ship Special vessel for offshore work or project cargo handling equipped with a heavy-lift crane of a maximum SWL of several thousand tonnes

Crewboat Fast ship for transporting personnel to drilling platforms and other offshore destinations

Cross trade Operation of liner carriers on routes distant from the home country of the shipping company

Crosshead Engine part on large two-stroke marine diesel engines, so-called long-stroke engines, for transmitting the piston movement to the crankshaft

Crossover (also named manifold) Distribution station with pipe connections on both sides of the main deck to the individual cargo tanks located at approximately half the deck length of a tanker

Cruise Voyage undertaken for pleasure on a passenger ship

Cruise liner Passenger ship for recreational voyages

Dangerous cargo Substances which nature may pose a danger to the ship, crew and environment

Deadweight (dwt) Load capacity (1 tonne = 1000 kg or tons = 1016 kg) Allowable payload, including fuel, supplies and people, to reach the low load line (freeboard mark)

Deck transporter Cargo ship transporting its entire cargo, mostly project cargo, on the weather deck

Depth In shipbuilding terms, the height of the ship measured from the inner edge of the keel to the lower edge of the main deck at its side

Diesel-electric propulsion A form of propulsion selected primarily on special ships and cruise liners, which uses the energy supplied by diesel generators or gas turbines to propel azimuthing propellers (pods) or conventional propellers via shaft motors

Discharge Unloading of the cargo

Displacement The weight in tonnes of water displaced by a ship and its cargo, which in freshwater is almost equal to the weight of the ship

Diving support vessel Specially equipped ship to support diving activities in the offshore industry

DNV GL Norwegian-German classification society, formed by the merger of Det Norske Veritas and Germanischer Lloyd

Dock Floodable pit closed by a gate (dock ship) for the construction or repair of a ship (dry dock, graving dock), or lowerable floating installation regularly unsuitable for shipbuilding, but for taking in a repair ship (floating dock)

Double hull Construction prescribed for tankers and common for most cargo ships featuring a double bottom and an inner plating formed by a longitudinal bulkhead next to the outer shell plating

Double bottom For safety reasons, a mandatory construction consisting of the ship's bottom (keel), the tank deck, floor plates and stringers, mostly serving as tank for water ballast

Down by the head Ship deeper in the water by the bow than aft

Draught (Draft) Distance from the waterline to the lower edge of the keel, depending on the loading condition of the ship and the physical properties of the seawater/freshwater, in its maximum size determined by the calculation of the freeboard

Dual fuel engine Technology for the alternative use of two types of fuel, such as heavy fuel oil/diesel oil and liquiefied natural gas (lng)

Dynamic Positioning (DP) Computer- and satellite-based system for positioning a ship by means of its propulsion and manoeuvring systems in conjunction with sensor elements and the Differential Global Positioning System (DGPS) to compensate for currents, wind and sea states, used since 1974 by special ships in the offshore industry and increasingly also on ferries, cruise liners and cargo ships

External identification Internationally prescribed letter/number combination for identifying a fishery vessel, applied to its bow in addition to the vessel name, e.g. H for Hull/UK, or NC for Niedersachen/Cuxhaven

Factory trawler Fishing vessel equipped with special fish processing and freezing machines and thus achieving a self-sustained sea endurance of up to three months, unlike fresh fish trawlers, which are at sea for about three weeks

Fairway A waterway identified by buoys

Feeder vessel (in short Feeder) Smaller vessel used for distributing and collecting containers or cars to and from secondary ports

Fender Element of wood, plastic or rubber for absorbing collision forces during docking manoeuvres

Ferry Ship carrying passengers and vehicles between two ports on marginal seas according to timetable

Fin Streamlined body attached to the hull or rudder to improve hydrodynamic conditions

Fishery protection vessel State-operated vessel for medical and technical assistance to the deep-sea fishing fleet and for fishery control tasks

Fishery research vessel Vessel similar to stern trawlers, usually state-operated for the purpose of exploring fish stocks and fixing of fishing quotas

Fish transporter Refrigerated vessel used for the transport of frozen fish, but not for other refrigerated cargoes, which need exact temperature control and are therefore regularly transported by reefer vessels

Fishing cutter Fishing vessel for use on rivers, marginal seas or the high seas (high-sea cutter)

Flag Rectangular piece of cloth as a sign of a ship's nationality, as a reference to the guest country (courtesy flag), or as a means of communication (signal flag, e.g. yellow quarantine flag)

Floating crane A crane mounted on a pontoon, self-propelled or moved by a tug for handling general, bulk, heavy lift and project cargo

Forepeak Forward of the ship, forward of the collision bulkhead, to receive ballast water

Frame Term in the design process of a ship and the name of a major structural element to strengthen the hull, traditionally running transversely, but also fitted longitudinally

Free a ship Emptying a tank or pumping out infiltrated water from the ship

Freeboard The minimum allowed freeboard is the distance from the upper edge of the deck to the lower edge of the freeboard line, determined by the classification society on the basis of the freeboard calculation and to be observed as an indication of the maximum permissible draught, with additional marks at different heights taking into account the shipping area, season and wooden deck load

Freight ferry Roll-on roll-off vessel for the transport of cargo loaded on trucks or trailers

Freight rate (or simply Freight) Price for the transport of cargo depending on the market, quantity, route and quality of the ship, to be determined by tariff or negotiated

Gas turbine Engine developed from modified aircraft turbines for ships with high speed requirements

General Average (Havarie grosse) An exclusive maritime law principle according to which in case of emergency the loss incurred by jettisoning or other measures for saving the ship is shared proportionately by the parties with a financial interest in the voyage.

General cargo Cargo of individual items transported in general cargo ships or container ships

Grain Indicates the cargo capacity for bulk cargo, often measured in cbf (1 cubic foot = 0.0283 m^3)

Handymax Designation for a bulk carrier between 40,000 and 50,000 dwt

Handysize Designation for bulk carriers with capacities up to 40,000 dwt, but also for tankers of corresponding deadweight

Hatch In a narrower sense describing, by forward starting counting, the openings in the weather deck and in the tweendecks for loading and unloading the cargo, in a broader sense the cargo hold compartment as a whole

Hatch cover By crane, chain or hydraulically moved cover on a hatch opening in the weather deck (watertight) or in the tween deck

Heavy Fuel Oil (HFO) Fuel predominantly used in shipping as a residual component of the refinery process, prohibited in SECAs (Sulphur Emission Control Areas) due to its high sulphur content, if not cleaned by scrubbers

Heavy-lift cargo vessel Equipped with especially strong, connectable heavy-lift cranes or booms of up to 1000 tonnes SWL for the transport of heaviest cargo

Heel compensation Carried out actively with powerful pumps or blowers by ballast transfer to compensate the ship's inclination during cargo handling or passively with the aid of roll damping tanks

High tide High water level caused by tidal impact or by storm

Hold Space for storing the ship's cargo

Hoistable deck Flexible loading decks installed on ro ro ships, which can be raised to provide greater height for the deck below

Homogeneous load Theoretical assumption of a uniform weight distribution of the cargo, e.g. of containers, as base for the calculation of loading cases to prove sufficient stability

Hopper dredger Dredger that dumps its dredged material through bottom flaps and thus quickly "hops" up

Hopper tank Tank with triangular cross-section between double bottom and outer skin, which is intended to facilitate the unloading of cargo on bulk carriers, and which serves as ballast tank

Ice class Classification of a ship with reinforced hull structures and greater engine power based on special ice class regulations of a classification society

Icepod Specially designed propeller for navigating in ice

Inland cruise vessel Passenger ship with cabins sailing on rivers for several days lasting voyages

Jetty Berthing place built at a distance from the coast into the sea for loading or discharging bulk, oil or gas cargoes

Kamsarmax Bulk carrier of a size class which dimensions still allow using the African bauxite export port of Kamsar

Knots abbr. kts, speed indication (1 kn = 1 nautical mile/h or 1.852 km/h)

Korrespondentreeder German term for a manager operating a vessel on behalf of a ship owner

Kort nozzle A propeller turning in a conical ring to improve the propulsion power and therefore particularly installed in tugs before the invention of new propeller systems, if rotatable the ring takes over the function of a rudder, named after its German inventor Ludwig Kort

Laker Bulk carrier which dimensions of 225.5 m length, 23.8 m breadth and maximum draught of 9 m have been designed to pass the locks of the St. Lawrence Seaway

Lashing Fastening of cargo, especially deck cargo, with the aid of lashing material consisting of chains, iron bars or belts

Leak safety Property of a ship to maintain buoyancy in the event of a leak

Length of the ship Overall ship length or length between perpendiculars (from the centre of the rudder stock to the point of intersection of the vertical with the design waterline)

Light ship Weight of an empty ship

Lightering Reducing the draught of a ship by partial discharge

Line Equator, short form of lat. *linea aequinoctialis*

Line of sight Line from the conning position in the wheelhouse to a point down to the waterline as a function of the ship's length, of particular importance for the allowed maximum height of the deck cargo and thus an important design condition of container ships, to be complied with regulations issued by the IMO, national and the Panama canal authorities

Liner shipping Unlike tramp shipping, the transport of cargo and/or passengers on fixed routes according to a schedule, which may vary in general cargo or break bulk shipping according to the volume of cargo and which is carried out at lower rates than in project-based shipping

Live fish transporter Vessel for transporting fish reared in farms to the processing factory

Livestock carrier A special vessel, often converted from a general cargo vessel, tanker or car carrier, for the transport of live animals such as sheep and cattle

Lloyd's Register Globally active company headquartered in London for the classification and supervision of ships according to its own regulations as well as for the certification and monitoring of technical installations

Lock Basin-like hydraulic structure with one or more gates, which can provide a constant water level in a harbour or compensate for differences in height on rivers and canals

Log carrier Bulk carrier suitable for the alternative transport of logs with cargo securing devices by means of fixed or detachable supports at the bulwark and lashing equipment

Loop A scheduled service route, also named string or line

Low tide (ebb) On the coast: low water level caused by tides or storms

Main dimensions The most important fixed shipbuilding characteristics of a ship, such as overall length, length between perpendiculars, moulded breadth, depth or design draught, whereas the maximum draught is a measure resulting from the calculation of the freeboard

Main engine A diesel engine used to propel the ship, as two-stroke engine driving the propeller shaft directly, or as four-stroke engine through a reduction gear

Measurement Determination of the tonnage of a ship by a national authority, recorded in the ship's tonnage certificate and forming the basis for the manning the ship and levying fees

Midship section One of the relevant constructional drawings showing the cross-section of a ship to determine the basic structure and material thicknesses

Membrane tank Construction of a non-self-supporting cargo tank in LNG tankers, its hull being manufactured of thin sheet metal membrane that can compensate for large temperature differences

Minibulker Bulk carrier with a deadweight capacity of about up to 10,000 tonnes

Model tank test Carried out in a water or ice tank with an exact hull model to optimise the lines of a ship and determine the required propulsion power

Monopile Monotube foundation pipe for offshore wind turbines

Moonpool Watertight passage through the hull of a ship for drilling in the seabed or carrying out underwater work

Mooring winch Winch for the handling of mooring lines, nowadays working automatically with a constant tension

Moss-Rosenberg Patent for the design of spherical, self-supporting tanks in LNG tankers

Motorway of the Seas EU transport concept to shift road transport to the sea for environmental reasons

Multi-purpose cargo vessel Cargo carrying ship type developed from the general cargo ship for the transport of general cargo, heavy goods, breakbulk and containers, equipped with its own cargo cranes

Nautical mile (nm) Distance given in shipping for a distance of 1852 m

Neopanamax Class of vessels with dimensions not exceeding 369 m length and 51.2 m breadth, the maximum dimensions for passage through the Panama Canal after its extension in 2016

Neopostpanamax Class of vessels with dimensions exceeding 369 m in length and 51.2 m in breadth, the maximum dimensions for passage through the Panama Canal after its extension in 2016

Northern Sea Route Shortened route from the Far East to Europe along the Siberian coast, which is increasingly easier for cargo ships to navigate due to reduced ice expansion caused by global warming

OBO carrier Type of ship for the alternating transport of oil, bulk or ore cargoes (Oil Bulk Ore), which was created for reasons of economy to avoid ballast voyages, but is now only rarely in service

Offshore industry A branch of the maritime industry practised near the coast for the exploitation of marine resources, more recently also for the production of energy from wind power stations

Open hatch vessels In container shipping vessels designed without hatch covers or partly without hatch covers, in bulk shipping ships with very wide hatch openings, which are yet closed by hatch covers

Operator Shipping company maintaining liner services, which may either own or charter the vessels employed

Panamax Vessel class with dimensions of maximum 294 m length, 32.3 m breadth and 12.8 m draught allowing them to pass the old Panama Canal locks before their extension in 2016

Partenreederei German term describing a corporate form in which a number of persons—including persons not involved in shipping—have shares (*Parten*) in the ownership of a ship and are therefore entrepreneurs

Petcoke Substance derived from petroleum and consisting mainly of carbon

Pier Berthing place for ships sticking out into the water, but as well used as term for a walled quay

Pilot Officially appointed nautical officer for the mandatory advice of the ship's command in a specified area (sea, river or canal pilot) and in port (harbour pilot) on board a ship, and during fog additionally by radar supervision from ashore

Pod Azimuthing drive unit (nacelle) combining the electric motor and the propeller

Pool Joint and therefore more economical deployment of ships from different owners, especially in tanker and bulk shipping

Postpanamax Vessels of a size class exceeding the former maximum dimensions of 294 × 32.3 m for passing through the Panama Canal before its extension in 2016

Product tanker Tanker transporting processed oil (products), usually also suitable for transporting edible oil or chemicals

Project Initiative for the construction of a ship, for which the project department of a shipyard produces initial design drawings and calculations according to the specifications of a shipowner or as an in-house development

Propeller Cast from bronze or steel (for navigating in ice), this is the part of the propulsion system that converts the rotary motion of the propeller shaft into propulsion power

Rail ferry Ferry for the exclusive transport of cargo or passenger trains or for combining them with other types of vehicles and passengers

Rate Daily price for chartering a ship as specified in a charter contract

Reefer container ship Cargo ship capable of cooling all or part of its cargo consisting of containers by electrical connections

Refrigerated vessel (reefer) Cargo ship for the transport of fruit, vegetables, meat and fish, which entire cargo space consists of insulated, temperature-controled rooms cooled down to -20 °C

Roll-on roll-off ship Vessel such as ferries, freight ferries, ro ro ships or car carriers equipped with doors and ramps for the horizontal handling of rolling cargo

Rudder Essential part of a ship's equipment consisting of the rudder shaft, trunk and blade and moved by the steering gear to steer the ship, designed as a spade, semi-balanced, or patent rudder

Saima canal Finnish inland waterway which locks determine the main dimensions of many European operated coastal cargo vessels (82 × 12.6 m)

Shaft motor Drive for mechanical transmission of electrically generated power to the propeller shaft

Shipowner Owner of a vessel operated for economical reasons

Sister ships Ships built to the same design

Suction dredger Dredger equipped with a lowerable suction pipe for deepening or maintaining the fairway

Sulphur tanker Special ship for the transport of liquid sulphur in coated tanks

Scrubber A device installed in the exhaust system to reduce sulphur in the emissions of main and auxiliary engines and boilers

Sea margin Regularly 15% of the engine power reserve included in the speed data

Sea-going inland vessel Low-built ship with a shallow draught and often a hydraulically liftable wheelhouse, which can be employed on inland waterways as well as in coastal waters

Seebäderschiff German term for passenger ships carrying tourists from the mainland to domestic islands

Self-discharging bulk carrier Cargo ship with an automated discharge system for unloading a bulk cargo such as grain, ore, stones or cement

Semi-submersible heavy lift vessel (dock ship) Special ship type that can be lowered like a floating dock to accommodate large cargo items, including ships, oil rigs etc.

Shaft generator Generator using mechanical energy for on-board operation from the main engine as economical source, usually via the Power-take off (PTO) of the reduction gear

Shell plating Outer hull of the ship made of plates of different thickness and at its inner side reinforced by frames and stringers

Shifting Feared cargo movement to one side of the ship, which can result in the loss of stability and capsizing of the ship

Shipyard Place or company with special facilities for the construction or repair of ships

Short sea shipping Shipping operations within a continent, in contrast to transoceanic connections (deep sea shipping)

Shuttle tanker Tanker with a special loading device for the transport of oil taken over at sea and carried from there to a refinery or to a transshipment location

Slow steaming Operation of ships at lower than service speeds to save fuel costs, requiring technical measures in engine operation, and, if the most weekly schedule scheme is to be maintained, either the cancellation of ports of call or an increase in the number of vessels, resulting in greater tonnage demand

Sponson Steel body welded to the sides of a ship to increase stability, especially fitted in the case of the lenghtening or raising of ship

Stability The ability of a vessel to maintain its regular—stable—buoyancy and thus avoiding capsizing, determined by the position of the centre of gravity and the shape of the vessel

Stack load Load on hatch covers per container stack, up to 100 tonnes for 20-ft and 140 tonnes for 40-ft containers

Stand-by Position of an auxiliary or emergency ship deployed for assistance, e.g. in the vicinity of offshore installations or as an emergency tug in the event of an accident

Stern Aft part of the vessel

Stern trawler Fishing vessel on which the catch is towed to the deck by means of a winch via a slope at the stern

Stern boom Double post at the stern of fishing, research and offshore vessels for hauling nets or gear

Stem (Bow) Forward end of the hull

Stowage factor Space requirement as ratio of volume to weight of the cargo

Supramax Bulk carrier with a deadweight of between 40,000 tonnes and 67,000 tonnes

Tanker Ship for the transport of liquid cargo of all kinds in separate tanks

Terminal Port facility for handling containers, vehicles or liquid cargoes as the end of the land-based logistics chain

Time charter Renting a ship from its owners for a certain time period, which can lasting from a few weeks to decades

Tonnage Describes the total size of a ship in gross tons, and its cargo carrying space in net tons

Top tank In cross-section a triangular tank between the main deck and the outer skin, which on bulk carriers is intended to prevent the cargo from shifting and to serve as ballast tank

Towing bridle A bridle with a hook in the center to which a towline is fastened, in order to distribute the pulling force, by this exercising a limited steering function on the towed object, or in order to tow two objects simultaneously

Transhipment Direct transfer of cargo from one ship to another, mostly carried out due to fairway restrictions by using the ship's own cargo handling equipment or a suitably equipped floating crane or crane ship

Transversal thruster A propeller rotating in a transversal tunnel in the fore or aft of the vessel, often referred to as a bow thruster

Trim Longitudinal floating position of a ship relative to the horizontal

Tugboat Special ship type of small size equipped with powerful engines for towing, assisting and the salvage of seagoing vessels, a special type is the pusher tug mainly used on inland waterways and in North America for pushing barges

Turret-Mooring Anchor system for positioning oil drilling vessels

Tween deck Deck or hold space on general cargo and refrigerated vessels between the weather deck and the tank deck, formed on today's multipurpose cargo vessels by removable pontoon-like hatch cover that can be positioned at different heights

Twistlock Corner connecting element for securing containers stacked on top of each other

Two-compartment status Requires the maintenance of the buoyancy even when two compartments of the ship are flooded

Two-island type Ship type with forward accommodation block and bridge and the engine room positioned separately aft

Ultramax designation for bulk carriers with a deadweight capacity from about 60,000 to 67,000 tonnes

Van Carrier (Straddle carrier) Wheeled vehicle used on terminals for the independent transport of containers between the container gantry bridge and the storage place, truck or train, able to stack up to four layers of containers

Vegetable oil tanker Tanker used for the transport of vegetable oil, either specially built or classed as chemical tanker

VLOC (Very Large Ore Carrier) Large cargo carriers from about 200,000 dwt to 400,000 dwt capacity for the exclusive transport of iron ore

VLCC (Very Large Crude Carrier) Large tankers from about 200,000 dwt to 400,000 dwt capacity for the exclusive transport of crude oil

Void space Unused space, e. g. space between the tanks and the hull, kept as small as possible for reasons of economy and, for example on LPG tankers, to improve leakage stability

Voith-Schneider propeller Established special propulsion technology with adjustable knives rotating around the vertical axis, which provides high manoeuvrability and is therefore installed in tractors in particular, despite lower tractive force development than with other types of propulsion

Water jet propulsion A pump-type propulsion system for ferries and yachts, which generates propulsion by the expulsion of water and enables high ship speeds

Well Stimulation Vessel Offshore special vehicle for increasing the yield of drilling sites, for example by introducing acid

Wide beam vessel Container ship design with a small length to width ratio developed in advance of the Panama canal extension, providing the ship greater stability and a smaller draught

Wine tanker Tanker equipped for the transport of alcohol with stainless or coated tanks

Woodchip Carrier Large-depth bulk carrier for the transport of wood chips, but also used for other bulk cargoes

X-bow Protruding bow shape developed by the Norwegian Ulstein Group, initially for offshore vessels and later also applied to container and passenger ships, to improve seaworthiness in heavy seas

Yacht A luxuriously equipped vessel primarily used for pleasure or, as a state yacht, for representative purposes

Index[1]

[1] Italics indicate an illustration.

R. Witthohn, *International Shipping*, https://doi.org/10.1007/978-3-658-34273-9

C

F

H

J

K

M

Q

W